Lecture Notes in Civil Engineering 494

Lecture Notes in Civil Engineering (LNCE) publishes the latest developments in Civil Engineering—quickly, informally and in top quality. Though original research reported in proceedings and post-proceedings represents the core of LNCE, edited volumes of exceptionally high quality and interest may also be considered for publication. Volumes published in LNCE embrace all aspects and subfields of, as well as new challenges in, Civil Engineering. Topics in the series include:

- Construction and Structural Mechanics
- Building Materials
- Concrete, Steel and Timber Structures
- Geotechnical Engineering
- Earthquake Engineering
- Coastal Engineering
- Ocean and Offshore Engineering; Ships and Floating Structures
- Hydraulics, Hydrology and Water Resources Engineering
- Environmental Engineering and Sustainability
- Structural Health and Monitoring
- Surveying and Geographical Information Systems
- Indoor Environments
- Transportation and Traffic
- Risk Analysis
- Safety and Security

To submit a proposal or request further information, please contact the appropriate Springer Editor:

- Pierpaolo Riva at pierpaolo.riva@springer.com (Europe and Americas);
- Swati Meherishi at swati.meherishi@springer.com (Asia—except China, Australia, and New Zealand);
- Wayne Hu at wayne.hu@springer.com (China).

All books in the series now indexed by Scopus and EI Compendex database!

José C. Matos · Paulo B. Lourenço ·
Daniel V. Oliveira · Jorge Branco · Dirk Proske ·
Rui A. Silva · Hélder S. Sousa
Editors

20th International Probabilistic Workshop

IPW 2024

Springer

Editors
José C. Matos
ISISE, Department of Civil Engineering
University of Minho
Guimarães, Portugal

Paulo B. Lourenço
ISISE, Department of Civil Engineering
University of Minho
Guimarães, Portugal

Daniel V. Oliveira
ISISE, Department of Civil Engineering
University of Minho
Guimarães, Portugal

Jorge Branco
ISISE, Department of Civil Engineering
University of Minho
Guimarães, Portugal

Dirk Proske
Department of Architecture, Wood and Civil Engineering
Bern University of Applied Sciences
Burgdorf, Switzerland

Rui A. Silva
ISISE, Department of Civil Engineering
University of Minho
Guimarães, Portugal

Hélder S. Sousa
ISISE, Department of Civil Engineering
University of Minho
Guimarães, Portugal

ISSN 2366-2557 ISSN 2366-2565 (electronic)
Lecture Notes in Civil Engineering
ISBN 978-3-031-60270-2 ISBN 978-3-031-60271-9 (eBook)
https://doi.org/10.1007/978-3-031-60271-9

This Springer imprint is published by the registered company Springer Nature Switzerland AG
The registered company address is: Gewerbestrasse 11, 6330 Cham, Switzerland

Preface

The imperative of managing uncertainty and facilitating informed decision-making renders evident the criticality of using probabilistic and reliability disciplines. This can be seen in the most recent advances in infrastructure maintenance and management, especially those related to safety under extreme events and security to non-conventional threats and dangers. Additionally, the escalating significance of climate change phenomena is incontrovertible, mostly affecting the likelihood and consequences of various natural hazards. Consequently, there is a need to develop deeper studies on probability and statistics for data science, as well as on its application to system analysis, combining these tools to face huge uncertainties.

The International Probabilistic Workshop (IPW) series started in 2003 as the Dresden Probabilistic Symposium at the Technical University of Dresden and repeated in 2004. In 2005, the 3rd edition held in Vienna was renamed as International Probabilistic Workshop. Subsequent IPWs took place in Berlin (2006), Ghent (2007), Darmstadt (2008), Delft (2009), Szcecin (2010), Braunschweig (2011), Stuttgart (2012), Brno (2013), Weimar (2014), Liverpool (2015), Ghent (2016), Dresden (2017), Vienna (2018), Edinburgh (2019) and Stellenbosch (2022).

The IPW 2020, planned to take place in September 2020 at the University of Minho, was postponed to May 2021 and transitioned to a digital format due to the global COVID-19 pandemic. IPW 2024 offers a renewed opportunity to gather the IPW Community in Guimarães, Portugal. Undoubtedly, IPW 2024 keeps the high-quality level of previous editions by bringing together high-calibre scientific contributions (51 papers spanning 22 countries) covering different applications and approaches to probabilistic-based methods.

The editors express their profound gratitude to all contributing authors, keynote speakers and attendees for their valuable contributions, to the members of the Scientific Committee for their meticulous work, and to the Workshop Secretariat for their dedicated teamwork.

May 2024

José C. Matos
Paulo B. Lourenço
Daniel V. Oliveira
Jorge Branco
Dirk Proske
Rui A. Silva
Hélder S. Sousa

Organization

Organizing Committee

José C. Matos (Chair)	University of Minho, Portugal
Paulo B. Lourenço (Chair)	University of Minho, Portugal
Dirk Proske	Bern University of Applied Sciences, Switzerland
Jorge M. Branco	University of Minho, Portugal
Hélder S. Sousa	University of Minho, Portugal

Scientific Committee

Daniel V. Oliveira (Chair)	University of Minho, Portugal
Abel Henriques (Vice-chair)	University of Porto, Portugal
Luís O. Santos (Vice-chair)	LNEC, Portugal
Rita Bento (Vice Chair)	University of Lisbon, Portugal
Rui A. Silva (Secretary)	University of Minho, Portugal

Members

Alfred Strauss, Austria
Andre T. Beck, Brazil
Ângelo Palos Teixeira, Portugal
Bernt J. Leira, Norway
Bram van den Eijnden, The Netherlands
Bruno Sudret, Switzerland
Carlo Pellegrino, Italy
Carmen Andrade, Spain
Chris Roth, South Africa
Christina McLeod, South Africa
Colin Caprani, Australia
Dan M. Frangopol, USA
Daniel Cantero, Norway
Daniel Straub, Germany
Dimitri Val, UK
Edoardo Patelli, UK
Eduardo Cavaco, Portugal
Emilio Bastidas-Arteaga, France

Eugene O'Brien, Ireland
Frank Coolen, UK
Hélder S. Sousa, Portugal
Hooman Ghasemi, USA
Ioannis Kougioumtzoglou, USA
Joan R. Casas, Spain
Joaquim Tinoco, Portugal
Jochen Kohler, Norway
John Sørensen, Denmark
Jorge M. Branco, Portugal
Jose C. Matos, Portugal
Jose Turmo, Spain
Laura Caldeira, Portugal
Leonardo Rodrigues, UK
Luís C. Neves, UK
Lukas Novak, Czech Republic
Marc Maes, Canada
Mariano Zanini, Italy
Maria Pina Limongelli, Italy
Mark Stewart, Australia
Matias Valenzuela, Chile
Maurício Sanchez, Colombia
Michael Beer, Germany
Michael H. Faber, Denmark
Michael Kraus, Switzerland
Michel Ghosn, USA
Miroslav Sýkora, Czech Republic
Mitsuyoshi Akiyama, Japan
Myrto Konstantinidou, Greece
Paula V. Ferreira, Portugal
Paulo B. Lourenço, Portugal
Peter Mark, Germany
Peter Tanner, Spain
Petr Konecny, Czech Republic
Pieter van Gelder, The Netherlands
Radomir Pukl, Czech Republic
Raphael Steenbergen, The Netherlands
Raquel Menezes, Portugal
Robby Caspeele, Belgium
Roman Wan-Wendner, Belgium
Ruben van Coile, Belgium
Thomas Brami, Germany
Tiago Ferreira, UK
Tiago Miranda, Portugal
Tom Lahmer, Germany

Vikram Pakrashi, Ireland
Xin Ruan, China

Sponsors

Workshop Organization

University of Minho
School of Engineering

Institute for Sustainability and Innovation in Structural Engineering

Supporting Entities

ESRA European Safety and Reliability Association

Ordem dos Engenheiros

Município de Guimarães

NORISK

Contents

Risk Assessment and Management

Keynotes

Life-Cycle Probabilistic Multi-objective Optimum SHM Planning

Dan M. Frangopol[1](✉) and Sunyong Kim[2]

[1] Lehigh University, Bethlehem, PA 18015-4729, USA
dan.frangopol@lehigh.edu
[2] Wonkwang University, Iksan, Jeonbuk 54538, South Korea

Abstract. Structural Health Monitoring (SHM) is widely utilized to detect potential damage and predict the structural performance of deteriorating structures. Timely damage detection and accurate structural performance prediction are crucial for ensuring structural safety and managing the service life of deteriorating structures. This paper addresses a life-cycle probabilistic multi-objective optimization for SHM planning. The life-cycle probabilistic multi-objective optimum SHM planning can be based on availability, damage detection, reliability, service life, life-cycle cost, and risk. When initial information is insufficient for reliable damage prediction, an availability-based objective is useful for optimum SHM planning. Otherwise, the damage detection delay, maintenance delay, reliability index, service life, life-cycle cost, and risk can be used for optimum SHM planning. They are formulated by integrating damage initiation and propagation, damage detection, and the effects of inspection and maintenance actions on service life, reliability, life-cycle cost, and risk. Approaches to solving the multi-objective optimization efficiently and multi-attribute decision-making to select the most appropriate solution among the Pareto optimal set are presented. Furthermore, an updating process incorporating information from SHM is outlined to enhance the accuracy of SHM planning.

Keywords: Availability · Damage Detection · Life-Cycle Cost · Maintenance · Multi-Objective Optimization · Reliability · Risk · Service Life · SHM planning

1 Introduction

Deteriorating structures demand maintenance interventions to sustain performance above the predefined threshold during their service life [1–3]. By implementing maintenance interventions, it's possible to delay the deterioration process, repair existing damage, enhance structural performance, and extend the service life of the structure [4–6]. For cost-effective maintenance, accurate assessment and prediction of the deterioration process and timely damage detection are crucial [7, 8]. However, they pose challenges for practical bridge management due to epistemic and aleatory uncertainties [9–11]. Consequently, the adoption of periodic inspections and maintenance has been a prevalent practice for several decades. To address the cost-effectiveness and uncertainty associated with management, it is necessary to adopt structural health monitoring (SHM) optimally [12–14].

J. C. Matos et al. (Eds.): IPW 2024, LNCE 494, pp. 3–11, 2024.
https://doi.org/10.1007/978-3-031-60271-9_50

In this paper, a life-cycle probabilistic multi-objective optimization for SHM planning is presented. The life-cycle probabilistic optimum SHM planning can be based on multiple objectives considering availability, damage detection, maintenance, reliability, service life, life-cycle cost, and risk. When initial information is insufficient for reliable damage prediction, objectives based on the availability of monitoring data are useful for optimum SHM planning. Otherwise, optimum SHM planning can be based on objectives considering damage detection, reliability index, service life, life-cycle cost, and risk. Their formulations require integrating damage initiation and propagation, damage detection, and the effects of inspection and maintenance actions on service life, reliability, life-cycle cost, and risk. In addition to formulating objectives for optimal SHM planning, this study presents efficient approaches for solving multi-objective optimization and employing multi-attribute decision-making to select the most suitable solution from the Pareto-optimal set. Furthermore, an updating process that incorporates information from SHM is outlined to enhance the accuracy and reliability of SHM planning.

2 Framework for Life-Cycle Probabilistic Multi-objective Optimum SHM Planning

The framework for life-cycle probabilistic multi-objective optimum SHM planning consists of three parts: (a) formulations of the objectives, (b) multi-objective optimization and decision making, and (c) updating process with the monitored data, as shown in Fig. 1. Through the presented framework, the monitoring starting time and duration can be optimized. Through the multi-objective optimization process, multiple well-balanced solutions are obtained, which provide flexibility for managers in determining the best Pareto optimal solution. The monitored data is utilized to update the parameters related to the formulations of objectives, and thereby reducing the uncertainties associated with SHM planning. The three parts of the framework are iteratively applied throughout the life-cycle of deteriorating structures.

3 Objectives for Probabilistic Optimum SHM Planning

The formulation of objectives for optimum SHM planning considers the availability of monitoring data, damage detection, reliability index, service life, life-cycle cost, and risk [15]. The availability of monitoring data can be used when the initial information for damage initiation and propagation is insufficient. If there is sufficient information for damage initiation and propagation, the objectives based on damage detection, service life, reliability, life-cycle cost, and risk can be applied for SHM planning [4, 16]. Table 1 summarizes the nine objectives (i.e., O1 to O9) and required estimations for the formulations of objectives.

The availability-based objectives are O1 = maximization of expected availability of monitoring data and O2 = minimization of expected monetary loss. The damage detection-based objectives include O3 = minimization of expected damage detection delay (or minimization of expected damage detection time), O4 = maximization of probability of damage detection, and O5 = minimization of expected maintenance delay.

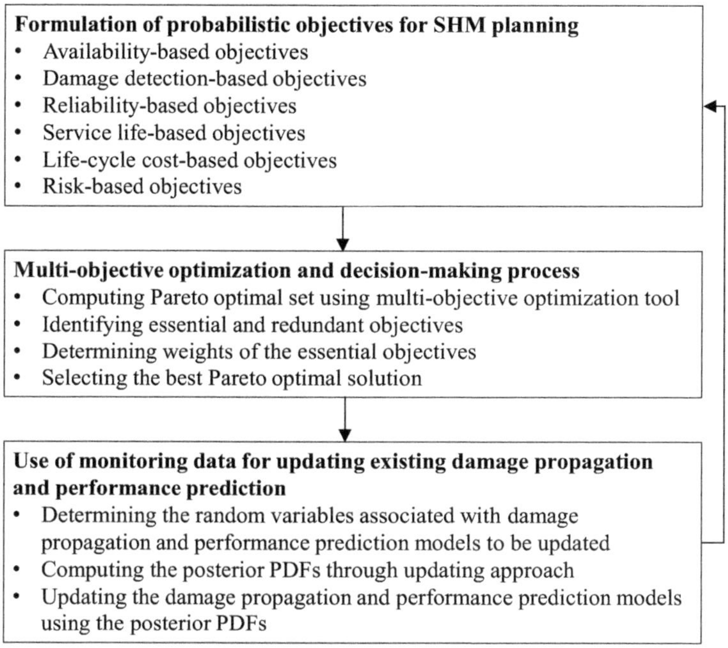

Fig. 1. Framework for life-cycle probabilistic multi-objective optimum SHM planning.

Table 1. Objectives and required estimations for their formulations

Objectives	Required estimations
O1 = maximization of expected availability of monitoring data	RE1
O2 = minimization of expected monetary loss	RE1 + RE2
O3 = minimization of expected damage detection delay (or minimization of expected damage detection time)	RE3
O4 = maximization of probability of damage detection	RE3
O5 = minimization of expected maintenance delay	RE3 + RE4
O6 = maximization of expected extended service life (or expected service life extension)	RE3 + RE4 + RE5
O7 = maximization of reliability	RE3 + RE4 + RE5
O8 = minimization of expected total life-cycle cost	RE3 + RE4 + RE5
O9 = minimization of risk	RE3 + RE4 + RE5 + RE6

Note: RE1 = monitoring cost during a given monitoring period; RE2 = monetary loss due to unavailable monitoring data; RE3 = damage initiation and propagation; RE4 = relation between the degree of damage and probability of maintenance; RE5 = effects of damage detection and maintenance application on reliability, service life, and life-cycle cost; RE6 = direct and indirect monetary loss due to structural failure

The objectives O6, O7, and O8 are based on service life, reliability, and life-cycle cost, respectively. Optimum SHM planning can also be based on O9 = minimization of risk. The detailed formulations of O1 – O8 are available in Kim & Frangopol [15] and Frangopol & Kim [4, 16].

The risk is the probabilistic performance indicator to integrate the occurrence probability of an adverse event (e.g., structural failure, malfunction) and the resulting monetary loss [17]. The risk due to the failure of the ith component of a system RS_i is expressed as [17, 18]

$$RS_i = RS_{dir,i} + RS_{ind,i} \tag{1}$$

The direct risk $RS_{dir,i}$ and indirect risk $RS_{ind,i}$ in Eq. (1) are estimated, respectively, as

$$RS_{dir,i} = p_{f,i} \cdot C_{dir,i} \tag{2a}$$

$$RS_{ind,i} = p_{f,i} \cdot p_{f,sub|i} \cdot C_{ind,i} \tag{2b}$$

where $p_{f,i}$ = probability of failure of the ith component; $p_{f,sub|i}$ = probability of system failure caused by the failure of the ith component; $C_{dir,i}$ = direct monetary loss; and $C_{ind,i}$ = indirect monetary loss. Using the damage detection time t_{det}, the probability of failure of the ith component $p_{f,i}$ can be computed as [4, 16]

$$p_{f,i} = P\left(t_{crt,i} - t_{det,i} < 0\right) \tag{3}$$

where $t_{crt,i}$ is the time to reach the critical degree of damage of the ith component, and $t_{det,i}$ is the damage detection time. Since the damage initiation and propagation are uncertain, $t_{crt,i}$, and $t_{det,i}$ can be treated as random variables. The damage detection time $t_{det,i}$ is estimated considering the relation among the damage initiation/propagation, monitoring starting time, and monitoring duration [4, 15, 16]. The direct monetary loss $C_{dir,i}$ is the required cost for the replacement of the ith component. The indirect monetary loss $C_{ind,i}$ includes operating cost, time-loss cost, and accident cost [17, 19]. The SHM planning can be optimized to minimize the risk caused by the failure of the monitored components.

4 Multi-objective Optimization and Decision Making

When more than two objectives for optimum SHM planning are considered simultaneously, the multiple-objective optimization problem can be solved using various methods such as genetic algorithms (GA), weight sum method, lexicographic method, and bounded objective, among others [20]. As a result, the Pareto optimal solutions can be obtained. The GA is widely used to find all Pareto optimal solutions regardless of the continuity or differentiability of the objective functions [16, 20]. The general formulation of the multi-objective optimization for SHM planning can be expressed as

$$\begin{aligned} \text{Find} \quad \mathbf{t_{ms}} &= \left\{t_{ms,1}, t_{ms,2}, \ldots, t_{ms,Nmon}\right\} \\ \mathbf{t_{md}} &= \left\{t_{md,1}, t_{md,2}, \ldots, t_{md,Nmon}\right\} \end{aligned} \tag{4a}$$

$$\begin{aligned}&\text{which minimizes } \Omega^{(-)} = \{O2, O3, O5, O8, O9\}\\ &\text{and/or maximizes } \Omega^{(+)} = \{O1, O4, O6, O7\}\end{aligned} \tag{4b}$$

$$\text{subject to } g_k(\mathbf{t_{ms}}, \mathbf{t_{md}}) \leq 0 \quad k = 1, 2, \ldots n_c \tag{4c}$$

$$\mathbf{t}^-_{\mathbf{ms}} \leq \mathbf{t_{ms}} \leq \mathbf{t}^+_{\mathbf{ms}} \text{ and/or } \mathbf{t}^-_{\mathbf{md}} \leq \mathbf{t_{md}} \leq \mathbf{t}^+_{\mathbf{md}} \tag{4d}$$

where $\mathbf{t_{ms}}$ and $\mathbf{t_{md}}$ are the vectors of design variables (i.e., monitoring starting times $t_{ms,i}$ and monitoring duration $t_{md,i}$), N_{mon} is the number of monitorings, $\Omega^{(-)}$ and $\Omega^{(+)}$ indicate the objective sets to be minimized and to be maximized, respectively, $g_k(\mathbf{t_{ms}}, \mathbf{t_{md}})$ is the kth inequality constraint among n_c constraints, and $\mathbf{t}^-$ and t^+ are the vectors of lower and upper bounds of the design variable $\mathbf{t}$.

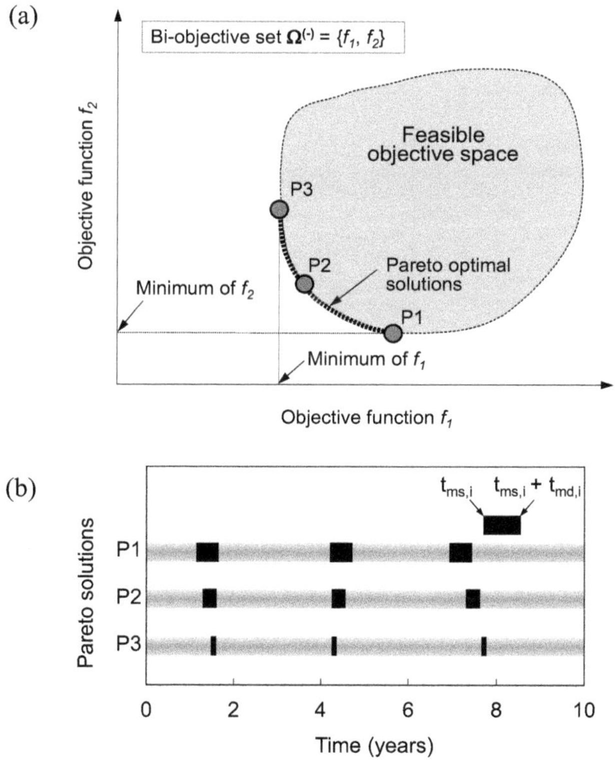

Fig. 2. Bi-objective optimization for SHM planning: (a) feasible objective space and Pareto optimal solutions; and (b) monitoring starting times and monitoring durations for the Pareto solutions P1, P2 and P3 presented in (a).

Figure 2(a) shows the feasible objective space and Pareto optimal solutions when two objectives to be minimized are considered. Any solution among the Pareto optimal solutions can be used for the optimum SHM planning. The monitoring starting times and monitoring durations of the representative Pareto solutions P1, P2 and P3 in Fig. 2(a) are illustrated in Fig. 2(b).

Decision making is the process to select the best solution from the computed Pareto optimal set, which determines the weight of the essential objectives, and estimates the overall assessment value of the Pareto optimal solutions [4, 15, 16]. The essential objective set indicates the minimal set of objectives that generates an equivalent Pareto front of the multi-objective optimization when compared to the original objective set. Through the dominance relation-based approach [21], the essential objective set can be identified. When depicting Pareto optimal solutions in multi-objective optimization with fewer than four objectives, the Cartesian coordinate system is useful. However, visualizing solutions with more than three objectives becomes challenging. To address this, the parallel coordinate system can be employed for the effective representation of Pareto optimal solutions. As shown in Fig. 3, in the parallel coordinate system, the vertical axes denote the values of the objective, and each Pareto solution is represented by a polyline connecting the corresponding values along the vertical axes.

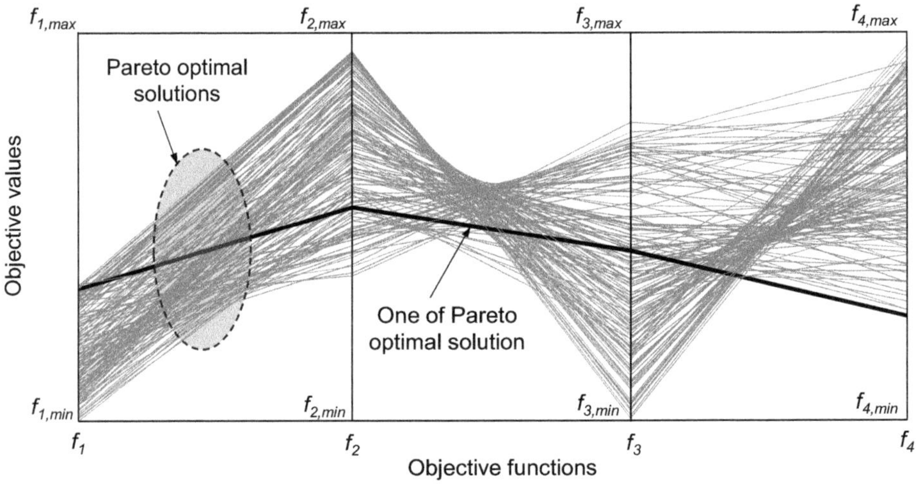

Fig. 3. Pareto solutions of a multi-objective optimization problem with four objectives in the parallel coordinate system.

The weights for essential objectives are calculated using various objective weight determination methods such as the standard deviation (SD) [22], criteria importance through the inter-criteria correlation (CRITIC) [23], and correlation coefficient and standard deviation (CCSD) [24] methods. Subsequently, multiple attribute decision-making approaches, including simple additive weighting (SAW) [25], technique for order preference by similarity to ideal solution (TOPSIS) [26], and elimination and choice expressing the reality (ELECTRE) [27] methods, are applied to estimate the overall assessment values of Pareto optimal solutions with the computed weights of essential objectives. The best optimal solution results in the largest overall assessment value.

5 Updating with Monitored Data

The optimum SHM planning is based on the multiple objectives considering damage propagation prediction, as shown in Table 1. Therefore, improving the accuracy of damage propagation prediction is crucial for enhancing the effectiveness of SHM planning. It can be achieved through the updating the damage propagation prediction with monitoring data [7, 8, 10, 16]. During the last two decades, various approaches for use of the monitoring data to update the damage propagation and structural performance prediction have been developed [4, 15, 16]. Most of updating approaches are based on the Bayesian theorem, which is generally expressed as [28]

$$f''_{\Theta}(\theta) = k \cdot L(\theta) \cdot f'_{\Theta}(\theta) \tag{5}$$

where $f''_{\Theta}(\theta)$ = updated probability density function (PDF) of the parameter θ; k = normalizing constant; $L(\theta)$ = likelihood function; and $f'_{\Theta}(\theta)$ = initial PDF of the parameter θ. The parameter θ is the involved in the PDF $f_X(x)$ of a random variable X. The likelihood function $L(\theta)$ with n monitored data associated with the underlying random variable X is expressed as

$$L(\theta) = \prod_{i=1}^{n} f_X(x_i|\theta) \tag{6}$$

The updated PDF of the underlying random variable X can be obtained as [29]

$$f''_X(x) = \int_{-\infty}^{\infty} f_X(x|\theta) \cdot f''_{\Theta}(\theta) d\theta \tag{7}$$

Markov Chain Monte Carlo (MCMC) methods are widely used in Bayesian statistics, particularly for updating posterior distributions with new information [31]. In the context of SHM, these methods allow for the integration of monitoring data to update the structural performance and SHM planning. One of the main strengths of MCMC methods is their ability to sample from complex, high-dimensional posterior distribution. It is useful in cases where direct computation of the posterior distribution is impractical or impossible due to its complexity [32]. MCMC methods generate a Markov chain that converges to the target distribution as the number of iterations increases. They are suitable for updating the parameters of complex damage propagation prediction. The accuracy and reliability of the updating can improve as the Markov chain converges to the true underlying distribution [33, 34]. Metropolis–Hastings, Slice Sampling, and Gibbs Sampling are well-known MCMC methods [35]. Metropolis–Hastings is a general algorithm applicable in various contexts, including those with high-dimensional spaces. Gibbs sampling is useful when dealing with multivariate distributions and can simplify computations. Slice sampling is a technique that can be effective when dealing with unnormalized distributions. The MCMC techniques allow for the sequential updating of the damage propagation, performance prediction, and optimum SHM planning as new SHM data becomes available [4, 8, 15, 16, 30].

6 Conclusions

This study presents a comprehensive framework for life-cycle probabilistic multi-objective optimum SHM planning. The framework effectively integrates various objectives based on availability, damage detection, maintenance, reliability, service life, life-cycle cost, and risk. Multi-objective optimization techniques can be utilized to efficiently identify Pareto optimal solutions. In the decision-making process, multi-attribute methods are employed to select the best SHM planning from the computed Pareto optimal solutions. Furthermore, the incorporation of SHM data updates the planning process, thereby enhancing its accuracy and reliability throughout the life cycle of the monitored structures. This presented framework can lead to improving structural safety but significant economic benefits by efficiently managing risks.

Acknowledgements. The support by grants from the US National Science Foundation (NSF) Award CMMI-1537926, and the National Research Foundation (NRF) of Korea NRF-2023R1A2C1007097 is gratefully acknowledged. The opinions presented in this paper are those of the a40uthors and do not necessarily reflect the views of the sponsoring organizations.

References

1. FHWA: Bridge Maintenance Reference Manual. Publication No. FHWA-NHI-14-050, Federal Highway Administration (FHWA), Washington, D.C. (2015)
2. Orcesi, A.D., Frangopol, D.M.: Optimization of bridge maintenance strategies based on structural health monitoring information. Struct. Saf. **33**(1), 26–41 (2011)
3. Biondini, F., Frangopol, D.M.: Design, assessment, monitoring and maintenance of bridges and infrastructure networks. Struct. Infrastruct. Eng. **11**(4), 413–414 (2015)
4. Frangopol, D.M., Kim, S.: Life-Cycle of Structures under Uncertainty: Emphasis on Fatigue-Sensitive Civil and Marine Structures. CRC Press, Boca Raton (2019)
5. Sánchez-Silva, M., Frangopol, D.M., Padgett, J., Soliman, M.: Maintenance and operation of infrastructure systems: review. J. Struct. Eng. **142**(9), F4016004 (2016)
6. Hurt, M., Schrock, S.D.: Highway Bridge Maintenance Planning and Scheduling. Butterworth-Heinemann, UK (2016)
7. Orcesi, A.D., Frangopol, D.M.: Bridge performance monitoring based on traffic data. J. Eng. Mech. **139**(11), 1508–1520 (2013)
8. Okasha, N.M., Frangopol, D.M.: Integration of structural health monitoring in a system performance based life-cycle bridge management framework. Struct. Infrastruct. Eng. **8**(11), 999–1016 (2012)
9. Biondini, F., Frangopol, D.M.: Life-cycle performance of deteriorating structural systems under uncertainty: review. J. Struct. Eng. **142**(9), F4016001 (2016)
10. Frangopol, D.M., Soliman, M.: Life-cycle of structural systems: recent achievements and future directions. Struct. Infrastruct. Eng. **12**(1), 1–20 (2016)
11. Mondoro, A., Frangopol, D.M., Liu, L.: Bridge adaptation and management under climate change uncertainties: a review. Nat. Hazard. Rev. **19**(1), 4017023 (2018)
12. Kim, S., Frangopol, D.M.: Cost-effective lifetime structural health monitoring based on availability. J. Struct. Eng. **137**(1), 22–33 (2011)
13. Kim, S., Frangopol, D.M.: Cost-based optimum scheduling of inspection and monitoring for fatigue-sensitive structures under uncertainty. J. Struct. Eng. **137**(11), 1319–1331 (2011)

14. Kim, S., Frangopol, D.M.: Multi-objective probabilistic optimum monitoring planning considering fatigue damage detection, maintenance, reliability, service life and cost. Struct. Multidiscip. Optim. **57**(1), 39–54 (2018)
15. Kim, S., Frangopol, D.M.: Computational platform for probabilistic optimum monitoring planning for effective and efficient service life management. J. Civ. Struct. Heal. Monit. **10**(1), 1–15 (2020)
16. Frangopol, D.M., Kim, S.: Bridge Safety, Maintenance and Management in a Life-Cycle Context. CRC Press, Boca Raton (2022)
17. Decò, A., Frangopol, D.M.: Risk assessment of highway bridges under multiple hazards. J. Risk Res. **14**(9), 1057–1089 (2011)
18. Zhu, B., Frangopol, D.M.: Reliability, redundancy and risk as performance indicators of structural systems during their life-cycle. Eng. Struct. **41**(8), 34–49 (2012)
19. Decò, A., Frangopol, D.M.: Life-cycle risk assessment of spatially distributed aging bridges under seismic and traffic hazards. Earthq. Spectra **29**(1), 127–153 (2013)
20. Arora, J.S.: Introduction to Optimum Design, 4th edn. Elsevier, UK (2017)
21. Brockhoff, D., Zitzler, E.: Objective reduction in evolutionary multi-objective optimization: theory and applications. Evol. Comput. **17**(2), 135–166 (2009)
22. Deng, H., Yeh, C.-H., Willis, R.J.: Inter-company comparison using modified TOPSIS with objective weights. Comput. Oper. Res. **27**(10), 963–973 (2000)
23. Diakoulaki, D., Mavrotas, G., Papayannakis, L.: Determining objective weights in multiple criteria problems: the critic method. Comput. Oper. Res. **22**(7), 763–770 (1995)
24. Wang, Y.-M., Luo, Y.: Integration of correlations with standard deviations for determining attribute weights in multiple attribute decision making. Math. Comput. Model. **51**(1–2), 1–12 (2010)
25. Yoon, K.P., Hwang, C.-L.: Multiple Attribute Decision Making: An Introduction. SAGE Publication Inc., London (1995)
26. Hwang, C.-L., Lai, Y.-J., Liu, T.-Y.: A new approach for multiple objective decision making. Comput. Oper. Res. **20**(8), 889–899 (1993)
27. Govindan, K., Jepsen, M.B.: ELECTRE: a comprehensive literature review on methodologies and applications. Eur. J. Oper. Res. **250**(1), 1–29 (2016)
28. Ang, A.H.-S., Tang, W.H.: Probability Concepts in Engineering: Emphasis on Applications to Civil and Environmental Engineering, 2nd edn. Wiley, New York (2007)
29. Bucher, C.: Computational Analysis of Randomness in Structural Mechanics. Structures and Infrastructures Series, vol. 3. CRC Press/Balkema/Taylor & Francis, Leiden, The Netherlands (2009). Frangopol, D.M. ed.
30. Soliman, M., Frangopol, D.M.: Life-cycle management of fatigue-sensitive structures integrating inspection information. J. Infrastruct. Syst. **20**(2), 04014001 (2014)
31. Robert, C.P.: The Bayesian Choice from Decision-Theoretic Foundations to Computational Implementation, 2nd edn. Springer, Berlin (2007)
32. Gilks, W.R., Richardson, S., Spiegelhalter, D.J.: Markov Chain Monte Carlo in Practice. Chapman and Hall/CRC, London, UK (1995)
33. West, M., Harrison, J.: Bayesian Forecasting and Dynamic Models, 2nd edn. Springer, NY (1997)
34. Brooks, S., Gelman, A., Jones, G., Meng, X.L.: Handbook of Markov Chain Monte Carlo. Chapman and Hall/CRC, London, UK (2011)
35. Casella, G., George, E.I.: Explaining the Gibbs sampler. Am. Stat. **46**(3), 167–174 (1992)

Toward a More Resilient Road Infrastructure

Rade Hajdin[1,2](✉)

[1] Department of Infrastructure Management, GmbH, Zurich, Switzerland
Rade.Hajdin@imc-ch.com
[2] Department of Civil Engineering, University of Belgrade, Belgrade, Serbia

Abstract. The transportation infrastructure serves as a cornerstone for economic prosperity, enabling market expansion and fostering a more efficient division of labor. Ensuring its continued functionality in the face of disruptive events is imperative. This paper introduces a practical method for assessing the functionality of road networks and outlines a procedure for estimating their resilience. While many of these disruptive events are stochastic in nature, such as natural hazards, planned maintenance interventions can also significantly impact the functionality of high-volume road networks. The proposed framework for resilience evaluation takes into account both stochastic events and planned interventions. It specifically examines the impact of recovery time on resilience, particularly in cases of stochastic disruptive events. Furthermore, the framework also separates resilience from the overall costs that are associated with achieving a certain level of resilience, facilitating a comparison between resilience levels and corresponding costs in decision-making processes. Additionally, the paper explores measures that target different aspects of resilience.

Keywords: Resilience · Road Infrastructure · Stochastic processes · Functionality · Disruptive event · Maintenance

1 Introduction

The term "resilient" can be traced back to as early as 1674, although its usage likely predates this. It originates from the Latin verb "resilire," which conveys the idea of rebounding or recoiling. In accordance with this definition, the adjective "resilient' emerged to describe an object's ability to endure a disturbance without suffering permanent deformation or rupture. The noun resilience has the same meaning as a property of objects to recover from deformation caused by mechanical forces.

The modern application of the term "resilience" gained prominence in the 1970 and 1980 by psychologist Emmy Werner. She conducted a 40-year study of 698 infants, the entire multiethnic birth cohort for the year 1955 on the Hawaiian island of Kauai. In her work [1] Werner employed the term "resilience" to describe "three kinds of phenomena: good development outcomes despite high-risk status, sustained competence under stress and recovery from trauma". The high-risk status means that these children were born in chronic poverty, had experienced perinatal stress and lived in a family environment

J. C. Matos et al. (Eds.): IPW 2024, LNCE 494, pp. 12–27, 2024.
https://doi.org/10.1007/978-3-031-60271-9_51

troubled by chronic discord, divorce, or parental psychopathology. By the age 10, two thirds of children have indeed developed serious learning and behavior problems or had delinquency records and mental health problems by the age of 18. However, one third of the children who had experienced four or more such risk factors defied the odds and matured into competent, confident and caring adults. Emmy Werner used the term "resilient" for these children.

The term "resilience" entered the engineering realm firstly in ecology [2], and then in structural engineering [3]. It quickly spread into fields of infrastructure and transportation Engineering. However, its rapid popularization also had a downside as it led to the emergence of multiple competing definitions. A clear delimitation to other related terms such as robustness, resistance, reliability and risk is missing.

World Road Association (PIARC) defines the resilience as follows [3]:

> *The ability of a system or systems to survive and thrive in the face of a complex, uncertain and ever-changing future.*

This is a quite generic definition applicable to all kinds of systems that need to be explained in detail to be useful in practice. Drawing from its usage in psychology, it's intuitively clear that resilient infrastructure can maintain its function during or restore its function after disruptive events. However, there's a significant difference compared to biological and ecological systems, where recovery from disruptive events occurs within the system itself. In the case of non-living matter such as road infrastructure, recovery after a disruptive event happens through deliberate action by society. Therefore, it's necessary to extend the system boundary for resilience evaluation to include the organization responsible for maintaining the functioning of the infrastructure.

In this paper the practice-oriented resilience evaluation for road infrastructure and the measures to improve the resilience of road infrastructure are discussed.

2 The Benefit of Transportation Infrastructure

The effect of road infrastructure or more generally of transportation infrastructure on economic prosperity was already recognized in the eighteenth century by Adam Smith [4]. He astutely identifies the division of labor or specialization as a key productivity driver and therefore essential for economic prosperity. Adam Smith recognized that the division of labor results in production surpassing the needs of the producer, leading to a surplus that must be traded for other goods. He emphasized that

> *"... it is the power of exchanging that gives occasion to the division of labour, so the extent of this division must always be limited by the extent of that power, or, in other words, by the extent of the market."*

Adam Smith found it self-evident that the extent of the market is determined by the capacity to transport products. Given the era in which he lived, he identified water transport as a means to expand the market.

Efficient utilization of transportation infrastructure facilitates market expansion and enhances the division of labor. Affordable transportation of goods enables the concentration of production facilities, leading to higher productivity. Additionally, it stimulates demand for niche products that require further specialization, thereby increasing productivity. These effects unfold gradually as the economy adjusts to improved transportation capabilities.

However, it's important to note that increased productivity isn't solely attributable to readily available transportation infrastructure. It is also influenced by factors such as technological innovation, education, and political stability. Therefore, accurately evaluating the long-term benefits of transportation infrastructure, akin to the benefits of education and political stability, is complex and multifaceted.

The lower limit of annual macroeconomic benefit of road traffic can be estimated by the contribution of road traffic to the Gross Domestic Product (GDP). It lies between 4% [5] and 10% [6] and [7], of Gross Domestic Product (GDP). The inconsistency in the contribution of road transportation to GDP stems from the differences in financing and accounting practices in different countries, as discussed in [8]. Apart from purely economic benefit, road infrastructure enables road users to be involved in various activities that yield other private, public, and social benefits [9] and [10]. In addition to these beneficial externalities, road traffic and indirectly the road infrastructure also have negative health, societal and environmental impact (e.g. accidents, noise, pollution), which is mostly difficult to ponder against the benefits. However, despite these detrimental externalities, which are indeed significant, they do not substantially diminish the overall benefit of mobility and, consequently, of road infrastructure to society.

It appears that measuring resilience based on the benefits of road transport, as proposed in [11] would be the most logical choice consistent with the preceding discussion. However, even if the benefit of road transport could be evaluated in sufficient granularity the portion attributable to road infrastructure is nearly impossible to estimate. Therefore, one resorts to proxies that serve as resilience measure.

3 Measuring Resilience

3.1 Functionality

One can assume that the benefit of road infrastructure, even if unknown, is fully exploited if the road network doesn't impose any impediments to the free flow of traffic, which is the primary purpose of road infrastructure. In the literature one associates this situation with a functionality[1] level of 100%. However, evaluating the functionality following a disruptive event, i.e., the reduction from 100%, is not straightforward. It is intuitively clear that this reduction is associated with network topology, traffic volume, and traffic composition. In the literature, there are suggestions to use network (or graph) properties such as connectivity [12] and centrality [13] as functionality measures. While these measures are quite useful, they do not adequately model the loss of benefit due to a disruptive event.

On the other hand, traffic-related models can capture the disbenefit of a disruptive event. For example, the disruption in the network, such as the loss of a road link between two junctions, will lead to a redistribution of traffic. The new traffic patterns can be determined using macroscopic or microscopic simulations, which are widely used in

[1] The terms "level of service" and "performance level" are also used with the same meaning.

practice for transport planning. Macroscopic simulations provide traffic equilibrium solutions before and after the disruption [14]. Assuming that the origins and destinations of road users remain unchanged, one can evaluate the additional travel time i.e. the difference between cumulative travel time before T_0 and after disruption T_{dis} for all vehicle categories.

This approach allows for the estimation of the throughput time for each origin and destination pair, as well as for each vehicle type, which can be compared between undisrupted and disrupted networks. The difference, i.e., the reduction in functionality, is illustrated in Fig. 1. The integral of the functionality reduction over time, depicted by the light grey area in Fig. 1, represents the loss of resilience.

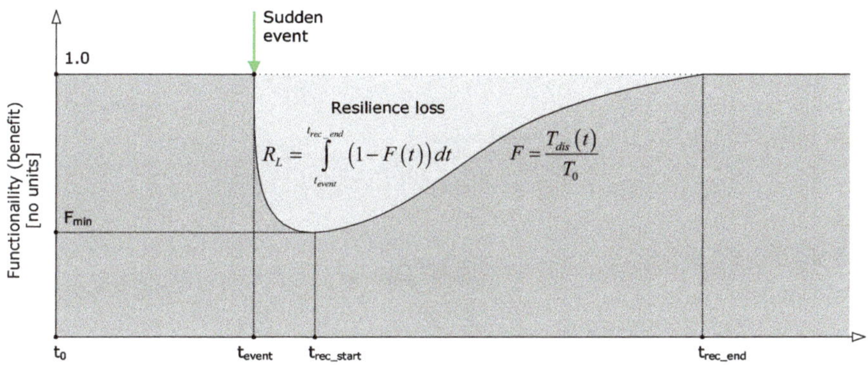

Fig. 1. Resilience on the basis of functionality.

Expanding upon the traffic composition, one can also assign monetary values to these additional travel times, resulting in a monetary disbenefit due to disruptions in the road network. Following this rationale, it can be argued that reintroducing the same link would result in an increased benefit equal to the disbenefit caused by the lost link. Therefore, this value can be assumed as the benefit of the said link [15].

Similarly, this reasoning applies if multiple links are affected simultaneously. However, it's crucial to note that this approach cannot be applied if the disruptive event severs the existing network into disconnected networks. In such cases, other transport modes need to be considered, or as a last resort, a macroeconomic analysis of the disconnected region needs to be conducted.

The functionality as defined in this chapter, is constrained to traffic throughput times, which may or may not be monetized. While this definition aligns with the concept of functionality, it raises questions about whether this should be the sole criterion when evaluating resilience.

3.2 Safety and Environmental Impact During Recovery

The redistribution of traffic due to disruptive event can in certain cases lead to change in accident rates. The detour on two-lane two-way roads with increased traffic density can lead to higher accident rates. Conversely, in some cases, slower traffic may result in lower or less severe accident rates. Considering that safety is one of the most critical performance indicators in Asset Management, the impact of safety should be included in resilience evaluations.

The same principle applies to environmental impact, as traffic redistribution resulting from disruptive events can lead to longer travel times, including congestion and higher gas consumption, thus contributing to negative environmental impacts. Similar to safety, environmental impact is an important performance indicator in Asset Management that should be included in resilience evaluations.

Evidently, once full functionality is restored, the original traffic regime, along with associated accident rates and environmental impacts, are also restored. If necessary, these effects can be monetized and added to monetized functionality.

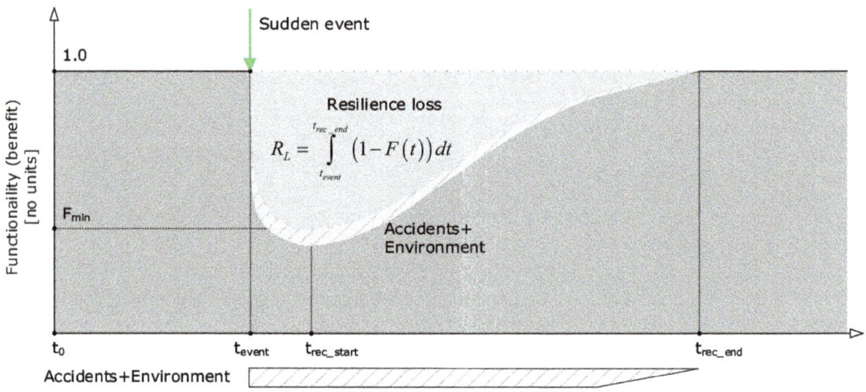

Fig. 2. Alternative consideration of safety and environmental impact.

Alternatively, in accordance with the chosen evaluation and decision-making framework, safety and environmental impact can be integrated into costs (see Sect. 3.5). Both alternatives are illustrated in Fig. 2. Accidents and environmental impact can be treated as the reduction of functionality (area with striped area in the upper part of the figure) or as separate performance indicator (striped area in the lower part of the figure).

3.3 Consequences of Disruptive Event

Disruptive events such as flooding or earthquakes not only abruptly alter the functionality of the road network but also have unrecoverable consequences. Loss of life, physical injuries, and damage to cultural heritage are examples of such consequences. While these consequences are typically considered in risk analysis [16], they are seldom included in resilience assessments due to their irrecoverable nature. These unrecoverable losses are depicted as a separate performance indicator in Fig. 3.

There are also consequences of disruptive events that generally can be alleviated or even eliminated, but the costs of doing so are disproportionally high. In rare cases, even the affected infrastructure can be abandoned altogether. A tragic example is the Lower Ninth Ward in New Orleans, which now has less than 35% of its pre-Hurricane Katrina population [17]. In such cases, there may be no need for road infrastructure recovery to the pre-disaster level. Indeed, for certain disaster events, the recovery level needs to be aligned with the demand for road infrastructure D, as depicted in Fig. 3.

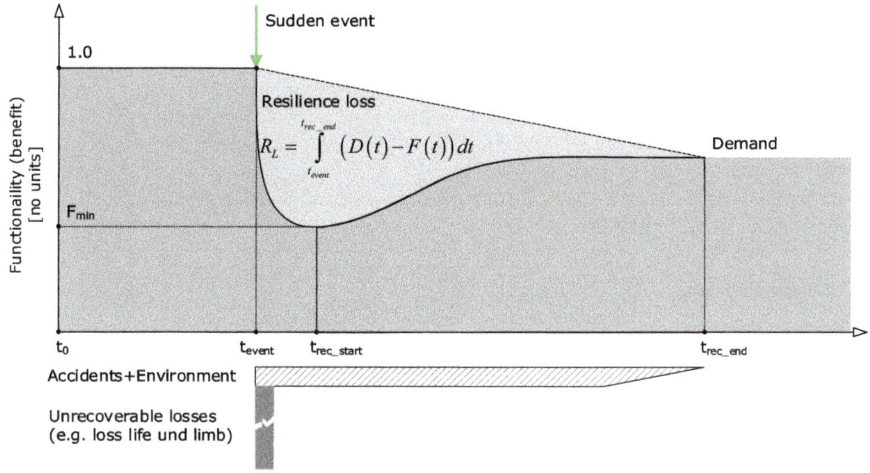

Fig. 3. Change in functionality demand after a disruptive event.

3.4 Maintenance and Improvement Interventions

In most developed countries, road networks operate at the limit of their capacity. Even minor incidents can result in impaired functionality. This also applies to planned maintenance and improvement interventions, as they too lead to capacity reduction and impaired functionality (Fig. 4).

The primary reason for this lies in the lack of road surface to establish a sufficiently large construction site. The only difference compared to natural hazards is that the occurrence time and impact of maintenance and improvement interventions exhibit a very low level of uncertainty. Otherwise, their impact on functionality, safety, and environmental impact is the same. Essentially, they can be treated in the same way as recovery measures after an unexpected disruptive event.

3.5 Costs

In general, the costs of preventative measures, emergency measures and recovery measures i.e. material and labor are not considered in the evaluation of resilience [18]. It is rather that these costs are associated with a disruptive event and associated scenario. This approach allows for a cost/resilience analysis to be conducted, enabling the selection of the optimal course of action with regard to measures aimed at improving resilience (Fig. 5).

Depending on the chosen evaluation framework also the monetized unrecoverable consequences (see Sect. 3.3), accidents and environmental impact (see Sect. 3.2) can be

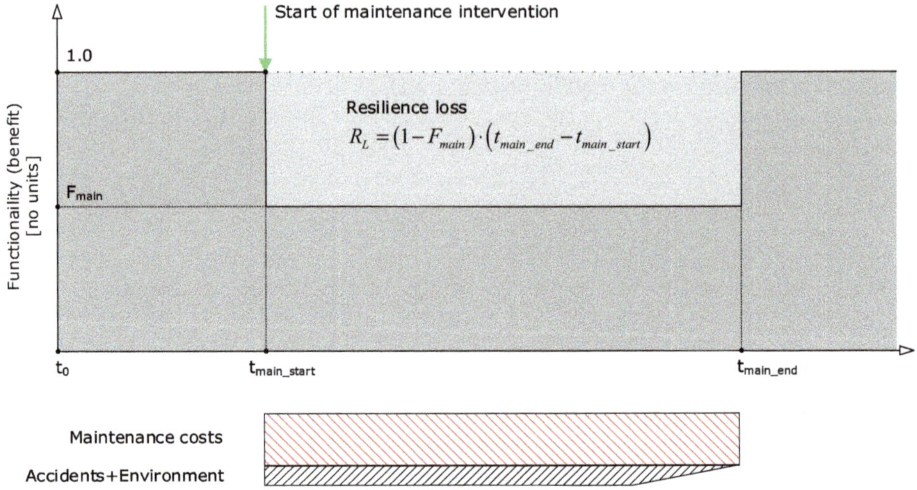

Fig. 4. Change in functionality due to maintenance intervention.

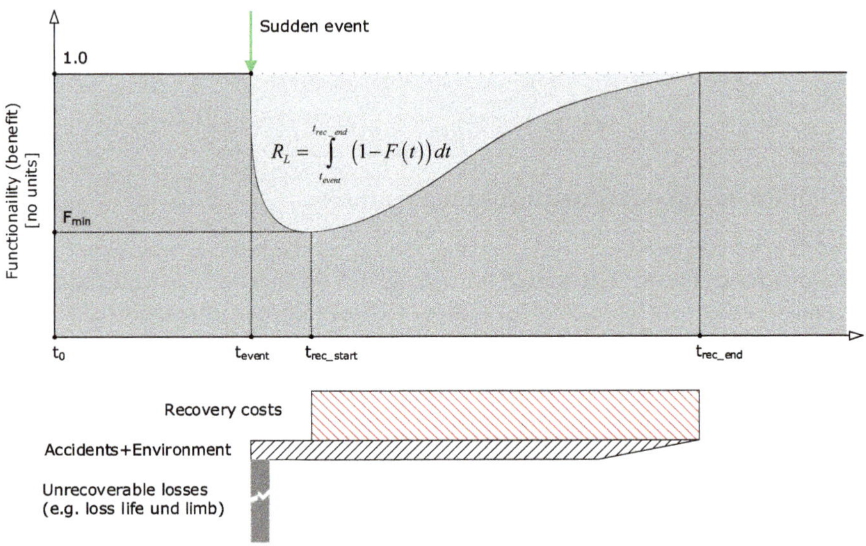

Fig. 5. Resilience and costs.

added to recovery costs. In this way the total cash-flow can be associated with resilience loss.

3.6 Uncertainty

In the preceding paragraphs, an event of a certain magnitude is assumed to have occurred, and consequently, the resilience loss can be regarded as a conditional resilience loss. Different events that can affect the functionality of road infrastructure are stochastic i.e.

random events with exception of planned maintenance measures (see Sect. 3.4). The occurrence rate of these events the probability of their magnitude, needs to be duly considered. To address this, the same procedure applies as for risk analysis.

Natural hazards are typically modeled as stochastic processes, often using Poisson processes, which can be either stationary or non-stationary. The intensity of these hazard events is also a stochastic variable with a certain probability distribution. An example of a realization of stochastic process is shown in Fig. 6.

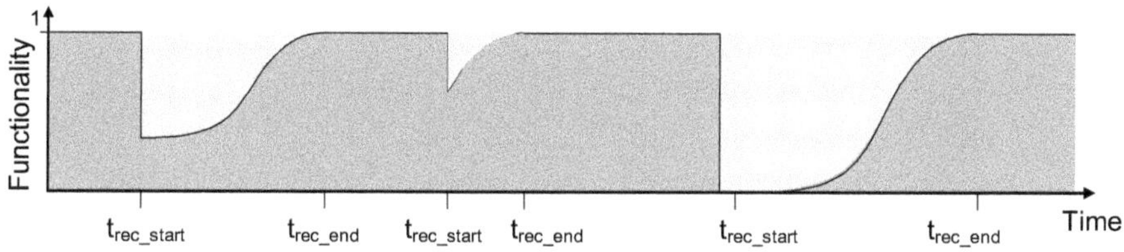

Fig. 6. Functionality due to multiple hazards events.

For certain stochastic processes, resilience, defined as the average functionality over time as depicted in Fig. 1, assumes a constant value. For illustration purposes it is assumed that hazard events follow a homogenous Poisson process and consequently the time interval between events τ are exponentially distributed with the parameter λ:

$$p_f(\tau) = \lambda \cdot e^{-\lambda\tau}$$

Furthermore, it is assumed the functionality loss L due to event is instant and constant, and the recovery time t_r is also constant. In this case the elementary functionality diagram is shown in Fig. 7.

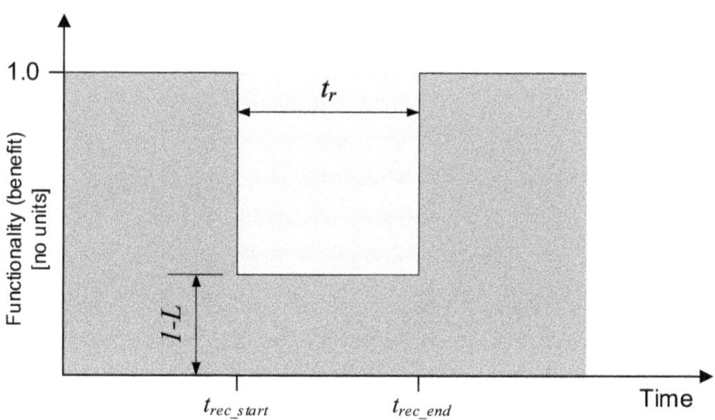

Fig. 7. Elementary functionality diagram.

To evaluate resilience, one needs to distinguish two cases: In the first case, where the random interval τ is smaller than recovery time t_r, and the resilience can be estimated as follows:

$$Re^{I} = \int_{0}^{t_r} \lambda e^{-\lambda\tau} \cdot (1 - L)\, d\tau = (1 - L) \cdot \left(1 - e^{-tr \cdot \lambda}\right)$$

In the second case the interval τ is larger than recovery time t_r, the resilience can be estimated as follows:

$$Re^{II} = \int_{t_r}^{\infty} \lambda e^{-\lambda\tau} \cdot \left(1 - \frac{t_r \cdot L}{\tau}\right) d\tau = e^{-t_r \cdot \lambda} - L \cdot t_r \cdot \lambda \cdot \Gamma(0, t_r \cdot \lambda)$$

where $\Gamma(0, t_r \cdot \lambda)$ is incomplete upper Gamma function. By adding the expressions for both cases one can obtain the full expression for resilience:

$$Re = 1 + \left(e^{-t_r \cdot \lambda} - 1\right) \cdot L - L \cdot t_r \cdot \lambda \cdot \Gamma(0, t_r \cdot \lambda)$$

For example assuming the average interval of 20 years i.e. $\lambda = 0.05$, one can obtain resilience diagrams as a function of recovery time for different functionality losses as presented in Fig. 8.

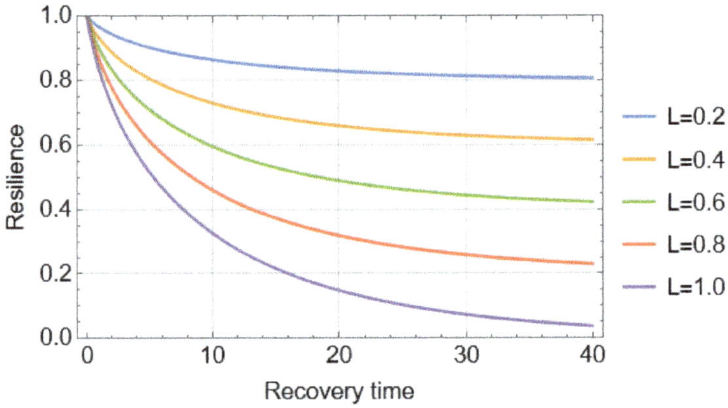

Fig. 8. Resilience due to stochastic events.

One realization of stochastic process is presented in Fig. 9. It should be noted that since the chosen stochastic process is stationary the convergence to the constant value of resilience is very fast. In Fig. 9, one can observe that the events can occur before the infrastructure is recovered. In severe cases this may lead to the "point of no return" i.e. that the road infrastructure may never recover its original functionality.

The presented procedure may also apply to more complex elementary resilience diagrams, reflecting the more realistic behavior of infrastructure affected by disruptive events.

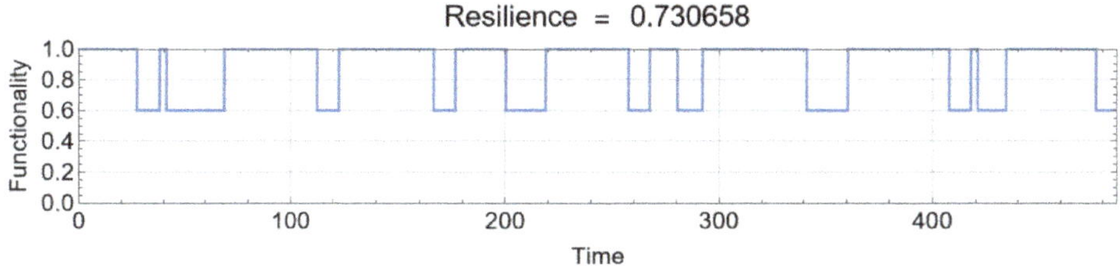

Fig. 9. Realization of stochastic process and corresponding resilience.

4 Measures to Improve Resilience

It is evident from the previous chapter that the resilience increases if the functionality loss due to a disruptive event and recovery time is reduced. In the practice the functionality loss in the road network can be achieved by increased robustness and redundancy. The rapid recovery incl. the emergency measures reduce self-evidently the time needed to restore full functionality and increases the resilience. Finally, all measures both prior to event and after the event are subject to resourcefulness. The knowledge and ingenuity of engineers i.e. their resourcefulness is indispensable to implement adequate measures. Clearly, also human, material and financial resource are necessary for preventative and recovery measures.

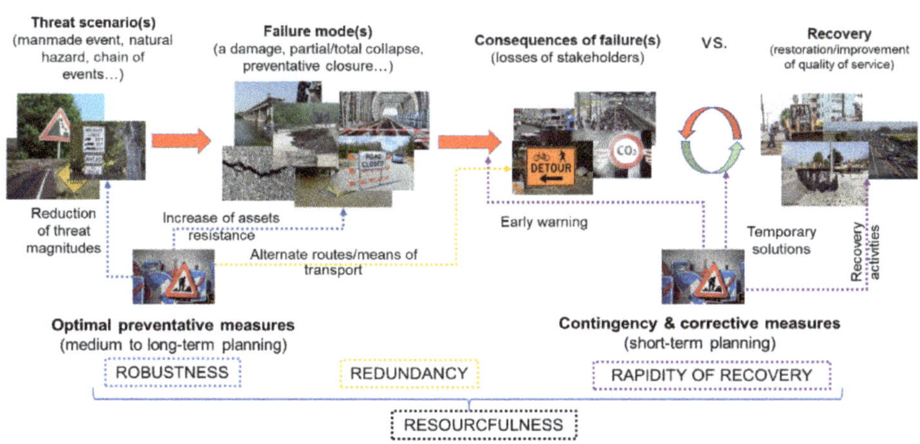

Fig. 10. Resilience framework and measures to improve resilience [19].

The resilience framework and the measures that affect different aspects of resilience i.e. robustness, redundancy, rapidity of recovery and resourcefulness are schematically presented in Fig. 10. The proposed framework is quite similar to one used in risk analysis with the significant difference that the recovery duration is not detailed within the risk analysis.

4.1 Measures to Improve Robustness

Traditionally, robustness refers to the ability of systems to withstand a disruptive event of a certain magnitude while experiencing little to no functionality loss. However, this concept can be broadened to encompass the system's ability to either mitigate the intensity of disruptive events or evade them entirely.

Measures to improve robustness are preventative measure that needs to be performed before a disruptive event occur. Typically they include

- the reduction of exposure to threats e.g. installation/upgrade of protection structures/devices, environmental adaptations at objects' locations and
- Increase of asset resistance to failures e.g. increasing resistance/displacement capacity of assets, controlling/triggering desired failure mode of assets.

Examples of measures to improve robustness are protective structure like avalanche galleries, rockfall protective nets and increasing load capacity of bridges.

It's important to recognize that robustness entails the capacity of road infrastructure to withstand a wide array of events, including unforeseen ones.

Within the quantified framework for resilience, robustness is defined as residual functionality after a disruptive event. This is however oversimplification as the residual functionality is also influenced by redundancy [19].

4.2 Measures to Improve Redundancy

Redundancy refers to the availability of substitutable elements in systems that can be activated in the wake of a disruptive event. If there is sufficient redundancy e.g., alternative routes with sufficient capacity the functionality of infrastructure will be moderately or not affected.

Implementing redundancy is a preventative measure that needs to be undertaken before a disruptive event occurs. Typically they include

- building alternate road routes with adequate capacity.
- increase of traffic mode interchangeability and the adequate capacity of alternative transport modes and
- provide enough road surface to accommodate maintenance measures (see Sect. 3.4).

A prime example of providing redundancy in a road network is the construction of a second tube of the Gotthard road tunnel. While the traffic capacity of both tubes is kept the same as having only one tube, the redundancy ensures that in instances such as maintenance interventions or emergencies, one tube can be utilized to its full capacity [19].

4.3 Measure to Improve Rapidity of Recovery

Rapidity of response refers to the ability to promptly address priorities and achieve goals in a timely manner in order to contain losses. The speed of response relies on the efficient utilization of available resources, thereby reducing recovery time and consequently enhancing resilience.

Preparations for recovery actions in the network may also be planned in advance, but in general need to be adapted in the wake of an disruptive event. Typically they include

- efficient repair/rebuild activity planning (prioritization) and execution,
- providing temporary roads and road structures,
- continuous training and improvements of the response teams and
- swift legal response (e.g., emergency funding)

Examples of measures to improve rapidity of recovery include rehabilitation of objects in manner that maximizes the increase of functionality or construction of temporary bridges that ensure sufficient functionality of road network [19].

4.4 Resourcefulness

Resourcefulness is the capacity to mobilize and apply material and human resources to achieve goals in the event of disruptions. The resourcefulness is a matter of careful resource planning and capacity building. If there is enough human, material resources in case of failure the recovery can proceed without any delays and hence increasing resilience. Typically they include

- the capacity to identify problems,
- expertise and experience of engineers to find optimum measure,
- prudent allocation of emergency budget, manpower and machinery,
- response planning that fosters simultaneous repairs of several assets and
- temporary solutions to provide minimum operational network performance.

The resourcefulness affects other aspects of resilience. The innovation and ingenuity of engineers, along with the availability of financial and material resources, are prerequisites for achieving the optimal balance between resilience and costs.

5 Case Study

A quantitative resilience framework has been implemented in a web-based standalone decision support tool (DST) prototype, developed within the EU Horizon 2020 project SAFEWAY [20]. The development, testing, and showcasing of the DST were based on a pilot project in the region of Santarém in Portugal. As a basis for resilience analysis, a threat from a 100-year flooding scenario is chosen. By a previous geographic coincident analysis it was estimated that 5 bridges and 28 road sections may be exposed simultaneously to this hazard as presented in Fig. 11.

This benefit, actually functionality is normalized assuming demand change in the future (linear traffic growth of 1%), which is factored into the evaluation of traffic redistribution due to flooding event.

In the assumed disruptive 100-year flooding event set to occurs in 2024, structural analysis indicates that three road bridges—Torto, Alviela, and Don Luis I—can simultaneously fail. In order to determine the optimum prioritization of recovery actions, as indicated in Sect. 4.3, two courses of action are considered:

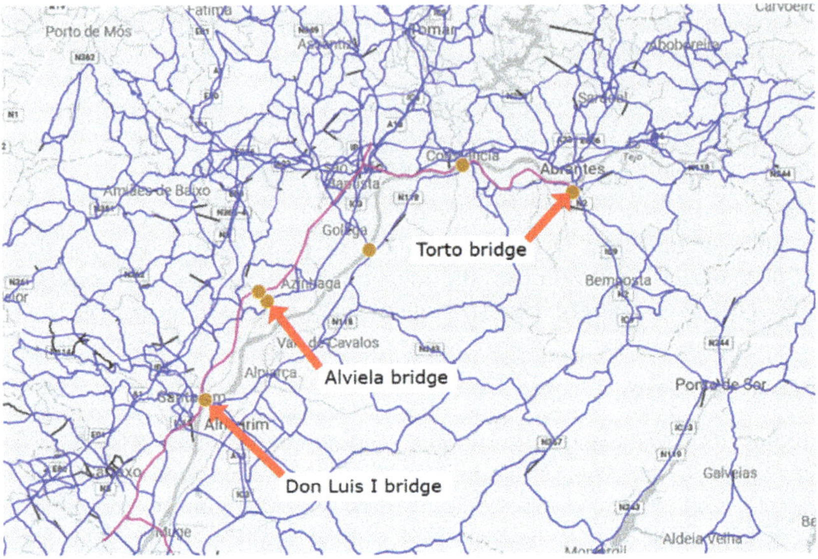

Fig. 11. Map of Santarém region in Portugal with bridge (orange dots) affected by 100-year flood.

- Torto bridge rehabilitation: – The rehabilitation of Torto bridge starts immediately and lasts for 300 days. The other failed two bridges are not rehabilitated. For this course of action the functionality is represented in Fig. 12(a)
- Dom Luis I bridge rehabilitation: The rehabilitation of Dom Luis I bridge starts immediately and lasts for 500 days. The other two failed bridges are not rehabilitated. For this course of action the functionality is represented in Fig. 12(b)

The associated resilience graphs depict the normalized benefit i.e. functionality on the y-axis (the baseline is taken for the year 2020), while the time is given on the x-axis (months). The cut-off of the analysis is adopted at year 2050.

It is seen in the analyzed case, that is better to repair the Dom Luis I bridge (Strategy 2) than the Torto bridge (Strategy 1), even if this repair lasts 200 days longer. It should be mentioned the recovery costs of the Dom Luis I bridge twice as high as those of Torto bridge. The decision on whether the higher costs associated with the Dom Luis I bridge justify the increase in resilience from 0.845 to 0.917 ultimately lies with the infrastructure operator.

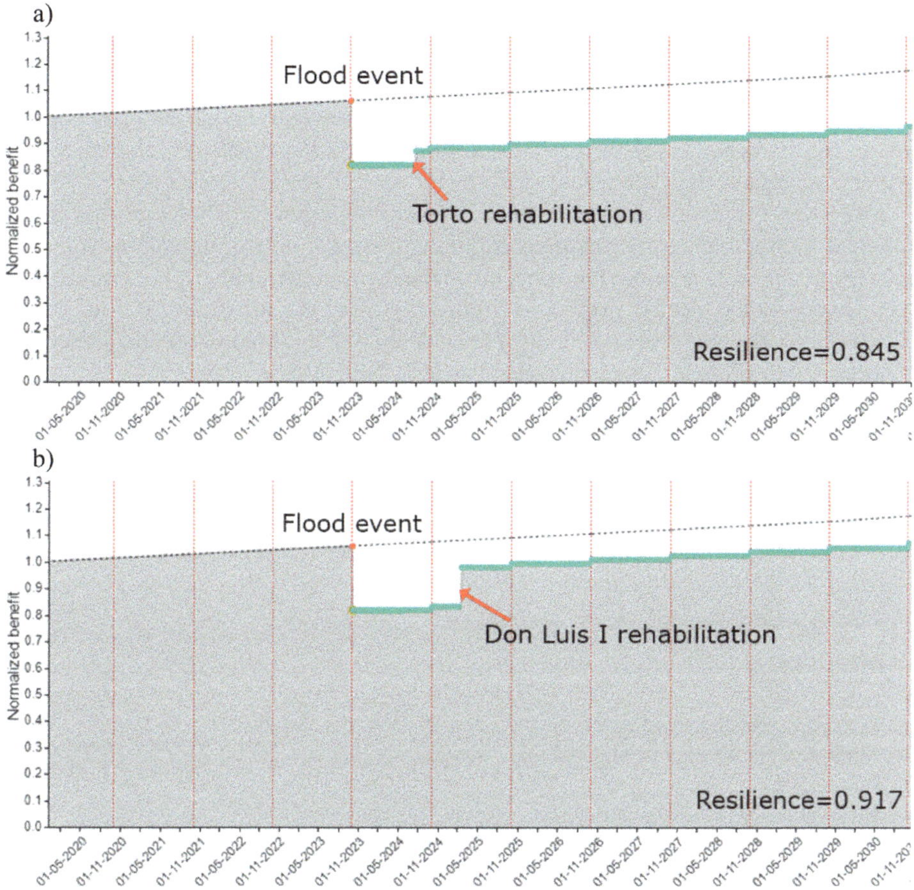

Fig. 12. Resilience graphs from DST.

6 Conclusion

The primary purpose of road infrastructure is to enable mobility of people and goods, thereby contributing to both economic and societal prosperity. This contribution is maximized when traffic flows freely without impediments, representing a functionality level of 100%. However, quantifying the functionality of a disrupted network poses a challenge. It seems logical to utilize changes in traffic flow to evaluate the functionality of disrupted network. This is also a pragmatic approach since the calculation of traffic flow i.e. travel times between any origin/destination pair is already widely used. Resilience, as proposed in existing literature, can be evaluated as the average functionality over time, with resilience loss understood as the additional travel time resulting from network disruptions.

The framework presented in this paper allow for consideration of collateral effects of traffic redistribution such as change in accident rates and environmental impact.

However, the issue of demand change needs further discussion and particularly the temporal misalignment between the development of demand and supply following a catastrophic event needs further research. The recovery time is as shown in the paper an essential parameter for resilience evaluation, yet reliable data for this parameter remain scarce, and its dependency on available resources poses a significant challenge. The further challenge is the dependency of recovery time on available resources.

Moreover, evaluating resilience solely for transportation infrastructure may not be sufficient. Disruptive events can affect multiple infrastructures simultaneously, and the recovery of one infrastructure may depend on the recovery of another [21]. For instance, the recovery of the electric grid can significantly influence the recovery time of road infrastructure. In some cases, adopting a holistic approach to community resilience may be necessary to address these interdependencies effectively.

Acknowledgements. The resilience framework presented herein is inspired by the work within the PIARC TC3.3. WG2: Measures for improving resilience of road network. The decision support tool for risk/resilience assessment mentioned in this paper was developed within European H2020 SAFEWAY project: "GIS-based infrastructure management system for optimized response to extreme events on terrestrial transport networks", grant agreement No. 769255

References

1. Werner, E.: Resilience in development. Curr. Dir. Psychol. Sci. **4**(3), 81–85 (1995)
2. Holling, C.S.: Resilience and stability of ecological systems. Annu. Rev. Ecol. Syst. **4**(1), 1–23 (1973)
3. Bruneau, M., et al.: A framework to quantitatively assess and enhance the seismic resilience of communities. Earthq. Spectra **19**(4), 733–752 (2003)
4. Smith, A.: Inquiry into the Nature and Causes of the Wealth of Nations. W. Strahan and T. Cadell, London (1776)
5. Departement of Transportation: Industry Snapshots: Uses of transportation. Departement of Transportation, Wahsington D.C. (2015)
6. Kurte, J., Esser, K.: ADAC study on mobility: Benefit of road traffic. ADAC, Munich (2008)
7. Federal Office for Spatial Development and Federal Roads Office: The benefit of transportation – Subproject: Contribution of transportation to value creation in Switzerland, Bern (2006)
8. Queiroz, C., Guatam, S.: Road Infrastructure and Economic Development: Some Diagnostic Indicators. World Bank Publications, Washington, D.C. (1992)
9. Frischmann, B.M.: Infrastructure - The Social Value of Shared Resources. Oxford University Press, New York (2012)
10. PIARC Technical Committee A.3 Road system Economics and Social Development: Approaches to evaluation of social impacts of road projects. PIARC (2012)
11. Faber, M.H.: On sustainability and resilience of engineered systems. In: Routledge Handbook of Sustainable and Resilient Infrastructure, pp. 28–49. Routledge, London (2018)
12. Gross, J.L., Yellen, J.: Handbook of Graph Theory. CRC Press (2004)
13. Bonacich, P.: Power and centrality: a family of measures. Am. J. Sociol. **92**(5), 1170–1182 (1987)
14. Wardrop, J.G.: Some theoretical aspects of road traffic research. Proc. Inst. Civ. Eng. **1**(3), 325–362 (1952)

15. Tanasić, N., Ilić, V., Hajdin, R.: Vulnerability assessment of bridges exposed to scour. Transp. Res. Rec. **2360**(1), 36–44 (2013)
16. Erath, A., Birdsall, J., Axhausen, K., Hajdin, R.: Vulnerability assessment of the swiss road network. Transp. Res. Rec J. Transp. Res. Board. **2137**, 118–126 (2009)
17. Greater New Orleans Community Data Center. https://www.datacenterresearch.org/data-resources/neighborhood-data/district-8/lower-ninth-ward/. Accessed 17 Mar 2024
18. Nikola, T., Rade, H.: Resilience-based decision support tool for management of transportation infrastructure. In: Pellegrino, C., Faleschini, F., Zanini, M.A., Matos, J.C., Casas, J.R., Strauss, A. (eds.) EUROSTRUCT 2021. LNCE, vol. 200, pp. 209–218. Springer, Cham (2022). https://doi.org/10.1007/978-3-030-91877-4_25
19. Hajdin, R., et al.: Measure to improve resilience of road networks. PIARC Technical report (2023)
20. Hajdin, R., Tanasić, N.: Deliverable 6.2: Decision Support System. H2020 SAFEWAY project (2021)
21. Blagojević, N., Stojadinović, B.: A demand-supply framework for evaluating the effect of resource and service constraints on community disaster resilience. Resilient Cities Struct. **1**(1), 13–32 (2022)

Resilience, Robustness and Redundancy of Infrastructure System

Aerodynamic Shape Optimization of Long-Span Bridges

Idham Ahraf[1](✉) and Gledson Rodrigo Tondo[2]

[1] Bauhaus-Universitat Weimar, Weimar, Germany
idham.ahraf@gmail.com
[2] Institut Für Konstruktiven Ingenieurbau (IKI), Bauhaus-Universitat Weimar, Weimar, Germany

Abstract. This work develops an optimization algorithm to select the most optimal section for a long-span bridge in terms of the minimum section weight by limiting the flutter limit. Parametric variations in the depth and width of the fairing shape are introduced as variables in the optimization process. Several methods for prediction, namely Polynomial, Extreme machine Learning, and Gaussian Process, are used to build the response surface model, to ensure that the design is reliable enough to always be on the safe side with the uncertainty in design variables, reliability-based design optimization (RBDO) is also introduced.

Keywords: Reliability-based design optimization · Aerodynamic Shape · Long-span Bridges

1 Introduction

Optimization process becomes an important factor nowadays in overcoming the limited resource, engineers want inexpensive and environmentally friend structures. Today, many structures tend to be built slenderer which are not only aesthetically pleasing but also structurally efficient. In the view of structural behavior, specifically in bridge structure, dead load is one of the governing factors in structural design since it is built longitudinally so that the bending moment produced depends on the length of the bridge deck. There are other factors that influence the behavior of the structure as a whole, but optimizing the volume of the structure can be a starting point for achieving good structural designs for the future.

One of the turning points that motivate many scholars to study the effect of wind phenomenon in Bridge design is the collapse of Tacoma Narrows Bridge in 1940. Prior to this event, many bridges were designed based on deflection theory, which considers the relationship between the deck and cable deflection and the required stiffness for increasing spans and without consideration of an environmental effect. The Tacoma Narrows Bridge collapsed primarily due to the aero-elastic flutter. However, these factors were not originally considered and only became part of the later forensics. This event raised awareness of Wind-Structure Interaction (WSI) and the aerodynamic behavior

J. C. Matos et al. (Eds.): IPW 2024, LNCE 494, pp. 31–40, 2024.
https://doi.org/10.1007/978-3-031-60271-9_1

of long-span bridge decks. Since then, aerodynamic and aero-elastic analysis became important for long span bridge design.

This case can be seen as an attempt to design a bridge with a longer span length by modifying the shape of the deck without adequate knowledge about the effect of wind phenomena on slender structures. The introduction of the H-shape section at Tacoma Bridge is seen as an optimization process to reduce the weight of the deck, where most of early generation suspension bridges used a frame section deck. Today more research about environmental load has taken a role in the development of design structure. Studies of slender structures need to be accompanied by a review of the effects of environmental loads on its behavior (Fig. 1).

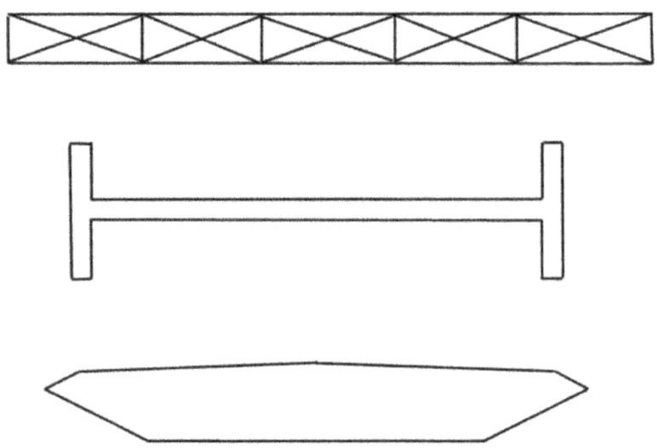

Fig. 1. Illustration of The Development Suspension Bridge Cross-section, from top to bottom: Frame section (Brooklyn Bridge, 1883), H-shape cross section (Tacoma Bridge, 1940), and Trapezoid shape cross section (Great Belt Bridge, 1998). **(Structural Research Laboratory and Washington Toll Bridge Authority,** 1949) **(Larsen,** 1993)

1.1 Flutter Instability Theory

Several phenomena occur to structure due to wind load, when the flow meets a bluff body such as bridge deck, the flow field around the body generates the flow-induced forces and these forces excite the structure and generate the flow induced vibrations that produce specific behavior in the structure.

These phenomena can be categorized based on limited and divergent amplitude wind-induced vibration. Phenomenon such as Vortex Induced Vibration (VIV) and buffeting, rain-induced vibration on cable is limited amplitude vibration whereas galloping, torsional and coupled flutter are considered as divergent-amplitude wind-induced vibration. Normally, Among the most common aerodynamic phenomena in long-span bridges are Flutter instability and VIV because these phenomena have been shown to cause failure or exceed the serviceability of the structure.

Flutter instability or aerodynamic instability is Instability of cross section indicate by increasing amplitude response due to loss of Total Structural damping and often divided into two categories, Single-Degree of Freedom (SDOF) flutter (for example, torsional flutter and galloping) and Two-Degree of Freedom (2DOF) flutter (also known as classical flutter). There are several approaches to calculate flutter instability, using Theodorsen theory that is based on flat plate airfoil, and Scanlan theory that used flutter coefficient considering a bluff body of the section (**Simiu, E. and Scanlan, R. H.,** 1996) (Theodorsen, 1935**).**

Long-span bridges are characterized by relatively light and flexible structures with low damping, which means that they are at risk of being excited into sizable motion by environmental forces. Designing deck cross-section shapes that are excited by vortex shedding neither in torsional nor vertical bending need to be carefully calculated to have a robust long-span bridge. There are two approaches to prevent aeroelastic instability, first by modifying the whole structure in the hope to change the modal frequency of structure. Normally engineers try to increase the torsional stiffness and separate the torsional and flexural mode as far as possible. The other way is by changing the properties of deck section, with a better section, the dragging force effect can be minimized so that the wind velocity where divergent amplitude occurs will shift. The later method is used mostly for the long span bridges when changing the structural system is not possible anymore.

2 Fundamental Studies

The actual structure considered in the calculation is Lillabelt Bridge in Denmark. The background in preparing a flutter analysis for the actual cross-section of the Lillabelt Bridge is to help understand the optimization process with the actual cross-section as a benchmark. Analytical and semi-analytic methods have been used in many practices, and these methods are quite reliable to predict critical wind velocity. Here, using section of Lilleabelt bridge the result critical wind velocity *Ucr* is 92 m/s indicates when the torsional mode damping become zero (Ostenfeld, 1970) (Fig. 2).

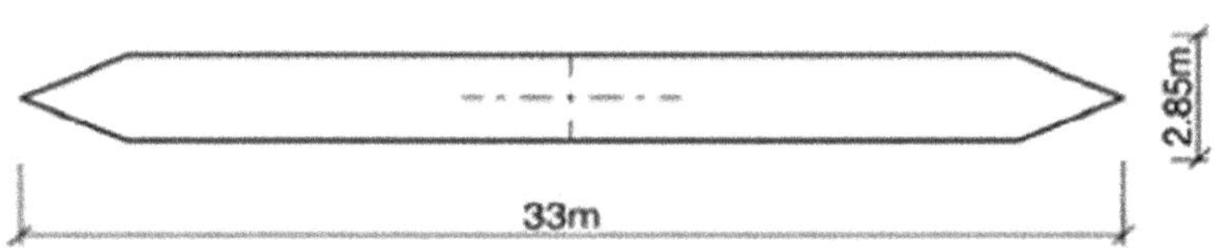

Fig. 2. Deck Cross Section Lilleabelt Suspension Bridge in Denmark 1970

In order to give a good illustration of how the modification of section can change the aerodynamic response of the structure, the variables of *Dbot* and *Wext* are introduced. *Dbot* is defined as the depth of bottom fairing and *Wext* is the width of the fairing, and by changing these variables, the angle of fairing is also changing where the width of the upper and bottom lane is fixed, while the top depth of fairing remained the same.

A sample field is created based on 25 samples of cross-section shapes, varying the *Dbot* from 1 m–3 m and *Wext* from 0 m–4 m, Fig. 3 gives an example of sections with a different value of *Dbot* and *Wext*. The section starts from a rectangular section (non-fairing section) and varies up to the maximum length of extension 4m, indicated by the last digit of the sample number 1–5.

These training sections are used to predict the response of the rest of the sections within the boundary by using the response surface method. Response surface models are more or less simple mathematical models which are designed to describe the possible experimental outcome (e.g., the structural response in terms of displacements, stresses, etc.) of a structural system. The response surfaces for the different methods are prepared from the training set *Ucr*, there are 25 points used as the training sample. Each of these points is calculated for the residual error of the prediction method so that the quality of the prediction can be measured.

There are three methods introduced, which are Polynomial, Extreme Learning Machine (ELM), and Gaussian Process (GP). The polynomial method is the most common regression method that has been widely used in statistics, other methods based on random processes will be compared to the polynomial method, which are Extreme Learning Machine and the Gaussian Process.

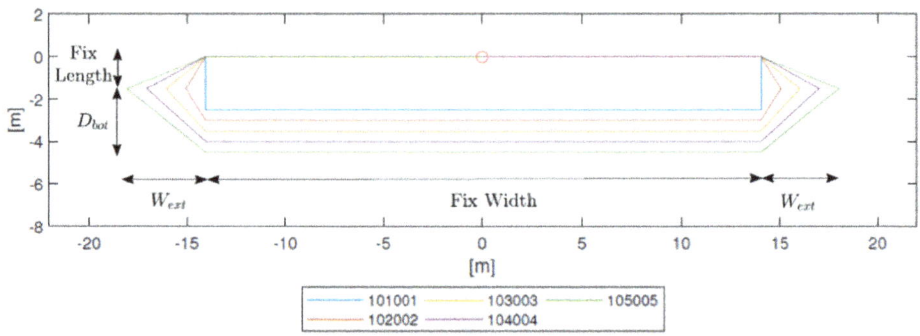

Fig. 3. Example of Parametric Variation Considered in the Deck Cross-Section for the Training Sections. The top and bottom width remain fix, and the variations are in *Dbot* and *Wext*

2.1 Stability Analysis of the Prediction Method

The first regression method is polynomial regression, which is presented in Fig. 4. This method is very common in many applications because of its robustness. This is a simple method and produces a polynomial function that is characterized by continuous and differentiated so that it is advantageous for a problem that can be solved analytically. However, the disadvantage of this method is that its accuracy cannot be adjusted to the desired limit. It shows that compared to other methods (ELM and GP), this method has the largest residual error in the training sample.

Due to random processes in the predictions of ELM and GP, the stability of the results is also a major factor in predicting the quality of the results. In the stability analysis, the flutter limit at *Ucr* 90 m/s was taken as the benchmark. Naturally, ELM always produces a different result because the random data in the matrix is not stored, and only considers the minimum weight function. On the other hand, GP calculates the properties of a random function and tries to minimize its properties and store them, thus naturally producing a stable result. However, both predictions still contain variations in the prediction results which can be seen in Figs. 5 and 6.

From this analysis, it can be concluded that areas with high uncertainty in prediction are indicated by high standard deviation values. The standard deviation represents the

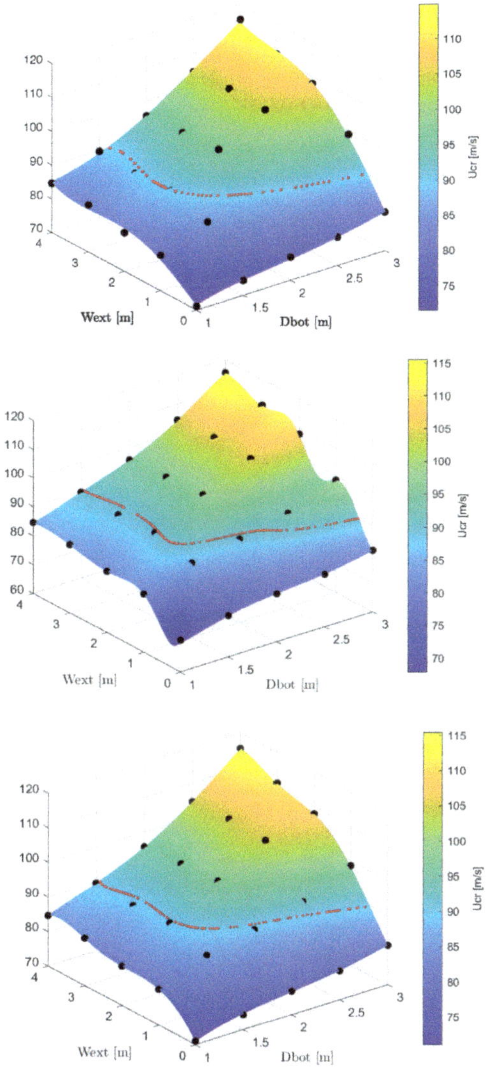

Fig. 4. Response Surface Critical Wind Velocity with Limit function at Flutter Limit $Ucr = 90$ m/s (...). From top to bottom: using Polynomial Order 4, ELM, and GP Method

spreading of *Ucr* at the specific position of variable *Dbot* and *Wext*. In the result of standard deviation for both methods, the data is presented in 2σ, so the confidence level for the respected upper and lower bound is about 95%. So the optimization result based on these methods is considered accurate if the optimized section lies in the area with a small standard deviation, because both methods show some area with high deviation.

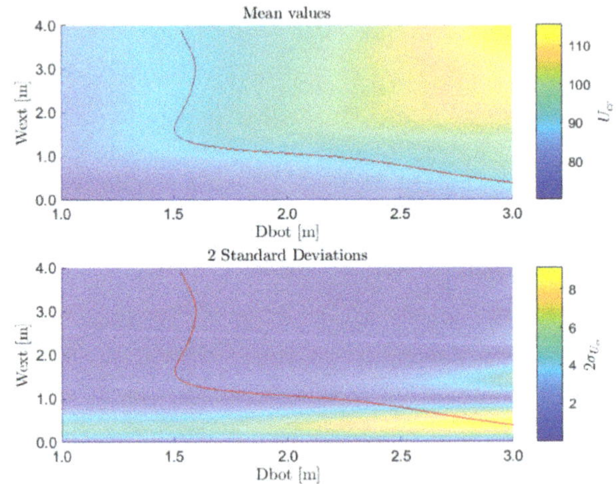

Fig. 5. ELM Regression Result, 20 times simulation. (---) benchmark $Ucr = 90$ m/s Top: Random Process of ELM, The mean values prediction. Bot: Standard Deviation of the prediction result

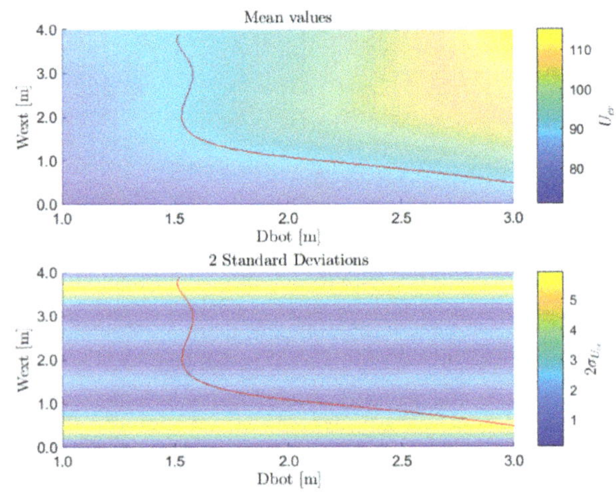

Fig. 6. GP Regression Result. (---) benchmark $Ucr = 90$ m/s Top: Random Process of GP, The mean values prediction Bot: Standard Deviation of the prediction result

3 Reliability-Based Design Optimization

Single objective optimization can be applied by defining the response surface of critical wind velocity to calculating the reliability index, and then the constraint of the minimization the section weight is applied directly to all sections at the limit function. The response surface for 25 sample points can be seen in Fig. 4. On this response surface, the flutter limit as the first objective is already stated, and all the sections located on the flutter limit are called the limit functions. With the definition of limit function, then the reliability index is calculated.

The Arbitrary points located near the limit function are defined and then random sample field for Monte Carlo Simulation (MCS) is generated for each of these points, these random sample represents the uncertainties of the variable *Dbot* and *Wext*. The uncertainties is modeled as a normally distributed with standard deviation of 0.1 as seen in Fig. 7.

The aim of this process is to calculate the actual failure probability and the reliability index of these arbitrary points and then shift these points to the designated target reliability index. This strategy has proven to be simple but requires careful observation. MCS can calculate the failure probability of each random sample if the number of samples itself is sufficient enough so that there are some samples that fall outside the limit function. It is important to note that the result of this simulation cannot give a zero-failure probability or negative reliability index values.

An alternative strategy can be implemented in calculating reliability index, the result of reliable points which is not completely dependent on MCS but also need to be compared with the analytical result from First-Order Reliability Method - Hasofer-Lind Rackwitz Fiessler (HL-RF) (**Choi, S.-K., Grandhi, R., and Canfield, R. A.,** 2006). The result of the shifted MCS points needs to be rerun again using algorithm from HL-RF. The result of HL-RF simulation is the actual reliability index for each arbitrary point and these values must match the predetermined reliability target. However, this strategy can only be used for response surface model using the polynomial method, where the limit function can be taken as a continuously differentiable function.

The other strategy applied in this process is to do MCS for each variable *Dbot* and *Wext*, which means double simulation for each arbitrary point. The problem with shifting the arbitrary points is the difficulty of finding the direction that is perpendicular to the angle of these points. The general solution is to first create an interpolation function for these points and then find the direction by differentiating this function. And it gives another consideration where sometimes the interpolation process also requires a good arrangement of points. By applying a double MCS simulation the step of calculating interpolation function can be skipped with the drawback of additional processing time. The direction of the shift can be calculated directly using the trigonometry principle with respect to each of the primary axes. The result of MCS for this strategy can be seen in Fig. 7 , the mean value is indicated as point B, and the MCS gives two reliability values, namely $\beta 1$ and $\beta 2$, and is illustrated as the distance AB and BC. The reliability index (β), which is the distance of BD, is calculated using the trigonometry principle. Finally, the direction to shift the arbitrary point is defined as βx and βy which are the scaling values for their respective primary axes.

MCS with a reasonable number of samples can be performed and the non-zero failure probability value (Pf) can be obtained. One thing to make sure of is the level of confidence of the result Pf, since zero Pf value may occur, each result of MCS simulation needs to be check, and also this method has twice more results compare to another alternative method. Moreover, the drawback of MCS, which is for a system with highly reliable components, the likelihood of observing a system failure during each Monte Carlo Simulation is very low. In consequence, there is a large variance in the estimation of the system reliability (**Hongzhou Wang, Hoang Pham,** 2006). If the number of samples is not sufficient or the deviation value too small for the point far from the limit value, the result *Pf* from MCS tends to become zero.

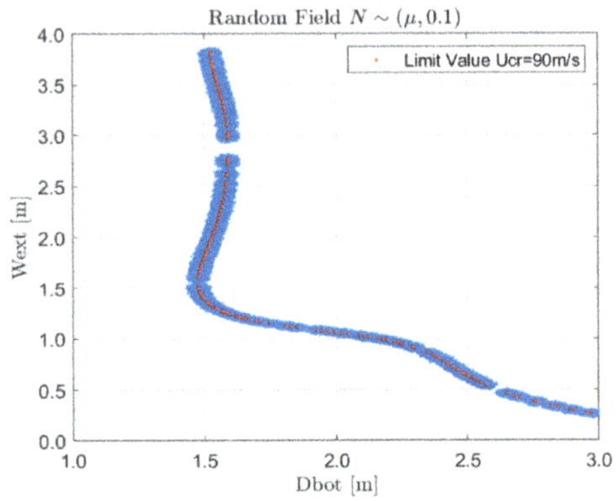

Fig. 7. Samples Random Field of the arbitrary points near the Flutter Limit 90 m/s for MCS Simulation. Using normal distributed with standard deviation σ 0.1, and number of samples 50000

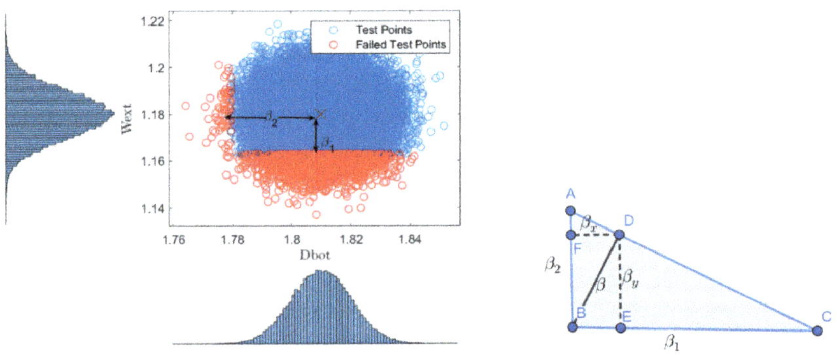

Fig. 8. Monte Carlo Simulation for 2 variables. *β1* and *β2* are the result of each MCS simulation, the actual reliability index (*β*) is calculated using trigonometry principle. The value *βx* and *βy* are the scaling factor to shift the arbitrary point on its primary axes.

4 Summary

To sum up, this study tries to find a more efficient approach with minimum number of simulations performed to examine the effect of aerodynamic phenomena by considering the parametric variations of the fairing on the cross-section represented by the variables *Dbot* and *Wext*. Critical Wind Speed (*Ucr*) objectives are introduced in the process to find the optimal solution in terms of structural response, and reliability-based design optimization was conducted to find the optimized section (Fig. 8).

The trade-off effect between computational effort and accuracy is explained in the use of this method. It was explained that with polynomial regression there is a faster and stronger method to calculate the reliability index, namely the First and Second Order Reliability Methods. Whereas random process regressions like ELM and GP require MCS to calculate the reliability index, but still the reliability results need to be compared with each other.

This result in Fig. 10 shows not only the optimized section with a good agreement with flutter limit objective but also the level of confidence of this section with the definition of the prescribed failure probability. The result of single objective shows a section that is similar to the original section, the comparison can be seen in Fig. 9.

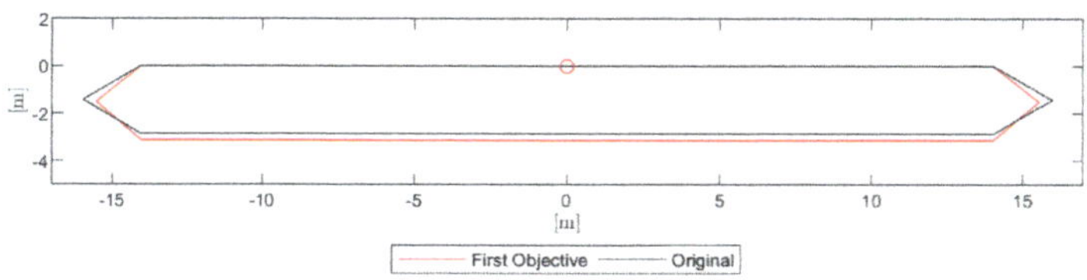

Fig. 9. Optimized cross-section for Single Objective. (---) Original Section as Reference

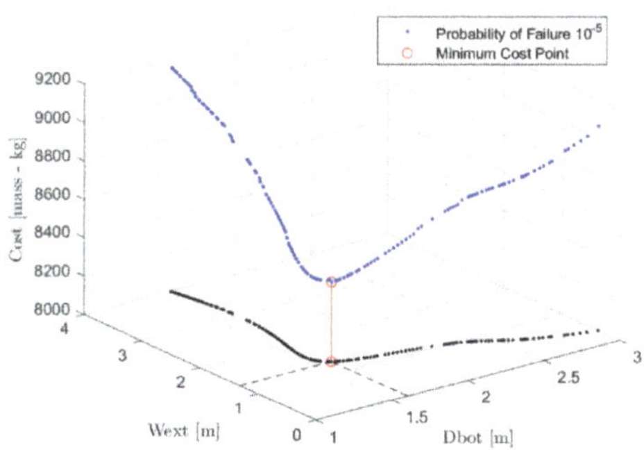

Fig. 10. Cost Optimization for Single Objective (Flutter Limit 90 m/s) with **prescribed failure probability = $\mathbf{10^{-5}}$**

4.1 Discussion

The MCS calculation presented in this thesis is based on the Performance Measure Approach (PMA) method, first is to prepare an arbitrary point near the limit function and then calculate the reliability index for each parameter *Dbot* and *Wext* separately. This method requires more computational effort because it has to perform MCS twice (as much as the variable defined). But the advantage of this method is that there is no need to do a regression analysis to find the direction of the shift in the mean value, because usually MCS is done by considering all variable uncertainties so that the reliability index (β) can be obtained directly. On the other hand, by calculating MCS separately, the shifting scale factor can be obtained as it calculates the projected reliability index to its principal axis.

Designing a structure with high target reliability can satisfy many constraints and increase the robustness of the structure, but it can generate a large cost leading to a more expensive structure. Therefore, the target reliability should be a decision variable that needs to be found to maximize the lifetime of the structure. It is noted that the sensitivity of failure probability needs to be accurately calculated using the procedure proposed in this study.

Acknowledgments. The authors thankfully acknowledge the support of Chair of Modelling and Simulation of Structures, Bauhaus-Universitat Weimar.

Reference

Choi, S.-K.G.: Reliability-Based Structural Design. Springer, London (2006). https://doi.org/10.1007/978-1-84628-445-8

Larsen, A.: Aerodynamic aspects of the final design of the 1624 m suspension bridge across the great belt. J. Wind Eng. Ind. Aerodyn. **48**(2–3), 261–285 (1993)

Ostenfeld, C.J.: Motorway Bridge Across Lillebælt: Design and Construction of the Bridge (1970)

Simiu, E., Scanlan, R.H.: Wind Effects on Structures, Fundamentals and Aplication to Design. Wiely (1996)

Structural Research Laboratory and Washington Toll Bridge Authority. Aerodynamic Stability of Suspension Bridges with Special Reference to the Tacoma Narrows Bridge: A Report of an Investigation, volume 1. University of Washington Press (1949)

Theodorsen, T.: General theory of aerodynamic instability and the mechanism of flutter. NACA (1935)

Wang, H., Pham. H.: Monte Carlo reliability simulation of complex systems. In: Reliability and Optimal Maintenance, by Hoang Pham Hongzhou Wang, pp. 275–294. Springer, London (2006) https://doi.org/10.1007/1-84628-325-6_12

Computational Investigations on the In-Plane Capacity of a URM Wall: Effect of Material Uncertainty

Bora Pulatsu[1(✉)], Semih Gonen[2], Luca Pela[2], and Paulo B. Lourenço[3]

[1] Department of Civil and Environmental Engineering, Carleton University, Ottawa, ON K1S 5B6, Canada
bora.pulatsu@carleton.ca

[2] Department of Civil and Environmental Engineering, Universitat Politècnica de Catalunya, C/ Jordi Girona, 1-3, 08034 Barcelona, Spain

[3] ISISE, Department of Civil Engineering, University of Minho, Campus de Azurém, 4800-058 Guimarães, Portugal

Abstract. Preserving the aging unreinforced masonry (URM) building stock and architectural heritage requires accurate predictions regarding their structural performance under various loading conditions to provide sustainable conservation plans. Often, the considerable variation in material properties and uncertainties regarding the as-is condition of the existing buildings forces us to perform conservative and deterministic analysis using various computational modelling strategies targeting the lower bound capacity of the structural systems. With recent advances in computational power, different probabilistic structural analysis approaches have been presented, often considering loading and material uncertainties within the adopted computational modelling framework. However, experimental data to support simulations is insufficient, and several assumptions often must be made regarding the probability distributions and distribution parameters of the variables. It is rarely investigated how these assumptions affect the behaviour and maximum load-carrying capacity of the analyzed URM structure. To this end, this study proposes a probabilistic analysis framework using the discrete element method (DEM) to investigate the effects of modelling parameters and their probability distribution on the in-plane lateral force capacity of URM walls subjected to combined compression-shear loading. The effect of material uncertainties is addressed by performing a large number of simulations with different bond and brick strength properties explicitly represented in the adopted DEM-based modelling strategy. The results show that the impact of the probability distribution of the bond frictional resistance for the examined URM wall may not significantly influence its behaviour and capacity. The adopted framework can be used in the decision-making processes, preserving cultural heritage and avoiding expensive repair costs.

Keywords: Computational Modelling · Material Uncertainty · Probabilistic Analysis · Discrete Element Method · Masonry

J. C. Matos et al. (Eds.): IPW 2024, LNCE 494, pp. 41–51, 2024.
https://doi.org/10.1007/978-3-031-60271-9_2

1 Introduction

The aging unreinforced masonry (URM) building inventory in Europe and North America, including residential, governmental, public, and cultural heritage, is quite substantial. Considering their aged and deteriorated conditions due to man-made and natural hazards, it is of the utmost interest to conserve existing building stock and accurately predict/assess their structural safety in a cost- and time-efficient manner and shed light on the need for future interventions under the given climate change crisis (such as increased frequency of extreme environmental actions). A considerable portion of old URM buildings suffer from localized cracks, severe deterioration in mortar joint and bond (unit-mortar interface) quality and masonry unit dislocations caused by differential soil settlement, seismic activity and exposure to various types of external loading. However, developing a systematic approach to capturing the short and long-term behaviour of old masonry buildings is a challenging task owing to the complexity of the problem and inherited uncertainties in the material properties, among other factors.

Typically, the nonlinear structural analysis of URM buildings or a component of a building (e.g., walls, pier-spandrel systems, arches, etc.) is performed via either continuum (e.g., [1–4]) or discontinuum-based analysis (e.g., [5–8]), which requires a large number of input parameters related to material properties and high confidence in the boundary and loading conditions. Due to the high computational cost of nonlinear structural analysis, the safety assessment of URM structures is performed using deterministic analysis where the input parameters of the models attain specific values. This approach has been widely used to optimize the available resources. Nonetheless, it is accepted by the scientific community that we can improve our predictions and make better-informed decisions if we can reduce the inherent uncertainties in our models or find a way to incorporate them into the analysis without exceeding the bounds of computational resources. In the former approach, the in-situ investigations (site surveys), data collection via remote sensing technologies, and structural health monitoring are beneficial to advance knowledge about the state of the structure and can be utilized to modify/update the computational models. However, the aleatoric uncertainties regarding the material properties of masonry composite can only be addressed by performing stochastic structural analysis. Therefore, with advances in computational power and parallel computing systems, probabilistic continuum and discontinuum-based analysis have attracted growing interest from researchers from various disciplines in the last decades.

The recent studies not only uncover the notable influence of material uncertainty on the force-displacement response of the analyzed URM structures but also show the diverse failure mechanisms of URM load-bearing systems explored through probabilistic analysis [9–12]. While it is preferable to include material uncertainty in the seismic analysis or safety assessment of URM buildings over the deterministic investigations, it brings additional costs. The main disadvantage of most probabilistic structural analysis approaches is their high computational demand and the absence of material testing (or input data) that can support the predetermined statistical parameters, such as the type of probability distribution and the mean and standard deviation. To this end, the present research investigates the effects of the assumptions we make in the probabilistic analysis of URM walls on the lateral load-carrying capacity. Here, a framework is presented for such an investigation, which exemplifies how the pre-defined probability distribution types of the input material parameters affect the outcome. In this research, a URM wall tested in a laboratory environment is utilized as a benchmark study to validate the proposed discontinuum-based modelling strategy. Once the adopted computational modelling approach is validated, nonspatial stochastic analysis is performed to determine the influence of input parameters on the lateral load-carrying capacity of the analyzed reference URM wall. Then, spatial stochastic analyses are conducted for the most influential parameter considering three different statistical probability distributions (i.e., normal, lognormal and Weibull), and important conclusions are drawn. Note that attention is given to a specific URM geometry and boundary conditions for which the proposed computational model is validated. However, the proposed framework can be utilized to investigate other conditions and parameters.

2 Discontinuous Presentation and Analysis of URM Walls

The URM walls inherently demonstrate significant strength and stiffness differences between the masonry constituents (i.e., unit, mortar and unit-mortar interface), yielding localized cracks and sliding failure predominately at the mortar joints and unit-mortar interfaces once they are subjected to in and/or out-of-plane loading. To capture these local failure mechanisms explicitly and address the effect of morphological features of URM walls, a block-by-block representation of masonry texture can be preferable over the homogenized continuum-based formulation of masonry [7, 13, 14]. In this study, a discontinuum-based analysis based on the discrete element method (DEM) – developed by Cundall [15] – is utilized, where each masonry unit is represented via two rigid blocks with a potential crack surface. The adopted modelling strategy is referred to as simplified micro-modeling or discrete rigid block analysis in the literature and is extensively used in the detailed structural analysis of URM walls [11, 16, 17].

The proposed DEM-based computational model relies on integrating equations of motion for each rigid block in the discontinuous system and predicts their translational and rotation velocities using the central difference method, which is further utilized to compute their relative displacements. In DEM, the mechanical interaction among the adjacent blocks is predicted via pre-defined three orthogonal springs (one in the normal and two in the shear directions) at the contact points, as shown in Fig. 1 Soft contact approach is adopted which allows inter-block penetration depending on the contact stiffness in the normal (k_n) and shear (k_s) directions (see Fig. 1). Note that each rigid block has 6 degrees of freedom – 3 translations and 3 rotations. The relative block displacements are used to obtain point contact forces that follow the pre-defined contact constitutive law. Throughout this study, a recently implemented contact constitutive model is used, which provides a bi-linear elasto-softening stress-displacement relationship for tension, compression and shear regimes, presented in [18]. The slope of the softening branch and the ultimate contact displacement are controlled by the mode-I, mode-II and compression fracture energies. The readers are referred to the reference study for further details regarding the mathematical modelling and computational procedure of the employed contact constitutive models [18].

The bond shear strength at the contact points is controlled by the Mohr-Coulomb failure criteria that require cohesion (c) and friction angle ($\theta_{0,res}$ subscript refers to

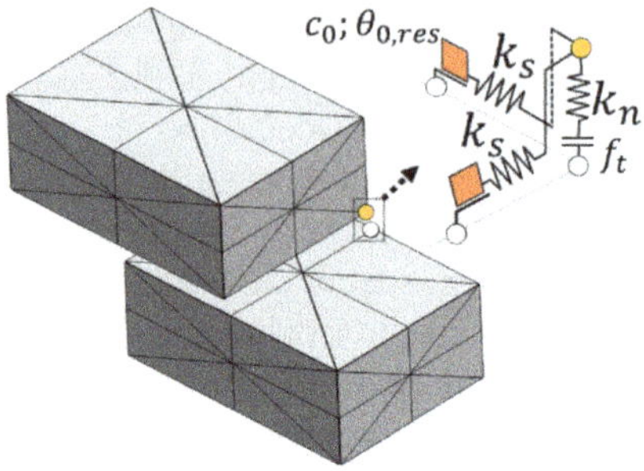

Fig. 1. Representation of two rigid blocks and a contact point.

initial and residual friction angle, respectively), whereas compression and tension cut-off yields functions are used in the normal direction governed by the bond tensile strength (f_t) and the composite compression strength of masonry (f_c). Following the standard predictor-corrector algorithm embedded in the explicit time-stepping solution scheme of DEM, the new stress state at the contact points is updated (if applicable) based on the determined failure criteria. Finally, the new contact forces are used in the equations of motion to predict the block velocities for the next time step. The quasi-static solutions are obtained using Cundall's local damping formulation [19]. It is worth noting that the adopted modelling strategy explicitly captures essential failure mechanisms at the bond – also referred to as unit-mortar interface – (such as cracking and sliding) and units as illustrated in Fig. 2 Throughout this study, a commercial three-dimensional (3D) discrete element code, 3DEC developed ITASCA, is used [20].

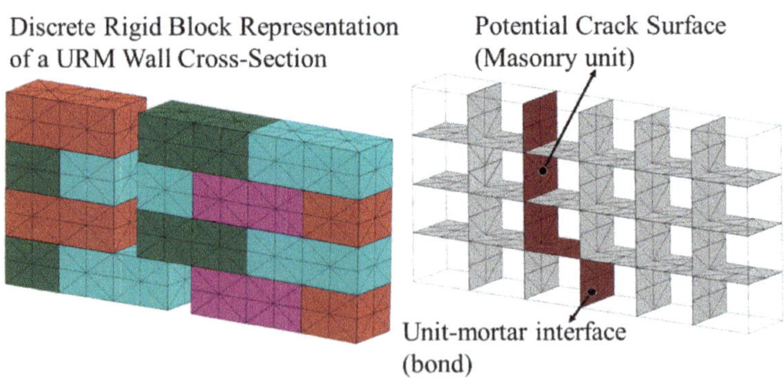

Fig. 2. Illustration of URM wall cross-section using discrete rigid blocks together with the associated contact planes corresponding to masonry unit potential crack surface and unit-mortar interface.

3 Validation of the Proposed Computational Modelling Strategy

First, the proposed DEM-based modelling approach is validated by comparing the results of the adopted computational modelling strategy against the recently performed experimental campaign on unreinforced calcium silicate (CS) brick masonry walls. In the benchmark study, a large-scale CS masonry wall (length × height × width – 4.0 × 2.760 × 0.102 m^3) was subjected to constant pre-compression vertical pressure (0.5 MPa) and lateral cyclic loading [21]. The authors of the reference study also conducted a comprehensive material characterization related to the masonry constituents (brick, mortar and bond) and composite (masonry prisms) that are utilized in the numerical model to assign input parameters (Table 1). Since the deformations are lumped at the contact planes within the proposed discontinuum-based modelling approach, the elastic properties of masonry (e.g., Young's modulus, E_m) are utilized to predict unit-mortar interface (or bond) stiffness (i.e., $k_{n,j} = E_m/h$, h is the vertical joint spacing). The missing input parameters and the mode-I (G_I), mode-II (G_{II}) and compressive fracture energy (G_c) values are assumed following the recommended values in the literature (e.g., [22–24]). Note that the subscripts j and u are assigned to the input parameters to denote that they refer to the bond and unit properties, respectively.

Table 1. Contact properties used in the computation model.

Unit-mortar interface			
$k_{n,j}$, $k_{s,j}$ (GPa/m)	$f_{t,j}$, c_j, f_c (MPa)	$\theta_{0,j}$, $\theta_{res,j}$ (°)	G_I, G_{II}, G_c (N/m)
38, 15	0.1, 0.14, 5.9	27.5, 27.5	2.1, 21, 9440
Masonry unit			
$k_{n,u}$, $k_{s,u}$ (GPa/m)	$f_{t,u}$, c_u (MPa)	$\theta_{0,u}$, $\theta_{res,u}$ (°)	G_I, G_{II} (N/m)
5kn,j, 0.4kn,u	1.4, 2.1	45, 45	21, 210

Following the test procedure, a cantilever boundary condition is considered in the discrete element model where the loading beam, defined above the URM wall, is only restrained in the out-of-plane direction. However, different from the reference study, a monotonic loading scheme is defined in the computational model, where the base shear and drift (i.e., lateral displacement at the monitoring point divided by the height of the wall) are monitored during the simulation. In Fig. 3a, the predicted force-drift behaviour (denoted as DEM) is compared against the backbone curve of the cyclic response of the wall obtained from the experiment. A reasonable match is noted in terms of ultimate lateral load and the location of the first crack presented in the benchmark study, shown in Fig. 3b Therefore, the given comparison is considered as the validation of the proposed modelling strategy, and it is further utilized to perform probabilistic analysis considering the spatial and nonspatial variation of the material properties in the following section.

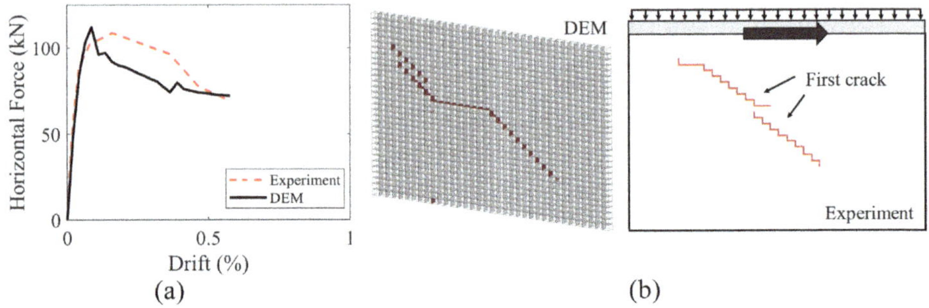

Fig. 3. (a) Comparison of force-drift curves computational model (DEM) and experimental results; (b) predicted the first crack vs benchmark study [21].

4 Data Preparation for Nonspatial and Spatial Probabilistic Analysis

In the probabilistic analyses, random variables and dependent parameters are considered for the bond and unit properties of masonry, based on the previous work of the authors [10, 25]. The mean values of the parameters and their standard deviations are taken from the experimental study [21] where available. Initially, for the nonspatial probabilistic analysis, a normal distribution is assumed for all random variables, as seen in Table 2. Considering the physically meaningful values that these parameters can attain, the distributions corresponding to joint cohesion, friction angle and brick stiffness are truncated within the bounds of [0.02 ∞], [15, 40], and [1 ∞], respectively. After establishing the statistical distributions and their parameters, the Latin Hypercube Sampling (LHS) approach is employed to generate sample values for Monte Carlo simulations [26]. A set of 200 simulations was conducted using the LHS method, and no correlations between the random variables were taken into account. The number of simulations is determined by observing the variation in the output maximum force, as illustrated in Fig. 4a.

The spatial stochastic analyses are performed considering the spatial variation of the most influential parameter determined in the previous phase. In addition to the normal distribution considered for the friction angle, lognormal and Weibull distributions (parameters A = 27.5 and B = 3) are defined; see Fig. 4b These two distributions are truncated in between 15–40 as well. Then, 200 simulations were conducted for each type of distribution, reaching a total of 600 simulations – excluding the previous 200 nonspatial analyses. In each of the 200 simulations, all random variables but the friction angle (θ_j) non-spatially varied and took the same values as they did in the nonspatial analyses, while the θ_j values changed spatially and attained different values for each of the 1801 unit-mortar interfaces (or contact planes) considered in the discrete element model. Interested readers are referred to the reference study for further information regarding the adopted approach used in assigning parameter values [25].

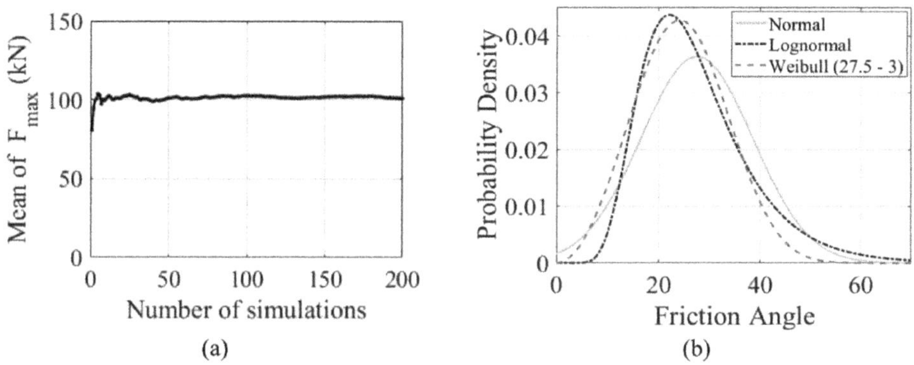

Fig. 4. (a) Variation of the mean of the maximum force with respect to increasing simulation number, (b) Probability distributions

Table 2. Input parameters together with their probability and statistical values.

Random Variable	Probability Distribution	Mean Values (μ)	Coefficient of Variation (CoV)
c_j	Normal	0.14 MPa	0.40
θ_j	Normal	27.5°	0.40
f_c	Normal	5.90 MPa	0.088
E_m	Normal	3.2 GPa	0.30
$f_{t,u}$	Normal	1.44 MPa	0.058
Dependent Variable		**Relationship**	
c_u		$2f_{t,u}$	
c_j		f_j	
G_I		$0.015f_{t,u}$	
G_{II}		$10G_I$	
G_c		$1.6f_c$	

5 Results of the Probabilistic Analysis

This section presents the results of the probabilistic analyses using the validated URM wall with identical load and boundary conditions. The given workflow offers to observe the effect of the pre-defined probability distribution of an input parameter.

5.1 Nonspatial Probabilistic Analysis

The nonspatial probabilistic analysis is performed considering 200 simulations following the statistical properties mentioned in the previous section. The results show a substantial correlation between the lateral load-bearing capacity of the URM wall and a plastic shear strength parameter (bond friction angle). In Fig. 5, a high correlation with a Pearson correlation coefficient value of 0.90 between the bond friction angle and the ultimate load can be noted compared to other parameters (Fig. 5), including cohesion, masonry compression strength, and brick tensile strength. These results suggest that most of the predicted failure mechanism includes sliding behaviour, mostly noted at the bed joints. It is also worth noting that pure sliding or mixed mechanisms are expected for non-slender (low aspect ratio) URM walls, and interaction between the mechanisms may occur during the analysis. To better understand the effect of the probability distribution type of bond friction angle, it is spatially varied, considering different statistical distributions in the next section.

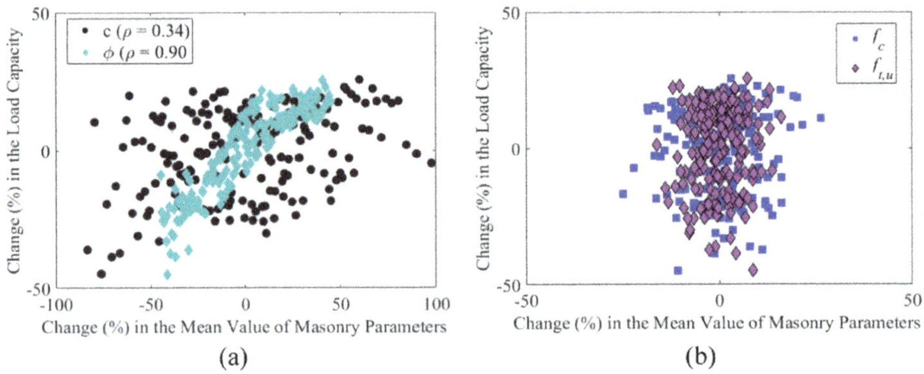

Fig. 5. Influence of the random parameters on the lateral force capacity of the wall in nonspatial probabilistic analysis.

5.2 Spatial Probabilistic Analysis

Unlike the previous analyses, the bond friction angle is spatially varied considering normal, lognormal and Weibull statistical distributions. Figure 6 shows a relatively consistent response regarding maximum force and macro force-drift behaviour in the analyzed URM wall compared to nonspatial analyses. Since similar force-drift curves are obtained from the pre-defined statistical distributions for the bond friction angle, only the normal distribution results are given and compared in Fig. 6 In Table 3, a 10% drop in the lateral load-carrying capacity and a significant decrease in its variation is also noticed when the nonspatial and spatial analyses are compared. This observation complies with the previous works of the authors.

As shown in Fig. 7 and Table 3, the effect of spatial variation of the friction angle is almost independent from the predetermined statistical distribution for the simulated benchmark study. The results regarding the maximum load (Fig. 7a) and displacement at the maximum load (Fig. 7b) are compared, where no significant effect is noticed. This outcome suggests that any of the adopted statistical distribution trends can be used in a similar discontinuum-based analysis setting to analyze URM walls, which predominantly fail in shear sliding.

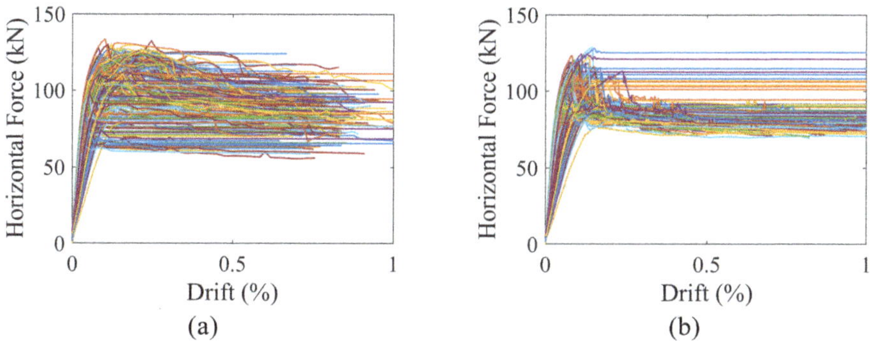

Fig. 6. Force-drift curves obtained from (a) nonspatial and (b) spatial analyses considering normal probabilistic distribution for the bond friction angle.

Table 3. Summary of the probabilistic analysis results.

Probabilistic Analysis	Mean F_{max} (kN)	CoV (%)
Nonspatial	101.3	15.3
Spatial (Normal)	94.3	10.8
Spatial (Lognormal)	91.4	11.6
Spatial (Weibull)	92.0	11.6

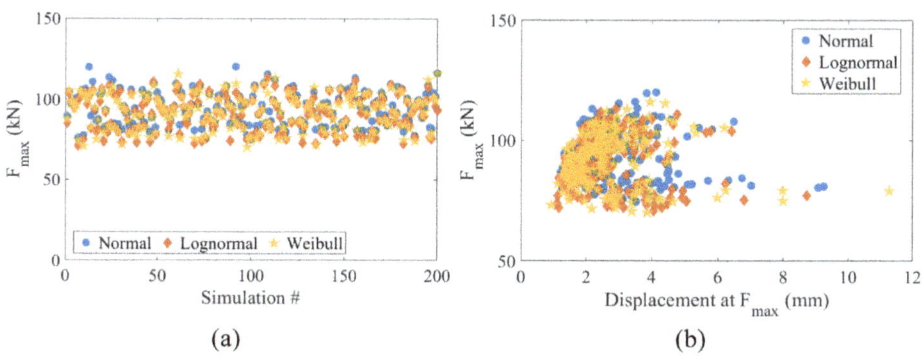

Fig. 7. (a) Maximum load and (b) displacement at maximum load comparison among the predetermined statistical distributions for the bond friction angle.

6 Conclusions

This study presents a framework for investigating the influence of inherent material uncertainty in the URM walls via a discontinuum-based computational modelling strategy based on the discrete element method. Through nonspatial probabilistic analysis, the validated numerical modelling approach is first utilized to understand the most significant input parameters on the ultimate load-carrying capacity of a reference URM wall. The results explicitly show that shear strength parameters (especially bond friction angle) have a considerable correlation with the lateral load carrying capacity under the given boundary and loading conditions of the analyzed URM wall. Then, spatial stochastic analyses are performed considering the bond friction angle with different statistical probability distributions (i.e., normal, lognormal and Weibull). No significant difference in ultimate lateral load is observed when the bond friction angle is spatially varied considering the given probability distributions; however, caution must be taken as the results also depend on the analyzed URM geometry and pre-compression vertical stress level. Similar future studies employing the presented framework should be carried out to investigate the effects of various material parameters, their distributions and the correlation between them, as well as different boundary conditions with varying restraints and vertical stress levels. Understanding these effects will help refine and optimize our numerical models and improve their accuracy.

Acknowledgments. The first author gratefully acknowledges the support of the Natural Sciences and Engineering Research Council of Canada (NSERC) grant RGPIN- 2022-04938.

References

1. Aguilar, R., Noel, M.F., Ramos, L.F.: Integration of reverse engineering and non-linear numerical analysis for the seismic assessment of historical adobe buildings. Autom. Constr. **98**, 1–15 (2019). https://doi.org/10.1016/j.autcon.2018.11.010
2. Funari, M.F., Hajjat, A.E., Masciotta, M.G., et al.: A parametric scan-to-FEM framework for the digital twin generation of historic masonry structures. Sustainability **13**, 11088 (2021). https://doi.org/10.3390/su131911088
3. Franchi, A., Napoli, P., Crespi, P., et al.: Unloading and reloading process for the earthquake damage repair of ancient masonry columns: the case of the Basilica di Collemaggio. Int. J. Archit. Herit. **16**, 1683–1698 (2022). https://doi.org/10.1080/15583058.2021.1904056
4. Lourenço, P.B., Rots, J.G., Blaauwendraad, J.: Continuum model for masonry: parameter estimation and validation. J. Struct. Eng. **124**, 642–652 (1998). https://doi.org/10.1061/(ASCE)0733-9445(1998)124:6(642)
5. Lourenço, P.B., Rots, J.G.: Multisurface interface model for analysis of masonry structures. J. Eng. Mech. **123**, 660–668 (1997)
6. Mendes, N., Zanotti, S., Lemos, J.V.: Seismic performance of historical buildings based on discrete element method: an adobe church. J. Earthq. Eng. **24**, 1270–1289 (2020). https://doi.org/10.1080/13632469.2018.1463879
7. Pulatsu, B., Bretas, E.M., Lourenço, P.B.: Discrete element modeling of masonry structures: validation and application. Earthq. Struct. **11**, 563–582 (2016). https://doi.org/10.12989/eas.2016.11.4.563
8. Gobbin, F., de Felice, G., Lemos, J.V.: Numerical procedures for the analysis of collapse mechanisms of masonry structures using discrete element modelling. Eng. Struct. **246**, 113047 (2021). https://doi.org/10.1016/j.engstruct.2021.113047
9. Gooch, L.J., Masia, M.J., Stewart, M.G.: Application of stochastic numerical analyses in the assessment of spatially variable unreinforced masonry walls subjected to in-plane shear loading. Eng. Struct. **235**, 112095 (2021). https://doi.org/10.1016/j.engstruct.2021.112095
10. Pulatsu, B., Gonen, S., Funari, M.F., Parisi, F.: Spatial stochastic D-RBA and limit equilibrium analysis of unreinforced masonry pier-spandrel structures. Eng. Struct. **296**, 116897 (2023). https://doi.org/10.1016/j.engstruct.2023.116897

11. Pulatsu, B., Gonen, S., Parisi, F., et al.: Probabilistic approach to assess URM walls with openings using discrete rigid block analysis (D-RBA). J. Build. Eng. **61**, 105269 (2022). https://doi.org/10.1016/j.jobe.2022.105269
12. Gonen, S., Pulatsu, B., Lourenço, P.B., et al.: Analysis and prediction of masonry wallette strength under combined compression-bending via stochastic computational modeling. Eng. Struct. **278**, 115492 (2023). https://doi.org/10.1016/j.engstruct.2022.115492
13. Lemos, J.V.: Discrete element modeling of masonry structures. Int. J. Archit. Herit **1**, 190–213 (2007). https://doi.org/10.1080/15583050601176868
14. Gonen, S., Pulatsu, B., Erdogmus, E., et al.: Quasi-static nonlinear seismic assessment of a fourth century A.D. Roman aqueduct in Istanbul, Turkey. Heritage **4**, 401–421 (2021). https://doi.org/10.3390/heritage4010025
15. Cundall, P.A.: A computer model for simulating progressive, large-scale movements in blocky rock systems. In: The International Symposium on Rock Mechanics, Nancy, pp. 47–65 (1971)
16. Roca, P., Cervera, M., Gariup, G., Pelà, L.: Structural analysis of masonry historical constructions. Classical and advanced approaches. Arch. Comput. Meth. Eng. **17**, 299–325 (2010). https://doi.org/10.1007/s11831-010-9046-1
17. Lourenço, P.B.: Computations on historic masonry structures. Prog. Struct. Eng. Mater. **4**, 301–319 (2002). https://doi.org/10.1002/pse.120
18. Pulatsu, B.: Coupled elasto-softening contact models in DEM to predict the in-plane response of masonry walls. Comput. Part. Mech. **10**(6), 1759–1770 (2023). https://doi.org/10.1007/s40571-023-00586-x
19. Cundall, P.A., Detournay, C.: Dynamic relaxation applied to continuum and discontinuum numerical models in geomechanics. In: Rock Mechanics and Engineering Volume 3: Analysis, Modeling and Design, pp. 45–90 (2017)
20. Itasca Consulting Group Inc.: 3DEC Three Dimensional Distinct Element Code (2013)
21. Messali, F., Esposito, R., Ravenshorst, G.J.P., Rots, J.G.: Experimental investigation of the in-plane cyclic behaviour of calcium silicate brick masonry walls. Bull. Earthq. Eng. **18**, 3963–3994 (2020). https://doi.org/10.1007/s10518-020-00835-x
22. Lourenço, P.B., Gaetani, A.: Recommended properties for advanced numerical analysis. In: Lourenço, P.B., Gaetani, A. (eds.) Finite Element Analysis for Building Assessment: Advanced Use and Practical Recommendations, pp. 209–320. Routledge, New York (2022)
23. Jafari, S., Rots, J.G., Esposito, R.: A correlation study to support material characterisation of typical Dutch masonry structures. J. Build. Eng. **45**, 103450 (2022). https://doi.org/10.1016/j.jobe.2021.103450
24. Pulatsu, B., Erdogmus, E., Lourenço, P.B., et al.: Simulation of the in-plane structural behavior of unreinforced masonry walls and buildings using DEM. Structures **27**, 2274–2287 (2020). https://doi.org/10.1016/j.istruc.2020.08.026
25. Gonen, S., Pulatsu, B., Erdogmus, E., et al.: Effects of spatial variability and correlation in stochastic discontinuum analysis of unreinforced masonry walls. Constr. Build. Mater. **337**, 127511 (2022). https://doi.org/10.1016/j.conbuildmat.2022.127511
26. Stein, M.: Large sample properties of simulations using Latin hypercube sampling. Technometrics **29**, 143–151 (1987). https://doi.org/10.1080/00401706.1987.10488205

Framework for Resilience Assessment as Decision Support for Traffic Infrastructure Operators

Christian Gasser[1](✉), Alois Vorwagner[1], Eva M. Eichinger-Vill[2], Robert Veit-Egerer[3], and Theresa Reimoser[3]

[1] Austrian Institute of Technology, Vienna, Austria
christian.gasser@ait.ac.at

[2] Vill Ziviltechniker GmbH, Vienna, Austria

[3] Vienna Consulting Engineers ZT GmbH, Vienna, Austria

Abstract. In the context of ageing transport infrastructure, it is crucial to allocate available funds for construction and maintenance in a way that minimizes the extent and duration of functionality loss. In simpler terms, the goal is to enhance the resilience. This contribution presents the outcomes of the REMAIN project (REsilient MotorwAy INfrastructure).

Within REMAIN, a framework was developed to assess the resilience of Austria's entire motorway infrastructure, considering five distinct asset categories: bridges, tunnels, noise barriers, retaining walls, and roadways. Resilience was assessed in the face of 18 hazards originating from natural, technical, and human factors. The framework is designed to offer the motorway operator a set of performance indicators exclusively focused on resilience. It is intended to serve as an additional decision-making tool for prioritizing infrastructure maintenance, alongside existing indicators like condition-based ones.

Particular emphasis was put on making the framework executable even in cases of limited data availability, with the flexibility to integrate additional data easily. The framework's ability to predict the resilience of an entire country's motorway infrastructure in the face of major hazards makes it unique in its comprehensive applicability.

While the framework was initially developed for motorway networks, its fundamental concept is universal and can be easily adapted for use in other infrastructure networks, such as secondary road networks and railway systems.

Keywords: Resilience · Motorways · Asset management · Robustness · Multi-hazard

1 Introduction

Malfunction of traffic infrastructure not only causes trouble and delays to users, but also constitutes a serious impediment to the economy. Various hazards of natural, technical, or human origin can severely disrupt network availability. Hence, it is highly desirable that

J. C. Matos et al. (Eds.): IPW 2024, LNCE 494, pp. 52–61, 2024.
https://doi.org/10.1007/978-3-031-60271-9_3

the road infrastructure is resilient. Resilience is the capacity to withstand or to recover quickly from impairments. The concept of resilience can readily be seen in Fig. 1, where the functionality Q of a system is shown as a function of time t. In the case of a traffic network, functionality is equivalent to the ability to *enable the flow of traffic.*

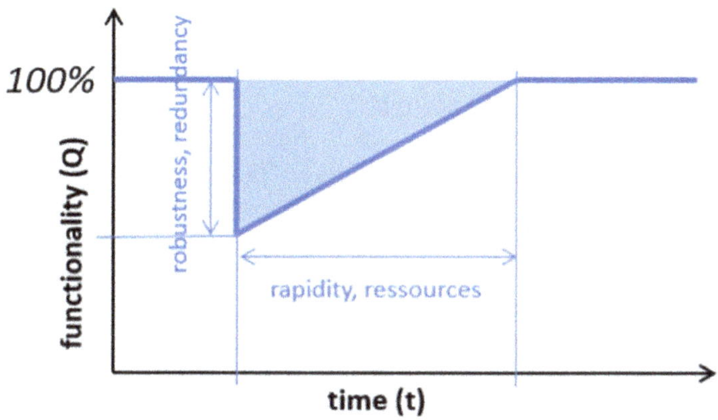

Fig. 1. Definition of resilience in terms of duration of reduced functionality: The smaller the blue area, the higher the resilience. While 'robustness and redundancy' influence the functionality (loss) of the system, 'resources and rapidity' influence the time (of restorage)

A reduction in functionality may be caused by continuous degradation (e.g., when a bridge must be closed to heavy vehicles because corrosion has reduced its bearing capacity), or by a sudden event, e.g., an earthquake or a flood. The focus of the project REMAIN was on sudden events since *resilience* is rather associated with sudden events (while *condition* is rather associated with continuous degradation).

On the other hand, time is crucial in the context of resilience. A resilient system is able the recover quickly from damage. In summary, according to Fig. 1, high resilience can be associated with a small blue area (integral of the functionality function).

It is widely recognised in research that there are four key aspects influencing resilience, cf. (Hajdin et al., 2021) or (Bruneau et al., 2003):

- **Robustness**: Refers to the strength or the ability of elements or systems to withstand a given level of stress or demand without suffering degradation or loss of functionality. Sufficiently high robustness will prevent failure and reduce functionality loss.
- **Redundancy**: Refers to the availability of alternate elements or systems that can be activated when disruptions occur. If there is sufficient redundancy, e.g., alternative routes with sufficient capacity, the functionality of infrastructure will be less affected.
- **Resources**: is the capacity to mobilise and apply human and material resources to conduct repairs after a disruptive event. Resourcefulness is a matter of careful resource planning and capacity building. If there are enough human and material resources in case of failure or disruption, restitution of functionality can be achieved faster.
- **Rapidity**: is the capacity to meet priorities and achieve goals in a timely manner in order to contain losses and avoid future disruption. The rapidity depends on efficient

use of available resources and can be considerably increased, e.g., by suitable rescue plans. By shortening the recovery time, resilience is increased.

In the context of the definition of resilience shown in Fig. 1, these four key aspects can clearly be attributed to one of the two diagram axes: Robustness and redundancy influence the functionality (loss) of the system, while resources and rapidity influence the time (of restorage).

2 Description of the Framework

The selection of the approach pursued to assess resilience in the project REMAIN was determined by specific requirements and boundary conditions. The major challenge in developing the framework was the need for comprehensiveness: The framework should be directly applicable to the whole motorway infrastructure of Austria, and for that reason it had to be based on data that is largely already available.

Five asset categories are considered individually:

- Bridges
- Tunnels
- Noise barriers
- Retaining walls
- Roadways

While in the case of the first four categories, the resilience evaluation was done for single structures, the category 'Roadways' is to be understood as a section of the motorway defined by its starting and ending chainage. 'Roadways' basically comprises all components of the motorway that are not covered by the other four categories (e.g., the road surface, drainage, gantries etc.).

The requirement to consider all possible hazards imposed a further challenge to the implementation of the framework. On the other hand, requirements regarding numerical precision were less strict. Given the primary goal of the research to develop a method for implementing resilience as an additional criterion in the asset management to prioritise maintenance measures for the national road administration (ASFINAG), it was essential to establish a framework allowing to *compare* different strategies. Accordingly, a procedure was developed which does not aim at an absolute quantification of resilience but rather allows systematic comparisons between different construction and maintenance strategies. For that reason, neither functionality loss nor the respective length of time is estimated quantitatively.

2.1 Hazards

In total, 18 hazards are considered in the framework:

- Storm
- Snow (drift), ice
- Extremely high temperature
- Earthquake

- Settlement
- Snow avalanche
- Mudslide
- Rockfall
- Flooding
- Undercutting, scouring
- Forest fire
- Windthrow
- Animals on the road
- Blackout
- Vehicle impact
- Explosion, fire
- Release of hazardous substances
- Overloading by vehicles

2.2 Influence Characteristics

For each of the five asset categories, a list was compiled containing the characteristics that were considered to have the most influence on the resilience of the respective asset category. All the characteristics were formulated as statements so that the user (who is assessing a specific asset) can state for each of them whether it applies or does not apply for the specific asset to be evaluated. The influence characteristics of one asset category, namely 'Bridges', are shown in Fig. 2. For each of the influence characteristics it was assessed whether its influence is,

- positive
- moderate or
- negative,

regarding:

- robustness/redundancy (decrease in functionality) as well
- resources/rapidity (recovery time).

This procedure is shown in Fig. 2 as an example. Figure 2 contains the assessments of some influence characteristic of the asset category 'Bridges' for the hazard 'Overloading by vehicles'. In case, for a hazard, an influence characteristic is irrelevant for one of the resilience dimensions, the respective cell is coloured grey.

The total number of assessments amounts to 8496 (236 influence characteristics, 18 hazards, 2 'dimensions').

The decisions about the assessments were made based on expert opinions. Naturally, this classification is subjective. To make the reasons for the decisions comprehensible also for outsiders, a list containing the explanations for each assessment was provided. In the following, a number of explanations are shown by way of example:

- Year of construction: Regarding many hazards, an early date of construction is generally disadvantageous. The overall quality of construction of bridges that were built in the post World War II period, but also in the 1960s and 1970s, often did not meet today's quality standards. Hence, these bridges are more susceptible to premature degradation phenomena.

Asset Category 1: Bridges		Robustness/Redundancy functionality, "y- axis"			Resources/Rapidity time, "x-axis"		
Influence characteristic		positive	moderate	negative	positive	moderate	negative
Year of construction	before 1960			x			
	1960 - 1980			x			
	1980 - 2000		x				
	after 2000	x					
Code used for dimensioning	Eurocode	x					
	according to older code			x			
Recalculation according to Eurocode or ÖN B 4008 available	yes	x					
	no			x			
Static system	statically determinate			x			
	statically indeterminate	x					
Building material	reinforced concrete						
	prestressed concrete						
	steel						
	composite						
Separate structures for each driving direction	yes	x			x		
	no			x			x
Camera surveillance	non-existent						x
	non-swivel					x	
	swivelling				x		
Traffic data sensors: jam detection sensors or similar available	yes				x		
	no						x
Structural health monitoring	available	x					
	not available			x			

Fig. 2. Assessment of some influence characteristics of the asset category 'Bridges' for the hazard 'Overloading by vehicles' regarding 'robustness/redundancy' and 'resources/rapidity' (grey cell if influence characteristic is irrelevant)

- Code used for dimensioning: Bridges designed according to Eurocode, generally have a higher robustness than bridges designed according to older national standards. Especially regarding durability of concrete, the relevant Eurocode document (Eurocode EN 1992–1-1, 2004) reflects a more advanced knowledge about the need to protect reinforcement by an appropriate concrete cover.
- Reassessment of load bearing capacity according to Eurocode or (ÖNORM B 4008–2, 2018) available: If it is available, a better estimation of the actual load-bearing capacity is possible since the reassessment considers both the as-built documents and the current condition of the structure.
- Static system: The evaluation strongly depends on the type of hazard. If the hazard under consideration has the character of an acting force (e.g., 'Vehicle impact', 'Explosion/fire', 'Overloading by vehicles', 'Earthquake'), a statically indeterminate system is more robust, since a force redistribution can take place. However, if displacements are imposed on the abutments ('Settlement', 'Undercutting, scouring'), constraining forces would arise in the case of statically indeterminate structures,

impairing the structural safety (robustness). An analogous situation applies for the hazard of 'Extremely high temperatures'.

- Camera surveillance: Regarding basically all hazards, a camera surveillance can be considered a plus in terms of 'Resources'. A swivelling system is even better for detectability.
- Traffic data sensors: jam detection sensors or similar available: Regarding basically all hazards, this can be considered a plus in terms of 'Resources' as information about functionality is reported in real time.
- Structural health monitoring: A structural health monitoring (SHM) system can indicate acting loadings and reaction stresses in the structure nearly in real time, making it easier to identify imminent situations of excessive stresses. Hence, the availability of a SHM can be considered favourable in terms of 'Robustness'.
- Travel time via alternative route: The availability of an alternative route renders an asset to a certain degree redundant. The related gain of resilience depends on the additional travel time via the alternative route as well as on its capacity. However, it is pointed out that the present resilience framework only considers the motorway network in Austria. Given that Austria is rather a small country, and the secondary road network is not taken into account, the redundancy via alternative routers is rather limited.

It should be noted that, due to the necessity for *general* applicability, also the assessments needed to be based on *general* facts.

The assignment of the influence characteristics to either 'positive', 'moderate' or 'negative' had to be done for the two dimensions 'functionality' and 'time' separately. Furthermore, the assignment had to be done for each hazard separately. This fact is illustrated in Fig. 3. In the example shown, 'hazard 2' (right image) causes a lower loss of functionality but leads to a longer recovery time than 'hazard 1' (left image).

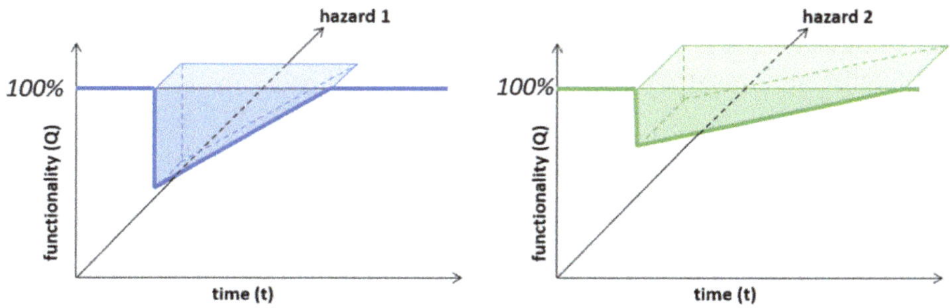

Fig. 3. Concept for assessing the resilience: In this example, 'hazard 2' (right image) causes a lower loss of functionality but leads to a longer recovery time than 'hazard 1' (left image). In addition, the risk from 'hazard 2' at the location under consideration is greater than from 'hazard 1'

A framework in which the resilience is assessed separately regarding each hazard also offers the advantage that the user can be provided with key figures that indicate the resilience of the examined object regarding each individual hazard (*partial resilience indicator*).

When deriving the *overall resilience indicator* that includes all the investigated threats, it is important to note that the hazard potential (i.e., the extent of the endangerment) must play a role.

In this respect, it is conceptually reasonable to introduce the hazard potential (HP) as a third dimension in the resilience assessment, as shown in Fig. 4. The resulting volume is a measure of the vulnerability of the asset regarding the hazard under consideration. The following applies: the smaller the resilience regarding the hazard and the greater the hazard potential, the larger the volume.

A concept for the evaluation of the hazard potential is described in the next section. The hazard potential is highly site-specific. For a number of hazards, it must be determined by the end user for the object to be examined individually.

2.3 Hazard Potential

The hazard potential (HP) is considered as the third dimension in the resilience diagram, see Fig. 4. The hazard potential is expressed on a 5 + 1 level scale where 'HP 5' is used for the highest risk and 'HP 1' for a very low risk. 'HP 0' means that the hazard does not exist in that site. The hazard potential is a measure for the occurrence probability (and hazard intensity). The number of 5 + 1 levels was chosen on purpose: For six of the considered hazards, hazard maps are available in Austria which are based on a 5 + 1 occurrence probability scheme.

For the other hazards, equivalent classifications were developed based on available relevant hazard maps, respectively based on specific site conditions. For example, the risk of 'Extremely high temperatures' is related to the number of days/year where the temperature exceeds 30 °C (for < 10 days/year: HP 1, for 10–30 days/year: HP 3, for > 30 days/year: HP 5). A somewhat different approach was chosen for the hazard 'Settlement': Here, the risk is determined according to the duration of the settlements. If they cease during construction phase, HP is equal to 1, if they cease in the first 5 years after construction, HP is equal to 3, and if settlements are present permanently or occur periodically, HP is equal to 5.

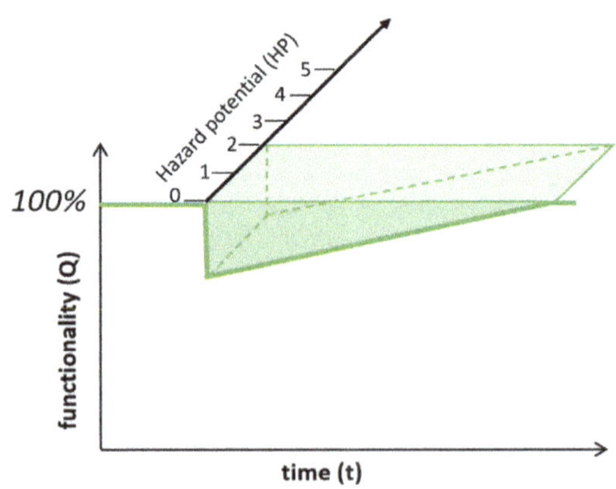

Fig. 4. The hazard potential may be interpreted as the third dimension in the diagram

2.4 Definition of Resilience Indicators

The evaluations from the procedure shown above are processed as follows: An influence characteristic statement rated with 'positive' assumes the value 1, if the rating is 'moderate' the value is 3, and if the rating is 'negative' the value is 5 (see Fig. 5). This five-step evaluation scheme is in accordance with the scheme of the condition-based assessment in Austria and many other countries.

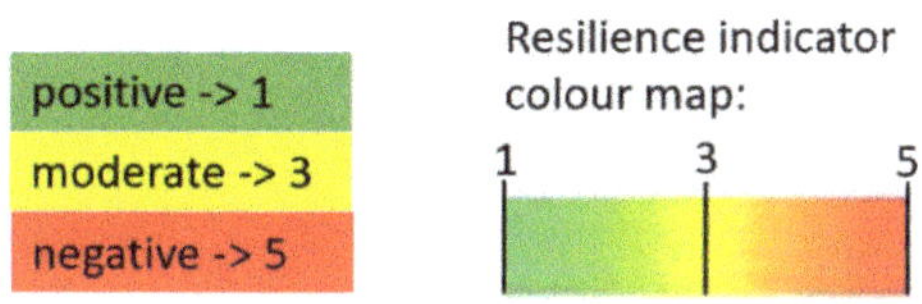

Fig. 5. Assignment of numerical values to the ratings

Partial resilience indicators $\mathrm{R}_{\mathrm{y,i}}$ and $\mathrm{R}_{\mathrm{x,i}}$ for an asset are calculated by summing up the corresponding assessments. The first index describes the dimension: y denominates the functionality dimension, x the time dimension. The second index i denotes the hazard the resilience indicator is referring to. The overall resilience indicator is calculated as:

$$R_{x,y,HP,i} = \left(\mathrm{R}_{\mathrm{x,i}} \cdot \mathrm{R}_{\mathrm{y,i}} \cdot \mathrm{HP}\right)^{1/3} \tag{1}$$

It includes all three dimensions: functionality, time, and hazard potential. The cubic root operation ensures that the resulting values are contained within the range between 1 and 5.

3 Visualisation

To visualise the results, Microsoft Power BI has been adopted. Business Intelligence (BI) is a suitable environment for visualising key figures of a system and to compare different strategies. A key element to illustrate the results of an assessment are spider charts. Figure 6 shows a spider chart illustrating the (*overall*) resilience before and after (envisaged) construction works of an asset for all the considered hazards. Values close to the centre indicate high resilience. Different (partial) resilience indicators (functionality, time, total) may be selected by a drop-down menu.

Figure 7 shows a resilience label for a bridge before and after execution of construction works. The resilience label shows the overall resilience indicator. Small values indicate high resilience.

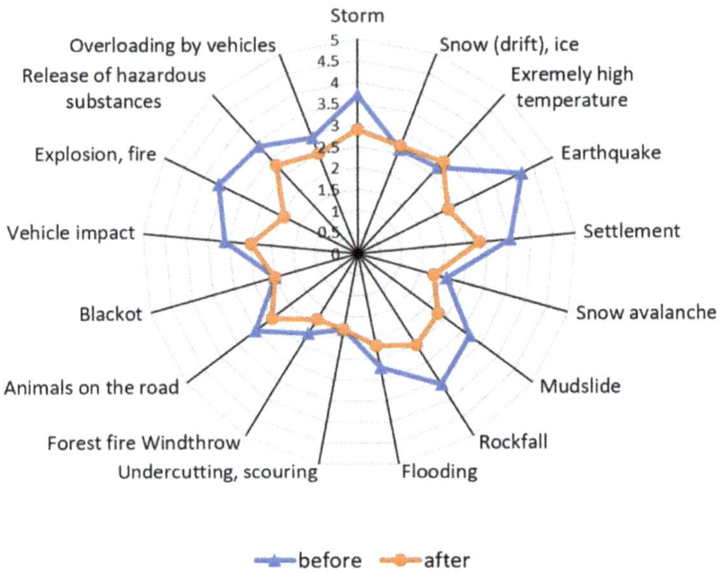

Fig. 6. Spider chart showing the resilience before and after construction works for all the considered hazards. Values close to the centre indicate high resilience

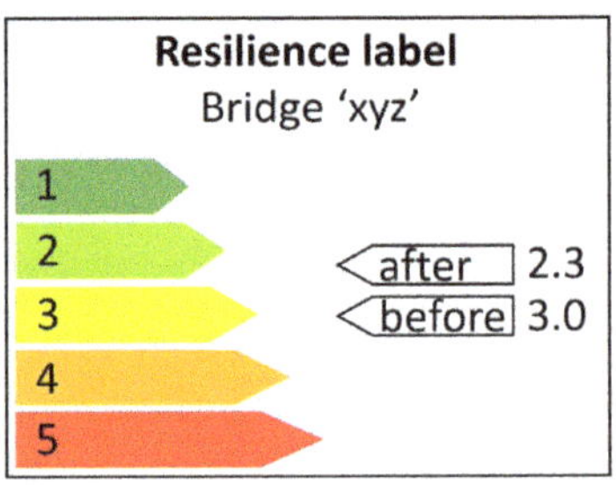

Fig. 7. Resilience label for a bridge before and after construction works. A small value indicates high resilience

4 Conclusion

A framework was developed which allows a straightforward assessment of the resilience of assets of a traffic infrastructure network. The specific implementation of the framework was carried out for the motorway network of Austria. The procedure allows comparisons between different construction/maintenance strategies. That way, it is suitable to provide performance indicators related to resilience, which can be used by the infrastructure operator as a decision basis, next to the existing condition-based indicators. Resilience indicators were developed referring to (1) single aspects, i.e., a single hazard and a single 'dimension' (functionality/time); (2) combined aspects like a single hazard and both dimensions; (3) all hazards and both dimensions, i.e., the overall resilience indicator.

The framework not only involves all major asset categories of a motorway network, but also considers all relevant hazards, 18 in total. The framework is immediately applicable because it relies on data of assets and hazards that, in general, are already available to the motorway operator, or can be collected easily. A scheme was developed to transfer information from available hazard maps and other hazard-related data bases consistently into the framework. In principle, any additional information increases the reliability of the results. More details can be found in [3] and [4].

Acknowledgement. The project REMAIN was conducted within the research program 'Mobility of the Future 2020' (contract no. 856137). The authors thank the Austrian Federal Ministry for Climate Action, Environment, Energy, Mobility, Innovation and Technology (BMK), ASFINAG, and the Austrian Research Promotion Agency (FFG) for their co-operation and support.

References

1. Hajdin, R., Hein, D.K., Honeger, C., Tanasić, N.: Improving Resilience of Road Networks—Framework. PIARC Routes/Roads, 389 (2021)
2. Bruneau, M., et al.: A framework to quantitatively assess and enhance the seismic resilience of communities. Earthq. Spectra **19**(4), 733–752 (2003). https://doi.org/10.1193/1.1623497
3. Gasser, C., et al.: Framework to assess the resilience of the motorway network in Austria. Struct. Infrastruct. Eng. (2023). https://doi.org/10.1080/15732479.2023.2260596
4. Eichinger-Vill, E.M., et al.: REMAIN Resilient Motorway Infrastructure – Ergebnisbericht. Österreichische Forschungsgesellschaft mbH (FFG) (2022). https://projekte.ffg.at/anhang/63d0ee2370a1d_REMAIN_Ergebnisbericht.pdf

Influence of Stud Shear Connectors Fatigue on the Entire Reliability of Composite Bridge Superstructure

Nikolai Kozak[1(✉)], Jose C. Matos[2], Hélder Sousa[2], Anton Syrkov[1], Dmitry Yaroshutin[1], and Vladimir Bystrov[1]

[1] Saint Petersburg State University of Architecture and Civil Engineering, Saint Petersburg, Russia
kozak.nickolay@gmail.com, nkozak@lan.spbgasu.ru
[2] ISISE, ARISE, Department of Civil Engineering, University of Minho, Guimarães, Portugal

Abstract. Based on multiple bridge inspections results, it was noticed that some composite steel–concrete bridges showed signs of a decrease in stiffness of their superstructures (in particular, additional sagging) after 20–30 years of operation. As one hypothesis, a reduction in the stiffness of the beam-slab joint elements can be considered. Most design codes (EN, AASHTO, among others.) are considered mandatory for fatigue check for joint elements, but in some codes (SP, Russia) fatigue check is not regulated. The purpose of this research is to assess the influence of fatigue damage accumulation in stud shear connectors on the performance of the entire superstructure and, in particular, on its reliability index. Based on the results of previous studies, a generalized model of the dependence of stud shear connector stiffness on the amount of its damage value has been proposed. Using this relationship model, a nonlinear finite element model of a simple composite beam has been developed that allows the calculation of the value of the load-bearing capacity objective function (by ULS criteria) depending on the number of load cycles. Using SOFiSTiK modules, the dynamics of declining the value of the load-bearing objective function were analyzed in both a deterministic formulation and using the probabilistic FORM method. To optimize the process of reliability index calculation, a search for an optimal model for clustering load cycles was performed. As a result of the research, the method for calculating the reliability index by considering the damage accumulation in stud shear connectors was proposed. This method can be used to substantiate the need for fatigue checks on joint elements numerically and determine reliability indices of bridge superstructures when combined with other methods. A series of calculations that simulate the operation of the superstructure model over various loading cycles was performed. The simulations revealed a significant decrease in quantitative reliability indicators as the number of loading cycles increased. This confirms the importance of ensuring the fatigue resistance of the joint elements in steel-reinforced concrete superstructures through reliability analysis methods.

Keywords: Bridge · Fatigue · Steel–concrete · Stud shear connector · Reliability

J. C. Matos et al. (Eds.): IPW 2024, LNCE 494, pp. 62–72, 2024.
https://doi.org/10.1007/978-3-031-60271-9_4

1 Introduction

Due to the strategic importance for the transport system, it is crucial that road bridges have an adequate reliability level. For the current widespread steel–concrete composite bridges, one of the least studied issue is the actual behavior and fatigue resistance of the structure joints between steel beams and concrete slab. In previous works, the authors contemplated the issues of the joint elements durability. An experimental study has shown that with high traffic intensity of heavy vehicles, there are significant risks of destruction of shear connectors due to the fatigue damage accumulation [1]. Next, numerical experiments were performed to simulate failures of discrete predefined groups of joint elements and for qualitative evaluation of these failures impact on design checks for ultimate limit state and serviceability limit state [2]. The current study progresses this topic by focusing on the modeling of the gradual failure of all joint elements (without predefined groups) and identifying quantitative factors related to structural reliability rather than qualitative ones.

The most common measure of structural reliability is currently the reliability index β (or associated failure probability index P_f) [3–7]. Modern European design standards, such as EN 1990 [8], specify basic target reliability indices depending on the reliability class of the structure, while separately stating that the minimum index for newly designed structures should be at least [β] = 3.8. Bridge structures, depending on the complexity of the structures and the length of the spans, belong to the RC2 (basic) class or to the RC3 (advanced) class, and for the base class the targets are highlighted separately for each limit state (Table 1).

Table 1. Reliability index targets (adapted from [8])

Reliability class	Limit states	Target index [β] for the period of	
		1 year	50 years
RC2	ULS	4.7	3.8
	Fatigue	-	1.5–3.8
	SLS	2.9	1.5
RC3	All	5.2	4.3

In the case of using the reliability index as a measure of the required reliability of the structure, the reliability condition can be represented in the form of in Eq. (1):

$$\beta_f \geq [\beta] \tag{1}$$

where $[\beta]$ is the target reliability index, β_f is the actual reliability index of the structure.

Various methods can be used to estimate the reliability of a structure. In [9], a classification of reliability estimation methods is proposed according to their level, to the accuracy of the results and to the used approximation algorithm. To tackle common engineering problems, an appropriate method is FORM (First Order Reliability Method), which allows to determine numerically the reliability indices β, with a sufficient accuracy, using current finite element analysis programs with excessive computational time.

The main goal of this research is to study the effect of possible failures (by fatigue) of the joint elements due to the accumulation of fatigue damage on the reliability level of composite superstructures by tracking its change in the reliability index.

To achieve the purpose of the study, a numerical experiment was planned and performed on a finite element model of a steel–concrete beam. The peculiarity of the model was the possibility to take into account the damage accumulation in the connectors, which results in a decrease in both the connector's stiffness and the overall stiffness of a superstructure. Based on the results of simulation at control points using reliability methods, design reliability indices were determined and how they change depending on the number of accumulated load cycles.

2 Model Analysis

2.1 Structural Model

To conduct a numerical experiment, a model of a simply supported steel–concrete superstructure with 33 m design span was considered (also used in [10]). The scheme and main dimensions of the superstructure are shown in Fig. 1. The superstructure is assumed to be designed for A14, H14 load (Russian codes SP 35.13330.2011 (rus. СП 35.13330.2011)). The main beam material is 15HSND (rus. 15ХСНД) steel (design yielding strength 295 MPa, elasticity modulus 206000 MPa), the main plate material is reinforced concrete of class B40 (design compression strength 40 MPa, modulus of elasticity 34500 MPa). The beam and plate are combined by means of shear stud connectors (height 180 mm, diameter 22 mm, steel 09G2S (rus. 09Г2С)), arranged evenly along the length of the beam spaced 0.30 m apart in pairs (total 220 studs per beam). Fatigue resistance characteristics for modeling the process of accumulation of fatigue damages in studs are taken according to EN 1994-2 and EN 1993-1-9 ($\Delta\tau_C = 90$ MPa, m = 8).

To optimize computing resources, it was decided to use a flat beam structural design. The finite element model of the superstructure beam was performed using the SOFiSTiK software package. The integration of a steel beam and a reinforced concrete slab of a superstructure was performed using elastic bonds with a predefined stiffness. To optimize the calculation, one common bond was set for each group of two pairs of studs (a total of 55 elastic bonds). Interval of these elastic bonds was set to 0.60 m. The initial stiffness of one stud in the longitudinal direction, in accordance with the results of the study [11], was taken equal to $K_{\text{stiff}} = 270{,}000$ kN/m. The friction of the metal-concrete contact was taken equal to zero.

For the loads acting on the superstructure, constant loads were considered (the self-weight of the superstructures, the self-weight of the bridge deck elements) applied as distributed, and a moving load for ULS AK_I (A14), including a car and a strip distributed part. The action of moving loads was determined using the transverse installation coefficient (*KPU*) calculated using the off-center compression method (Fig. 2). The values of unfactored loads and design loads, including moving ones for ULS, are given in Table 2. The design values for moving loads are calculated taking into account the load partial safety factor y_f and dynamic impact coefficients $(1 + \mu)$.

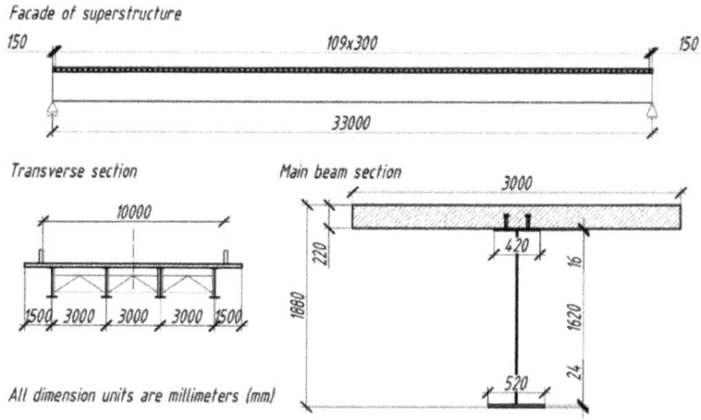

Fig. 1. Schematic and basic dimensions of the superstructure model (adapted from [10])

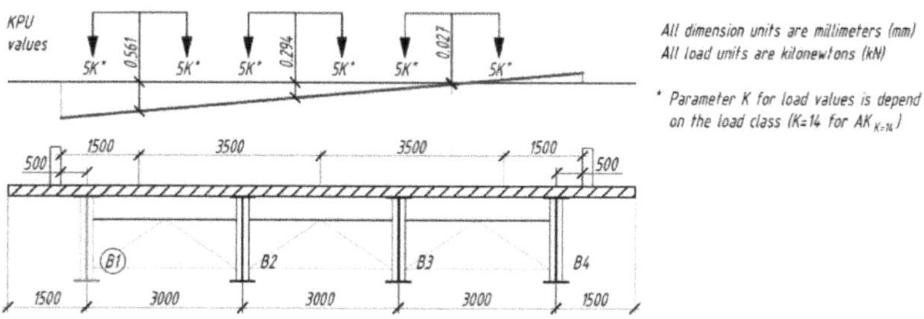

Fig. 2. The scheme of the transverse arrangement of the load AK_I (adapted from [10])

Table 2. Unfactored and design loads on the beam

Load	Unfactored value	Design value
Span construction dead load	–	2.77 kN/m
Bridge bed dead load	–	12.04 kN/m
AK_I (A14, wagon)	$140 \cdot KPU = 140 \cdot 0.745 =$ $= 104.30$ kN	$104.30 \cdot \gamma_f \cdot (1 + \mu) =$ $= 104.30 \cdot 1.50 \cdot 1.40 = 219.03$ kN
AK_I (A14, distributed)	$14 \cdot KPU = 14 \cdot 0.745 =$ $= 10.43$ kN/m	$104.30\gamma_f \cdot (1 + \mu) =$ $= 104.30 \cdot 1.15 \cdot 1.00 = 11.19$ kN

2.2 The Dependence of the Joint Elements Stiffness on the Number of Loading Cycles

Previous studies have shown that during the operation of a bridge structure under the action of transport loads, the operating mode of shear stud connectors is characterized by a significant number of alternating cycles [1]. Such a mode of operation contributes to the accumulation of fatigue damage in joint elements thus it affects their stiffness. For the current study it is proposed a model of dependence of shear stiffness on the index of fatigue damage accumulation (resource usage indicator). This model is primarily based on the results of the study [12] and also take into consideration [13, 14], where behavior of similar stud shear connectors was investigated. This model considers the gradual decrease of the joint element stiffness and includes 3 sectors, as shown in Fig. 3. The damage D accumulated by the element can be determined according to the Palmgren–Miner linear damage accumulation hypothesis.

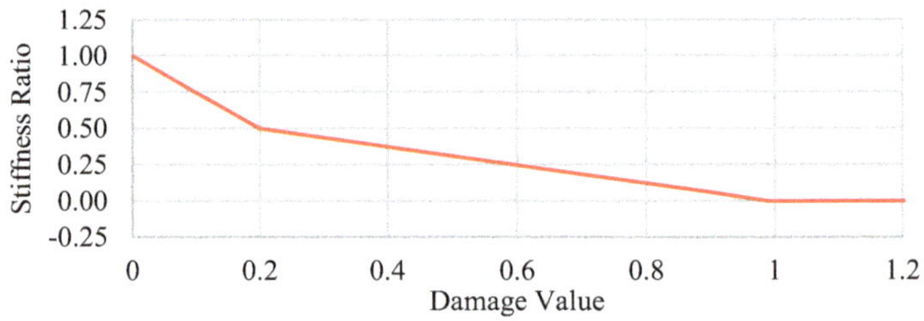

Fig. 3. The approximate model of dependence of shear stiffness on resource utilization index (model is proposed by authors and based on the results of research [12])

2.3 Equivalent Load Determination Method

An important issue in modeling the accumulation of fatigue damage in a structure is the value of the load used in the loading cycle. The use of the design load AK_I with class $K = 14$ in each cycle seems incorrect since its main purpose is modeling the maximum possible load, but not cyclic one.

Methods that include sets of loads with random transverse distributions and load values (models of real traffic flows) are widely used in tasks of fatigue assessment [15, 16]. However, for current research tasks, these methods appear unreasonable because they introduce new statistical uncertainties of a different nature, which depend not only on the object's characteristics but also on the time.

As a solution, one may consider the use of equivalent fatigue loads AK_{fat} based on AK loads with an equivalent class $K = K_{fat}$. This class can be determined by selection based on the condition $D[AK_{fat}] > D_{ave}(N)[TrafficFlow]$, where $D[AK_{fat}]$ is damage in the particular element for 1 pass of the AK_{fat} load (an AK car with an equivalent class $K = K_{fat}$), and $D_{ave}(N)[TrafficFlow]$ is the average damage from the effects of N loads from the traffic flow model. At the same time, the number of considered impacts of N loads should be significant ($N \geq 10{,}000$).

Models based on traffic composition data can be considered as a traffic flow model. In the absence of such data, it is proposed to use a flow model based on the AK load, taking into account the parameters of its car by random values. The distributed load, at the same time, remains similar to that used in AK_I, but taking into account $KPU = 0.499$, calculated for two lanes without considering of movement along safety lanes. Table 3 shows the parameters for modeling random variables $V = V_{perm} + K \bullet V_{prob}$ loads AK.

All values are assumed for this study. The uncertainty values for the axial load P were chosen under the condition that the value of the axial load is not greater than the value of the standard AK load model in more than 1% of the cases ($\gamma_f = 1.4$). The values for the dynamic impact coefficient $(1 + \mu)$ were chosen based on the experimental results [1]. The values for KPUs were determined based on the geometric parameters of the bridge deck (mainly in the middle of the road line, possible displacement towards the road fence, it is impossible to enter the opposite lane).

Table 3 Parameters for modeling random variables of the AK load

Load parameter	Distribution	V_{perm}	K	V_{prob}	
				MV	SD
P (kN)	lognormal	0	140	1	0.2
$(1 + \mu)$		1	0.4	1	0.5
KPU_1 (left lane)		0.329	0.076	1	0.3
KPU_2 (right lane)		0.169	−0.076	1	0.3

Modeling of the structure loading with AK loads with random parameters using the Monte Carlo method was performed using the RELY module in the SOFiSTiK software package. 10,000 simulations were performed, based on the results of which the average value of maximum damage per load pass was determined $D_{ave}(N)[\text{TrafficFlow}] = 1.462 \cdot 10^{-7}$. Based on the results of its comparison with the dependence curve of $D[\text{AK}_{fat}]$ on K_{fat} [10] the equivalent class $K_{fat} = 9.3$ was determined. In further calculations, the AK_{fat} load will be used to simulate the process of damage accumulation in the joint elements.

2.4 Algorithm of Analysis and Clustering of the Loading Process

The difference between the design yielding strength of the steel R_y and the design stresses σ in the lower fiber of the steel beam in the middle of the span from the design load AK_I (ULS condition, Eq. 2) was chosen as the target function PF_1 to determine the reliability indicator:

$$PF_1 = R_y - \sigma \tag{2}$$

Figure 4 shows a schematic flowchart of the algorithm for determining quantitative indicators of structural reliability.

Obviously, the calculation with direct modeling of each *j* cycle from the many million ones in the history of loading *S*, will require significant time and computing resources that are not currently available. In order to optimize the process, it seems rational to consider in each iteration of the nonlinear calculation a certain cluster of N_{load} loads, assuming an identical operating mode within the cluster. To determine the optimal cluster size, a series of nonlinear calculations of beam structures for the effect of an equivalent fatigue load of AK_{fat} over 200 million load cycles were performed, with a breakdown of cycles into clusters of various sizes. The results of the study showed that the use of clusters up to $2.5 \cdot 10^6$ cycles makes it possible to obtain quite similar curves of variation of the PF_1 parameter throughout the cyclic loading process. It was concluded that it is rational to use the cluster with $N_{load} = 2.5 \cdot 10^6$ cycles.

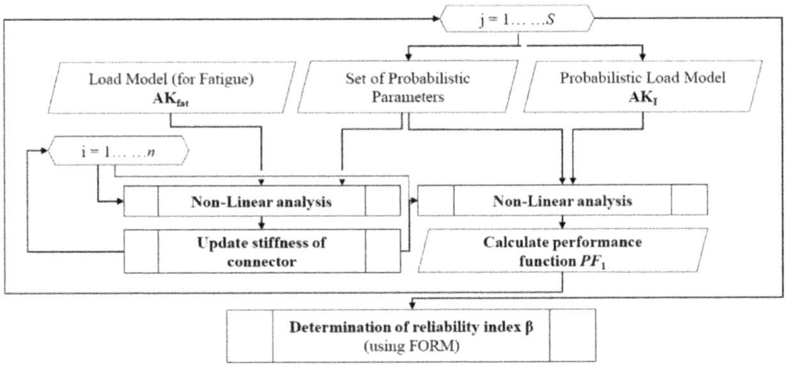

Fig. 4 Schematic flowchart of the algorithm for determining quantitative indicators of structural reliability

2.5 Parameters of the Probabilistic Calculation

During the study, two variations of a probable design parameters set were considered. In the first (approximate) set A, only the probabilistic parameters of steel yielding strength and the probabilistic parameters of the AK_I moving load were considered. In the second (advanced) set B, probabilistic parameters for the initial stiffness of the studs, the fatigue resistance characteristics of the studs, as well as for the fatigue failure criterion were also considered. Individual parameters were set for each of the 55 joint elements of the calculation model. A significant disadvantage of set B is the increased resource intensity of calculations. The parameter values were adopted according to the recommendations [17] and are shown in Table 4.

For the probabilistic calculation, the RELY module of the SOFiSTiK was used. The probabilistic parameters (from sets A or B) were updated with each iteration and transferred to the main FE model for nonlinear calculation. As a result of the analysis the value of the target function PF_1 was obtained. To determine the reliability index β, the FORM method was used. In total, for each number of cycles, 4 iterations were performed totally with 841 samples of parameter value sets.

Table 4 Parameters for modeling random variables of characteristics

Parameter	Set	Distribution	V_{prob}	
			MV	SD
Steel yielding strength R_y, MPa	A, B	lognormal	363	25.4
Coefficient for P in AK_I car	A, B	lognormal	1	0.172
Stud stiffness $K_{stiff,i}$	B	normal	270,000	13,500
Number of cycles N_i if $\Delta\tau_C$	B	normal	2,000,000	100,000
Failure criterion $D_{lim,i}$	B	lognormal	1	0.3

3 Results and Discussion

Figure 5 shows the results of calculating reliability indices depending on the number of accumulated loading cycles for two variations of structural parameters set for a range up to $4 \cdot 10^8$ million cycles (a) and a segment of the "operating" range up to $0.8 \cdot 10^8$ million cycles (b).

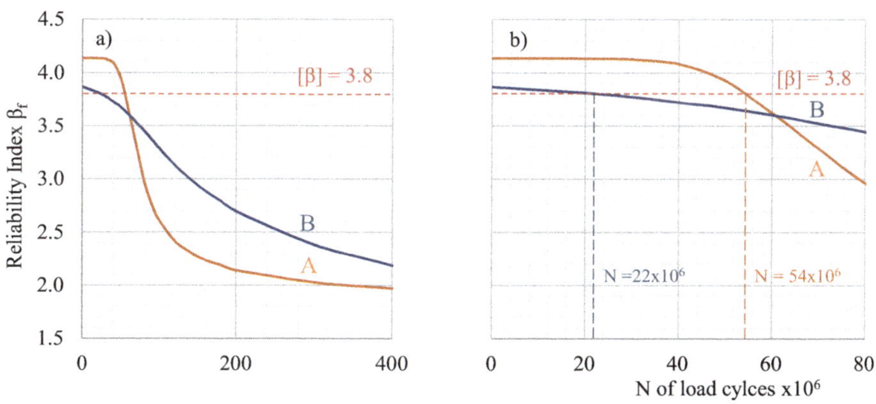

Fig. 5 Diagrams of the dependence of the structural reliability index β_f on the number of loading cycles *N*

Based on the results of the numerical experiment, it is possible to confirm the hypothesis that considering the damage accumulation in the joint elements allows for refining the definition of the reliability index β. It is also clearly confirmed inverse relationship between the number of loading cycles and the reliability index β. The following results of the analysis should be highlighted:

1. Comparing the β_f-N curves (Fig. 5) showed that during the first 40 million load cycles when using parameter set A, the reliability index remained almost unchanged. In this case, it also should be noted that 40 million cycles is almost the complete life cycle of most bridges. For comparison, when using set B, the index constantly gradually

decreases throughout the entire period of operation. Thus, it is important to note that the choice of parameter set significantly influences the accuracy of the reliability index definition.

2. For reliability class RC2 (basic), the target reliability index in the first year should be at least 4.7, and within 50 years – at least 3.8. The comparison shows that the initial calculated reliability index of the selected structure (designed according to the Russian codes) is significantly below the target value. At the same time, in the long term, the target index $[\beta] = 3.8$ is provided when using a set of parameters A for 54 million cycles, when using a set of parameters B – for 22 million cycles.
3. The requirements for ensuring the target reliability index $[\beta] \geq 3.8$ for 50 years are thus met at a traffic intensity of less than 1205 heavy vehicles per day (51 heavy trucks per hour) in the case of calculation using the parameter set B and at an intensity of less than 2959 heavy vehicles per day (124 heavy trucks per hour) for the A set.
4. A comparison of the reliability indices β-curves and the value of the ULS PF_1 function depending on the number of loading cycles (Fig. 6) shows that the decrease in the actual reliability index occurs earlier than the failure according to ULS ($PF_1 \leq 0$), which may indicate the need to refine the parameters of the design load AK_I.
5. Despite the significant resource intensity of calculations with a set of parameters B, it seems recommendable to use it to determine the reliability indices of structures, taking into account the accumulation of fatigue damage.

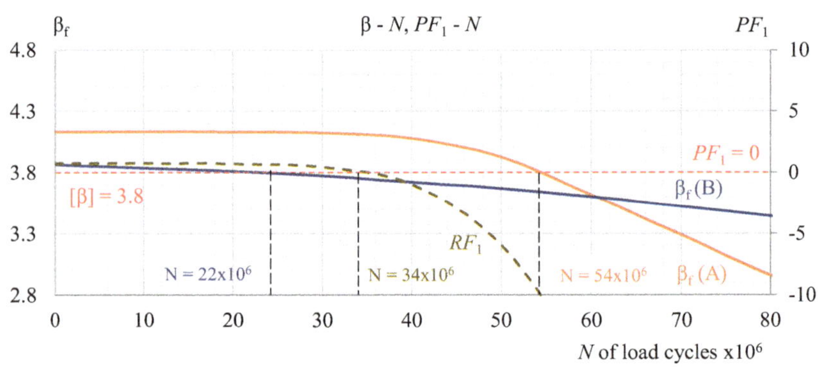

Fig.6 Comparison of diagrams of dependence of reliability index β_f and parameter PF_1 on the number of loading cycles

4 Conclusion

In this study, the authors developed and tested a method for reliability analysis of a steel–concrete composite beam, considering the fatigue damage accumulation in the stud shear connectors. The results of a numerical experiment confirmed the inverse relationship between the number of load cycles and the overall reliability index of structure β. An analysis of the experimental results indicated that with an expanded (more complete) set of probabilistic parameters, the actual reliability index might decrease less than the target one after 22 million load cycles, equivalent to 50 years of operation with an average traffic intensity of 1205 trucks per day. The authors are supposed to adjust the parameters of the AK traffic load (Russian design standards) in accordance with the required levels of reliability. It also seems advisable to take into account, when designing superstructures, not only the fatigue resistance of the joint elements but also to ensure their stiffness does not fall below the limit value for the entire period of design operation. Further development of the study is expected in the clarification of sets of probabilistic parameters used in reliability analysis, as well as in the application of this method to the superstructures of existing road bridges.

References

1. Kozak, N., Bystrov, V., Yaroshutin, D., Daliaev, N.: Improving of fatigue assessment method of stud shear connectors using experimental data from studs' test of existing road bridge. In: Proceedings of IABSE Symposium Prague 2022: Challenges for Existing and Oncoming Structures. Prague, Czech Republic, pp. 224–234 (2022)
2. Kozak, N., Syrkov, A., Bystrov, V., et al.: Determination of quantitative indicators of reliability of steel-reinforced composite beam considering fatigue damage accumulation in stud shear connectors (2023). (in Russian). https://doi.org/10.13140/RG.2.2.19503.92322
3. Das, P.C.: Safety Concepts and Practical Implications. In: Safety of Bridges, pp. 29–35. Thomas Telford Limited, London (1997)
4. Matos, J.C., Moreira, V.N., Valente, I.B., et al.: Probabilistic-based assessment of existing steel-concrete composite bridges –application to Sousa River Bridge. Eng. Struct. **181**, 95–110 (2019). https://doi.org/10.1016/j.engstruct.2018.12.006
5. Kim, S., Ge, B., Frangopol, D.M.: Optimum target reliability determination for efficient service life management of bridge networks. J. Bridge Eng. **25**, 04020087 (2020). https://doi.org/10.1061/(ASCE)BE.1943-5592.0001623
6. Ghasemi, S.H., Nowak, A.S.: Target reliability for bridges with consideration of ultimate limit state. Eng. Struct. **152**, 226–237 (2017). https://doi.org/10.1016/j.engstruct.2017.09.012
7. Eidsvig, U., Santamaría, M., Galvão, N., et al.: Risk assessment of terrestrial transportation infrastructures exposed to extreme events. Infrastructures **6**, 163 (2021). https://doi.org/10.3390/infrastructures6110163
8. EN 1990 (2002) (English): Eurocode - Basis of structural design
9. Melchers, R.E., Beck, A.T.: Structural Reliability Analysis and Prediction. Wiley, Hoboken (2018)
10. Kozak, N.V., Syrkov, A.V., Bystrov, V.A., et al.: Quantitative reliability measures determination of a steel-reinforced concrete beam taking into account fatigue damage accumulation in flexible pin stops (2023). (in Russian). https://doi.org/10.15862/05SATS423
11. Kozlov, A.V., Kozlov, V.A.: Experimental study of SSS of a steel concrete beam taking into account the rigidity of the connecting seam. J Phys: Conf Ser **1902**, 012010 (2021). https://doi.org/10.1088/1742-6596/1902/1/012010
12. Xu, C., Sugiura, K., Su, Q.: Fatigue behavior of the group stud shear connectors in steel-concrete composite bridges. J Bridge Eng **23**, 04018055 (2018). https://doi.org/10.1061/(ASCE)BE.1943-5592.0001261
13. Wang, B., Huang, Q., Liu, X.: Comparison of static and fatigue behaviors between stud and perfobond shear connectors. KSCE J. Civ. Eng. **23**, 217–227 (2019). https://doi.org/10.1007/s12205-018-1303-0

14. Hassanin, A.I., Shabaan, H.F., Elsheikh, A.I.: The effects of shear stud distribution on the fatigue behavior of steel-concrete composite beams. Arab. J. Sci .Eng. **45**, 8403–8426 (2020). https://doi.org/10.1007/s13369-020-04702-4
15. Croce, P.: Impact of road traffic tendency in Europe on fatigue assessment of bridges. Applied Sciences **10**, 1389 (2020). https://doi.org/10.3390/app10041389
16. Hao, Q., Zhihua, N., Pingming, H.: Research on random fatigue load model of highway bridge by vehicle traffic based on GMM. Teh vjesn **27**, 961–969 (2020). https://doi.org/10.17559/TV-20190629100555
17. JCSS Probabilistic Model Code. Part III – Material Properties. Joint Committee on Structural Safety (2002). https://www.jcss-lc.org/jcss-probabilistic-model-code/. ISBN: 978-3-909386-79-6

Surrogate Model Approaches for the Evaluation of Structural Safety of Suspended Ceilings

Nikolaos Mellios[1,2](✉), Agnetha Jörg[1], Panagiotis Spyridis[3], and Alfred Strauss[1]

[1] Institute of Structural Engineering, BOKU Wien, Gregor-Mendel-Straße 33, 1180 Vienna, Austria
nikolaos.mellios@tu-dortmund.de

[2] Chair of Fastenings Engineering, TU Dortmund University, 40476 Dortmund, Germany

[3] Chair of Concrete Construction, Rostock University, Justus-Von-Liebig-Weg 2, 18059 Rostock, Germany

Abstract. Suspended ceiling systems are widely used in construction as a solution especially for public buildings and offices. By installing a secondary ceiling system, the room height can be regulated and the utility systems of a structure can be accommodated. There are many methods to fix the suspended ceiling but one of the newest is the use of direct fastening. This method can significantly increase the installation speed of a ceiling and therefore reduce the cost of installation. Based on previous own laboratory tests conducted at the Institute of Structural Engineering BOKU and the use of a digital ceiling system the reliability of the fixing was evaluated. The results of the experimental investigation were used to feed the FE digital system on the basis of a Monte Carlo simulation. To extend our knowledge of the structural reliability of the system for more configurations the use of less computationally demanding surrogate models was implemented. Based on the surrogate models the optimized system configuration in terms of structural safety can be identified.

Keywords: Direct fastening · Structural safety · Suspended ceilings · Stratified sampling · Surrogate model

1 Introduction

1.1 General

Suspended ceiling systems are widely used as an architectural solution in construction. These systems can be fastened to the structure with the use of direct fasteners i.e. nails. This fastening method can increase the installation speed drastically therefore reducing the installation cost. Extensive research is required to evaluate the safety and reliability of the suspended ceilings with this fastening method. This study considers the experimental results from an extensive experimental program that evaluates the load distribution of a suspended ceiling and the entire life load-bearing resistance of the nails. The experimental results are implemented in numerical models to evaluate the reliability of the

J. C. Matos et al. (Eds.): IPW 2024, LNCE 494, pp. 73–83, 2024.
https://doi.org/10.1007/978-3-031-60271-9_5

system. This study aims to develop a series of surrogate models based on the computationally heavy numerical models and the experimental results. These models can be implemented as input to evaluate the reliability of suspended ceiling systems with different configurations with minimal computational costs.

1.2 Suspended Ceilings

A suspended ceiling, also known as a false ceiling, creates a space between the statically effective ceiling and the room itself. This space can be used to accommodate various devices, cables or insulation. These include, for example, air conditioning systems, fans, cables and pipes, lights, heat and sound insulation [1]. Secondary ceiling systems not only have the advantage of visually concealing installations but can also cover a wide range of protective functions. These include fire protection, moisture protection, cold protection, and sound insulation. The choice of ceiling can influence the acoustics of the room according to specific requirements. It is also possible to use a suspended ceiling as a room design element. In practice, such a ceiling not only fulfills one of the purposes mentioned but is required for several functions at the same time [1].

A typical example of a suspended ceiling is presented in Fig. 1. The first component of a suspended ceiling is the supporting structure which can consist of different materials such as concrete or steel. To the supporting structure, the top profile is attached by means of mechanical fasteners. The third component is a hanger which is attached to the top profile and connects to the base profile transferring the loads to the supporting structure. The base profile is an element of the substructure and in some cases, it can connect directly with a hanger to the supporting structure. Finally, the cover layer components are placed between the base profiles. The cover layer components are a part of the ceiling and remain visible.

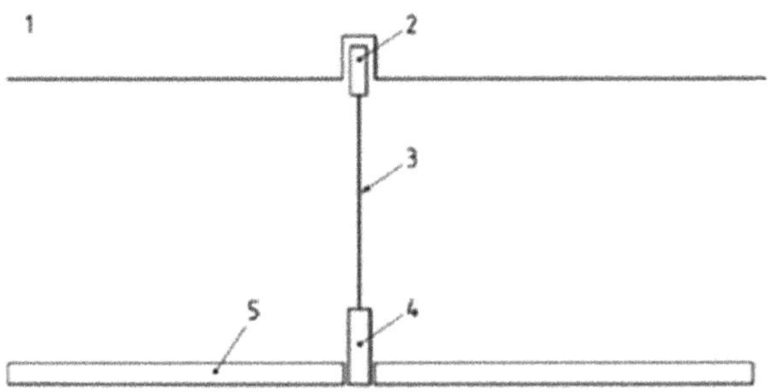

Fig. 1. Components of a suspended ceiling. (1) Support structure, (2) top profile, (3) hanger, (4) base profile, (5) cover layer components.

For the design of a suspended ceiling, the first step is to select a specific thickness of the cladding. With this information and the information about the additional loads from the insulation, the final load class can be determined in kN/m^2. Basically, four classes are differentiated. The highest load class is between 0.50 and 0.65 kN/m^2. Another class includes loads between 0.30 and 0.50 kN/m^2. The next smaller class considers loads from 0.15 to 0.30 kN/m^2. Finally, all structures with a load lower than 0.15 kN/m^2 are grouped together in the lowest class [2]. In this study, the load classes 0.50, 0.30, and 0.15 are considered. Finally, to ensure that the installation of the load-bearing component is cost-effective, the fastening of the top profile should be maintenance-free during the entire service life of the structure. For this reason, it is important to fulfill the durability limit state.

1.3 Direct Fastening

Direct fastening technology is an efficient method in which a stud or nail is driven into a hard material like concrete or steel using a powder-actuated tool. Whether the nail is in a good or poor setting condition it can be influenced by the properties and the specific condition of the substrate base material. This influence is more prominent when the base material is concrete due to the composite nature of the material and the low tensile strength. If the conditions are unfavorable the nail may fail during installation. Therefore the load-bearing capacity of a direct fastener is largely dependent on the substrate base material. Experimental studies have shown that the nail load-bearing capacity has a continuous mixed distribution and usually a very high relative scatter of over 50% [1]. A high distribution density can already occur from a value of zero. In principle, the statistical distribution of the load-bearing capacities varies greatly and depends on the properties and conditions of the selected reinforcement. While the anchors are being set on a concrete substrate, the installation forces can overcome the tensile strength of the surface and lead to concrete spalling. This can lead to the serviceability limit state not being met.

1.4 Redundant Fastening

As mentioned above the use of direct fastening for ceiling suspension faces some difficulties due to installation failure and the significant number of nails with low load bearing capacity. Further nail failures can occur due to the formation of cracks and crack movement [1]. Therefore it is not possible to install a hanger directly with a single nail since the failure of a hanger is not permitted. Therefore the use of direct fasteners must be implemented in a redundant fastening system. With a redundant fastening system the safety level of the suspended ceiling is not affected. It is also ensured that the fastening is not the weakest component in the construction.

For the direct fastening system sheet metal profiles are fixed on the support structure. The top profiles can be mounted separately or assembled as a frame. The hangers are attached to the top profile independently from the position of the nails. The number of nails and hangers can be varied. This means that the number of nails per hanger can vary. An example of the redundant fastening system is presented in Fig. 2. The static model of the system is a continuous beam. In this system, nail failures can occur and support forces can be developed against the anchoring base. In the event of a nail failure, the load can be redistributed to the remaining nails of the system [1].

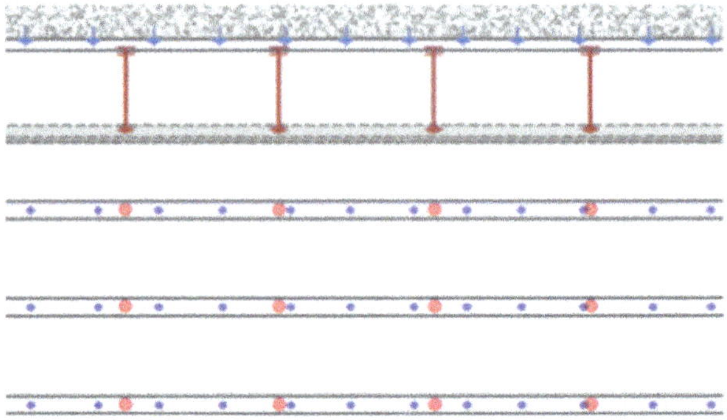

Fig. 2. Redundant direct fastening system. Nails are shown with blue color and hangers are shown with red.

1.5 Stratified Sampling Methods

One of the most widely implemented probabilistic methods is the Monte Carlo Method also known as Monte Carlo (MC) simulation. It is commonly used due to its easy implementation and the ability to handle multi-dimensional problems. For MC simulation a random number generator is used to generate random numbers of the distribution function used for the task. The associated realizations x_{ik} are determined on the basis of the values and the distribution function [3]. The MC method can be used for example to make statements about the probability of failure for a known limit state function. When the MC method is implemented in a given limit state function all realizations that are less than zero are considered failure cases. From the number of failures z_0 and the total number of realizations z the probability of failure can be determined according to Eq. (1).

In civil engineering problems due to the significantly small probability of failure, more advanced probabilistic methods have to be implemented. Such a method is the stratified sampling. In this method instead of implementing the computationally heavy limit state function for the total number of realizations z the population is separated into subpopulations (strata) with specific characteristics. The limit state function is then evaluated only for the subpopulations that are probable to lead to a failure. The total number of failures z_0 is recorded. Equation (1) can be implemented to evaluate the probability of failure.

$$p_f \approx \frac{z_0}{z} \tag{1}$$

1.6 Surrogate Model

Surrogate models, also known as metamodels, are being used more and more frequently. Their purpose is to obtain satisfactory results for the problem under investigation on the basis of a developed model using fewer resources. Various techniques, primarily from the field of machine learning, are used to create a surrogate model [4–7]. When investigating costly and time-consuming simulations a surrogate model can be developed to obtain a fast mathematical approximation. Once the surrogate model is created it is possible to implement it to update the source model in an iterative method in order to improve the accuracy of the results. These models can also be used for data mining. In this process, insight into the relationships between the variables of a problem and their significance can be gained.

2 Materials and Methods

2.1 Experimental Setup

A test rig for the heavy suspended ceiling was developed at the Institute of Structural Engineering at BOKU Vienna. The nail failures were simulated and the resulting redistribution effects were analyzed. The collected data were used as the basis for the creation of a suitable probabilistic model and its verification. The test stand has dimensions of approximately 4 by 4 m. The fastening system of the setup consists of 6 profiles. Each profile is fixed with 10 nails and has a total of 6 hangers. Figure 3 shows the assembly of the experimental suspended ceiling. More detailed information for the test setup and the measurement methodology can be found in [8].

(a) (b)

Fig. 3. Experimental setup of suspended ceiling. a) top profile, b) base profile

From the test rig, the applied load to each hanger was estimated. Nail failures were simulated and the redistribution of the load to the hangers was measured statistically. The scatter of the suspension loads depends on various influencing factors. These include the bending stiffness, the cover slab thickness, and the assembly sequence. Based on these experiments the distribution function of the suspension hangers was estimated. The loads can be described very accurately by a Weibull distribution in Eq. (2).

$$E(Z) = 1.165 \cdot ln\left(\frac{1}{1-\frac{Z}{1.001}}\right)^{0.304} \quad (2)$$

The permanent nail load-bearing capacity was determined experimentally in [8]. For the experimental investigation, the Hilti X-P 27 MX nails with an embedment depth of

27 mm and diameter of 4 mm were tested under realistic conditions. The nails were inserted into a concrete ceiling from below i.e. overhead. On the ceiling, traffic loads were simulated as cyclic loads. As a result, cracks were formed on the ceiling with different crack widths. The nail load-bearing capacity with a crack width from 0.00 to 0.36 mm can be described by a Weibull distribution. The distribution function is valid for concrete grade C20/25.

2.2 Numerical Modeling

Finite Element (FE) models of the suspended ceiling were created with the structural analysis software Sofistik. The input in the numerical models was the experimentally derived distribution functions for the suspension load of hangers and the load resistance of the nails. The FE models are used as the basis for a probabilistic calculation. The models are calculated for a high number of realizations according to the MC method. If the corresponding number of nails and their group load-bearing capacity is overwhelmed by the suspension loads from the hangers then the system fails. The results are then evaluated statistically and the failure probabilities are determined based on Eq. (1). From the calculated failure probabilities the the reliability index β is estimated. The suspended ceiling under investigation is assigned a reliability class of RC2. Therefore the target reliability index for the load-bearing capacity is $\beta = 3.8$ and for the serviceability $\beta = 1.5$. For the models investigated the number of hangers per nail group and the number of nails per nail group were varied.

All the FE models investigated have a profile length of 4 m, and the hangers are arranged with a spacing of 700 mm. Three different nail spacings are considered 450, 360, and 280 mm. The smaller the nail spacing the more nails are required for the profile. A further investigation was conducted that considered the reaction force applied to the profile by the ceiling support. Figure 4 shows the static system with a nail spacing of 450 mm. The nails are shown with red color the hangers with brown and the profile with blue.

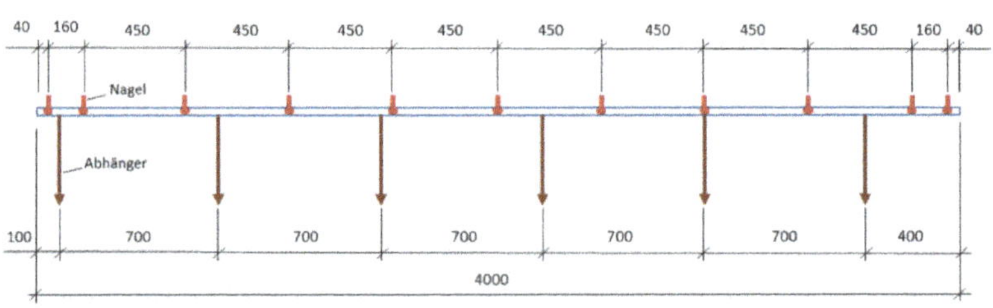

Fig. 4. Suspended ceiling system with nail spacing of 450 mm and hanger spacing of 700 mm

Some preliminary results of the probabilistic FE model were derived for evaluation purposes. These results were derived for a limited number of realizations i.e. 200000. The reliability index was calculated based on the probability of failure. The aimed reliability index was $\beta = 3.8$. Additional simulations were carried out with reduced hanger loads, these are 80% and 60% of the original load class 0.50 kN/m^2. From the early evaluations, it was identified that by reducing the nail spacing or the load the reliability of the system increases. It is also evident that the reaction force from the ceiling to the profile influences negatively the reliability of the system. The concentrated preliminary results are presented in Table 1.

Table 1. Concentrated preliminary reliability indices

β-reliability index	With ceiling support			Without ceiling support		
	60%	80%	100%	60%	80%	100%
280 mm spacing	3.35	3.15	2.92	4.08	3.88	3.65
360 mm spacing	3.21	2.85	2.60	3.46	3.10	2.80
450 mm spacing	2.76	1.96	1.16	-	-	-

2.3 Simplified Model

Due to the high computational cost of the probabilistic FE model, a simplified probabilistic model was developed. To develop the simplified model the full MC simulation of the FE model was evaluated for the 100% of the load class 0.50 kN/m^2 for 1 million realizations. The failure cases from the FE model were selected and their corresponding input data i.e. suspension loads, and nail capacities. The results were examined more closely and therefore patterns between the actions and resistances were highlighted and therefore certain failure patterns were recognized. A zip type of failure was identified which could lead to an overall system failure. With the use of this information, the input data that could lead to failure cases are separated into subpopulation (strata). Based on this knowledge new simplified models and an advanced stratified sampling can be created.

From the analysis of the results, 3 critical combinations were identified resulting in 3 failure criteria. The first failure criterion is the high utilization of the total nail load-bearing capacity. The other two criteria are related to the zip type failure. Half the profile is considered, in order to represent that the zip type failure begins from one side. The moment ratio between the nail resistance and the applied suspension loads is estimated. If a high ratio of moments is identified in one half of the profile that means that the zip type failure can initiate in this side.

3 Results and Discussion

In this chapter, the results of the probabilistic calculations are presented. The simulations cover a wide area of suspended ceiling configurations with different load classes, hangers, and nails per group. A notation system was developed for all the simulated configurations the first component SC stands for suspended ceiling the second component is the load class multiplied by 100 e.g. 0.50 kN/m^2 is represented by 50. The following factors represent the profile length in centimeters the number of hangers and finally the number of nails.

In the initial step, the results of the heavy SC-50 and the medium SC-30 load class with a profile length of 4 m and a hanger distance of 700 mm were investigated. The reliability index β of these systems was derived from the FE model with 1 million realizations for each configuration. The input values of the realizations that failed were evaluated and the subpopulation of input combination that led to failure was identified. With the use of this information, the same levels of reliability were achieved with only 200,000 realizations. In a second iteration of analyzing the properties of the input of the failed scenarios, it was possible to further reduce the subpopulation that leads to failure. As a result, the desired probability of failure or reliability index can be achieved with less than 20,000 realizations. A visual representation of the stratified sampling interations is presented in Fig. 5. For the calculation of the probability of failure, a combination of the probabilistic models described in Sect. 2 was used. For the remaining simulations, the simplified model developed is considered. For these models, the stratified sampling method developed from the previous models is implemented. In Table 2 the results from the two load classes SC-50 and SC-30 and the corresponding reliability indices are concentrated.

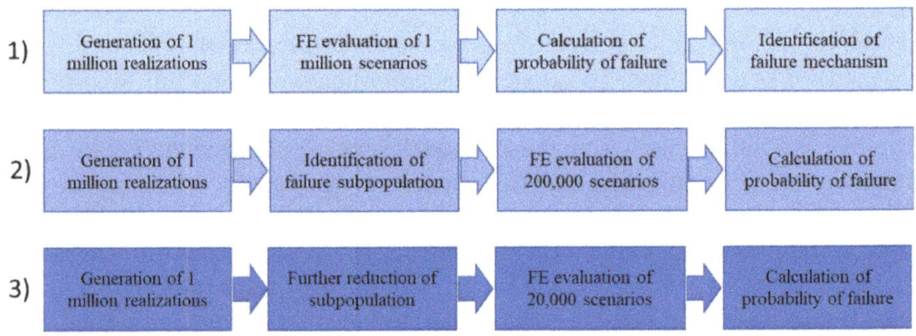

Fig. 5. Flowchart of the 3 interations of the optimization method for the stratified sampling

The combined results of the two probabilistic methods in terms of reliability index are illustrated in Fig. 6. The results are presented in relation to the nail spacing. The β-values determined by the probabilistic FE for the ceilings with hanger spacing of 700 mm and profile length of 4 m are presented with blue circles. These results are derived after 1 million realizations. The corresponding results for medium load class SC-30 are presented with yellow circles. The results from the simplified probabilistic models are presented with triangles and the same color code is used for the load classes. It can be seen that for smaller nail spacing a higher reliability of the system can be achieved.

It was decided as a next step to develop a surrogate model based on the existing results. The aim is to create a soft computing model that reflects the relationship be-tween nail spacing, hanger spacing, and reliability index. Such models have to be created for each evaluated ceiling system. Three load classes 0.50, 0.30, 0.15 kN/m^2 are considered. A variation of the nail spacings is considered and three hanger spacings are 700, 800, and 1000 mm. These configurations lead to a total of 9 different systems. A new notation system was implemented for the surrogate models the first term is SC for suspended

Table 2. Tested configurations and reliability indices

	Designation	Profile length (m)	Nail spacing	Number of hangers	Number of Nails	β-index
Load class 0.5 kN/m^2 hanger spacing 700 mm	SC-50 40-6-18	4	240	6	18	3.75
	SC-50 25-4-16	2.5	160	4	16	3.62
	SC-50 13-2-10	1.3	140	2	10	4.05
	SC-50 6,4-1-5	0.64	140	1	5	3.41
Load class 0.3 kN/m^2 hanger spacing 800 mm	SC-30 40-5-15	4	300	5	15	3.71
	SC-30 40-5-16	4	280	5	16	3.8
	SC-30 20-3-13	2	160	3	13	3.73
	SC-30 6,4-1-5	0.64	140	1	5	3.66

ceiling. The second term is the load class multiplied by 100, the third term is the hanger spacing in mm. The final term is SIM for the simplified model, MC for the FE model, and GAUSS for the Gaussian function.

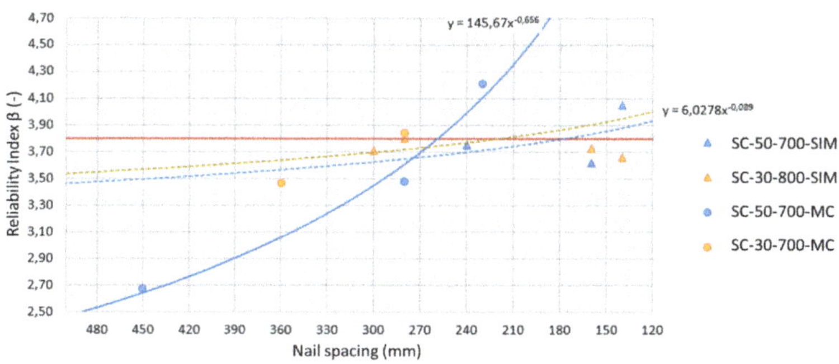

Fig. 6. Development of the reliability index values β for the SC-50 and SC-30 ceiling systems depending on the nail spacing. The aim reliability index of 3.8 is highlighted with a red line

For the surrogate model, a 3rd-order Gaussian function was adapted to the probabilistic FE simulations and the simplified probabilistic model results. In order to draw conclusions about the ceilings with hanger spacing of 800 and 1000 mm, the action variables were set in relation to the actions of the ceiling with 700 mm hanger spacing. The same was also done for the resistance. The resulting action was then subtracted from the value of the resistance. Figure 7 presents the Gaussian functions of reliability to nail spacing for the heavy ceiling 0.50 kN/m^2. It can be seen that only the ceiling system with 700 mm hanger spacing reaches the limit of 3.8 at a nail spacing of 255 mm.

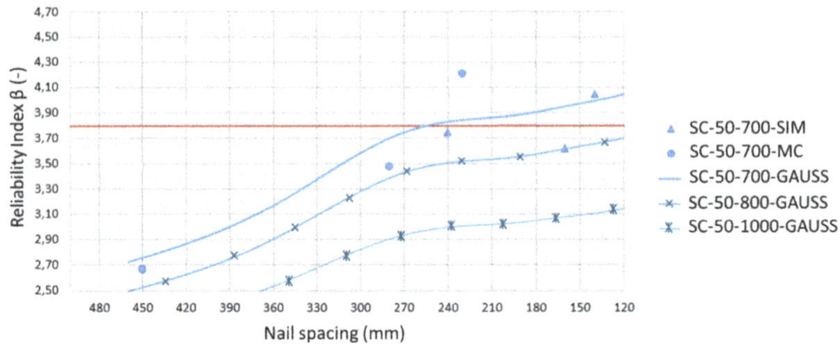

Fig. 7. 3rd-order Gaussian function of the reliability index versus the nail spacing for the heavy suspended ceilings 0.5 kN/m^2.

4 Conclusions

This study aims to develop some initial surrogate models to describe the reliability level of suspended ceiling systems fastened with direct fasteners. The investigated configurations are regarding different load classes, nail spacings, and hanger spacings. The surrogate models are developed based on the reliability levels that are derived by an iterative probabilistic FE simulation. The probabilistic FE models use as input the experimentally derived distribution functions for the nail resistance and the hanger loads. The most important outcomes of the study are presented below.

- A digital model of the suspended ceiling system that was tested in the laboratory of BOKU was developed and verified.
- An extensive MC simulation was conducted implementing the digital model to estimate the reliability of the system and determine the failure mechanism.
- Based on the identified zip-type failure mechanism a stratified sampling methodology was developed that reduced the number of simulations from 1 million to a few tens of thousands.
- Simplified 2D models that can describe the zip-type failure were developed to further reduce the computational cost.
- Surrogate models that can represent the reliability of a variety of suspended ceiling systems were developed by combining both the extensive and th simplified models

References

1. Block, K., Dreier, F.: Zusammenfassender Bericht 18-115-2 zum Forschungsvorhaben "Befestigung von redundanten nichttragenden Systemen an Betonkonstruktionen mit Hilfe der Direktmontagetechnik"- abgehängte Decken in direktem Befestigungsverfahren ohne Vorbohrung -"Dortmund" (2019)
2. Knauf Gips KG. www.knauf.de 11 1015. https://www.knauf.de/profi/sortiment/systeme/plattendecken.html#vpDetails. Zugriff 23 Aug 2022
3. Schneider, J.: Sicherheit und Zuverlässigkeit im Bauwesen - Grundwissen für Ingenieure, vdf Hochschulverlag AG an der ETH-Zürich, Zürich (2018)

4. Forrester, A.I.J., Sóbester, A., Keane, A.J.: Engineering Design via Surrogate Modelling: A Practical Guide. Wiley, Chichester (2008)
5. Gorissen, D., Couckuyt, I., Demeester, P., Dhaene, T., Crombecq, K.: A surrogate modeling and adaptive sampling toolbox for computer based design. J. Mach. Learn. Res. **11** (2010)
6. Razavi, S., Tolson, B.A., Burn, D.H.: Review of surrogate modeling in water resources. Water Resour. Res. **48**(7), 31 (2012)
7. Ferreira, J., Pedemonte, M., Torres, A.I.: A genetic programming approach for construction of surrogate models. Comput. Aided Chem. Eng. **47** (2019)
8. Hauser, M., Schwenn, M., Zeman, O.: Prüfbericht Nr. 875-19-020: Untersuchungen zum Lastabtragungsverhalten von abgehängten redundanten Deckensystemen. Wien (2019)

Understanding Corrosion in Restored Concrete Zones Through Sensor Data Analysis

Luis F. Rincon[1,2(✉)], Emilio Bastidas-Arteaga[2], José C. Matos[1], Yina M. Moscoso[1], and Elsa Eustaquio[3]

[1] ISISE, ARISE, Department of Civil Engineering, University of Minho, Guimarães, Portugal
luis.rincon_prada@univ-lr.fr

[2] Laboratory of Engineering Sciences for the Environment (LaSIE - UMR CNRS 7356), La Rochelle University, La Rochelle, France

[3] Laboratório Nacional de Engenharia Civil, Lisboa, Portugal

Abstract. Despite reinforced concrete bridges being highly durable, they remain susceptible to natural ageing and disasters, potentially impairing their performance over time. This drives an increase in maintenance, rehabilitation, and repair expenditure, accounting for half of the construction sector spending in developed nations. Structural and durability monitoring becomes crucial to establish long-term maintenance schedules and ensure safety throughout their lifespan. Typically, maintenance schemes rely on sensor systems that periodically or continuously collect information, utilizing various sensors such as electrical, optical, chemical, and sound sensors. Sensor reliability is subject to environmental factors, its own durability and potential power outages, which can temporarily or permanently interrupt data acquisition making difficult or even impossible damage detection. This study focuses on analyzing the evolution of concrete electrical resistivity over time in five repaired concrete zones instrumented with durability sensors to estimating corrosion progression. Based on a compiled and filled database by the authors, this research utilizes degradation models and correlation indices to estimate concrete cover deterioration. Leveraging over fourteen years of sensor data from a reinforced concrete structure, the study focuses on comprehending concrete resistivity evolution in repaired zones to gain insights into the loss of protective concrete cover characteristics to reinforcement corrosion. The study deliberates its findings and extends recommendations for applying this approach to other sensor analysis, enabling a more profound understanding of concrete structure health and longevity.

Keywords: Structural health monitoring · Concrete Resistivity Sensors · Corrosion · Degradation models

1 Introduction

Ensuring the longevity and reliability of reinforced concrete bridges stands as a paramount concern amidst the persisting vulnerabilities to natural disasters. Despite their inherent durability, these critical infrastructures confront ongoing challenges that

J. C. Matos et al. (Eds.): IPW 2024, LNCE 494, pp. 84–91, 2024.
https://doi.org/10.1007/978-3-031-60271-9_6

impact their performance over time, necessitating substantial expenditure, accounting for nearly half of the construction sector's spending in developed nations [1, 2]. To address these issues, structural and durability health monitoring emerges as a pivotal strategy, essential for establishing comprehensive maintenance schedules and safeguarding their sustained safety throughout their lifespan.

Typically reliant on sensor systems, these monitoring approaches engage diverse sensors, encompassing electrical, chemical, optical, and sound sensors, to gather essential structural information periodically or continuously. However, the reliability of such sensors faces susceptibility to environmental influences, durability constraints, and potential power disruptions, posing risks of temporary or permanent interruptions in data acquisition.

Degradation models serve as pivotal tools in unraveling structural decay and forecasting the long-term performance of assets, crucial for crafting effective maintenance schedules amidst diverse environmental and societal influences. These models, spanning mechanistic, probabilistic, statistical, and metaheuristic approaches, offer distinct perspectives on structural deterioration. Mechanistic models, seen in works by Villain et al. [3] and Nguyen et al. [4], meticulously establish links between influential variables and degradation outcomes, utilizing complex simulations to understand phenomena like chloride penetration into concrete. In contrast, probabilistic models, as evidenced by studies from Kwon et al. [5], and Bastidas-Arteaga et al. [6], deal with inherent variability, considering stochastic elements impacting corrosion and structural fatigue. Markov Chain (MC) models, a standout in statistical approaches, discretize degradation states to probabilistically estimate transitions, showcased in methodologies by Bastidas-Arteaga and Schoefs [7] and Zhang et al. [8], integrating inspection data and chloride penetration models. While each model category offers unique insights, mechanistic models demand high costs, probabilistic models navigate complexity, statistical models lack resistance quantification, and metaheuristic models, notably Artificial Neural Networks (ANNs), show promise but rely heavily on available data and opaque decision processes.

This study undertakes a comprehensive analysis of five repaired concrete zones, delving into the longitudinal evolution of concrete electrical resistivity over time and scrutinizing reinforcement corrosion progression. By using a recently processed database, which includes a new methodology for imputing missing data in an SHM system [9], this research aims to estimate concrete deterioration employing degradation models. Spanning over fourteen years of sensor data from a reinforced concrete structure, this study meticulously scrutinizes the evolution of concrete resistivity within repaired zones, offering profound insights into corrosion progression. The paper is structured as follows: Sect. 2 outlines the methodology employed; Sect. 3 presents the application's information. Section 4 describes the results and discussion, and finally, the research conclusions are presented in Sect. 5.

2 Methodology

This chapter presents the methodology employed to assess corrosion in a repaired reinforced concrete structure by monitoring electrical resistance as a key indicator. Specialized sensors were installed in the structure, enabling continuous measurement of resistivity over time. Utilizing these resistivity data, stochastic degradation models based on

Markov chains were proposed to predict the evolution of corrosion in the structure. This chapter details the significance of electrical resistance in concrete and the application of Markov degradation models.

2.1 Electric Resistance as an Indicator of Durability

Estimating the deterioration of a structure due to corrosion can be achieved through the utilization of various indicators linked to the durability of materials against the phenomenon of reinforcement steel corrosion. Concrete resistivity, among other parameters, like corrosion potential, emerges as a pivotal parameter in the durability assessment of reinforced concrete structures. Due to the simplicity of the measure, it is use in degradation models and as it gauges the electrical resistance of concrete. Making it, a crucial marker for the presence of cracks and chloride ions that may trigger the initiation of steel reinforcement corrosion within the structure [10].

Nevertheless, resistivity sensors might be influenced by the concrete's degree of saturation and temperature. A higher degree of concrete saturation substantially lowers the determined electrical resistivity [11, 12]. Regarding the temperature's influence, an increase in temperature augments ion mobility and consequently reduces concrete resistivity [12, 13]. Hence, the analysis of electrical resistivity measurements in concrete should be conducted with special care because of environmental variations but also because they are affected by temporary or permanent interruptions in data acquisition particularly detrimental to information analysis.

2.2 Stochastic Degradation Model

There are many models for calculating the deterioration of reinforced concrete over time. The calculation of deterioration can be proposed through empirical or physical models. Generally, models that consider resistivity focus on calculating the initiation period and in very few cases the corrosion propagation period. Models such as those of Andrade and D'Andrea [14] or EHE-08 [15], are robust models that make use of resistivity together with other variables such as the resistivity of concrete saturated in water and cured for 28 days or the corrosion rate. These models can be difficult to implement especially in older structures where resistivity is taken from recently added sensors and little complementary information has been collected.

On the other hand, statistical models rely on analyzing a repository of established failure knowledge or historical records to predict future deteriorations. They find common application in assessing the extended performance of structures such as bridges and operate through deterministic or stochastic methodologies [16].

The Markov Chain (MC) models describe a system's evolution through defined states within a transition period. These models dictate that the probability of the future state relies solely on the present state, excluding dependence on past states [17]. Within a transition period, the system holds a probability of moving from one state to another, symbolized by transition probabilities P_{ij} from state S_i to future state S_j [18]. The transition probability of each state is calculated using historic information data using

$$P_{i,j} = \frac{S_{i,j}}{S_i} \tag{1}$$

These probabilities construct a transition probability matrix (See Eq. 2), with P_{ij} denoting the likelihood of transition between states i and j. Such algorithms offer insight into structural degradation as an ongoing, natural process, crucial for understanding deterioration trends without external intervention [19, 20].

$$P_{i,j} = \begin{bmatrix} P_{1,1} & P_{1,2} & \cdots & P_{1,m} \\ P_{2,1} & P_{2,2} & & \\ \vdots & & \ddots & \\ P_{m,1} & & & P_{m,m} \end{bmatrix} \quad (2)$$

Once this probability matrix representing structural degradation is established, it enables the prediction of bridge performance over a specified period. Subsequently, using the initial database, degradation curves are modeled to span the lifespan of these structures, typically extended up to a period of 100 years.

3 Application Description

A database of eight resistivity sensors installed in a repaired concrete area of reinforced concrete bridge located in Portugal near the sea. This database was previously populated by a study conducted by the authors to have a complete database of 14 consecutive years [9]. The concrete resistivity was measured using two-graphite electrode sensors with a daily measurement of the bridge, as shown in Fig. 1.

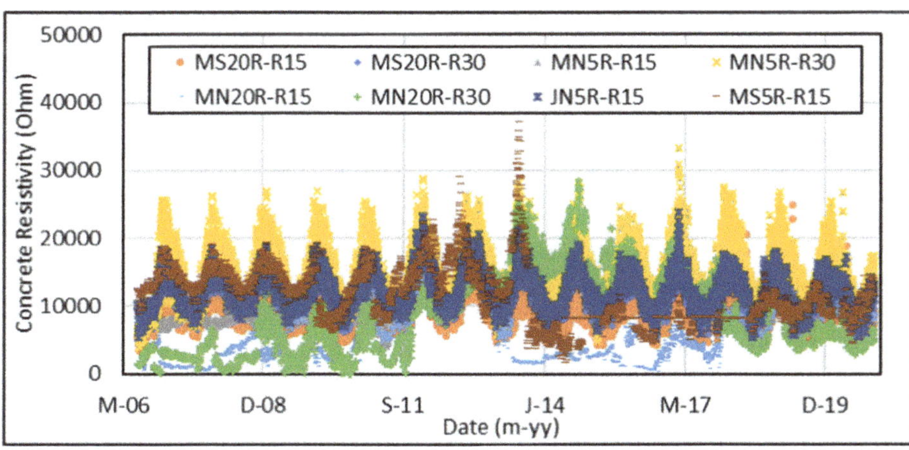

Fig. 1 Concrete resistivity sensors installed in the structure. Adapted from [9]

4 Results and Discussion

For the development of MC, it is necessary to establish sensor condition states. Table 1 presents the levels set for this analysis. The initial condition state represents a high risk of corrosion, the following two states represent a moderate risk of corrosion, and the remaining states signify a passive condition.

Table 1 Condition states establish for the MC model

Condition state	Resistivity
1	< 100
2	100–500
3	500–1000
4	1000–3000
5	3000–6000
6	6000–20,000
7	> 20,000

In civil engineering the MC models have primarily been applied to bridge deterioration based on visual inspections [19]. Simple MC models consider deterioration as a degenerative system; therefore, structures in the future can only remain in the current state or deteriorate. Using this basis, the probability transition matrix for repaired concrete of the analyzed structure is calculated (See Eq. 3). From this matrix, the degradation curve is established for a structure with an initial sensor condition of 5 and compared with the daily average of the sensors (See Fig. 2).

$$P = \begin{bmatrix} 1 & 0 & 0 & 0 & 0 & 0 & 0 \\ 0.06 & 0.94 & 0 & 0 & 0 & 0 & 0 \\ 0.01 & 0.03 & 0.96 & 0 & 0 & 0 & 0 \\ 0 & 0 & 0.01 & 0.99 & 0 & 0 & 0 \\ 0 & 0 & 0 & 0.02 & 0.98 & 0 & 0 \\ 0 & 0 & 0 & 0 & 0.01 & 0.99 & 0 \\ 0 & 0 & 0 & 0 & 0 & 0.30 & 0.70 \end{bmatrix} \tag{3}$$

With this model, it is estimated that sensors will begin detecting a high risk of corrosion after 30 years of measurements. However, significant differences are observed when considering only the degradation of the environment detected by the sensors, rendering the model unreliable.

As mentioned earlier, sensors are influenced by external factors, primarily temperature and humidity. Thus, high temperature and low humidity can lead to decreased water content in concrete, improving sensor measurements and altering the periods of corrosion initiation or propagation. Therefore, a second transition probability matrix is calculated to account for the deterioration and improvement of sensor readings over time (See Eq. 4).

$$P = \begin{bmatrix} 0{,}39 & 0{,}17 & 0{,}06 & 0{,}39 & 0 & 0 & 0 \\ 0{,}05 & 0{,}76 & 0{,}09 & 0{,}07 & 0{,}03 & 0 & 0 \\ 0{,}01 & 0{,}02 & 0{,}86 & 0{,}11 & 0 & 0 & 0 \\ 0 & 0 & 0{,}01 & 0{,}78 & 0{,}21 & 0 & 0 \\ 0 & 0 & 0 & 0{,}02 & 0{,}75 & 0{,}23 & 0 \\ 0 & 0 & 0 & 0 & 0{,}01 & 0{,}98 & 0{,}01 \\ 0 & 0 & 0 & 0 & 0 & 0{,}30 & 0{,}70 \end{bmatrix} \tag{4}$$

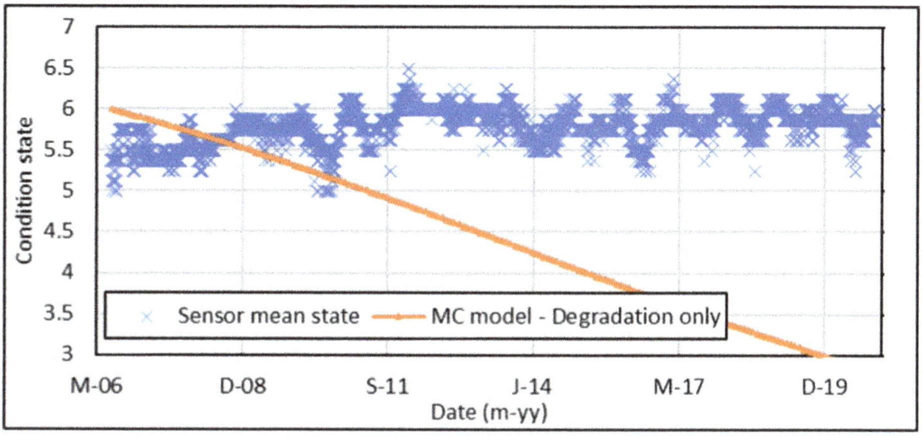

Fig. 2 MC model considering deterioration only

Figure 3 displays the outcomes of sensor data prediction, starting with an initial state of 5. It can be observed that the system converges closer to the daily average of the sensors. Nonetheless, a stationary distribution is evident in the model with a value of 5.96. Consequently, in the long term, the variance outside this value is minimal, and no deterioration will be observed using this model.

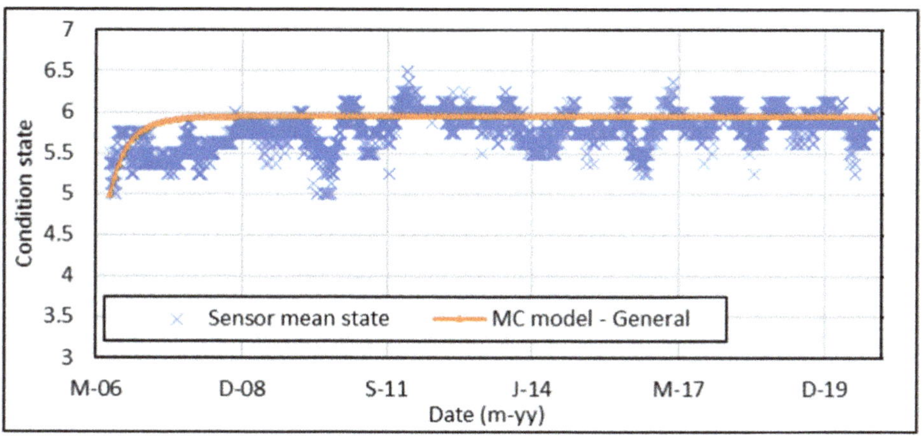

Fig. 3 MC model considering deterioration and improving of the sensors

5 Conclusions

The present paper explores two stochastic Markov Chain models to calculate concrete degradation over time. An eight-sensor database from a repaired concrete area on a bridge in Portugal is utilized to demonstrate the practical application of these models in predicting degradation and structural lifespan.

The single deterioration Markov Chain (MC) model fails to accurately represent the data evolution. The comprehensive MC model, accounting for both deterioration and improvement, yields a stationary distribution, resulting in a singular value over time, thereby complicating the estimation of future degradation in this specific case. Consequently, both analyzed MC models are deemed invalid for predicting the future deterioration of concrete based on the examined sensors.

Acknowledgement. This work is financed by national funds through FCT - Foundation for Science and Technology, under grant agreement 2021.05862.BD attributed to the first author.

This work was partly financed by FCT / MCTES through national funds (PIDDAC) under the R&D Unit Institute for Sustainability and Innovation in Structural Engineering (ISISE), under reference UIDB / 04029/2020 (doi.org/https://doi.org/10.54499/UIDB/04029/2020), and under the Associate Laboratory Advanced Production and Intelligent Systems ARISE under reference LA/P/0112/2020.

References

1. Pereira, D.D.: Lifetime evaluation of maritime structures: application to the bridge pier at the Leixões Seaport (In portugueses). Universidade do Minho (2017). http://hdl.handle.net/1822/49829
2. Bastidas-Arteaga, E., Schoefs, F.: Sustainable maintenance and repair of RC coastal structures. Proc. Inst. Civ. Eng. Eng. **168**(4), 162–173 (2015). https://doi.org/10.1680/jmaen.14.00018
3. Villain, G., Sbartaï, Z.M., Dérobert, X., Garnier, V., Balayssac, J.-P.: Durability diagnosis of a concrete structure in a tidal zone by combining NDT methods: Laboratory tests and case study. Constr. Build. Mater. **37**, 893–903 (2012). https://doi.org/10.1016/j.conbuildmat.2012.03.014
4. Nguyen, P.T., Bastidas-Arteaga, E., Amiri, O., El Soueidy, C.P.: An efficient chloride ingress model for long-term lifetime assessment of reinforced concrete structures under realistic climate and exposure conditions. Int. J. Concr. Struct. Mater. **11**(2), 199–213 (2017). https://doi.org/10.1007/s40069-017-0185-8
5. Kwon, S.J., Na, U.J., Park, S.S., Jung, S.H.: Service life prediction of concrete wharves with early-aged crack: probabilistic approach for chloride diffusion. Struct. Saf. **31**(1), 75–83 (2009). https://doi.org/10.1016/j.strusafe.2008.03.004
6. Bastidas-Arteaga, E., Bressolette, P., Chateauneuf, A., Sánchez-Silva, M.: Probabilistic lifetime assessment of RC structures under coupled corrosion-fatigue deterioration processes. Struct. Saf. **31**(1), 84–96 (2009). https://doi.org/10.1016/j.strusafe.2008.04.001
7. Tesfamariam, S., Bastidas-Arteaga, E., Lounis, Z.: Seismic retrofit screening of existing highway bridges with consideration of chloride-induced deterioration: a Bayesian belief network model. Front. Built Environ. **4** (2018). https://doi.org/10.3389/fbuil.2018.00067
8. Zhang, Y., Kim, C.W., Tee, K.F., Lam, J.S.L.: Optimal sustainable life cycle maintenance strategies for port infrastructures. J. Clean. Prod. **142**, 1693–1709 (2017). https://doi.org/10.1016/j.jclepro.2016.11.120
9. Rincon, L.F., et al.: Enhancing corrosion assessment in structural health monitoring: missing data estimation method for concrete structures. Struct. Heal. Monit. - An Int. J., p. Under review (2024)
10. Azarsa, P., Gupta, R.: Electrical resistivity of concrete for durability evaluation: a review. Adv. Mater. Sci. Eng. **2017**, 1–30 (2017). https://doi.org/10.1155/2017/8453095

11. Gjørv, O.E., Vennesland, Ø., El-Busaidy, A.H.S.: Electrical Resistivity of Concrete in the Oceans (1977). https://api.semanticscholar.org/CorpusID:137164502
12. d'Andréa, R.: Predicción de la durabilidad del hormigón armado a partir de indicadores de corrosión: aplicación de la resistividad eléctrica. Universidad Politécnica de Madrid (2010)
13. Hope, B.B., Ip, A.K., Manning, D.G.: Corrosion and electrical impedance in concrete. Cem. Concr. Res. **15**(3), 525–534 (1985). https://doi.org/10.1016/0008-8846(85)90127-9
14. Andrade, C., D'Andrea, R.: Electrical resistivity as microstructural parameter for the calculation of reinforcement service life. Microstruct. Relat. Durab. Cem. Compos. **1**, 1483–1490 (2008)
15. Ministerio de Fomento. Instrucción Española de Hormigón Estructural (2008)
16. Rincon, L., Moscoso, Y.M., Hamami, A., Matos, J.C., Bastidas-Arteaga: Degradation models and maintenance strategies for reinforced concrete structures in coastal environments under climate change: {A} review
17. Jiang, Y.: Application and comparison of regression and Markov chain methods in bridge condition prediction and system benefit optimization. J. Transp. Res. Forum. **49**(2), 210 (2010)
18. Setunge, S., Hasan, M.S.: Concrete bridge deterioration prediction using markov chain approach. Digital Library, University of Moratuwa, Sri Lanka (2011)
19. Moscoso, Y.F.M., Rincón, L.F., Leiva-Maldonado, S.L., Campos e Matos, J.A.S.C.: Bridge deterioration models for different superstructure types using Markov chains and two-step cluster analysis. Struct. Infrastruct. Eng. 1–11 (2022). https://doi.org/10.1080/15732479.2022.2119583
20. Rincon, L.F., Muñoz, Y.F., Leiva, S.L., Matos, J.C.: Stochastic degradation model analysis for prestressed concrete bridges. In: IABSE Symposium Prague, Prague (2022)

Risk Assessment and Management

Application of Structure-From-Motion Multi-view Stereo (SfM-MVS) Photogrammetry in First Response Inspection of Distressed Buildings

Zvonko Sigmund[1] and Ivan Marović[2](✉)

[1] Faculty of Civil Engineering, University of Zagreb, Zagreb, Croatia
[2] Faculty of Civil Engineering, University of Rijeka, Rijeka, Croatia
ivan.marovic@gradri.uniri.hr

Abstract. The digital transformation of built environments into 3D computer models via photogrammetry for inspection, documentation, or remote mapping has evolved over the last few decades. Today, the appropriateness of different cameras and laser scanners gives even more opportunity for engineers in different situations to capture and create very accurate and high-detailed 3D models. The Structure-from-Motion Multi-View Stereo (SfM-MVS) photogrammetry is a prominent way of reducing capturing time even more by using a 360-degree camera. In situations of buildings in distress, especially those exposed to extreme environmental hazards, such a method for rapid photogrammetric reconstructions seems to be the most suitable one for everyone, especially for first responders. Therefore, the application of SfM-MVS photogrammetry in first response inspection of distressed buildings is presented in a case study from a recent earthquake series that happened during the year 2020 in Croatia. The goal of this paper is to bring insights into the possibilities of applying SfM-MVS photogrammetry in the built environment and to present such digital models in a way that can help first responders make informative decisions for inspecting distressed buildings.

Keywords: rapid photogrammetry · inspection · distressed buildings · 360-degree camera · crisis management

1 Introduction

1.1 The Strong Motion Events of 2020

Croatia's seismicity is generally moderate, with infrequent instances of strong events—a characteristic often observed in regions experiencing intraplate seismicity. Despite not being the most earthquake-prone area, Croatia faces significant seismic vulnerability, primarily influenced by its economic and political standing. The northwest part of the Croatian mainland, constituting most of mainland Croatia, accommodates 45% of the country's population and contributes 55% of the Croatian gross national product [1]. Still, recently in 2020, Croatia, on the verge of the COVID-19 pandemic, was struck

J. C. Matos et al. (Eds.): IPW 2024, LNCE 494, pp. 95–104, 2024.
https://doi.org/10.1007/978-3-031-60271-9_7

with a possibly unconnected series of earthquakes: Zagreb earthquake that struck in March 2020 (M5.0), just after the Croatian government had issued a complete lockdown due to the COVID-19 pandemic; and the Petrinja (about 50 km from Zagreb) earthquake (M6.4) in December 2020 [2].

The aftermath of the Zagreb earthquake affected the city center the most, and early reports suggested significant damage. Numerous cultural and historical architectural landmarks in Zagreb suffered considerable harm, including severe damage to the south tower of Zagreb Cathedral. Approximately 16,555 residential, public, and monumental structures were affected. Here, state institution buildings, religious buildings, museums, galleries, the Mirogoj Cemetery, the Rectorate, faculties, schools, and theatres suffered a significant amount of damage as the majority of them were situated in old city center buildings [3].

The Petrinja earthquake affected the Sisak – Moslavina county mainly, but the damages resulting from this earthquake and the geological formations were reported even in Karlovac and Krapina - Zagorje country, which is almost 100 km from the epicenter. About 50.000 damaged buildings (Fig. 1) were assessed after the earthquake [4].

Fig. 1. Damage to heritage buildings in Petrinja region

1.2 Motivation

To the point of restoration of life, as was, engineers were and still are involved in numerous activities like:

- immediate post-disaster response activities supporting first-responders
- damage assessments to assess the safety of damaged buildings
- rapid damage and needs assessments (RDNA) as a support to the government
- post-disaster recovery regulatory framework development support
- but nowadays, they still have an important role – a crucial role in the building-back-better process

To understand the possible contribution of technological innovation and use, the application of Structure-from-Motion Multi-View Stereo (SfM-MVS) photogrammetry in post-disaster activities, which is the aim of this paper, one must understand the function of engineers and the decision-making in the recovery process.

According to the organizational structure of the Croatian platform for disaster risk reduction, the civil protection units are under the immediate jurisdiction of the Croatian government [5]. Here, the civil protection units can and did include expert working groups, which consisted of the scientific community [2] as well as a large number of structural engineers from the private sector [6].

The immediate task of engineers in the post-disaster activities was to support the first responders in search and rescue activities, which were only conducted immediately after the major earthquake on 29th December 2020, but also to assess the safety and stability of the buildings that were reported damaged [4]. Here, during a survey conducted by on-site structural engineers in Sisak-Moslavina County after the M6.2 Petrinja earthquake in 2020, vulnerability classes and damage grades were established in accordance with the EMS-98 scale to assess building damage [7]. These assessments are nowadays used as a decision-making platform for the Croatian Ministry of Physical Planning, Construction, and State Assets for the recovery financing options [8].

In the first years after the Petrinja earthquake, about 1,600 engineers were involved, tasked with damage assessments [9]. Throughout the immediate post-disaster response phase and the recovery period inevitable mistakes were made:

- As Petrinja earthquake had a rather unsignificant pre-shock that happened on 28th December [7], a certain number of volunteers aided to help speed up the assessment process of the Petrinja area. A relatively small number of engineers were professionally trained in behaving in cases of an emergency, which resulted in a small but still existing number of injuries caused by the assessment teams being entrapped in the collapsed building after the earthquake on 29th December.
- The post-earthquake damage assessment procedures were established just after the Zagreb earthquake, eight months prior to the Petrinja earthquake [10]. Such resulted in high unreliability of damage assessments made by the experts.
- Trained assessment experts were tasked by the Ministry of Physical Planning, Construction, and State Assets to conduct the Digital Living Network Alliance necessary for the help from the European Solidarity Fund, the World Bank, and others [11, 12]. The RDNA was conducted based on the damage assessments conducted on the field. However, due to the unreliability of the damage assessments conducted in the field, the RDNA could just marginally rely on the damage classification from the assessments.

Using technological advancements like SfM-MVS photogrammetry, several goals can be achieved: reducing the unreliability of data received, increasing the safety of the first responders and supporting engineering teams, and creating a database for reliable, unbiased decision-making platform based on the real data on buildings. In addition, the collected footage would nowadays serve as concrete, real, and realistic media for the education of first responders as well as for the education of young or inexperienced engineers on damage assessments and post-disaster safety concerns.

2 SfM-MVS in Use

2.1 Education Based on the SfM-MVS Collected Examples

To make disaster education programs more effective for community members, target groups and individuals should be identified [13, 14]. The protocol and educational methods are different for everyone in the community, with different levels of authority, education, and information.

In the post-earthquake event, certain types of expertise levels and areas are required. Besides the expected managerial and human resource management skills, on-site, where real danger lurks while there is a requirement to assess the situation and objectively report on the observed damages, make hard decisions, and give solid suggestions, it is beneficial for the involved parties to have at least met the real situation. Training processes must be as challenging and realistic as possible so that the training group can experience being in a real situation. Using simulation systems replicating the actual situation in the field is one of the most effective and economically viable ways to train personnel involved in crisis management. The increasing power of IT systems and the faster turnover of information are elevating the capabilities of simulation systems to previously unimaginable heights, but simulation systems have their limitations. They will never be able to fully replace training activities at specialized training sites or training ranges [15]. Here, in the scientific community, the use of virtual and augmented reality is increasing in importance and is ever more used for experience building for a particular use: first responders, medical, construction, and other experts, etc. [16–18]. Still, realistic education materials would benefit from realistic models collected on-site, like from actual damaged buildings in real environments.

2.2 Assessments Supported by SfM-MVS

Ambiguities or reliability of assessed data can be reduced or eliminated entirely only by trained expert users assessing the damages and assessment methods that rely on something other than expert opinion or decision-making. Here, a helpful approach is to have a recording of data based on which a decision has been made. The SfM-MVS processes are undoubtedly the definite way the assessment process will develop in the near future. With the rapid development and availability of machine learning and AI, researchers have already started harvesting the benefits of automated processes that can be developed or implemented.

Improving building damage identification in the aftermath of disasters, especially when labeled samples are limited, and there is a high similarity between partially damaged and undamaged buildings. The authors finally demonstrated the effectiveness of their automated damage detection using SfM-MVS through experiments and the emphasis on the importance of category-level alignment in complex image classification tasks [19]. Others proposed [20] a comprehensive practice-oriented approach for UAS-based survey and assessment of bridge infrastructures, focusing on mission planning design and optimization of data acquisition and processing. In addition, the authors aim to build a cooperative, autonomous, self-recharging drone system for autonomous inspections of transport infrastructures. Additionally, remote sensing is vital for conducting damage assessment after a disaster, and building/structural damage assessment is crucial for supporting rescue operations and recovery planning [21].

Currently, human support is still necessary and inevitable, but the way toward safer and more reliable decision-making for first-responder support and renovation planning measures is clear. The rapid development of SfM-MVS and the availability of technological solutions for reality capturing make the automation of processes even more interesting and achievable.

3 Experience in Post-Disaster Educational Materials Collection

UAVs can be utilized for gathering spatial data related to the condition of buildings and cultural heritage, as well as obtaining spatial information over larger areas for assessing risks such as damage detection and mapping, particularly in the aftermath of events like earthquakes. These platforms are outfitted with state-of-the-art sensors for determining in-flight positions accurately. These sensors encompass GNSS (Global Navigation Satellite System), RTK (Real-Time Kinematic), and IMU (Inertial Measurement Unit) systems, enabling real-time positioning with centimeter accuracy. To facilitate high-precision photogrammetric measurement of existing structures' safety and seismic vulnerability, the systems are equipped with sensors that gauge obstacle distance, allowing for close-proximity flights to buildings using technologies like microradar and LiDAR. The generated photorealistic 3D models and products, derived from sources such as orthophotos, can be employed for the identification and mapping of structural damage in buildings.

3.1 General Overview of the On-Site Situation

On a broader scale, SfM-MVS data acquisition processes can be employed to obtain valuable information, particularly for swiftly assessing damage across city blocks. In this context, UAVs can assist first responders and enhance their immediate emergency planning efforts following a disaster. However, a rapid zoning of city blocks is essential to pinpoint higher-priority zones within the city. Notably, this activity does not necessitate post-processing; a visual data acquisition approach, achieved by flying over different zones, is sufficient.

The main limitation of this procedure lies in the level of detail in the analysis. While examining drone footage, one can easily identify significant faults or damages, ranging from minor issues like the destruction of a chimney to complete building destruction (Fig. 1), without the need for post-processing or closer scrutiny. However, identifying more significant damages that did not result in complete destruction may be challenging without additional post-processing or closer inspection. Still, this information can be a precious asset in decision-making during immediate post-disaster activity and resource planning.

3.2 More Detailed Insight to the On-Site Situation

Obtaining more detailed information about buildings necessitates a close examination of the particulars. Such becomes challenging due to the perilous conditions surrounding buildings affected by earthquakes. Aftershocks, smaller seismic events that ensue after the initial quake, can further destabilize already damaged structures, increasing the risk of additional collapses or falling debris that pose a significant threat of injury. The structural integrity of these buildings is often compromised, rendering them unsafe and susceptible to sudden collapses. Furthermore, there are hidden dangers, such as gas line leaks and the potential for fires or explosions. Compromised electrical wiring introduces additional risks, including possible electrical discharges, shocks, and fires.

Additionally, older buildings may harbor hazardous materials like asbestos or lead, which can be released into the air when the structures are damaged. Hence, maintaining a safe distance from earthquake-damaged buildings and adhering to the guidelines provided by emergency response teams is paramount for ensuring safety. Engineers can use various photogrammetry techniques to conduct more advanced assessments in such situations. Drones can navigate areas otherwise inaccessible or too hazardous for human access, particularly in densely populated urban zones or areas where buildings are prone to collapse. Fitted with high-resolution cameras, UAVs capture detailed images and videos of damaged structures, enabling a more precise and rapid assessment of the extent of damage compared to ground-based observations. Such is especially true when postprocessing units, such as capable computers with adequate software support and Artificial Intelligence (AI) tools, are utilized.

Our primary reliance here is on the analysis of photogrammetric data. The quality of the acquired data may vary depending on the specific objectives, but the outcomes can yield photorealistic results with high accuracy (Figs. 2 and 3). These data acquisition protocols were developed as part of the NORMENG project (https://normeng.eu) but are versatile and can be applied for various purposes.

Fig. 2. Photo of the photogrammetry experiment site (NORMENG project)

Fig. 3. Postprocessing (left - photorealistic 3D model, right - display of photogrammetry point cloud results focusing on data scattering) (NORMENG project)

3.3 More Detailed Insight to the On-Site Situation

All UAV data acquisition processes are dependent, in one way or another, on the availability of GPS. While photogrammetry does not necessarily rely on GPS, it can offer significant advantages when GPS is accessible. Nonetheless, photogrammetry demands a sufficient amount of photogrammetric data (photos, images, screens, etc.) to make the process viable. In this context, LIDARs and similar technologies surpass photogrammetry in speed as well as quantity of gathered data.

As mentioned earlier, damaged buildings are not safe environments, but 3D data and interior damage intensities for engineering purposes and analyses are just as crucial as external information. Therefore, we directed part of our research toward interior data acquisition.

Given that LIDAR scanning is time-consuming, requiring at least 2 – 3 min for each scanning position and additional scans for transitions between rooms, we conducted experiments with 360-degree camera data acquisition. To test this approach, we selected a secure location: an undamaged storage area in an area less susceptible to seismic activities. The storage area was primarily scanned from the inside using LIDAR as a reference for control measurements and a 360-degree camera for interior scanning. As evident from the interior scans presented in Figs. 4 and 5, the LIDAR scan maintains its superiority over the 360-degree camera scan. However, the 360-degree camera exhibits its advantages in other aspects.

The entire scanning process, starting with 20 min using LIDAR for those ten scans, lacks color information and cannot be utilized for crack detection. In situations where time is crucial, such as recording in an endangered zone, the 360-degree camera demonstrates its superiority.

The recording using the 360-degree camera takes approximately 5 min to accumulate sufficient data for photogrammetry results. It is noticeable that the overall dimensions do not deviate significantly. The general dimensions are accurate within 0.5% ($\pm$ 5 cm in this case), but all geometric features are available, allowing for the straightforward recreation

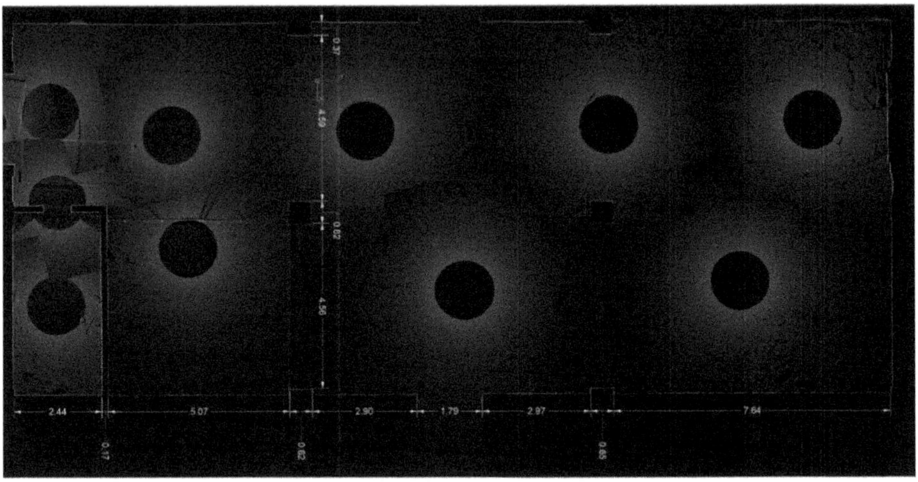

Fig. 4. LIDAR scan of the interior

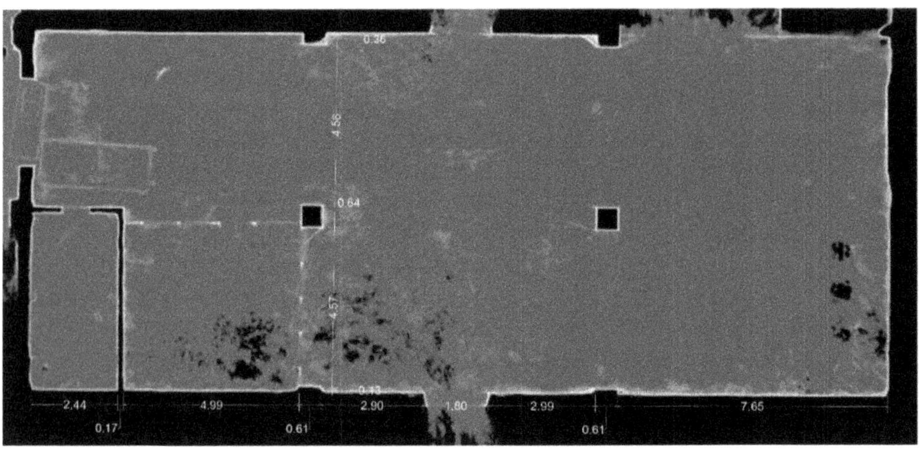

Fig. 5. 360-degree camera scan of the interior

of visible construction features. In addition to minimizing exposure to potential risks (these recordings were made using a UAV-mounted 360-degree camera), drone footage can be dissected and analyzed independently of point cloud results.

4 Conclusion

This paper presents the capabilities of SfM-MVS protocols with a focus on research gaps and needs in post-disaster activities, using the example of the Republic of Croatia, which experienced the 2020 earthquake series. The focus of the research was to aid the engineer and first responders by eliminating the unnecessary exposure of intervention teams while at the same time helping the engineers with their responsibilities when objective decision-making is required in stressful situations.

The SfM-MVS processes have become very advanced, and the tools for data capturing are both advanced and relatively accessible. With the advancements in computer technology, especially in AI, machine learning, and computer visioning, it is just a mere question of will and time until all these become united and specialized enough to support all the post-disaster activities. Still, with or without those helpers, education of the target groups like first responders, search and rescue teams, engineers, and decision-makers is of the essence. Here, SfM-MVS photogrammetry presents a fast-paced tool that can help reduce ambiguities in the engineering tasks, help decision-making in the long term during the post-disaster recovery, and straight at the disaster scene help all engaged forces in decision-making and saving lives of both the rescuers as well of those that were endangered by the disastrous event.

Still, one must be aware of the current technological limitations that are presented by the capabilities of our tools and that data collection for different purposes requires different data collection protocols. The main concern is that data processing takes time, which is of absolute essence in the immediate post-disaster time. On the other hand, tasks like a general overview of the disastrous area may require flying, while other tasks like rapid assessments may benefit from earth-bound technologies. The proper choice of technology and technique is still the most challenging decision to make in first response inspection of distressed buildings.

Acknowledgement. This work has been fully supported by the University of Rijeka project uniri-iskusni-tehnic-23-65 "Development of performance management model for construction projects based on soft computing methods (PerfMAN)" and NORMENG project KK.01.1.1.07.0057 co-financed by the European Union from the European Regional Development Fund.

References

1. Herak, D., Herak, M., Tomljenović, B.: Seismicity and earthquake focal mechanisms in North-Western Croatia. Tectonophysics **465**(1–4), 212–220 (2009)
2. Sigmund, Z., Radujković, M., Atalić, J.: The role of disaster risk governance for effective post-disaster risk management—case of Croatia. Buildings **12**(4), 17 (2022)
3. Markušić, S., Stanko, D., Korbar, T., Belić, N., Penava, D., Kordić, B.: The Zagreb (Croatia) M5.5 earthquake on 22 March 2020. Geosciences **20**(7), 21 (2020)
4. Atalić, J., et al.: The December 2020 magnitude (Mw) 6.4 Petrinja earthquake, Croatia: seismological aspects, emergency response and impacts. Bull. Earthq. Eng. **21**, 5767–5808 (2023)
5. Ministry of the interior: Ministarstvo unutarnjih poslova; Ravnateljstvo civilne zaštite. Retrieved 19.01.2024., from Hrvatska platforma za smanjenje rizika od katastrofa (2024). https://civilna-zastita.gov.hr/hrvatska-platforma-za-smanjenje-rizika-od-katastrofa/80
6. United Nations Office for Disaster Risk Reduction: Regional Assessment Report on Disaster Risk Reduction 2023: Europe and Central Asia. Geneva, Switzerland: United Nations Office for Disaster Risk Reduction (2023)
7. Markušić, S., et al.: Destructive M6.2 petrinja earthquake (Croatia) in 2020—preliminary multidisciplinary research. Remote Sens. **13**(6), 29 (2021)

8. Croatian Pairlament: Law on the Reconstruction of Buildings Damaged by Earthquake in the Area of the City of Zagreb, Krapina-Zagorje County, Zagreb County, Sisak-Moslavina County, and Karlovac County. NN 21/23. Official Gazette (2023)
9. Zamolo, M.: Inženjeri dnevno obave i do 500 pregleda kuća stradalih u potresu na Banovini. tportal.hr (2021). https://www.tportal.hr/vijesti/clanak/inzenjeri-dnevno-obave-i-do-500-pregleda-kuca-stradalih-u-potresu-na-banovini-20210213
10. Uroš, M., et al.: B. Procjena oštećenja građevina nakon potresa - B.2. Postupak provođenja pregleda zgrada na terenu. In: Uroš, M., Todorić, M., Crnogorac, M., POTRESNO INŽENJERSTVO - Obnova zidanih zgrada, pp. 89–114. Zagreb: Građevinski fakultet Sveučilišta u Zagrebu (2021)
11. Government of Croatia; World Bank: Croatia Earthquake Rapid Damage and Needs Assessment. The World Bank, Washington, DC (2020)
12. Government of Croatia; World Bank: Croatia December 2020 Earthquake—Rapid Damage and Needs Assessment. The World Bank, Washington, DC (2021)
13. Faber, M.H., Giuliani, L., Revez, A., Jayasena, S., Sparf, J., Mendez, J.M.: Interdisciplinary approach to disaster resilience education and research. Procedia Econ. Financ. **18**, 601–609 (2014)
14. Lopes, R.: Community Partnerships in Education: Dimensions, Variations and Implications, pp. 15–20. The University of Hong Kong, Senegal (1999)
15. Šimić, M., Sigmund, Z.: Simulation systems in support of the organization and implementation of crisis management exercises. In: 2nd Croatian Conference on Earthquake Engineering – 2CroCEE, pp. 910–916. Zagreb: Građevinski fakultet Sveučilišta u Zagrebu (2023)
16. Ahir, K., Govani, K., Gajera, R., Shah, M.: Application on virtual reality for enhanced education learning, military training and sports. Augmented Hum. Res. **5**, 1–9 (2020). https://doi.org/10.1007/s41133-019-0025-2
17. Migliorini, M.L.: Virtual and augmented reality for disaster risk reduction. In: 1st Croatian Conference on Earthquake Engineering, p. 8. Zagreb: University of Zagreb Faculty of Civil Engineering (2021). https://doi.org/10.5592/co/1crocee.2021.201
18. Kman, N.E., et al.: First responder virtual reality simulator to train and assess emergency personnel for mass casualty response. J. Am. Coll. Emerg. Physicians Open **4**(1), 8 (2023)
19. Lin, D., Wang, J., Li, Y.: Unsupervised building damage identification using post-event optical imagery and variational autoencoder. IEICE Trans. Inf. Syst. **E104.D**(10), 1770–1774 (2021). https://doi.org/10.1587/transinf.2021EDL8034
20. Mandirola, M., Casarotti, C., Peloso, S., Lanese, I., Brunesi, E., Senaldi, I.: Use of UAS for damage inspection and assessment of bridge infrastructures. Int. J. Disaster Risk Reduction **72**, 21 (2022)
21. Ghaffarian, S., Kerle, N.: Towards post-disaster debris identification for precise damage and recovery assessments from UAV and satellite images. Int. Arch. Photogrammetry Remote Sens. Spat. Inf. Sci. **XLII-2/W13**, 297–302 (2019). https://doi.org/10.5194/isprs-archives-XLII-2-W13-297-2019

Comparison of Earthquake Design Regulations for Buildings in Iran and Germany

Regine Ortlepp(✉) and Reinhard Schinke

Leibniz Institute of Ecological Urban and Regional Development, Dresden, Germany
r.ortlepp@ioer.de

Abstract. The contribution addresses a comparison of normative building codes for earthquake design between Germany and Iran. The underlying objective is to assess the possibilities and limitations of transferring building typological approaches for risk assessment and building materiality analysis from the European to the Iranian context. Research in building codes is closely related to risk analyses of natural and anthropogenic hazards, considering uncertainties in predicting their intensity and frequency. On this basis, building codes bridge the gap between insights from hazard research and the implications or requirements for structures. They involve determining relevant parameters and their effects on the structural design, required load-bearing capacity, and usability of individual components, taking into account the entire building. In a comparative study involving the Iranian earthquake design standard "Standard 2800" and the "DIN EN 1998" (Eurocode 8 and German National Annex) regulating seismic design in Germany and the EU, the primary focus is to assess the codes' applicability in risk assessment and their potential contribution to an integrated perspective, particularly from a resources perspective. The comparative study provides a brief overview of most important contents and compares key parameters of seismic design. Specifically, the study compares seismic zones characterizing the impact, subsoil types significantly responsible for the transmission and damping of earthquake waves as well as the importance factor, which characterises the social significance of the buildings as a kind of affectedness of the people living in them. The normative comparison reveals some similarities but also differences. While both codes essentially define the importance factor in the same way, geological conditions differ, resulting in distinct approaches to design parameters, among other aspects. Overall, both the Iranian and European seismic codes present a distinct perspective on content. While the Iranian code focuses on seismic design, the Eurocode 8 incorporates additional information on specific seismic requirements related to the use of building materials in structural planning.

Keywords: building codes · earthquake design · risk assessment · building types · building materials

1 Introduction

The research on building codes is closely related to risk analyses of natural and anthropogenic hazards, considering uncertainties that arise in the prognosis of their intensities and frequencies. In this context, the overall objective is to elaborate a set of rules to

J. C. Matos et al. (Eds.): IPW 2024, LNCE 494, pp. 105–114, 2024.
https://doi.org/10.1007/978-3-031-60271-9_8

ensure, in particular, the safety of buildings and the health of people against these hazards. On this basis, the building code provides a bridge between findings on hazard research and the implications or requirements for buildings. It includes the identification of relevant parameters and their effects on building construction providing the required load capacity and the usability of individual components considering the entire building.

Based on a comparative study on the Iranian and European seismic code, the focus here is to identify the usability of the code in risk assessment and a possible contribution in multi hazard/risk approaches. It gives a brief overview about the main contents and shows some similarities and differences. Geographical aspects and case examples - calculating the load assumptions for a virtual building - illustrate this.

Beginning here with the development of the "Iranian code of practise for seismic resistant design of buildings. Code No. 2800", it started with a first edition in 1987/88. Fundamentals for this code are mainly related to the US Building Officials and Code Administrators International (BOCA). In addition there are some references to the (i) National Building Code of Canada (NBC), 1970; (ii) the Building Standard Law (BSL) of Japan, 1960; and (iii) BSL of France [1]. In order to improve the safety level of buildings, the second edition includes elements of the new Uniform Building Code at that time (UBC, 1994). This was followed by revisions of the Code in 2005 (3rd edition) and in 2013 (4th edition) in order to include all relevant research output over time. The development steps characterise [2] related to the 1st edition – as "pre-code" while he named the second edition as "moderate-code", and the following editions as "high-codes".

Before a uniform European system of standards existed, regulations were developed and used in a national context. In Germany, the first seismic regulation was established in 1957, especially in consequence of the 1951 earthquake with 11 fatalities and many damaged buildings. The revised version of 1981 considered already dependencies of subsoil and the vibration characteristics of buildings [3]. The harmonization process with the European Code began with the update in 2005. The version of 2011 of the German National Annex (NA) [4] implements the EC 8 [5] regulations and determines parameter values for German context. Important changes are the transition from deterministic approaches describing seismic impacts (which was based on a deterministic specification of maximum intensities) to a semi-probabilistic concept with partial safety factors. In addition, the reassessment of the national Annex in the last years is characterized by a paradigm shift from intensity-based to magnitude-based assessment. Due to this shift, it gives much more detailed and reliable information considering uncertainties in models and assumptions. However, at the time of the study, the recently released update from November 2023 was not yet available, so our comparison is based on the 2011 version of the German NA.

In order to harmonise the standards within the European Union, the first generation of European standards – related to structural design of buildings – were adopted between the end of 1990s and mid 2000s [6]. It includes 10 different standards beginning with fundamentals of the structural design and load assumptions. Six of the 10 regulations related to different building materials with its specific requirement for the dimensioning of building components. One relates to geotechnical issues and another separate code (EC 8 [5]) include all aspects of seismic design of buildings.

In the view of both, the Iranian and the European earthquake codes, it gives a different perspective to the content. While the Iranian code focuses on the seismic design, the Eurocode EC 8 gives additional information to specific – seismic related – requirements with regard to the use of building materials in the structural design. Fardis [7] (P.1020) explains this as a convenience for countries with very low seismic hazard to dispense the application of the EC 8 [5] altogether.

The following sections provide insights into seismic analyses for buildings in Iran and Germany in order to find interlinkages between regulation and risk assessment. The basis for this comparison are the (i) 4th edition of Iranian code of praxis (IS 2008) [8], (ii) the EC 8 [5], as well as the German NA [4].

2 Code Comparison

2.1 General

Earthquakes are the shaking of the subsurface caused by a sudden release of energy, generating seismic waves. This energy signifies an impact, resulting in forced movements (displacements) and deformations of building structures. Given the significance and utility of buildings, the primary goal of seismic codes is to absorb and resist seismic energy impacts, thereby preventing building losses, preserving building functions, and, most importantly, protecting human life. The overarching concept involves verifying resistance (stability, usability) against these impacts, with both aspects characterized by a distribution of values. In this context, it is essential to ensure a sufficient distance between these distribution functions. Codes usually describe seismic effects through zoning of earthquake hazard and the properties of the subsurface.

2.2 Seismic Zones Characterising the Impact

One key element of seismic regulations is directed toward a simplified characterization of the impacts. Potential seismic events with a return period of 475 years are used as the design parameter, corresponding to a 10% probability of occurrence within 50 years. Both the Iranian and the European codes are related to this probability of occurrence, but the zonation uses slightly different formulations.

According to IS 2800 [8], seismic zones in Iran are divided into four classes using design base accelerations *g*. Similarly, in EC 8, the reference peak ground acceleration a_{gR} is used, which is associated with an intensity range I of the seismic zone. The National Annexes regulate the classification itself and, in the German context, relate to DIN EN 1998–1/NA [4]. The reference ranges characterizing the impacts in Iran and Germany are, of course, quite different, as illustrated by the use of different lines in the table. Both seismic codes include nationwide differentiation in space using tables where the zones are related to administrative boundaries and maps that are reflected in Fig. 1.

IS 2800			**DIN EN 1998-1 NA**		
Zone: Description	g [m/s²]	A [m/s²]	I acc. to EMS*)	a_{gR} [m/s²]	Zone: Description*)
I: zone with very high risk	0.35	3.43			
II: zone with high risk	0.30	2.94			
III: zone with medium risk	0.25	2.45			
IV: zone with low risk	0.20	1.96			
			$6 \leq I < 6.5$		0: very low seismicity
			$6.5 \leq I < 7$	0.4	1: low seismicity
			$7 \leq I < 7.5$	0.6	2: low seismicity
			$7.5 \leq I$	0.8	3: low seismicity

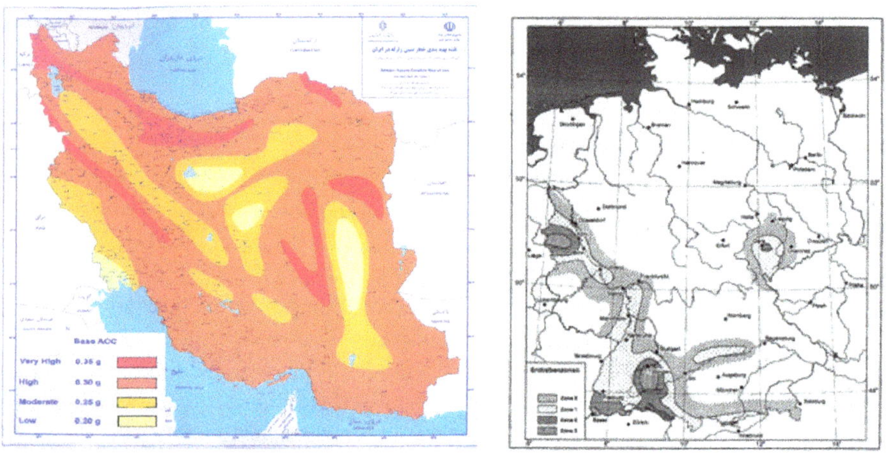

*) EMS = European Macroseismic Scale

Fig. 1. Comparison of seismic zones in IS 2800 [8] and DIN EN 1998–1/NA [5].

2.3 Ground Types Characterising the Transmission of Seismic Waves

Another central parameter reflects the properties of the subsoil, as these are decisively responsible for the transmission and damping of earthquake waves. Table 1 reflects the classes of ground types embedded in the Iranian (Fig. 2a) and European seismic code (Fig. 2b).

It can be seen that both standards have a comparable classification system for the subsoil. The supplementary classes of the Eurocode (E, S_1, S_2) specify the classes A to D regarding the associated mean shear wave velocity and/or the layer thicknesses in the stratigraphic profile.

Concerning Germany and its national annex, the ground types are further differentiated. Here, a distinction is made between subsoil classes and geological ground types. For relevant combinations, corresponding parameters are given for the response spectrum. Behind the definition of classes, the German annex includes a map with the spatial distribution of the geological subsoil classes as one of the two parts that describe the ground type (whereby the map is restricted to the seismic zones). Such a map is not

Table 1. Comparing the characteristics of ground types in IS 2008 [8] and EC 8 [5].

IS 2800		EN 1998–1	
Description of stratigraphic profile	$v_{s.30}$	$v_{s.30}$	Description of stratigraphic profile
I: Rock and semi-rock includes igneous rocks, sedimentary, metamorphic rocks, very strong stiff soil with a maximum of 5 m weaker materials to ground level	≥ 750	> 800	**A:** Rock or other rock-like geological formation, including at most 5 m of weaker material at the surface
II: Very dense soil or loose stone, includes dense sand, very hard clay with thickness more than 30 m which its mechanical properties improve by increasing depth loose igneous, sedimentary rocks like tuff or aerated rock	375 to 750	360 to 800	**B:** Deposits of very dense sand, gravel, or very stiff clay, at least several tens of metres in thickness, characterised by a gradual increase of mechanical properties with depth
III: Dense to medium soil includes dense to medium sand or hard clay with thickness more than 30 m	175 to 375	180 to 360	**C:** Deep deposits of dense or medium- dense sand, gravel or stiff clay with thickness from several tens to many hundreds of metres
IV: Medium to soft soil. Non-sticky soil layers or with a little sticky soil with medium to low density, the completely sticky, soft soil to firm layers	< 175	< 180	**D:** Deposits of loose-to-medium cohesionless soil (with or without some soft cohesive layers), or of predominantly soft-to-firm cohesive soil
			E: A soil profile consisting of a surface alluvium layer with vs values of type C or D and thickness varying between about 5 - 20 m, underlain by stiffer material with $v_s > 800$ m/s
		< 100 (indi-cative)	**S1:** Deposits consisting, or containing a layer at least 10 m thick, of soft clays/silts with a high plasticity index (PI > 40) and high water content
			S2: Deposits of liquefiable soils, of sensitive clays, or any other soil profile not included in types A – E or S1

included in the Iranian code. However, Farahani et al. [9] produced a soil classification map for Iran, using the method of Allen and Wald [10]. In contrast to the code, the site classes have slight differences related to $v_{s.30}$[1]: B > 760; C: 300 … 760; D: 240 … 300; E: < 180 [m/s].

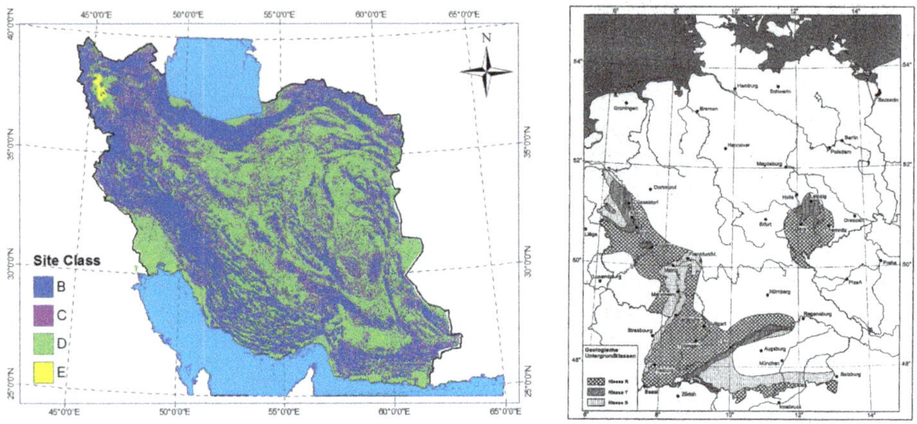

a) Iran **b)** Germany

Fig. 2. Maps with (geological) Ground types in Iran [9] and Germany [4].

2.4 Importance Factor Characterising the Social Significance

The importance factor classifies buildings in terms of their social significance and focuses on the consequences of a building failure. It considers how many people would be affected, which basic social functions (water, energy, health care) have to be ensured, but also the impact on national heritage. Both codes use the same classes with the same importance factor and nearly the same descriptions (the differences are more or less differences in translation). Table 2 reflects the importance factor for both codes.

2.5 Standard Design Response Spectrum Characterising the Building's Response

Characterizing the ground motion at a specific point on the surface, the standards define so-called 'elastic ground acceleration response spectra'. The response spectra are determined by various influencing factors, especially the ground type and the seismic level (ag). Figure 3 shows the response spectra for different combinations of influencing factors in the Iranian and European codes, as well as the spectra in the German National Annex.

The seismic standard IS 2800 defines the response spectrum B1 for various ground types and seismic levels (Fig. 3a). Additionally, IS 2800 increases the design spectrum values by the spectral correction factor N (Fig. 3b), which depends on the vibration period of the structure and the degree of seismicity in the construction area [11].

[1] $V_{S.30}$ is the average value of propagation velocity of S waves in the upper 30 m of the soil profile at shear strain of 10^{-5} or less (see EC 8).

Table 2. Comparing the importance factor in IS 2008 [8] and EC 8 [5].

IS 2800		EN 1998–1	
Description of importance	I	γ_I	Description of importance
I: Very high importance Essential buildings: Special importance for usability after earthquake (e.g. water supply, power plant, airport, telecommunication, military, help centers) Risky buildings: Facilities that causes long term effects for environment due to toxic and harmful materials	1.4	1.4	**IV:** Vital importance Buildings whose integrity during earthquakes is of vital importance for civil protection, e.g. hospitals, fire stations, power plants, etc
II: High importance A: Buildings which damage cause a lot of casualties (e.g. schools, mosques, theater, …, > 300 people) B: Buildings which damage cause the loss of national wealth (museums, libraries …) C: Industrial facilities which cause environmental pollutions or widespread fire	1.2	1.2	**III:** Buildings whose seismic resistance is of importance in view of the consequences associated with a collapse, e.g. schools, assembly halls, cultural institutions etc
III: Medium Importance Ordinary buildings, not belonging in the other categories (residential buildings, administrative, commercial, industrial buildings, …)	1.0	1.0	**II:** Ordinary buildings, not belonging in the other categories
IV: Low Importance A: Buildings which cause minor losses and minor human causalities B: Temporary buildings (< 2yr)	0.8	0.8	**I:** Minor importance Buildings of minor importance for public safety, e.g. agricultural buildings, etc

For better comparability, the response spectra of EC 8 and German NA are normalized by (a_g) and shown in Fig. 3c,d. The response spectra are defined by expressions illustrating dependencies on the soil factor S, the peak ground acceleration on ground type A (a_{gR}), and the importance factor (γ_I) by Eq. (1). Additionally, it considers 5% damping by the factor $\eta = 1$, which can further specify.

In a general view, the diagrams illustrate that softer subsoil classes lead to response spectra with longer plateau areas and more damped functions.

$$a_g = \gamma_I \cdot a_{gR} \tag{1}$$

with: a_{gR} … Reference peak ground acceleration for soil class A,
a_g … Design value of ground acceleration for soil class A,
γ_I … coefficient of significance; γ_I is set to 1 for the reference values.

In relation to the EC8's German NA, the differentiation of geological ground types and subsoil classes leads to a broad differentiation of elastic response spectra (Fig. 3d). The graphic clearly illustrates that – with a view of the same subsoil classes – the response

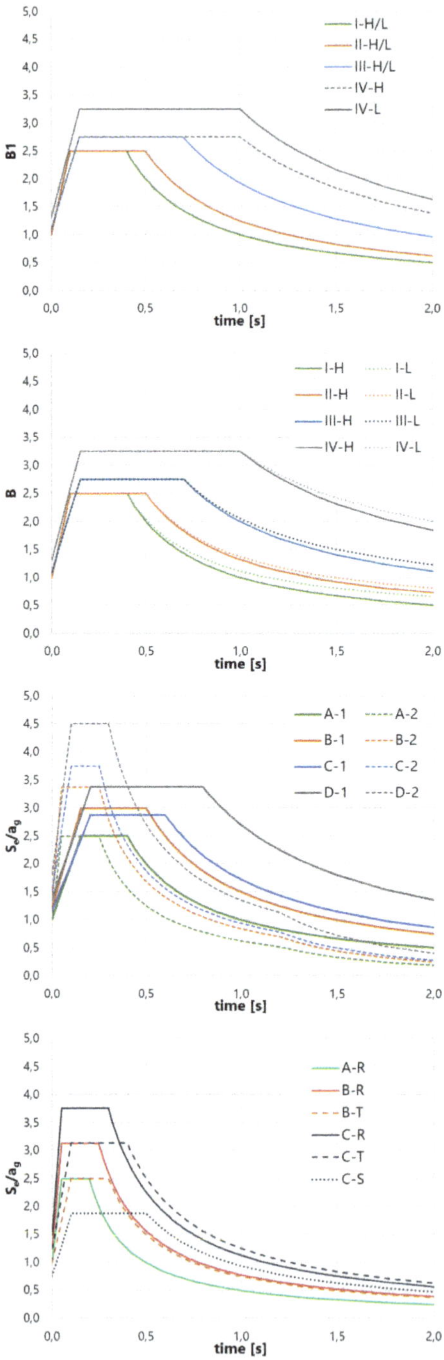

a) IC 2800: B1

For Parameter B1, the differences between (i) high and very high risks ("H") and (ii) relatively low and medium risks ("L") are only relevant for ground type IV.

b) IC 2800-4: B

Building response factor *B* considers the correction factor *N* by equation $B = B1 \times N$

c) EC 8

Spectra normalised by a_g shown for different ground zones (A-D) and seismic types (Type 1: $M_S \leq 5.5$; Type 2: $M_S > 5.5$)

d) EC 8, German NA

Spectra normalised by a_g shown for relevant combination of subsoil and geological ground zones (Type 1)

Fig. 3. Elastic response spectra based on IC 2800 [9], EC 8 [5] and the German NA [4].

spectra of the geological ground type R (rock, solid lines) lead to short response times (narrow spectra) and high spectral values in the plateau region compared to types T (dashed lines) and S (dotted line).

3 Conclusion

Building codes serve as a crucial link between hazard research and structural impacts, taking into account uncertainties in predicting hazard intensity and frequency. The contribution conducts a comparative analysis of normative building codes for earthquake design in Germany and Iran, aiming to evaluate the feasibility of transferring building typological approaches from the European to the Iranian context for risk assessment and building materiality analysis.

The comparative study, which focuses on various parameters related to seismic hazard, geological conditions and social importance in terms of the consequences of building failure, shows both similarities and differences. While the importance factor, for example, which includes social significance, is defined similarly in both codes, the geological conditions lead to different approaches to the design parameters. As the codes have a considerable number of similarities, the transferability of approaches from the German context to the Iranian context is possible in principle.

Nevertheless, there are differences that result primarily from the geological conditions. Thus, the intensity of earthquake-induced damage hinges on various factors, encompassing the proximity to the epicenter, building design, soil composition, and localized conditions. These parameters exhibit disparities between Iran and Germany, exerting significant influence on the magnitude and characteristics of seismic impact considered in design standards. Distinctive features differentiating Iran and Germany include the type of soil prevalent in each region – Germany often characterized by solid rock substrates, in contrast to the softer soil layers found in Iran. Furthermore, the frequency and intensity of seismic events differ, with Iran, located in a seismically active zone, experiencing more frequent and higher-magnitude earthquakes compared to less seismic-prone Germany. The depth of the earthquake epicenter also varies, with Iranian seismic activities generally occurring at deeper crustal levels than those in Germany. Additionally, the types of earthquake waves, such as P-waves (Primary waves) and S-waves (Secondary waves), exhibit differences. Regions near tectonic plate boundaries, like Iran, witness both wave types, while Germany, positioned farther away, predominantly experiences P-waves. Recognizing these distinctions is pivotal for tailoring seismic design and engineering practices to the specific geological and seismological conditions prevalent in each region. These aspects are taken into account in the various response spectra as defined in the compared codes.

These differences in geological conditions emphasize the need for careful consideration when transferring risk assessment approaches from the German to the Iranian context, indicating the potential necessity for specific adaptations to account for geological differences.

Acknowledgments. We acknowledge support by the Federal Ministry of Education and Research (BMBF) within the project "Inclusive and Integrated Multi-Hazard Risk Management and Engagement of Volunteers to INCREASE Societal Resilience in Times of Climate Change".

References

1. Imashi, N., Massumi, A.: A comparative study of the seismic provisions of Iranian seismic code (standard no. 2800) and international building code 2003. Asian J. Civil Eng. **12**(5), 579–596 (2011)
2. Sadeghi, M., Ghafory-Ashtiany, M., Pakdel, N.: Developing seismic vulnerability curves for typical Iranian buildings. Proc. Inst. Mech. Eng.Part O J. Risk Reliab. **229**(6), 627–640 (2015)
3. Baden-Württemberg, W. (ed.): Erbebensicher Bauen: Hinweise für das Bauen in Erdbebengebieten Baden-Würtembergs, 6th edn. Wirtschaftsministerium Baden-Württemberg, Stuttgart (2008)
4. DIN EN 1998–1/NA: National Annex – Nationally determined parameters – Eurocode 8: Design of structures for earthquake resistance - Part 1: General rules, seismic actions and rules for buildings, 01/2011 (2011)
5. EN 1998–1: Eurocode 8: Design of structures for earthquake resistance – Part 1: General rules, seismic actions and rules for buildings, 01/2009 (2009)
6. Tsompanakis, Y.: Earthquake Return Period and Its Incorporation into Seismic Actions. In: Beer, M., Kougioumtzoglou, I.A., Patelli, E., Au, S.-K. (eds.) Encyclopedia of Earthquake Engineering, pp. 818–851. Springer, Berlin Heidelberg (2015). https://doi.org/10.1007/978-3-642-35344-4_116
7. Fardis, M.N.: European structural design codes: seismic actions. In: Beer, M., Kougioumtzoglou, I.A., Patelli, E., Au, S.-K. (eds.) Encyclopedia of Earthquake Engineering, pp. 1020–1038. Springer, Berlin Heidelberg (2015). https://doi.org/10.1007/978-3-642-36197-5_119-1
8. IS 2800–4: Iranian Code of Practice for Seismic Resistant Design of Buildings, 4th edn. Building and Housing Research Center (BHRC), Tehran (in Persian) (2014)
9. Farahani, S., Behnam, B., Tahershamsi, A.: Macrozonation of seismic transient and permanent ground deformation of Iran. Nat. Hazard. **20**(11), 2889–2903 (2020)
10. Allen, T.I., Wald, D.J.: On the use of high-resolution topographic data as a proxy for seismic site conditions (VS 30). Bull. Seismol. Soc. Am. **99**(2), 935–943 (2009)
11. Rouhi, H., Kontoni, D.-P.N.: Calculation of the spectral correction factor for the design of reinforced concrete frames equipped with steel plate shear wall under near-field earthquakes. Asian J. Civil Eng. **24**, 2363–2373 (2023)

Deriving Analytical Fragility Curves for Masonry Churches Based on Stochastic Nonlinear Analyses

Federica Del Carlo[1](✉), Silvia Caprili[1], Tiago Miguel Ferreira[2], Pere Roca[3], and Marco Uzielli[4]

[1] University of Pisa, Largo Lazzarino 1, 56122 Pisa, Italy
federica.delcarlo@phd.unipi.it
[2] University of the West of England (UWE Bristol). Frenchay Campus, Coldharbour Lane, Bristol BS16 1QY, UK
[3] Technical University of Catalonia, Jordi Girona 1-3, 08034 Barcelona, Spain
[4] University of Florence, Via di Santa Marta 3, 50139 Florence, Italy

Abstract. This paper proposes an approach for deriving analytical fragility curves for religious masonry buildings based on stochastic static nonlinear analyses to support seismic vulnerability assessment and risk mitigation. A portfolio of churches located in the Tuscany region (Italy) was created and statistically analyzed to identify recurrent typologies. Given its architectural simplicity and greater frequency within the dataset, the one-nave type was selected as a pilot case. Using clustering algorithms, the geometrical properties of the one-nave churches sample were analyzed to obtain prototype buildings whose material properties were treated as random variables varying within predefined ranges. A building prototype was then selected as the reference structure to define the FE model. Geometrical properties were considered deterministic, while the variability of the mechanical properties was accounted for in the nonlinear analyses. The Latin Hypercube Sampling (LHS) method was used to generate a suitable set of input variables from the cumulative distribution functions of the mechanical parameters, and nonlinear static (pushover) analyses were conducted to derive capacity curves and the probability density function of selected damage states. Conversely, the N2 method provided the displacement demands corresponding to different levels of ground motion. Through the convolution of the complementary cumulative distribution of demand with the probability density function of each damage state, the method allowed the estimation of the probability of reaching or exceeding selected damage states and, consequently, to derive the respective fragility curves.

Keywords: Fragility curves · Masonry churches · Building typologies · Stochastic nonlinear analysis · DIANA-FEA

1 Introduction

Masonry is one of the most widely employed structural systems in historical buildings worldwide, especially religious ones. The desire to preserve and restore unreinforced masonry (URM) churches resulted in the development of several experimental

J. C. Matos et al. (Eds.): IPW 2024, LNCE 494, pp. 115–124, 2024.
https://doi.org/10.1007/978-3-031-60271-9_9

and numerical tools aimed at assessing their vulnerability to seismic events [1, 2]. It is commonly recognized that due to their unique structural and architectural features, these structures are more likely to respond to earthquakes through case-specific "local" mechanisms rather than with a uniform, "global" behavior [3]. For this reason, post-earthquake damage survey campaigns have played a pivotal role in the development of this field [4, 5] through the identification and systematization of damage mechanisms [6], which are now integrated into the Italian form for the post-earthquake survey of churches [7]. Survey data collected from large samples of masonry churches have also been extensively used to build empirical fragility curves. Typically, the procedure entails identifying the most recurrent typologies of churches among the surveyed sample. Subsequently, damage probability matrices for homogeneous classes of churches at various damage levels are compiled through the analysis of the observed damage, and the corresponding fragility curves in terms of PGA are implemented [8–10]. While empirical fragility curves derived from post-earthquake damage surveys have significantly contributed to our understanding of the seismic vulnerability of unreinforced masonry (URM) churches, there are inherent limitations to this approach. Reference [11] performed an extensive review of the state of the art in the construction of empirical fragility curves. Among other relevant aspects, the authors report significant differences in the damage observation databases used, the functional forms chosen to fit the data, the selected ground motion intensity measures, and the statistical modeling technique employed. Analytical fragility curves, on the other hand, aim to define a direct mechanical relationship between construction characteristics, structural response to seismic action, and damage effects [12]. This alternative approach proved extremely useful when damage data was unavailable [13]. In this study, a methodology for obtaining analytical fragility curves for classes of churches is proposed. Nonlinear static analysis and the correlation between acceleration/displacement capacity curves and spectral responsum curves [14, 15] are used to derive probability distributions of capacity and demand accounting for material variability. The identification of fragility points was made possible by convolution of the complementary cumulative distribution of displacement demand at different PGA levels with the probability density functions of displacement capacity for a specific damage condition. The points were subsequently fitted by lognormal distributions to obtain analytical fragility curves. The proposed approach is illustrated via its application to an archetype URM church, representative of the one-nave typology, which is common in Tuscany (Italy). On the one hand, the suggested approach allows for the consideration of the material mechanical parameters variability when determining the capacity of a building typology. On the other hand, it is an effective strategy for identifying fragility curves when no empirical data is available.

1.1 The Developed Methodology

The proposed methodology for deriving analytical fragility curves relies on conducting nonlinear analyses on a set of archetype buildings representative of relevant categories of religious building typologies. The definition of a population of archetype buildings and the clustering of buildings into archetypal typologies allowed the hypothesis of the intra-cluster geometrical similarity. Consequently, geometry can be considered deterministically in the modeling and analysis of each archetype. Differently, the mechanical properties of masonry were modeled as random variables. The resulting fragility curves thus account for the geometric and material variability inherent in the chosen typology, therefore considering the two most important sources of variability involved in the problem. Three main typologies, one-nave, three-nave, and Latin cross, were identified from a portfolio of seventy-one surveyed buildings. Due to its simplicity, the one-nave typology was chosen as the pilot typology to test the proposed methodology. Representative geometric parameters were then selected through statistical approaches – these included the length (L) and width (S) of the nave, the height of the perimeter walls (H_1), and the thickness of the walls (t_w). To this purpose, various clustering techniques were applied to identify subpopulations within the typology and the respective representative values. Geometric properties' values were combined two by two, resulting in fourteen combinations that, together, represent the archetypes of the one-nave typology.

Concerning material properties, four materials were identified from the surveys. The distribution of these materials within the single-nave typology was studied using a descriptive statistical approach involving the calculation of frequency histograms. Two independent mechanical properties, the compressive strength (f_C) and the Young's modulus (E), were selected to account for the material variability along the nonlinear static analyses. The choice of these two properties is a function of the constitutive model adopted to describe the masonry behavior, as discussed later. The two properties were then described through lognormal distributions using the range, mean, and standard deviation values provided by the Italian Code [16] since no experimental data were available. The Latin Hypercube Method (LHS), which is computationally less expensive than Monte Carlo simulation, was used to sample values from the lognormal distributions by subdividing the range of a random variable into equal-probability intervals and subsequently selecting a single value from each interval. Finally, to account for the frequency of each material in the one-nave portfolio, weights have been assigned to the sampled property values so that materials with a higher frequency have a higher likelihood of being selected during the stochastic nonlinear analysis. After defining the geometrical characteristics of the model and the mechanical properties values to use, nonlinear static pushover analyses were run to estimate specific damage states, including the overturning of the main façade's gable (DS1), the diagonal cracking of the longitudinal walls (DS2), the cracking of the apse vault (DS3) and the overturning of the back façade gable (DS4). Within this study, the application of the method is reported for DS4. Accordingly, the probability distribution functions were defined for the capacity displacement points associated with DS4.

On the other hand, the N2 method [15] – implemented in Eurocode 8 [17] and the Italian Building code [18] – was used to evaluate the seismic performance of the analyzed structures. Specifically, the method was used here to determine the displacement demand point through the intersection between the code-based acceleration spectrum defined for different Peak Ground Acceleration (PGA) levels and the idealized elastic/perfectly plastic force–displacement relationship. Four seismic zones could be identified over the Italian territory according to [19]. Five code-based acceleration spectra were selected, so as to cover the four Italian seismic zones. The cumulative distribution function of the displacement demand imposed on the building for different levels of PGA was obtained from the displacement demand points.

Convolution of the complementary cumulative distribution of demand and the probability density functions of each damage state allows us to derive fragility points. The points were fitted using a lognormal distribution to obtain fragility curves associated with different damage states and levels of PGA.

1.2 Material Sampling Method

To consider the material variability of the population of buildings in the one-nave typology, the mechanical properties of masonry were assumed as random variables. The Latin hypercube sampling (LHS) method was used to generate samples for the mechanical properties from their associated cumulative distribution functions. The compressive strength (f_c) and Young's modulus (E) were considered unknown parameters. Therefore, a log-normal cumulative distribution function (CDF) was assigned to each variable based on ranges of variation provided by the Italian Code [16], as mentioned earlier. The Latin hypercube sampling [20] procedure operates by dividing the range of each variable (i.e., the x*j*) into *n*S disjoint intervals of equal probability, and then one value is selected randomly from each interval. In a two-dimensional problem, as in this case, the *n*S values obtained for x_1 are paired randomly without replacing them with the *n*S values obtained for x_2. It's important to note that the individual variables x_j must be independent for this sampling technique to achieve the desired results.

2 Application to the Case Study

2.1 Building Archetype Description

The archetype chosen as a pilot case represents one of fourteen possible geometric variations within the single-nave typology. It has a dimension in plan of 5.40 m × 15.3 m and a height of 7.50 m. The masonry-bearing walls are 60 cm thick. The geometry of the structure has been considered deterministic. Regarding material characteristics, since this building is an archetype that represents a class of similar buildings, they have been assumed to be random variables, as mentioned earlier.

2.2 Archetype Finite Element (FE) Model

A finite element (FE) model was adopted and implemented in the DIANA-FEA (DIANA FEA BV, 2023) software. Quadratic (CQ40S) and triangular (CT30S) shell elements were used to model the masonry walls, the apse, and the roof. Roof shell and gamble shell were disconnected, thus allowing the activation of out-of-plane mechanisms. Roof timber beams were modeled using the 3D beam element (CL18B, composed of three nodes and six degrees of freedom for each node). The boundary conditions at the base of the building were assigned as totally fixed by restricting both translational and rotational movements.

Masonry was simulated as a continuum material with properties corresponding to the average response of the composite material. Its nonlinear behavior was acknowledged using the Total Strain Fixed Crack model implemented in DIANA-FEA software.

Compression behavior was assumed to follow a parabolic softening curve, while tensile behavior was associated with an exponential softening curve under tension. The shear behavior after cracking was described through a constant shear stiffness reduction with a shear retention factor of 0.01. The tensile strength was assumed to be equal to 2% of the compressive strength. The value of the compressive fracture energy has been defined as a function of the compressive strength, Eq. (1), as per [21]:

$$G_c = df_c \tag{1}$$

where d = 1.6 mm. The tensile fracture energy was assumed to be 50 N/m, as widely accepted in literature. Timber members were modeled as linear elastic. Both geometrical and mechanical nonlinearities were considered in the analyses.

2.3 Mechanical Parameters' Ranges

From the in-site surveys, four different materials were observed. Following the nomenclature in Table 1, such materials show a sample distribution of 53% for MAT1, 26% for MAT2, 12% for MAT3, and 9% for MAT4. *A priori* log-normal distributions were derived by the interval of values proposed in the commentary to the Italian Code [16] for the compressive strength (f_c) and Young's modulus (E). To capture the variability inherent in the detected materials, the Latin Hypercube Sampling (LHS) method was employed to obtain representative samples for each material. For each material category (MAT1 to MAT4), the LHS method was applied to derive 40 samples of compressive strength (f_c) and Young's modulus (E) data pairs within the specified intervals. Subsequently, the distribution percentages were used to select a total number of 40 data pairs ($f_c - E$) to be used in the analysis that could reflect the material distribution observed in the surveyed buildings. The goal was to ensure that the resulting statistical models adequately capture the variability present in the surveyed building population.

Table 1. Four most recurrent materials detected for the one-nave typology.

Rubble masonry (MAT1)	Partially dressed stone masonry (MAT2)	Fully dressed stone masonry (MAT3)	Squared stone masonry (MAT4)
		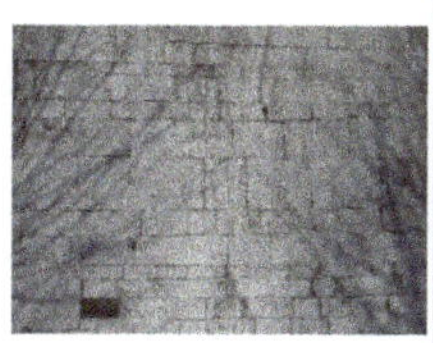	
Compressive strength f_c [MPa]			
1.0 – 2.0	2.0	2.6 – 3.8	5.8 – 8.2
Young's modulus E [MPa]			
690 - 1050	1020 - 1440	1500 - 1980	2400 - 3300

3 Nonlinear Static Analysis

The analyses have been carried out only in the positive longitudinal (Y) direction. Gravity was applied in the first loading step. Then, seismic forces proportional to the structure's mass were incremented until the analysis stopped due to the overturning of the back façade gable (DS4). The capacity curves have been obtained using a control point located at the top of the gable of the back façade, as illustrated in Fig. 1. A total of 40 pushover analyses have been performed on the selected archetype building according to the 40 data pairs ($f_c - E$) selected. Figure 2 shows the pushover curves obtained from the set of pushover analyses. As can be seen, there was a large variability in the results due to the variation in the material properties. Since the analysis stopped due to the overturning of the back façade gable (DS4), the maximum displacement values were considered to define the probability density functions associated with DS4. Finally, the Kernel density estimation method [22] was used to obtain a smooth empirical pdf based on sample data. The seismic demand due to different code-based acceleration spectra was obtained using the N2 method. Assuming to consider the churches as an SDOF system, the curves obtained from the pushover analysis were bilinearised, based on the principles of energy equivalence, i.e., that the areas below the capacity and bilinear curve are equal. The resulting curves, showing the elastic/perfectly plastic force–displacement relationship, where intersected with the seismic demand spectra on an Acceleration Displacement Responsum Spectrum (ADRS), obtaining the displacement demand points, for five different PGA levels, respectively 0.277 g, 0.26 g, 0.195 g, 0.13 g, 0.072 g. The empirical complementary cumulative distributions of the displacement demand points were obtained using the Kernel density estimation method once more.

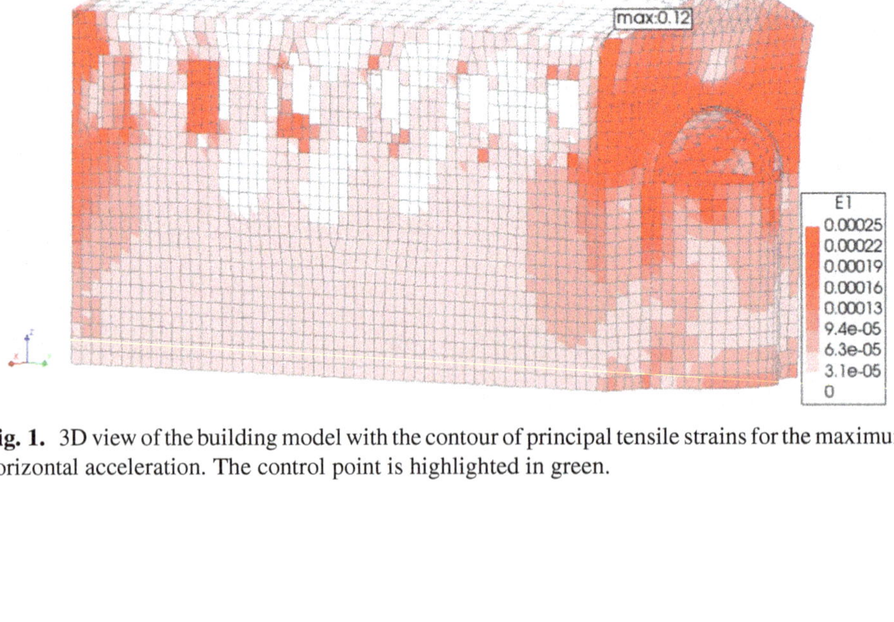

Fig. 1. 3D view of the building model with the contour of principal tensile strains for the maximum horizontal acceleration. The control point is highlighted in green.

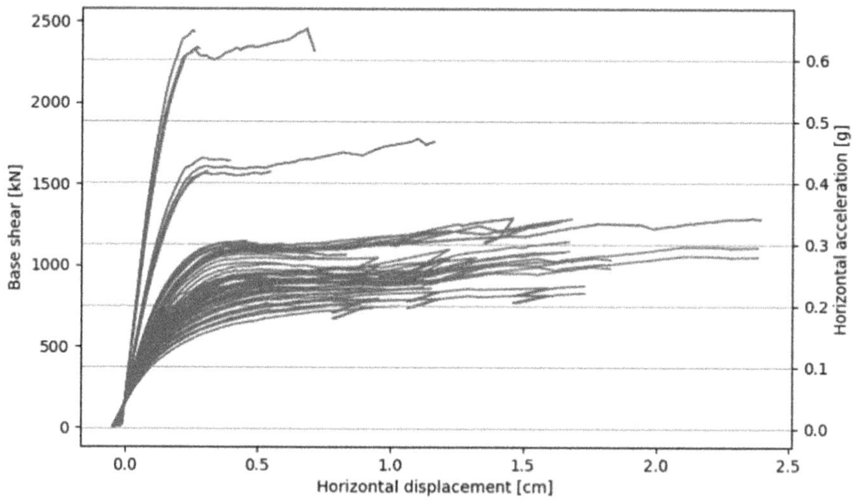

Fig. 2. Results of the set of pushover analyses in the longitudinal (Y) positive direction.

3.1 Derivation of the Fragility Curve

Once the probability density functions of the considered DS4 and the cumulative distribution function of the displacement demand corresponding to different levels of PGA were obtained, it was possible to convolve the two results to get fragility points, which were subsequently used to construct the fragility curve. Following the approach proposed by [23], the probability of failure associated with a predefined limit state was defined as Eq. (2), where $F_{PGA}(\alpha)$ is the cumulative density function (cdf) of the demand for a given level of PGA and f_C is the is the probability density function (pdf) of the capacity for DS4. Figure 3 graphically illustrates the convolution between the complementary cumulative distribution function of the demand (dash-dotted line) for a PGA of 0.195g with the probability density functions of the DS4 (continuous line).

$$Pf(PGA) = \int_0^{\infty} [1 - F_{PGA}(\alpha)]f(\alpha)d\alpha \tag{2}$$

The result of the convolution is highlighted using the dashed line. The integral on the area below the dashed line corresponds to the probability of exceedance of DS4 for a PGA of 0.195 g. The same procedure was repeated for all the PGA values considered, obtaining five fragility points representing the probability of reaching or exceeding DS4 for each PGA value. The points were then fitted using a lognormal probability density function. The points obtained and the lognormal curves fitting them are plotted in Fig. 4, together with the parameters (μ and σ) of the lognormal distributions.

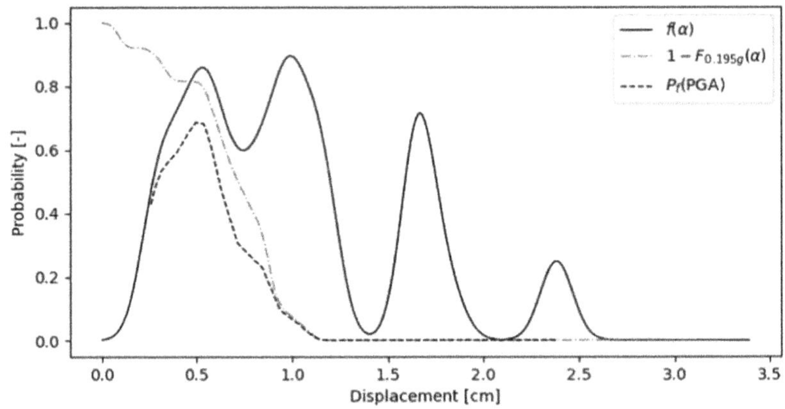

Fig. 3. Convolution of pdf of DS4 with and complementary CDF of demand for PGA 0.195 g.

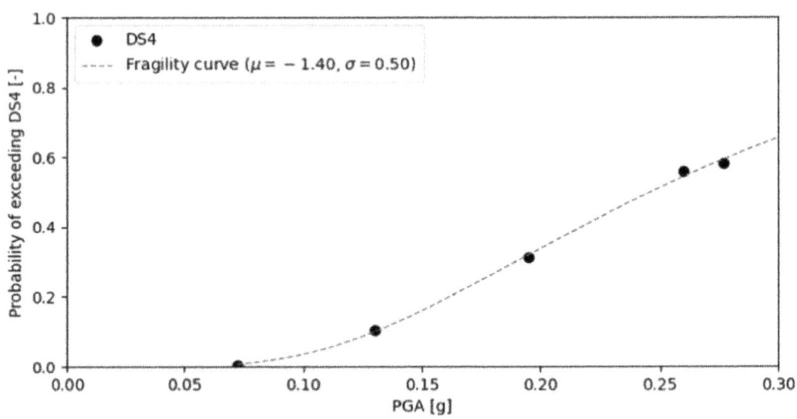

Fig. 4. Lognormal fragility curve fitting the analytically derived points.

4 Conclusion

This paper presents an initial study aimed at defining a methodology for obtaining analytical fragility curves for masonry churches using nonlinear static analysis. The proposed approach is based on the results of nonlinear stochastic analyses of an archetype building selected among all the archetypes considered representative of a church typology. The masonry compressive strength and Young's modulus are assumed to be random variables with which normal probability distributions are associated based on realistic ranges of variation. Using the LHS method, 40 data pairs of (fc – E). Are selected and assigned to the FE model of the archetype church to perform advanced nonlinear analyses. Four damage states were obtained from the pushover curves, and DS4 was selected to be used as a pilot case. The probability density function of DS4 was then determined based on displacement capacity points obtained through the pushover analyses. The cumulative distribution function of the displacement demand was obtained using the N2 method.

Fragility points were obtained through the convolution of the complementary cumulative distribution of demand obtained for different PGA values and the probability density function of the displacement capacity. Finally, the lognormal fragility curve for the DS4 was obtained by fitting the fragility points.

It is important to note that the proposed approach relies on a reasonable yet substantial simplification of the architectural configuration and structural behavior of churches. Consequently, its use is appropriate for an initial structural assessment at the level of the building type and not at the level of the individual building. However, the proposed methodology has general validity and can be easily extended to other church typologies or types of buildings. Moreover, this pilot study focused on one of the geometric configurations identified within the single-nave church typology. Nonetheless, the aim is to extend the methodology to all geometric configurations identified to also consider the variability of geometric attributes that characterize the plan and sectional configuration of the church. Finally, it should be noted that the code-based responsum spectra used to identify displacement demand were taken according to the most common rock soil type of the selected area. Hence, a possible improvement of the proposed methodology could address different soil conditions.

References

1. Cundari, G.A., Milani, G., Failla, G.: Seismic vulnerability evaluation of historical masonry churches: proposal for a general and comprehensive numerical approach to cross-check results. Eng. Fail. Anal. **82**, 208–228 (2017)
2. Lourenço, P.B., Roca, P., Modena, C., Oliveira, D.V.: Reducing the seismic vulnerability of cultural heritage buildings (2006)
3. Lagomarsino, S.: On the vulnerability assessment of monumental buildings. Bull. Earthq. Eng. **4**, 445–463 (2006)
4. Canuti, C., et al.: Post-earthquake damage and vulnerability assessment of churches in the marche region struck by the 2016 Central Italy seismic sequence. Int. J. Archit. Heritage **15**(7), 1000–1021 (2021). https://doi.org/10.1080/15583058.2019.1653403
5. Ruggieri, S., Tosto, C., Rosati, G., Uva, G., Ferro, G.A.: Seismic vulnerability analysis of Masonry Churches in Piemonte after 2003 Valle Scrivia earthquake: post-event screening and situation 17 years later. Int. J. Archit. Herit. **16**, 717–745 (2022)
6. Lagomarsino, S., Podestà, S.: Seismic vulnerability of ancient churches: II. Statistical analysis of surveyed data and methods for risk analysis. Earthq. Spectra **20**(2), 395–412 (2004). https://doi.org/10.1193/1.1737736
7. PCM-DPC Mibac (Presidenza Consiglio dei Ministri—Dipartimento della Protezione Civile Ministero Beni e Attività Culturali): Model A-DC Scheda per il rilievo del danno ai beni culturali—Chiese (2006)
8. Cescatti, E., Salzano, P., Casapulla, C., Ceroni, F., da Porto, F., Prota, A.: Damages to masonry churches after 2016–2017 central Italy seismic sequence and definition of fragility curves. Bull. Earthq. Eng. **18**, 297–329 (2020)
9. Sisti, R., Argiento, L.U., Ceroni, F., da Porto, F., Prota, A., Casapulla, C.: Empirical fragility curves for masonry churches and their macro-elements using a large database: Proposal of a new likelihood function. Structures **57**, 105164 (2023). https://doi.org/10.1016/j.istruc.2023.105164
10. De Matteis, G., Brando, G., Corlito, V.: Predictive model for seismic vulnerability assessment of churches based on the 2009 L'Aquila earthquake. Bull. Earthq. Eng. **17**, 4909–4936 (2019)

11. Rossetto, T., Ioannou, I., Grant, D.N.: Existing empirical fragility and vulnerability relationships: compendium and guide for selection. GEM Found. Pavia (2013)
12. Rossetto, T., D'Ayala, D., Ioannou, I., Meslem, A.: Evaluation of existing fragility curves. In: Pitilakis, K., Crowley, H., Kaynia, A.M. (eds.) SYNER-G: Typology Definition and Fragility Functions for Physical Elements at Seismic Risk: Buildings, Lifelines, Transportation Networks and Critical Facilities, pp. 47–93. Springer Netherlands, Dordrecht (2014). https://doi.org/10.1007/978-94-007-7872-6_3
13. Battaglia, L., Ferreira, T.M., Lourenço, P.B.: Seismic fragility assessment of masonry building aggregates: a case study in the old city centre of Seixal. Portugal. Earthq. Eng. Struct. Dyn. **50**, 1358–1377 (2021). https://doi.org/10.1002/eqe.3405
14. FEMA: HAZUS-MH Technical Manual. Federal Emergency Management Agency, Washington, DC (2003)
15. Fajfar, P.: A nonlinear analysis method for performance-based seismic design. Earthq. Spectra **16**, 573–592 (2000)
16. Italian Ministry of Infrastructure and Transport: Circolare 2 febbraio 2009, n. 617. Istruzioni per l'applicazione delle «Nuove norme tecniche per le costruzioni» di cui al decreto ministeriale 14 gennaio 2008 (in Italian). Accessed 11 May 2023
17. Code, P.: Eurocode 8: Design of structures for earthquake resistance-part 1: general rules, seismic actions and rules for buildings. Bruss. Eur. Comm. Stand. (2005)
18. D.M.17 gennaio: Norme tecniche per le Costruzioni (NTC 2018). Gazzetta Ufficiale, n. 42 del 20/02/2018, Supplemento ordinario n.8 (2018)
19. OPCM 28 aprile 2006 n. 3519: Ordinanza PCM 3519 del 28/04/2006. Criteri generali per l'individuazione delle zone sismiche e per la formazione e l'aggiornamento degli elenchi delle medesime zone, G.U. n.108 del 11/05/2006 (in Italian) (2006)
20. Helton, J.C., Davis, F.J.: Latin hypercube sampling and the propagation of uncertainty in analyses of complex systems. Reliab. Eng. Syst. Saf. **81**, 23–69 (2003)
21. Lourenço, P.B.: Recent advances in masonry modelling: micromodelling and homogenisation. Multiscale Model. Solid Mech. 251–294 (2009). Portugal
22. Węglarczyk, S.: Kernel density estimation and its application. ITM Web Conf. **23**, 00037 (2018). https://doi.org/10.1051/itmconf/20182300037
23. Rota, M., Penna, A., Magenes, G.: A methodology for deriving analytical fragility curves for masonry buildings based on stochastic nonlinear analyses. Eng. Struct. **32**, 1312–1323 (2010)

Discussion of Consequence Parameters for Risk Assessment of Bridges and Retaining Structures

Dirk Proske[1] (✉), Dimitrios Papastergiou[2], and Herbert Friedl[3]

[1] Bern University of Applied Sciences, 3400 Burgdorf, Switzerland
dirk.proske@bhf.ch
[2] Federal Roads Office FEDRO, Pulverstrasse 13, 3003 Bern, Switzerland
[3] Swiss Railway SBB, Asset Management and Infrastructure, 3000 Bern, Switzerland

Abstract. Numerous risk studies address the consequences of bridge and retaining structure failures within infrastructure networks. Although there are general recommendations for the selection of consequence and damage parameters, the risk studies show considerable variation in both the number of consequence parameters and the selection of such parameters. Of course, not only theoretical considerations play an important role, but also the availability of data on the parameters and the possibility of using surrogates. For example, some risk studies use only a few outcome parameters, while others use up to fifty. Few risk studies consider potential fatalities and injuries, while others use surrogates such as traffic volume on bridges. This paper reviews the most common consequence parameters and provides recommendations for their practical application. Ultimately, a trade-off between feasibility, data availability and risk analysis requirements must always be made in practice.

Keywords: Damage · Consequences · Bridges · Risk Assessment

1 Introduction

The increasing age of the built infrastructure and limited financial resources lead to increasing demands on infrastructure management. One possible way to ensure the availability and safety of infrastructure networks with constant or decreasing financial resources is to apply risk-based maintenance concepts. These approaches consider not only the probability of failure of a single infrastructure element, such as a bridge or a supporting structure, but also the consequences of failure and collapse. Risk-based approaches therefore take into account the importance of individual elements to the functionality of the networks and the impact on society.

However, while highly detailed standards and guidelines exist for determining the load-bearing capacity (e.g., Eurocodes) and failure probability (e.g., JCSS model code) of individual structures, there are often only general recommendations for determining risk and considering consequences. Furthermore, the recommendations often differ both in the number of individual consequence parameters and in the accuracy of the individual parameters (Western et al. 2016). Some examples are provided below:

J. C. Matos et al. (Eds.): IPW 2024, LNCE 494, pp. 125–136, 2024.
https://doi.org/10.1007/978-3-031-60271-9_10

"These costs are usually evaluated as the sum of direct costs and indirect costs. Direct costs include the costs of performing a repair, performing an inspection, monitoring, building, or replacing bridges, etc. Indirect costs are associated with the cost of detours and delays due to traffic interruption, as well as compensation for the emergency cancellation of public transportation. Consequences related to the operation such as damaged vehicles or loss of human lives are not considered here." (Wang and Leander 2023).

"They (consequence parameters) should be chosen in a clear, differentiated, and relevant manner, as too high a resolution neither appears economical nor necessarily provides greater insight but increases the effort of the evaluation and calls into question the meaningfulness of the statement." (Rielage 2016).

"The consequences of failure can include costs of repair or replacement, limited use/operation during repair or replacement, public inconvenience due to delays and detours, a higher risk of accidents, increased insurance premium, and loss of human life and limb… legal costs, such as liability in case of injuries." (Nowak et al. 2009).

"In bridge management, the primary concern is disruption of expected or designed service levels, which may cause injuries or property damage, loss of mobility, loss of environmental sustainability, and immediate expenditures or long-term excess costs." (Western et al. 2016).

The determination of the individual consequence parameters of risk-based maintenance management is significantly influenced by the availability of data, but also by political framework. For political reasons, estimating the number of fatalities that can of course occur in bridge collapses is often avoided. In addition, data on traffic volumes, diversion routes or environment impacts of bridge are often unavailable or not of sufficient quality.

This paper discusses the most common damage and consequence parameters used in risk-based maintenance management of bridges and retaining structures and discusses their appropriateness.

2 State of the Art

2.1 Introduction

If there is uncertainty in the selection of consequence parameters, the question is how to select the best parameters in terms of sensitivity, robustness, relevance, and availability. This question may be best answered by the observed performance of the risk parameters. A successful application of risk parameters for risk-based maintenance of bridges should result in either *lower financial resources* needed to maintain the bridges and the network compared to the original maintenance strategies, or *lower damage frequency and damage consequences* compared to the original strategy in the long run. Such a comparison is not known to the authors because in most cases mixed strategies are used under real world conditions, such as condition and risk-based maintenance. Furthermore, real-world constraints such as limited schedules for track closures for maintenance must also be considered and would affect the outcome of such comparisons.

Comparison in terms of ranking, such as bridges or retaining walls, can also be a measure, but the target criteria are either unknown or difficult to apply. For example, developing a risk ranking of all bridges within a network and comparing it to a list of bridges based on other criteria may or may not show reasonable differences. During the risk analysis of Swiss railway bridges, a comparison with an independent study showed that about 60% of the 100 highest-risk bridges were in agreement (Proske et al. 2022). Three of the top 100 risk bridges are now closed due to safety concerns.

2.2 Codes for Risk Analysis

There are a variety of codes and recommendations for performing risk analysis in different areas and for different structures. Such codes include ONR 49000, ISO 310010 (2009) and SIA 269/8 (2017). SIA 269/8, as a structural code for seismic evaluation of existing structures, includes recommendations such as.

- 10.2.3: Risk reduction shall at least consider the risk to people.
- 10.2.4: If structures, other valuables, and operation are of high value, they should be considered.
- 10.2.6: Protection of the environment and cultural heritage should be considered.
- E.3.1. The risks to other valuables should be considered if the value is comparable to the value of the structure.
- E.4.1. Loss of operation should be considered if the loss exceeds 20% of the value of the structure.

These are very specific recommendations that are often missing in other codes. Such recommendations can help in the selection of consequence parameters. However, the selection may require detailed information on certain costs. Furthermore, it is difficult to understand why operational losses should be considered if they exceed 20% of the value of the structure, while other values should be considered if they reach the value of the structure.

2.3 Comparison of Different Risk Parameters

Various risk parameters can be used in risk analysis, such as simple risk plots (relative risks) and risk matrices, mortality, *F-N* plots (frequency of events versus fatalities), years of life lost, etc. Some of these risk parameters are domain specific, e.g., different risk parameters in medicine such as relative risk, other risk parameters can be applied across domains. Comparisons of the results of different risk parameters can be found in the literature (Proske 2022a). For example, the results of the risk analysis of different modes of transportation show clear differences in risk based on the number of trips or distance traveled. Differences are also well known for the major hazards to human life, based either on fatalities or years of life lost (Proske 2022a).

2.4 Comparison of Risk Parameters in Other Fields

There are several empirical studies that show that companies with a better risk-return profile perform significantly better (see, for example, Gleissner et al. 2022). However, as mentioned above, it is much more difficult to determine whether the good risk position can be directly attributed to formal enterprise risk management or whether other decisions, such as strategic decisions, are decisive. This argument can be resolved by considering any activity that deals with risk as risk management (see Gleissner 2019, Horvey and Odei-Mensah 2023).

Technical and natural risks in nuclear power plants are considered in large probabilistic safety analyses (PSA). Especially after the Fukushima nuclear accident, there have been some strong criticisms of PSA considering certain limitations such as multi-unit sites, correlation, cutoff, and importance factors. Dorfman et al. (2013) stated, "*Given the degree of uncertainty and complexity... in... Nuclear risk assessment... Risk-based probabilistic assessment has proven to be very limited.*" Consequences are considered in three types (Level 1, 2, and 3), such as core damage, release of radioactive material, and loss of life and injuries. Conclusions cannot be drawn directly from one level alone. For example, a nuclear power plant may have a higher frequency of core damage, but it may have a filtered containment venting system with a cleaning system that keeps the amount of material released low.

2.5 Comparison of the Robustness of Risk Results

Proske et al. (2022) conducted a sensitivity study on the risk assessment of bridges in the Swiss railway network. Another study is currently underway for the bridges of the Swiss motorway network. The sensitivity calculation for the Swiss rail-road bridges was performed in two steps. First, four risk parameter combinations were calculated for approximately 4,000 bridges, and second, approximately 30 risk parameter combinations were calculated for 30 bridges.

Figure 1 shows the results of the second sensitivity study. Each column is a bridge and each point per column is a damage and consequence parameter combination. The black dot is the selected base case. Only the values of the consequence parameters were varied, not the number of consequence parameters. However, one can clearly see that the risk ranking of the bridges would change significantly with the changing numbers.

The range of the probability of failure of the bridges can be between 10^{-2} and 10^{-7} per year, which covers about five orders of magnitude. The range of consequences can be from \$ 100,000 to \$ 1 billion, also covering about five orders of magnitude. Since the largest contributions come from the consideration of hypothetical fatalities, up to

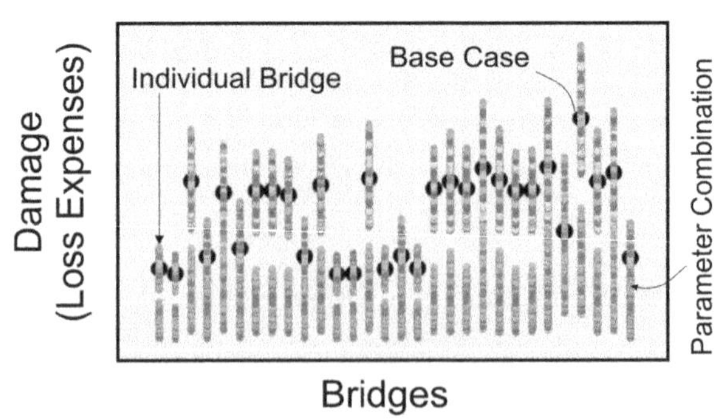

Fig. 1. Risk sensitivity study of app. 30 bridges.

200, and operational losses, both must be considered in a high-quality risk assessment. Furthermore, if highly detailed codes exist for calculating probabilities, a comparable approach should be available for calculating consequence parameters.

2.6 Conclusion

In this section we have shown that.

- Most risk codes give general recommendations; some give more specific rules for considering consequences.
- Different risk parameters can lead to different risk rankings.
- In other areas, attempts have been made to prove the superiority of risk analysis.
- The robustness of risk results has been investigated for bridges and retaining walls.

There is still a large uncertainty in the whole risk analysis process, considering the choice of the risk parameters, the relevant consequence parameters, the proof of superiority and the robustness of the results. However, first codes with detailed rules have been introduced.

3 Discussion of the Consequence Parameters

3.1 Introduction

According to Smith (1999), the consequences of failure and collapse can be divided into.

- Direct and indirect
- Tangible and intangible
- Primary and secondary

The direct, tangible, and primary consequence parameters are easier to consider because they can often, but not always, be measured or quantitative data is available. For example, damage or replacement costs are direct, tangible, and primary consequences that are well known to operators. Indirect, intangible, and secondary costs are very difficult to obtain, require information from third parties, and are therefore extremely difficult to consider.

Naik and Prasad (2022) distinguish the risk contributions to businesses into pure risks (such as extreme natural events), market risks (such as interest rate, exchange rate, or price changes), operational risks (such as fraud, theft, cybersecurity), and strategic and reputational risks (such as litigation). Although these are hazard and risk classes, they give rise to different consequence classes.

Figure 2 shows different types of risks considered in the nuclear industry. The risks shown in the orange rectangle are considered in large probabilistic safety analyses (PSAs). Such computational analyses result in different consequence classes (level 1, 2 and 3) as mentioned before. Other risks are often estimated on the basis of expert judgement and are considered more as economic and operational losses.

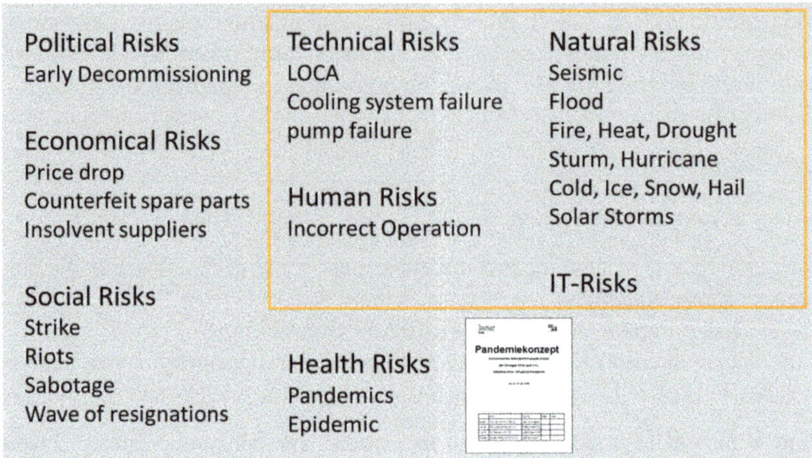

Fig. 2. Risks considered in nuclear power plants. (LOCA: Loss of coolant accident)

3.2 Number of Consequence Parameter

The number of consequence parameters in different bridge risk studies varies from very few to a maximum of 50 parameters. Only a few studies with more than 15 consequence parameters are known. The number of consequence parameters used in practice is usually less than 10, sometimes less than 5. Figure 3 shows the relative frequency of the number of different damage and consequence parameters in about twenty risk analyses of bridges. It seems that the determination of more than 10 consequence parameters is not feasible

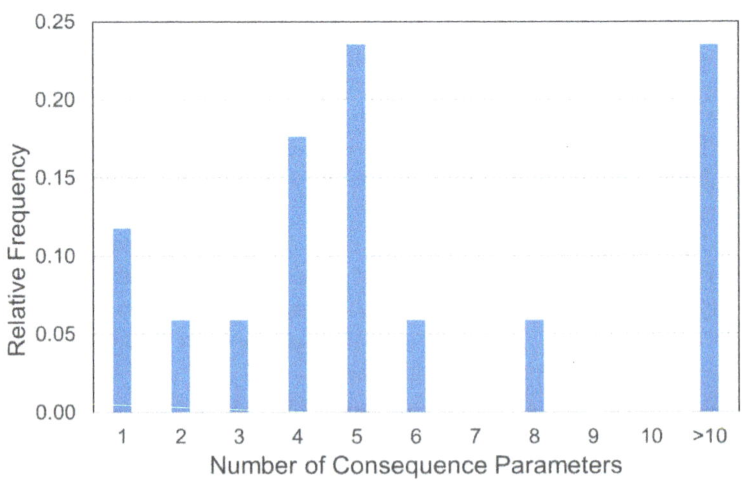

Fig. 3. Relative frequency of the number of consequence parameters in approximately twenty risk analysis of bridges. Please be aware, that the column > 10 includes publications with up to 50 consequence parameters.

in practice for the operators of large bridge portfolios since the majority of analyses uses less consequence parameters.

3.3 Most Common Consequence Parameter

While some consequence parameters can be standardized, e.g., the replacement costs of bridges depending on the bridge area or the detour route can be determined automatically today, some consequence parameters have to be determined individually for each bridge and have to be provided manually. Some of the data is also confidential.

The most common consequence parameters for bridges and retaining walls are.

- Traffic volume
- Detour length
- Replacement value

Traffic volume in terms of Average Daily Traffic (AADT) and Average Daily Truck Traffic (AADTT) is the most commonly used parameter in studies, with detour length and replacement value used in about 2/3 to half of the cases where traffic volume is used. All other parameters are used much less frequently.

3.4 Appropriateness of Consequence Parameter

Detailed risk analyses of bridges are available and can be used to determine the importance and contribution of each consequence parameter. Such values are shown in Table 1. It is clear that fatalities and injuries are a consequence parameter that cannot simply be neglected. The low value of fatalities found by Imhof (2004) is based on the failure of road bridges on minor roads. The high value of Proske et al. (2022) is related to railroad bridges rather than road bridges. The value of 8.5 in Schubert (2023) is a summary value of business losses, fatalities, and accident damages. The comparable value in Proske et al. (2022) would be 11. The table clearly shows that using only the replacement cost of bridges leads to a significant underestimation of the consequences.

Table 1. Normalized consequence parameter contributions for bridge collapses. The normalization is based on the bridge replacement costs.

Reference	Bridge	Operational Loss	Fatalities	Means of Transport/ Accident Damage	Sum
Proske et al. (2022)	1.0	6.5	4.0	0.5	12.0
Schubert (2023)	1.0	->	8.5	<-	9.5
Imhof (2004)	1.0	–	1.5	5.0	7.5
Al-Wazeer (2007)	1.0	–	6.0	4.5	12.0
Saydam (2013)	1.0	–	5.0	1.0	8.0

Values for the cost of bridges per square meter are given in Table 2. There is a clear difference between road and railway bridges. The median values are in the range of

3,000 CHF/m^2 for road bridges and about 10,000 CHF/m^2 for railway bridges. Such differences may be related to different bridge equipment requirements, different maintenance requirements, and different serviceability limit state requirements. In addition, this table does not distinguish prices due to different bridge spans and lengths, different construction technologies, and different construction materials.

Table 2. Values for the costs of bridges per square meter.

Reference	Road bridge in CHF/m^2	Railway bridge in CHF/m^2
Netherlands	3,400	
COMBRI (France)	2,000	
Saxony (Germany)	2,300	
SN 640 907 (Swiss)	2,500	
USA	500–3,200	
Swiss Railway	2,800	
Kuhlmann (Germany)	1,600–2,400	
Iordanescu (Rumania)	2,500–3,300	
Arndt (Germany)	1,000–7,300	
EconoMe (Switzerland)	3,000–4,000	
Swiss Highway	5,000–6,000	
Edwin (International)		4,300–12,900
Swiss Railway		12,500–13,400
EconoMe (Switzerland)		50,000–80,000

The cost of constructing new retaining walls also varies widely, ranging from a few hundred to 2,000 CHF/m^2. In particular, reconstruction after a collapse can be even more expensive, reaching up to CHF 4,000/m^2.

If traffic volume is used as a proxy for fatalities and other types of consequences, the correlation between traffic volume and fatalities should be studied. In fact, there are studies that examine the number of fatalities in bridge collapses. Such studies show that a low number of fatalities is related to both the cause of the bridge collapse, such as floods, and the class of the road, such as major or minor roads. In other words, bridge collapses due to floods on minor roads result in either no or very low number of fatalities whereas bridge collapses due to overload on major roads may result in a significant number of fatalities (Montalvo and Cook 2019, Hingorani et al. 2020, Spector and Gifford 1986, Imhof 2004). However, Hingorani et al. (2020) could not show statistical evidence between the number of fatalities and daily traffic. On the other side operational losses will be strongly correlated with traffic volume.

Other proxies, such as bridge span and length, have also been studied (Imam and Chryssanthopoulos 2012, Shirole and Holt 2005, Caspeele, Steenbergen, Sykora et al.

2016). However, studies on the relationship between bridge length or maximum span and the number of fatalities give conflicting results.

4 Discussion

Operators of large infrastructure networks are looking for better management criteria. As a result, the development and use of risk-based maintenance approaches has increased in recent years. On the other hand, operators often do not have sufficiently accurate data or the necessary data for all structures. Therefore, a compromise must be found between the level of detail in the description of possible consequences and the available data. For such simplifications, the importance of the consequence contributions can be used. Table 3 shows the consequences related to Table 1, the cause of collapse based on Proske (2022a), and the relative risk. It shows that the highest relative risk for bridges is associated with collapses caused by impacts. In other words, the frequency and consequences of impacts should be studied with the highest effort and detail.

Table 3. Normalized consequence parameter for different bridge collapse causes.

Cause of Collapse	Consequence	Relative Frequency	Relative Risk
Flood	1–7.5	0.5	0.5–3.7
Impact	8–12	0.3	2.4–3.6
Overload	8–12	0.1	0.8–1.2

If the AADT or the AADTT are used as substitute for the fatalities and the operating loss, then the factor for the AADT should be increased to correspond to the results from the risk analysis using hypothetical fatality and injury numbers respectively and operating loss.

It is interesting to note that the AADT and AADTT respectively is also a very frequently used substitute measure for the bridge condition (Proske 2022b).

The same concepts may be applied to nuclear power plants. However, here the definition of consequences is extremely difficult. For example, is the decommissioning of eight nuclear power plants that are not exposed to tsunamis and earthquakes in Germany after the Fukushima accident a consequence of the accident or a consequence of certain political interests? Furthermore, the discussion of the consequences of the accidents is extremely dominant compared to the discussion of the consequences of the mitigation measures taken after the accident, e.g., resettlement. Again, as an example, the operational losses of shutting down and decommissioning nuclear power plants in Japan are either greater than or at least equal to other costs, such as cleanup costs (Barnard 2018) or the costs of preventing loss of life, even taking into account the sad loss of suicides (Expert Workshop 2016) due to resettlement. Own estimates of operational losses are in the order of a trillion dollars. Moreover, recent studies have shown that mitigation measures have cost significantly more lives than they have saved (Neidell et al. 2019). This follows the imbalance in German news coverage after the accident (Kepplinger and

Lemke 2012) or the evaluation of the necessary refurbishment of the Beznau nuclear power plant based on the Life Quality Index (Proske 2022a). Such imbalances are also known for the avoidance of air travel after 9/11 and the use of cars instead (Gaissmaier and Gigerenzer 2012). Such consequences are indirect, intangible, secondary or tertiary, and extremely difficult to quantify. Risk codes and regulations should define a cut-off point at which such consequence classes are not considered anymore (Table 4).

Table 4. Risk Analysis Levels for Nuclear Power Plants and Bridges.

Level	Nuclear	Bridges	Retaining Walls
1	Core Damage Frequency	Probability of Failure (repairable)	Probability of Failure (repairable)
2	Release Frequency	Collapse (not repairable)	Collapse (not repairable)
3	Fatalities		
4a		Simplified Consequence consideration (substitute measure such as AADT)	Simplified Consequence consideration (substitute measure such as AADT)
4	All Consequences (very important, not common)	All Consequences	All Consequences (perhaps fatalities can be neglected)

The study comes to the following conclusions:

1. Risk analysis should be standardized in a much stricter way, as is the case for the load-bearing capacity of building materials or components. Some approaches can be found e.g. at FEDRO and SBB, in EconoMe, in the new VKF fire protection guideline or in SIA 269/8. In Switzerland, some input parameters are also defined by government authorities, such as the value of the Implied Cost Averting a Fatality, Value of a Statistical Life, Willingness to Pay (ICAF/VSL/WTP). Such detailed recommendations are essential for widespread and successful application of risk analyses.
2. The determination of the relevant consequence parameters should be based on a multi-step, at least two-step, procedure. The first step should be to use the data available from the major infrastructure operators. In a second step, all necessary data should be identified, collected, and used by the major infrastructure operators.
3. The results of the risk analysis and assessments must be checked against observations. Such observations can be collapses, but also damages or the development of long-term financial costs. In addition, loss data from other countries can be checked for representativeness and possibly included. For example, the consequences of earthquakes in comparable regions, such as the United States and the Soviet Union, have been compared (Proske 2022a).

References

Al-Wazeer, A.A.-R.: Risk-based Bridge Maintenance Strategies. Dissertation, University of Maryland (2007)

Barnard, M.: Fukushima's Final Costs Will Approach A Trillion Dollars Just For Nuclear Disaster, Cleantechnica (2018). https://cleantechnica.com/2019/04/16/fukushimas-final-costs-will-approach-one-trillion-dollars-just-for-nuclear-disaster/

Caspeele, R., Steenbergen, R., Sykora, M.: Fib Bulletin 80: Partial Factor Methods for existing Concrete Structures. Recommendation, Task Group 3.1, December 2016, Lausanne (2016)

Gaissmaier, W., Gigerenzer, G.: 9/11, Act II: a fine-grained analysis of regional variations in traffic fatalities in the aftermath of the terrorist attacks. Psychol. Sci. **23**(12), 1449–1454 (2012)

Dorfman, P., Fucic, A., Thomas, S.: Late lessons from Chernobyl, early warnings from Fukushima, In: Late Lessons from Early Warnings: Science, Precaution, Innovation, Summary, European Environment Agency, EEA Report 1/2013, Copenhagen (2013)

Workshop, E.: Fukushima Five Years on: Legal Fallout in Japan, Lesson for the EU, 4–5 March 2016. Darwin College Cambridge, Cambridge (2016)

Gleissner, W.: Cost of capital and probability of default in value-based risk management. Manag. Res. Rev. **42**(11), 1243–1258 (2019)

Gleissner, W., Günther, T., Walkshäusl, C.: Financial sustainability: measurement and empirical evidence. J. Bus. Econ. **92**, 467–516 (2022)

Hingorani, R., Tanner, P., Syrkov, A., Proske, D.: Development of Models für Estimation of Consequences to Persons due to Bridge Failures, IABSE TG1.5 (2020)

Horvey, S.S., Odei-Mensah, J.: The measurements and performance of enterprise risk management: a comprehensive literature review. J. Risk Res. **26**(7), 778–800 (2023)

Imam, B.M., Chryssanthopoulos, M.K.: Causes and consequences of metallic bridge failures. Struct. Eng. Int. **22**(1), 93–98 (2012)

Imhof, D.: Risk Assessment of Existing Bridge Structures. University of Cambridge, Dissertation, Kings College (2004)

ISO 310 010: Risk management, Risk assessment techniques, ISO/TC 262, Geneve (2019)

Kepplinger, H.M., Lemke, R.: Die Reaktorkatastrophe bei Fukushima in Presse und Fernsehen in Deutschland, Schweiz, Frankreich und England, Jahrestagung 2012 der Strahlenschutzkommission. 15. März 2012, Hamburg (2012)

Montalvo, C., Cook, W.: Retrospective analysis of hydraulic bridge collapse. J. Perform. Constructed Facil. **34**, 04019111 (2019)

Naik, S., Prasad, C.: Risk and risk management: a historical review and research agenda. Int. J. Bus. Continuity Risk Manage. **12**(3), 244–262 (2022)

Neidell, M., Uchida, S., Veronesi, M.: Be Cautious with the Precautionary Principle: Evidence from Fukushima Daiichi Nuclear Accident, IZA DP No. 12687, IZA – Institute of Labor Economics, Bonn (2019)

Nowak, A.S., Kozikowski, M., Lutomirska, M.: Risk Mitigation for Highway and Railway Bridges. University of Nebraska, Final Reports & Technical Briefs from Mid-America Transportation Center. 1. Lincoln (2009)

ONR 49 000: Risikomanagement für Organisationen und Systeme, Austrian Standards Institute (2014)

Proske, D.: Katalog der Risiken: Risiken und ihre Darstellung. Springer Fachmedien Wiesbaden, Wiesbaden (2022). https://doi.org/10.1007/978-3-658-37083-1

Proske, D.: Überführbarkeit der Eingangsgrößen von Lebenszykluskostenanalysen bei Massivbrücken. Beton- und Stahlbetonbau **117**(3), 167–175 (2022b)

Proske, D., Payeur, J.-B., Girardin, B., Friedl, H.: Risk based maintenance for swiss railway bridges: concept, implementation and first experiences. In: IPW 2022, South Africa, Acta Polytechnica CTU Proceedings, vol. 36, pp. 167–174 (2022)

Rielage, T.: Risikoanalyse und Risikomanagement, Verschiedene Definitionen von Risikoanalysen, Multirisikoanalyse und Darstellung des Risikomanagements für die Bevölkerung. Rheinische Friedrich-Wilhelms-Universität Bonn, Geographisches Institut, Bundesamt für Bevölkerungsschutz und Katastrophenhilfe, Bonn (2016)

Saydam, D.: Reliability and risk of structural systems under progressive and sudden damage. Thesis and Dissertation, Paper 1616, Lehigh University, Lehigh Preserver (2013)

Schubert, M.: Engineering methods and practical examples … and beyond. In: JCSS & IABSE Workshop on Risk Acceptance Criteria in Civil Engineering Decision Making, Trondheim, 20.6.2023, presentation (2023)

Shirole, A.M., Holt, R.C.: Planning for a comprehensive bridge safety assurance program. Transp. Res. Rec. **1290–005**, 39–50 (2005)

SIA 269/8: Erhalten von Tragwerken – Erdbeben, Zürich, SIA (2017)

Smith. D.I.: Floods. Physical Processes and Human Impacts In: Smith, K., Ward, R. John Wiley, Chichester 1998. Earth Surface Processes and Landforms, 24, 13, p. 1261 (1999)

Spector, A., Gifford, D.: A computer science perspective of bridge design. Commun. ACM **29**(4), 268–283 (1986)

Wang, R., Leander, J.: Risk-based decision-making for system-level fatigue assessment of bridges. Eng. Fail. Anal. **153**, 107533 (2023). https://doi.org/10.1016/j.engfailanal.2023.107533

Western, J., Bye, P., Valeo, M., Thompson, P.D., Frazier, E.: Assessing Risk for Bridge Management. Final Report, prepared for: AASHTO Standing Committee on Highways, NCHRP 20–07/Task 378 (2016)

Evaluation of Fault Tree Analysis Algorithms for Probabilistic Risk Assessment: A Systematic Comparative Study

Nailah Afshan[1(✉)], Saran Srikanth Bodda[1], Abhinav Gupta[2], and Kevin Han[1]

[1] Department of CCEE, North Carolina State University, Raleigh, NC 27695, USA
nafshan@ncsu.edu

[2] Center for Nuclear Energy Facilities and Structures, North Carolina State University, Raleigh, NC 27695, USA

Abstract. Civil infrastructures are constantly exposed to external hazards, such as earthquakes, flooding, and storm surges. These hazards underscore the need to estimate the risk of critical facilities due to catastrophic events. Given uncertainties associated with hazards, their propagation, and the complex interdependent nature of civil infrastructures, probabilistic risk assessment (PRA) has been gaining popularity for risk-informed decision-making, both under normal and extreme scenarios. One of the key steps of PRA is systems analysis, where fault and event trees are used to propagate the fragilities of structures, systems, and components (SSCs) to estimate the system-level risk. This paper presents the development of selective scenarios with increasing complexity from small to large realistic networks to conduct a comparative analysis among different fault tree analysis (FTA) algorithms. Some of these application scenarios can serve as benchmarks for future studies. Computational efficiency and accuracy are quantified and compared for a suite of selected algorithms. The results show that the performance of these algorithms is affected by the structure of the fault tree under consideration. In a few cases, the key sources of complexity in algorithms that influence the speed and memory/storage requirements are also identified.

Keywords: Fault Tree Analysis · MOCUS Algorithm · ZBDD Algorithm · BDD Algorithm · Probabilistic Risk Assessment

1 Introduction

Probabilistic risk assessment (PRA) is a systematic and comprehensive approach to evaluating risks and improving the safety and security of critical industrial facilities, structural systems, and the environment during extreme natural hazards. It has become a rigorous technical approach to manage design, operation, and maintenance vulnerabilities, enhance safety, and minimize costs of complex facilities, such as nuclear power plants [1]. The safe and reliable operation of nuclear power plant structures and systems is critical to protect both personnel and the surrounding environment. However, these

J. C. Matos et al. (Eds.): IPW 2024, LNCE 494, pp. 137–146, 2024.
https://doi.org/10.1007/978-3-031-60271-9_11

structures are subject to numerous risks, including natural disasters, equipment malfunctions, and human error. PRA allows the identification and quantification of these risks, enabling safer and more reliable design and operation of the facility [2]. PRA assesses potential hazards, estimates the likelihood of their occurrence, and evaluates the consequences of failures related to these hazards. By quantifying the risks associated with nuclear structures, it is possible to design safety systems and procedures to mitigate these risks [2].

Fault tree analysis (FTA) is one of the most used computational approaches in PRA. Various FTA algorithms with different approaches have been proposed, such as cut set-based, Boolean function-based, and compressed truth tables-based. Previous studies have shown that most of the existing FTA algorithms can incur significant computational expenses [3–5].

In this paper, a comparative analysis of a suite of selected FTA algorithms is conducted by developing selective scenarios with increasing complexity, from small to large realistic networks. These scenarios can serve as benchmarks for future studies. Four algorithms are considered for analysis, including the method of obtaining cut sets (MOCUS), Binary Decision Diagram (BDD), Zero-suppressed Binary Decision Diagram (ZBDD), and Compressed Truth Table (CTT) algorithm [3–5]. CTT is a recent addition to FTA algorithms and is based on the concept of compressed truth tables, which has shown promising results [4]. A rigorous comparative analysis is done to compare the algorithms with respect to their computational efficiency and accuracy of probability calculations, and some critical cases are identified.

To conduct comparative analysis in this paper, an open-source Command-line Risk Analysis Multi-tool (SCRAM) is used to do the analysis for the first three (MOCUS, BDD, and ZBDD) of the above-mentioned algorithms, while CTT is run through standalone C++ command line [6]. The SCRAM code is modular and uses different preprocessing algorithms to optimize the efficiency of its FTA algorithms. However, CTT has not yet been integrated with SCRAM, and its current code lacks appropriate optimization or efficient preprocessing approaches. Therefore, only the first three FTA algorithms of SCRAM are compared for computational efficiency while accuracy is compared among all four algorithms. A unified platform to enable comparative analysis of all these algorithms will enhance the comparative evaluation for complex fault trees [7, 8].

2 Theoretical Background

2.1 Scram

The present work uses SCRAM, an open-source tool for probabilistic risk analysis. The tool is primarily written in C++ with some Python scripts available for generating random fault trees. SCRAM can perform different PRA analyses such as event tree analysis (ETA), static FTA, and common cause failure models analysis, among others. The tool can also manage non-coherent fault trees that include NOT logic. SCRAM is a useful package for different types of PRA analyses, but there are some missing features that can be incorporated to enhance the tool's utility. The GUI front-end is not fully developed, which hinders the visualization and manipulation of risk analysis models and reports. Some other features that can be added include sensitivity analysis, dynamic fault tree analysis, systems with loops, Markov chains, and reliability block diagrams. SCRAM also lacks advanced optimization techniques such as ETA shadow-variables, variable ordering, and reordering heuristics for BDD and cut-offs for ZBDD. These features can improve the complexity and probability calculations of the FTA algorithms used in SCRAM. Additionally, more tests are needed for preprocessing techniques and expressions in coding [6].

2.2 MOCUS Algorithm

MOCUS was first proposed by Fussell and Vesely (1972). It is a traditional fault tree reduction algorithm that generates minimal cut sets using a gate substitution technique. It is widely recognized as one of the most popular and easy-to-understand FTA techniques. MOCUS supports approximations and is suitable for analyzing relatively smaller fault trees. However, it faces difficulties when analyzing large and complex fault trees leading to exponential computational time [10]. The algorithm is sensitive to the way formulae are written and uses complemented events that slow down the Boolean reduction process. There are several suggested improvement techniques for MOCUS, including preprocessing techniques, rewriting heuristics, and using shadow variables [11].

2.3 BDD Algorithm

BDDs are canonical and efficient data structures used for encoding and manipulating Boolean functions [12]. These are based on Shannon's decomposition theorem, which enables them to represent logical functions as canonical and compact graphs [13]. BDDs follow a bottom-up approach that is mature and finely tuned for logical applications. BDDs are primarily used in design and synthesis of logic circuits, formal verification, cryptography, Bayesian reasoning, product configurations, and FTA in PRA [14–16]. BDD is a rooted, directed acyclic graph with two terminal nodes labeled as '0' and '1' representing an operational or a failed state of the system, respectively. Each non-terminal node is associated with a Boolean variable 'x' and has two outgoing edges called as 1-edge and 0-edge representing the failure and operation of the corresponding component, respectively. BDD has variants depending on the application and the type of Boolean function being analyzed, such as ordered BDD (OBDD), free binary decision diagrams (FBDDs), reduced OBDD (ROBDD), zero-suppressed BDD (ZBDD), multi-terminal BDDs (MTBDDs), and interval BDD (IBDD). In SCRAM, BDDs are constructed from propositional directed acyclic graphs (PDAGs) for analysis and are transformed into ZBDD for computing minimal cut sets [6].

2.4 ZBDD Algorithm

The ZBDD algorithm is recognized as a more effective substitute for cu test-based algorithms that relies on conventional Boolean algebra [17]. The ZBDD algorithm is a compact variant of the BDD algorithm used for efficiently encoding minimal cutsets in fault trees. It outperforms traditional cut set-based algorithms by generating numerous minimal cutsets and efficiently solving large fault trees. The algorithm is built on a set of operation formulae for ZBDD data structures, which are more straightforward and concise than those used in traditional cut set-based algorithms. However, the application of Boolean operators with the ZBDD decomposition requires additional computations when compared to the BDD method [17]. To reduce the complexity of ZBDDs for highly complex fault trees, different techniques have been proposed, including parallel algorithms and heuristic approaches for implementing probability truncation and Boolean simplification [18]. In SCRAM, all FTA algorithms produce ZBDD data structures that can work directly with PDAGs [6].

2.5 CTT Algorithm

This approach of solving a fault tree using tree data structure was proposed quite recently [1, 4]. In this approach, a fault tree is represented by a tree data structure using the concept of compressed truth tables (CTTs). This bottom-up algorithm calculates the failure probability of the top node by first calculating the height of internal nodes, sorting them in ascending order, and then using logic gates to calculate binary state probabilities for each internal node. To handle dependent nodes, Boolean algebra is used to find the intersection of child nodes. The CCT approach has low computational requirements and gives exact probability calculations but has not been tested on highly complex and dynamic fault trees. In this study, a C++ code is used to implement CTT algorithm which does not use any optimization facilities. Hence, it is not compared with SCRAM FTA algorithms for computational efficiency but evaluated for accuracy only.

3 Analysis Methodology

This section outlines the analysis methodology used in the study. A range of scenarios with varying complexity are developed to compare different FTA algorithms, with variations in the fault tree's geometry, connectivity, gate type, and number and distribution of dependent nodes.

First, fixed height fault trees are analyzed. This analysis uses the fault tree structure given in Fig. 1. The number of intermediate gates (N) is fixed and the number of basic

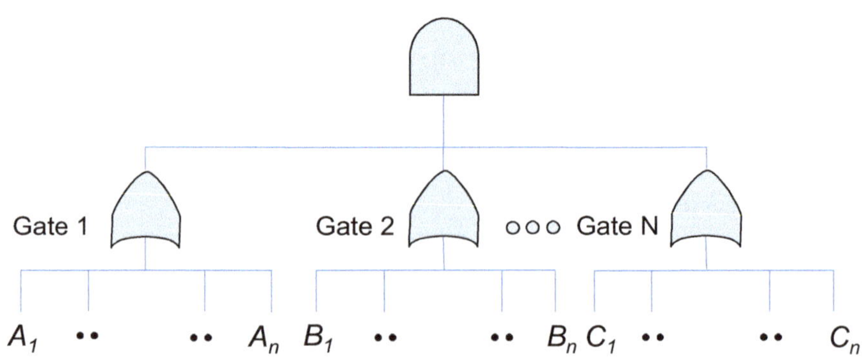

Fig. 1. Fault tree structure having fixed height

events (n) for each N is the same in this case. Next, more complex fault trees with varying geometric configurations are created and analyzed by varying their width, height, number of intermediate gates (N), and the number of basic events (n) as given in Table 1. To study the impact of dependent nodes on efficiency and accuracy of the FTA algorithms, fault trees with different number of dependent nodes are created from the structure given in Figure 1 which are listed in Table 2. Lastly, fault tree generator (FTG) code of SCRAM tool is used to create different complex fault trees with varying widths, heights, dependent nodes, and number of basic events. All SCRAM FTA algorithms are implemented for analysis and identification of some critical cases.

Table 1. Cases representing fault trees of varying geometric configurations

Case No	Width	n	Height	N
1	100	2031	6	107
2	2002	40071	6	2011
3	5002	100071	6	5011
4	103	2097	12	119
5	3003	60097	12	3019
6	5003	100097	12	5019
7	10003	200097	12	10019
8	100	2114	25	140
9	1000	20114	25	1040
10	5000	100114	25	5040

Table 2. Cases representing fault trees with dependent nodes

Case No	n	Dependent nodes
1	10	10
2	100	100
3	500	160
4	1000	230

4 Results and Discussion

In this section, the efficiency and accuracy of different FTA algorithms is discussed. The efficiency of algorithms.is determined in terms of their computational run times and the accuracy is analyzed in terms of their calculation of the top node failure probability. This comparison is conducted for different cases discussed in the previous section.

4.1 Efficiency

The fixed height fault tree structure shown in Fig. 1 is used to calculate the computation time of SCRAM FTA algorithms for various values of n, ranging from 10 to 1000. The resulting time values (t) are plotted against n in Fig. 2. It is observed that t increases with increasing n for all algorithms, with MOCUS and ZBDD having a larger increase than BDD. BDD takes the least computation time which is in the range of microseconds even when n is increased to 1000. Based on this analysis, it can be inferred that in a fault tree with fixed height and a large number of basic events but a fixed number of intermediate gates, BDD is more efficient than MOCUS and ZBDD.

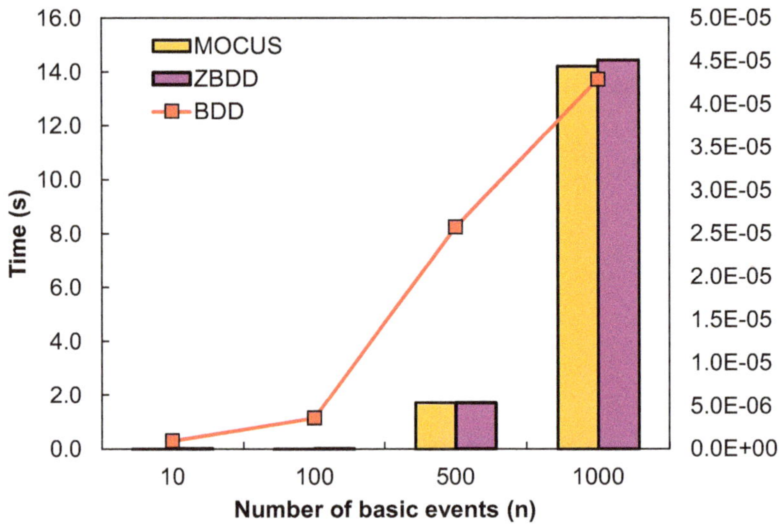

Fig. 2. Time complexity for fault trees with fixed height and varying n

When SCRAM FTA algorithms are implemented for different cases in Table 1, BDD is found to be the most efficient one as can be seen in Fig. 3. However, all three algorithms display a similar trend of variation in efficiency with change in geometric configuration. MOCUS and ZBDD have similar t values for all the cases and hence, exhibit comparable efficiency. It is observed that efficiency decreases for all the algorithms when n and N are increased for same height of the fault tree in case 1 and case 2. But for case 3 where the height is same as in previous cases, the efficiency remains unaffected when n and N are increased further. When height is increased to 12 in case 4, there is improvement in efficiency which then remains same for all other cases with same height but varying and increased number of other geometric parameters. On increasing height to 25 in case 8, efficiency decreases initially but then remains constant even if other parameters are changed. It is evident from this exercise that efficiency changes when height of the fault tree changes but no significant change is observed in efficiency by changing width, number of basic events and number of intermediate gates when height is fixed.

For different cases of Table 2, BDD is much more efficient than MOCUS and ZBDD as is evident from Fig. 4. However, the trend of performance change is similar for all

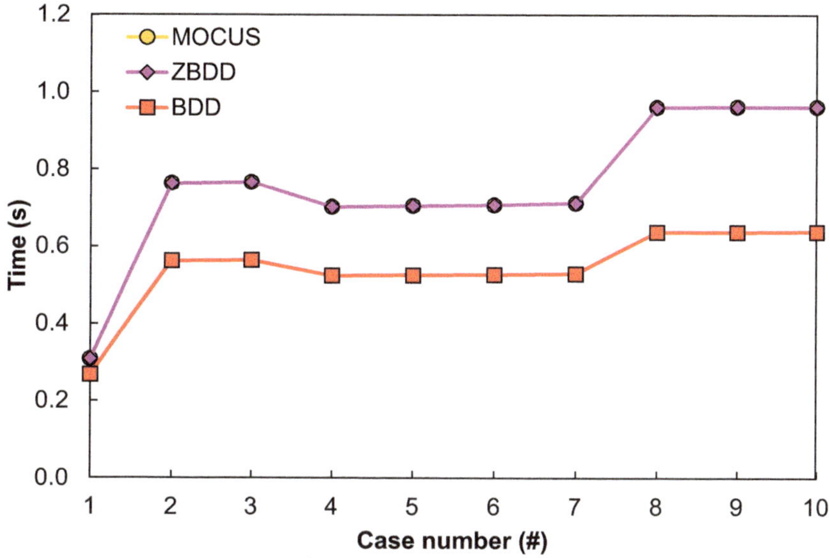

Fig. 3. Time complexity for cases with varying geometric configurations

the three algorithms. MOCUS and ZBDD have similar efficiency. The two algorithms perform well till case 2 when *n* and number of dependent nodes is small. As *n* increases beyond 100 with increase in the number of dependent nodes as well, the efficiency reduces. This analysis indicates that for a fault tree of fixed height and width, efficiency decreases with increase in the number of basic events and the number of dependent nodes.

Lastly, the complex fault trees of different geometric configurations generated from the FTG code of SCRAM are analyzed by increasing *n*. It is observed that MOCUS and ZBDD become computationally more efficient when complexity is increased by increasing *n* beyond 3000. BDD takes more computational time for some simpler fault trees and when *n* is increased to generate more complex fault trees, the algorithm fails to run.

4.2 Accuracy

To analyze the accuracy of all the four algorithms, it should be noted that approximations are used by MOCUS and ZBDD for probability calculations while BDD and CTT follow exact computations. From this study, it is observed that MOCUS and ZBDD produce similar results in terms of top node failure probability values in all the cases irrespective of any changes in the geometric configurations and the overall complexity of fault trees Due to approximations, both the algorithms fail to calculate the top node failure probability when the basic events have higher probability values. Table 3 shows the top node failure probability values for fixed height fault tress with varying number of basic events. On the other hand, BDD and CTT perform well in such cases and give similar and accurate

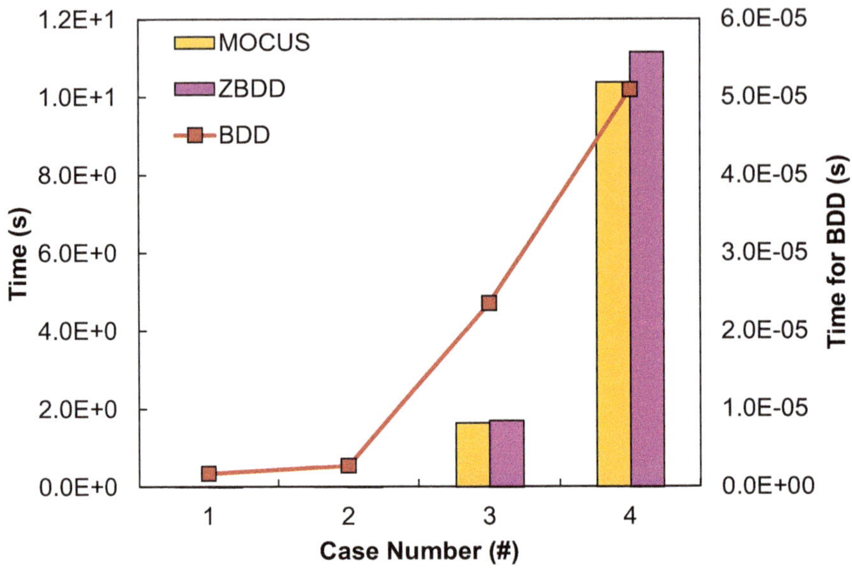

Fig. 4. Time complexity for cases listed in Table 2

results for probability values for all the cases considered in this study. This analysis indicates that CTT is a promising algorithm in PRA.

Table 3. Top node failure probability for fixed height fault trees with varying basic events

n	MOCUS	ZBDD	BDD	CTT
10	0.002231	0.002231	0.002057	0.002057
100	3.62E-06	3.62E-06	3.58E-06	0.000004
500	0.000773	0.000773	0.000682	0.000682
1000	0.006504	0.006504	0.004995	0.004995

5 Conclusions

This study conducts a comparative analysis of several FTA algorithms needed in probabilistic risk assessment (PRA), evaluating their computational efficiency and accuracy for a set of realistic network scenarios with incremental complexity. The results demonstrate that the performance of the algorithms is dependent on the structure of the fault tree being analyzed, with different trade-offs between speed, memory, and accuracy. Notably, BDD stands out as the most efficient algorithm for fault trees with fixed height and width, with increasing efficiency as the number of basic events increases. However, for highly complex fault trees generated using FTG code, BDD can exhibit memory or storage issues and fail to run successfully, while MOCUS and ZBDD become more efficient as the number of basic events and dependent nodes increases. The analysis also shows that the efficiency of the algorithms decreases with increasing height, number of dependent nodes, and number of basic events for fault trees of fixed height and width. Overall, these findings can serve as a reference for future research on FTA algorithms and highlight the need for carefully selecting the appropriate algorithm based on the characteristics of the fault tree being analyzed. Moreover, further investigation is necessary to address the trade-offs between efficiency and accuracy and develop more robust and scalable algorithms for large-scale fault trees, and CTT algorithm appears to be promising in this end.

References

1. Vaishanav, P., Gupta, A., Bodda, S.S.: Limitations of traditional tools for beyond design basis external hazard PRA. Nucl. Eng. Des. **370**, 110899 (2020). https://doi.org/10.1016/j.nucengdes.2020.110899
2. USEPA. Risk Assessment Forum White Paper: Probabilistic Risk Assessment Methods and Case Studies (2014)
3. Rauzy, A.: Toward an efficient implementation of the MOCUS algorithm. IEEE Trans. Reliab. **52**(2), 175–180 (2003). https://doi.org/10.1109/TR.2003.813160
4. Vaishanav, P.: Uncertainty Quantification and Propagation for External Hazard Probabilistic Risk Assessment (2021). https://repository.lib.ncsu.edu/handle/1840.20/38984. Accessed 14 Mar 2023
5. Jung, W.S., Han, S.H., Ha, J.: A fast BDD algorithm for large coherent fault trees analysis. Reliab. Eng. Syst. Saf. **83**(3), 369–374 (2004). https://doi.org/10.1016/j.ress.2003.10.009
6. SCRAM. Probabilistic Risk Analysis Tool (2016). https://github.com/rakhimov/scram. Accessed 14 Mar 2023
7. Afshan, N., Bodda, S.S., Gupta, A., Han, K.: Development of an integrated platform for probabilistic risk assessment using fault tree analysis. In: 2023 ASCE Engineering Mechanics Institute Conference (2023)
8. Afshan, N., Bodda, S.S., Gupta, A., Han, K.: SCRAM+: a unified probabilistic risk assessment platform for improved decision-making in reliability analysis. In: Transactions of the 27th International Conference on Structural Mechanics in Reactor Technology (2024)
9. Fussell and Vesely. A new methodology for obtaining cut sets for fault trees. Trans. Am. Nucl. Soc. **1**(15), 262–263 (1972). https://inis.iaea.org/search/search.aspx?orig_q=RN:4058963 Accessed: Feb. 15, 2023
10. Ruijters, E., Stoelinga, M.: Fault tree analysis: a survey of the state-of-the-art in modeling, analysis and tools. Comput. Sci. Rev. **15–16**, 29–62 (2015). https://doi.org/10.1016/J.COSREV.2015.03.001
11. Russell, K.D., Rasmuson, D.M.: Fault tree reduction and quantification—an overview of IRRAS algorithms. Reliab. Eng. Syst. Saf. **40**(2), 149–164 (1993). https://doi.org/10.1016/0951-8320(93)90105-8
12. Bryant, R.E.: Graph-based algorithms for Boolean function manipulation. IEEE Trans. Comput. **C–35**(8), 677–691 (1986). https://doi.org/10.1109/TC.1986.1676819
13. Xing, L., Amari, S.V.: Binary decision diagrams and extensions for system reliability analysis (2015)
14. Tan, S.X.D.: Symbolic analysis of analog circuits by Boolean logic operations. IEEE Trans. Circuits Syst. II Express Briefs **53**(11), 1313–1317 (2006). https://doi.org/10.1109/TCSII.2006.882356

15. Jensen, R.M.: CLab: A C++ library for fast backtrack-free interactive product configuration. In: Wallace, M. (ed.) CP 2004. LNCS, vol. 3258, pp. 816–816. Springer, Heidelberg (2004). https://doi.org/10.1007/978-3-540-30201-8_94
16. Lipmaa, H.: First CPIR protocol with data-dependent computation. In: Lee, D., Hong, S. (eds.) ICISC 2009. LNCS, vol. 5984, pp. 193–210. Springer, Heidelberg (2010). https://doi.org/10.1007/978-3-642-14423-3_14
17. Jung, W.S.: ZBDD algorithm features for an efficient probabilistic safety assessment. Nucl. Eng. Des. **239**(10), 2085–2092 (2009). https://doi.org/10.1016/j.nucengdes.2009.05.005
18. Wang, J., et al.: Fault-tree-based instantaneous risk computing core in nuclear power plant risk monitor. Ann. Nucl. Energy **95**, 35–41 (2016). https://doi.org/10.1016/j.anucene.2016.02.024

Exoskeletons Development: Military Load Effort Analysis in an Operational Environment

Luís Quinto[1,2], Joni Santos[1], Paulo Fernandes[1], Rui Lucena[1], Miguel Silva[2], and Paula Simões[1,3](✉)

[1] Military Academy Research Center - Military University Institute (CINAMIL), Lisboa, Portugal
{luis.quinto,rui.lucena,paula.simoes}@academiamilitar.pt, {santos.jg,fernandes.pjl}@exercito.pt
[2] Mechanical Engineering Institute (IDMEC), Instituto Superior Técnico, University of Lisbon, Lisboa, Portugal
miguelsilva@tecnico.ulisboa.pt
[3] NOVA MATH - Center for Mathematics and Applications, NOVA University of Lisbon, Caparica, Portugal

Abstract. Military personnel are subject to carrying loads that have physiological impacts on their bodies, which may compromise their ability to perform tasks effectively during operations.

Exoskeletons can support soldiers carrying loads and reduce fatigue. Challenges in developing military exoskeletons include the specific nature of the equipment carried by soldiers, the diversity of specific movements and the physical environment.

Integrated into the ELITE project, which aims to develop exoskeletons to support human movement, this research analyses the effort to transport military loads within the infantry squad in an operational environment.

The sample consisted of 180 soldiers from the Portuguese Army, belonging to the 6th to 12th National Forces, deployed to the Central African Republic from 2020 to 2023. Military personnel were grouped according to their roles within the Infantry squad to evaluate their overload level and compare different functions during military operations.

Statistical Inference techniques and probability models were used to evaluate the effort of soldiers and understand their proportion under stress/workload during peacekeeping operations in an International environment.

Results indicate that exoskeleton development should consider different roles in military operations according to the specific tasks and load carried since these factors vary significantly and imply different requirements.

Keywords: Exoskeleton · Military Load · Soldier's Combat System · Military Requirements · Statistical Inference

J. C. Matos et al. (Eds.): IPW 2024, LNCE 494, pp. 147–156, 2024.
https://doi.org/10.1007/978-3-031-60271-9_12

1 Introduction

Military operations involve a high degree of physical strain, taking place under adverse conditions, frequently under extreme circumstances, leading to soldier's exhaustion and increasing the risk of injury. In addition, the complexity of the current operational environment also led to the soldier's need to transport more equipment, limiting their mobility and agility. Performing long approach marches while carrying significant loads (57% to 73% of body mass) can result in a lack of cognitive control and awareness and ultimately lead to exhaustion [1]–[4]. These factors combined can result in the reduction of the operational readiness level of such units, as well as increased absenteeism, both of which represent a significant concern [2, 5, 6]. Several studies have indicated that during operations, the load transported by soldiers (Assault Load) should not exceed 30% of the body weight (BW) [4, 7, 8]. In some cases, the transported load can reach 40% of the BW, with some limitations in mobility and increased injury likelihood [9, 10].

Similarly, situations of extreme effort can also be found in non-military contexts, for instance, law enforcement agents, firefighters, industrial or construction workers, or nurses and carers, resulting in diverse health challenges and representing a potential burden for global health systems, social security and society in general [2, 11, 12].

Exoskeletons can help reduce the physical effort exerted in several activities, improving performance while reducing the risk of injury, ultimately improving performance and the readiness level of its user. These dual-purpose systems can also help people with different physical disabilities or the elderly regain their independence or perform rehabilitation tasks [2, 5, 13]–[17].

Exoskeletons can be characterised as external structures that support or increase their user's ability to perform specific tasks. They are composed of three core systems: 1) structure, which links the system to the user and supports the force or weight distribution; 2) force elements, which provide additional strain to the user when he needs it most; and 3) control system, which defines when/how the exoskeleton should support its user. Exosuits differ from exoskeletons, presenting a complacent structure rather than a rigid one [18]–[20].

A significant set of limitations to the use of exoskeletons and exosuits in military operations is related to the specific equipment that soldiers must carry, the diversity of movements that must be viable, and the environmental factors that must be accounted for when developing such system [21, 22].

To that end, the aim of this study is the identification and characterisation of the effort of soldiers in the operational environment faced by the Portuguese Armed Forces in the Central African Republic, integrated into the United Nations Multidimensional Integrated Stabilization Mission in the Central African Republic (MINUSCA).

This work is integrated into the ELITE – Enhancement Lite Exoskeleton project, developed by the Portuguese Army and Instituto Superior Técnico, aiming to develop dual-purpose solutions to improve soldiers' performance and reduce injury likelihood.

2 Methods

This research focuses on understanding the impact that load can have on military personnel at an operational level and the importance of developing military exoskeletons to overcome these physiological impacts. In this way, the initial research was based on articles related to load transport's impacts and exoskeletons' application in an operational environment. A questionnaire survey was designed and implemented to collect the necessary data to characterise the load carried by military personnel in an operational environment.

Considering the weight associated with equipment and weaponry, by functions performed during military operations, the aim is to analyse and compare the proportion of weight carried compared to the weight of the soldier, as well as whether there are significant differences between the different functions, characterising the associated level of overload. Such an approach takes into account that carrying backpacks with excessive weight significantly reduces the efficiency and speed of the soldier in the performance of his work, increasing the risk of injury and that the load should not exceed 30% of the soldier's body weight during operations (Assault Load) [4, 7, 8].

The sample consisted of 180 soldiers from the Portuguese Army, belonging to the 6th to 12th National Forces deployed to the Central African Republic from 2020 to 2023 (Fig. 1). Military personnel were grouped according to their roles within the Infantry Squad. The roles are Commander (or Leader), Designated Marksman, Automatic Rifleman, Driver and Rifleman.

Fig. 1. The Central African Republic is the theatre of operations (TO) under consideration (adapted from: exercito.pt).

The variables of interest are defined as the percentage of weight carried (PWC) compared to the weight of the soldier and the proportion of soldiers under stress (P). The data was analysed using various statistical methods, implemented using IBM SPSS Statistics 28 software, namely descriptive statistical analysis, essentially in the information summary phase to characterise the load carried by military personnel according to their functions. Statistical inference techniques and probability models were considered to evaluate the effort of soldiers and understand their proportion under stress/workload during peacekeeping operations in an international environment. Point estimation, interval estimation and implementation of various hypothesis tests for the parameters of interest (mean value, median value, and proportion) are considered, taking into account the enunciated problem, with a significance level of 5% [23]. Resorting to parametric statistical methods for hypothesis testing development, considering the T-Student test and also non-parametric statistical methods, using the Kruskal-Wallis test, with multiple pairwise comparisons, the study is performed in three main phases [23]. First, considering the average percentage of weight carried (PWC) by soldiers, according to their function, during peacekeeping operations in an international environment, and second, a comparison, in terms of medians, between these different functions follows [5]. Then, the overall characterisation of the force is also considered to understand the actual proportion (P) of military personnel under stress/workload, with a threshold of 40% of their weight carried.

3 Results and Analysis

The 180 Portuguese Army soldiers that constitute our sample are all male, who had served in the 6th, 7th, 8th, 10th and 11th National Detached Forces (NDF), the majority of whom were non-commissioned officers (84.5%), sergeants (10.5%) and officers (5%). All the respondents have more than one year of service, with between 2 and 25 years of service. Regarding participation in the NDF, most respondents were part of the 8th NDF (64.1%). Regarding the number of NDF participations, most respondents were only present in one NDF (61.9%).

Regarding the function in the NDF, most participants were as follows: Rifleman (28%), Automatic Rifleman (18.2%); Designated Marksman (18.2%), Driver (15.5%) and Commander (Company, Platoon, Squad, Team) (13.3%).

The participants were between 20 and 44 years old, with an average of 25.28 years and a standard deviation (SD) of 4.19 years, with most respondents aged between 22 and 25 years, corresponding to 64.1% of the entire sample. Weight was between 63kg and 110kg, with an average of 80.08kg and a SD of 8.99, with most participants weighing between 72 kg and 80 kg (52.5%). The height of the soldiers ranged from 163 cm to 190 cm, with an average of 175.58 cm and an SD of 5.53 cm, with most participants being between 175 cm and 180 cm (38.7%). Regarding years of service, there was a sample range of 23 years, between 2 and 25 years of service, corresponding to an average of 5.27 years with a SD of 4.14 years, with the majority of participants having four years of service (33.7%).

3.1 Effort Characterisation

The effort is characterised for each function in terms of the weight of equipment and weaponry in each role. An example of the weight carried is given for the Rifleman in Fig. 2.

Based on the data collected through the questionnaire survey, Table 1 was drawn up to analyse the equipment associated with each function, which provides the total weight carried by the soldiers (see Fig. 3).

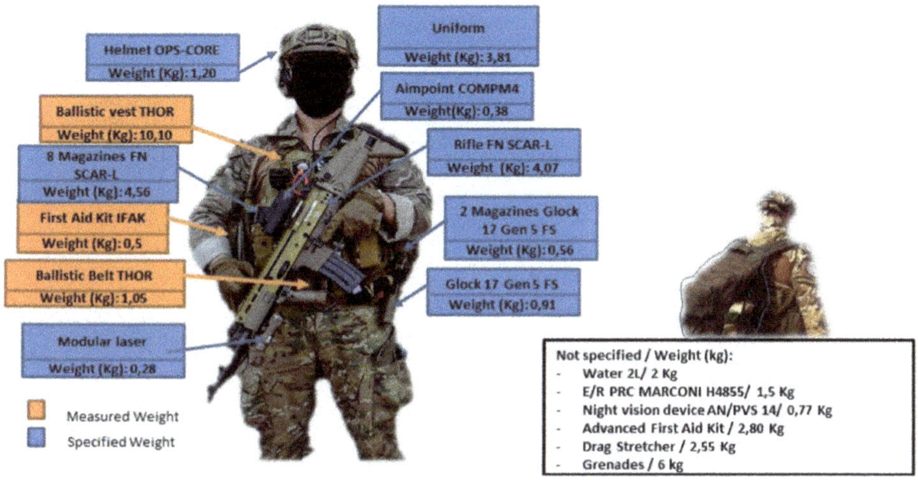

Fig. 2. Weight carried by a Rifleman.

3.2 Military Load

The load that military personnel are subjected to has been increasing over time, but the load should not exceed 30% of the military person's body weight.

According to different functions, the carried weight varies between 29,65 kg and 41,48 kg, with a soldier's mean weight of 80,1 kg. The mean weight of soldiers in each military function varies between 79.36 kg and 82.68 kg (Figs. 3 and 4). According to the study's first objective and taking into account the hypotheses H0: The mean percentage of weight carried (PWC) equals 30% vs H1: The mean percentage of weight carried (PWC) is higher than 30%, the overload level is analysed.

Resorting to T-student test, the null hypothesis is rejected ($t\_obs = 24{,}656$; p-value $< 0{,}001$) and the mean percentage of weight carried in global (PWC) is considered higher than 30%. When the T-student test is applied considering each of the military functions, the conclusion is similar. The results are presented in Table 2 including the 95% confidence interval for the mean percentage of weight carried in each function. Note that in the conducted test, the assumptions of normality of distributions were validated.

Considering the non-parametric statistical test for the null hypotheses H0: The median value of the percentage of weight carried is the same across functions; this one is rejected (p-value $< 0{,}01$), and when we go through the pairwise comparisons between functions stand out that the Automatic Rifleman function and the Rifleman function have a higher level of overload and the other three functions (Commander, Driver and Designated Marksman) are similar to each other (Fig. 5). However, it turns out that the level of overload within the infantry squad in an operational environment is high in all the functions.

Table 1. Equipment associated with each function.

Equipment	Designated Marksman (n= 33)	Automatic Rifleman (n=38)	Driver (n=28)	Commander (n=28)	Rifleman (n= 53)
Security	Ballistic helmet; Ballistic waistcoat; Glock 17 Gen 5 FS pistol; Ballistic belt; Communications equipment.	Ballistic helmet; Ballistic waistcoat; Glock 17 Gen 5 FS pistol; Ballistic belt; Communications equipment.	Ballistic helmet; Ballistic waistcoat; Glock 17 Gen 5 FS pistol; Ballistic belt; Communications equipment.	Ballistic helmet; Ballistic waistcoat; Glock 17 Gen 5 FS pistol; Ballistic belt; Communications equipment	Ballistic helmet; Ballistic waistcoat; Glock 17 Gen 5 FS pistol; Ballistic belt; Communications equipment.
Armament	FN SCAR-H Assault rifle	ML FN Minimi 5.56 MK3	FN SCAR-L rifle	FN SCAR-L rifle	FN SCAR-L rifle
Backpack material	Magazine/Belt; Zip ties; Batteries	Magazine/Belt; Zip ties; Batteries; Offensive grenades; Smoke grenades	Magazine/Belt; Zip ties; Batteries; First Aid Kit; Flashbang Grenades	Magazine/Belt; Zip ties; Batteries	Magazine/Belt; First Aid Kit; Offensive grenades; Smoke Flashbang grenades
Magazine	5	5	5	5	8

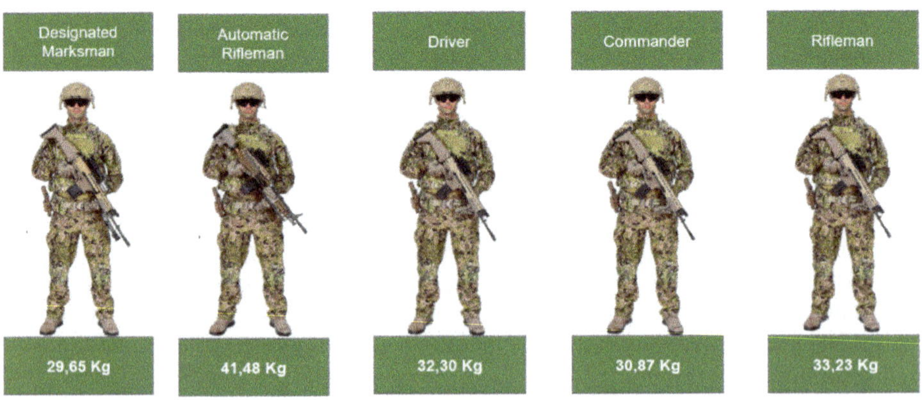

Fig. 3. Carried weight in each military function (image adapted from: warriors.pt).

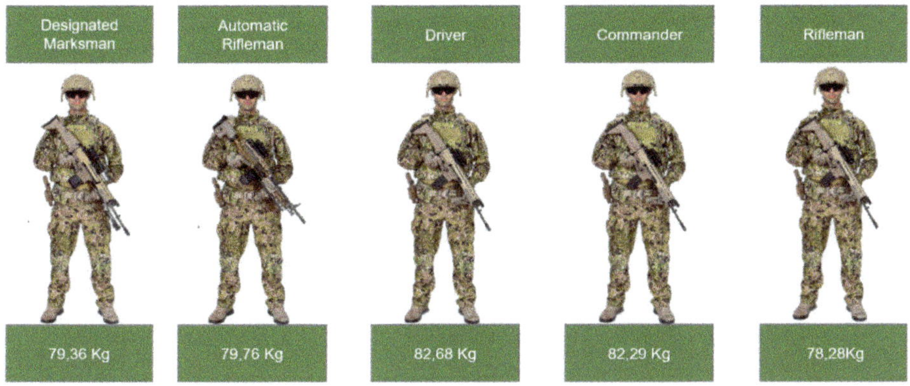

Fig. 4. Mean weight of each soldier in each military function (image adapted from: warriors.pt).

Table 2. T-Student test associated with each function.

Equipment	Designated Marksman (n = 33)	Automatic Rifleman (n = 38)	Driver (n = 28)	Commander (n = 28)	Rifleman (n = 53)
T value/p-value	t_obs = 9,9; p < 0,001	t_obs = 26,9; p < 0,001	t_obs = 11,9; p < 0,001	t_obs = 10,7; p < 0,001	t_obs = 22,3; p < 0,001
PWC Mean	37%	52%	39%	38%	43%
Sd Deviation	4,6%	5,1%	4,2%	3,9%	4,2%
95% Confidence Interval	**36,3–39,6%**	**50,4–53,7%**	**37,9–41,2%**	**36,5–39,5%**	**41,8–44,1%**

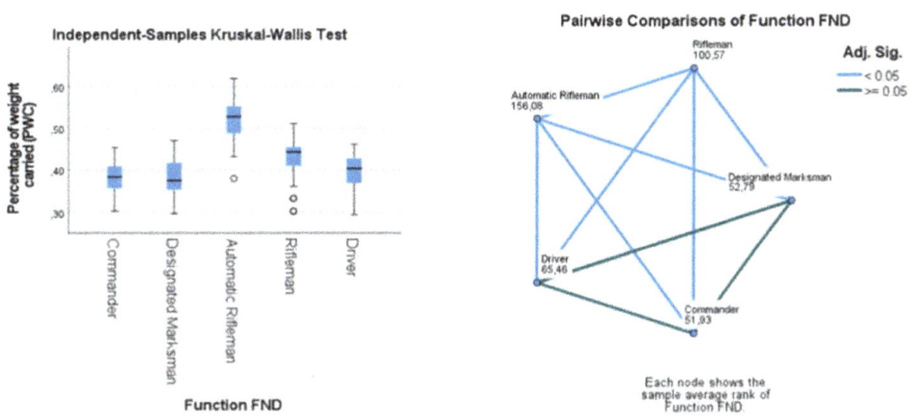

Fig. 5. Boxplot relative to the percentage of weight carried by function (left) and pairwise comparisons between functions (right).

3.3 Military Workload

The overall characterisation of the National Force was also taken into account in order to understand the real proportion of soldiers under stress/ workload, with a threshold of 40% of the soldier's body weight in relation to the weight carried.

Regarding the proportion of soldiers under strain, 65,6% of analysed military soldiers from the Portuguese Army are in stress/workload.

The confidence intervals for this parameter, according to each function, are presented in Table 3.

Table 3. Proportion of military workload associated with each function.

Equipment	Designated Marksman (n = 33)	Automatic Rifleman (n = 38)	Driver (n = 28)	Commander (n = 28)	Rifleman (n = 53)
Sample Proportion	33%	**97%**	54%	32%	**87%**
95% Confidence Interval	17%-49%	**92%-100%**	35%-72%	15%-49%	**78%-96%**

4 Conclusions

As for the equipment transported by each soldier, the most significant differences are related to the weapon, wish varies for each function, and ammunition. Based on this, the Automatic Rifleman and the Rifleman exert the most significant effort, with 52% and 43%, respectively. In fact, all soldiers are transporting a load above the load limit for the developed operations (under 30% of BW) and are very close to the limit defined for load bearing for trained personnel during specific tasks (under 40%). This increased load influences soldiers' mobility and awareness, increasing the risk of injury and implying substantial limitations to the development of the Armed Forces mission, and can become a burden to society in the long term.

Therefore, the development of lower limb exoskeletons that can enhance the capabilities of military personnel to meet all operational needs becomes evident. Some of the highlighted needs are associated with load bearing both during walking and during manual handling of loads. These essential tasks are way ahead of the development of lower limb exoskeletons, capable of supporting their user's transported load during operations in any environment or conditions. Two of the most essential features of such systems are the ability to maintain mobility and range of motion and its integration with the soldier system.

Future research should analyse other deployed Contingents since the operational environment and mission typologies significantly vary and may impact the exoskeleton's needs and requirements.

Acknowledgements. This work is funded by national funds through the project ELITE2 – Enhancement LITe Exoskeleton, an Army research and development project of the Military Academy Research Center.

References

1. Giles, G., Hasselquist, L., Caruso, C.M., Eddy, M.D.: Load carriage and physical exertion influence cognitive control in military scenarios. Med. Sci. Sports Exerc. **51**(12), 2540–2546 (2019). https://doi.org/10.1249/MSS.0000000000002085
2. Orr, R. et al.: Soldier load carriage, injuries, rehabilitation and physical conditioning: An international approach. Int. J. Environ. Res. Public Health **18**(8) (2021). https://doi.org/10.3390/IJERPH18084010
3. Schuh-Renner, A., et al.: Risk factors for injury associated with low, moderate, and high mileage road marching in a U.S. Army infantry brigade. J. Sci. Med. Sport **20**, S28–S33 (2017). https://doi.org/10.1016/J.JSAMS.2017.07.027
4. Bachkosky, J., et al.: Lightening the Load. Technical Report NRAC-07–02 (2007). https://www.onr.navy.mil/-/media/Files/history/nrac-reports/2007_rpt_lightening_the_load.ashx?la=en
5. US Army Research Laboratory, Methodologies for Evaluating the Effects of Physical Augmentation Technologies on Soldier Performance (2018). https://apps.dtic.mil/sti/pdfs/AD1057611.pdf
6. Knapik, J.J., Reynolds, K.L., Harman, E.: Soldier load carriage: historical, physiological, biomechanical, and medical aspects. Mil. Med. **169**(1), 45–56 (2004). https://doi.org/10.7205/MILMED.169.1.45
7. Andersen, K. A. , Grimshaw, P. N. , Kelso, R. M. and Bentley, D. J.: Musculoskeletal lower limb injury risk in army populations. Sport. Med. – Open. **2**(1) (2016). https://doi.org/10.1186/S40798-016-0046-Z
8. Department of Defense, MIL STD 1472 G - Design Criteria Standard - Human Engineering (2019). http://everyspec.com/MIL-STD/MIL-STD-1400-1499/MIL-STD-1472G_39997/
9. Birrell, S.A., Hooper, R.H., Haslam, R.A.: The effect of military load carriage on ground reaction forces. Gait Posture **26**(4), 611–614 (2007). https://doi.org/10.1016/J.GAITPOST.2006.12.008
10. Knapik, J., Reynolds, K., Santee, W.R., Friedl, K.E.: Load Carriage in Military Operations a Review of Historical, Physiological, Biomechanical, and Medical Aspects
11. World Health Organization, World report on Ageing and Health (2015). https://apps.who.int/iris/bitstream/handle/10665/186463/9789240694811_eng.pdf?sequence=1
12. World Health Organization, 'International Perspectives on Spinal cord injury, World Health Organization (2013). http://www.who.int/mediacentre/factsheets/fs384/en/
13. Farris, D.J., et al.: A systematic literature review of evidence for the use of assistive exoskeletons in defence and security use cases. Ergonomics **66**(1), 61–87 (2023). https://doi.org/10.1080/00140139.2022.2059106
14. Mudie, K., Billing, D., Garofolini, A., Karakolis, T., LaFiandra, M.: The need for a paradigm shift in the development of military exoskeletons. Eur. J. Sport Sci. **22**(1), 35–42 (2022). https://doi.org/10.1080/17461391.2021.1923813
15. Proud, J.K., et al.: Exoskeleton application to military manual handling tasks. Hum. Factors **64**(3), 527–554 (2022). https://doi.org/10.1177/0018720820957467
16. Proud, J.K. et al.: Exoskeleton Application to Military Manual Handling Tasks. Hum. Fact., 18720820957467 (2020), (Accessed Jan 06 2022). http://www.ncbi.nlm.nih.gov/pubmed/33203237

17. Quinto, L., Pinheiro, P., Goncalves, S.B., Roupa, I., Simões, P., Tavares da Silva, M.: 'nalysis of a passive ankle exoskeleton for reduction of metabolic costs during walking, Def. Technol. (2023). https://doi.org/10.1016/J.DT.2023.11.015
18. Dollar, A.M., Herr, H.: Lower Extremity exoskeletons and active orthoses: challenges and State-of-the-Art. IEEE Trans. Robot. **24**(1), 144–158 (2008). https://doi.org/10.1109/TRO.2008.915453
19. Pons, J.L.: Wearable Robots: Biomechatronic Exoskeletons. Wiley, Hoboken, N.J (2008)
20. de la Tejera, J.A., Bustamante-Bello, R., Ramirez-Mendoza, R.A., Izquierdo-Reyes, J.: Systematic review of exoskeletons towards a general categorisation model proposal. Appl. Sci. **11**(1), 1–25 (2021). https://doi.org/10.3390/app11010076
21. Crowell, H.P., Park, J.-H., Haynes, C.A., Neugebauer, J.M., Boynton, A.C.: Design, evaluation, and research challenges relevant to exoskeletons and exosuits: a 26-year perspective from the u.s. army research laboratory. IISE Trans. Occup. Ergon. Hum. Factors **7**(3–4), 199–212 (2019). https://doi.org/10.1080/24725838.2018.1563571
22. Quinto, L., Pinheiro, P., Gonçalves, S., Ferreira, R., Roupa, I., da Silva, M.T.: Development and functional evaluation of a passive ankle exoskeleton to support military locomotion. Adv. Military Technol. **17**(1), 79–95 (2022)..https://doi.org/10.3849/aimt.01536
23. Casella, G., Berger, R.: Statistical inference. Duxbury. Pacic Grove, CA (2002)

Flood Vulnerability of Transport Infrastructure: Classification of Construction Types, Structural Condition and Adaptation Measures to Reduce Vulnerability

Regine Ortlepp(✉) and Frank R. Beyer

Leibniz Institute of Ecological Urban and Regional Development, Dresden, Germany
r.ortlepp@ioer.de

Abstract. The study aims to develop a method for characterizing the vulnerability of transportation infrastructures such as roads and railways, starting with an analysis of damage patterns from flooding events. It classifies damage patterns based on exposure considerations and systematically categorizes types of damage. The vulnerability assessment involves classifying various construction types, emphasizing the inclusion and evaluation of structural conditions. Relevant factors such as terrain connection are also taken into account. Initially, the common building types were categorized according to their susceptibility to flooding and assigned to technical hazard classes. Construction-type-specific criteria were developed to assess structural conditions, enabling assignment to corresponding classes. Vulnerability and condition classes are methodically related, resulting in the assignment to a vulnerability class. The combined analysis of construction-type susceptibility and structural condition reveals that both an unfavorable construction type and impairments to the structural condition can increase vulnerability. Mapping vulnerability classes provides insights into potential damage locations, addressing the need to reduce vulnerability. Beyond selecting less susceptible construction types in the future, the study emphasizes the potential to avoid condition impairments. The findings offer specific recommendations for action.

Keywords: transport infrastructures · construction types · vulnerability assessment · susceptibility · flood

1 Introduction

Effective flood protection for transportation infrastructures such as roads and railway lines is crucial to ensure the functionality and safety of these vital systems. The escalating challenges posed by climate change and more frequent extreme weather events have underscored the need to develop appropriate protective measures. Despite this urgency, there is currently limited knowledge, particularly concerning the vulnerability of these infrastructures to flooding events. Small and medium-sized municipalities are particularly affected by this problem, as they often lack financial and human resources.

J. C. Matos et al. (Eds.): IPW 2024, LNCE 494, pp. 157–166, 2024.
https://doi.org/10.1007/978-3-031-60271-9_13

A comprehensive systematic literature review on the topic of increasing transportation infrastructure resilience to flooding events was conducted by Watson and Ahn [1] in the year 2022. The main research categories identified in this literature review predominantly focus on the analysis of flood risk, implementation of real-time flood forecasting, investigations into the impacts of floods on transportation infrastructures, vulnerability analyses of transportation infrastructures, and measures for responding to and preparing for flood events. However, limited attention has been directed towards vulnerability analyses specifically for roads, with most studies primarily considering larger geographical scales. In the context of transportation infrastructures, finer-grained information such as construction type and road condition is crucial for assessing vulnerability, as these factors significantly influence the extent of damage. However, the spatial resolution of studies reported in the literature is generally insufficient for capturing such detailed information. Consequently, the models employed in these studies are to be regarded as highly generalized. Very isolated regional studies have provided initial findings on damage patterns and risk factors [2, 3]. In these studies, characteristic damage patterns and associated damage mechanisms were identified and systematized, and damage types were classified on this basis. These are dependent on the intensity of the impact and the specific erosion stability of the construction, particularly of the surface layers. Unfortunately, there is currently no recognizable systematic and uniformly regulated procedure for data collection and the data sets are scattered decentrally, often inaccessible or difficult to access and therefore not quantifiable without further ado [2]. As rivers do not stop at national borders, the problem of data availability and data heterogeneity becomes even greater when planning cross-border adaptation measures. However, a comprehensive method for the systematic characterization of vulnerability that is applicable both across borders and for small and medium-sized communities with limited resources and data availability is lacking. Therefore, a simplified methodology is needed that allows for quick and cost-effective pre-scoring of vulnerability hotspots based on visually gathered qualitative data. A more precise and more costly engineering analysis using also higher data quality then only needs to be carried out locally for the hotspots identified in this way.

To address this gap, this study is dedicated to developing such a method for capturing the diverse influences on vulnerability with a focus on a simplified classification. In this context, it is crucial to formulate a methodology that not only addresses damage patterns but also focuses on the structural and preservation condition. The aim is to highlight particularly vulnerable areas to identify locational needs for intervention. Because the necessary quantitative data - e.g. on the thickness of the layers - is often not available, the pre-scoring method must be capable of operating on the basis of a qualitative visual assessment. To achieve this, we had to identify key influencing parameters and to link them methodically. Thereby, we needed to consider various influencing factors, such as construction type, condition characteristics, and the nature of terrain connectivity. This method enables the identification of areas with increased vulnerability or heightened damage potential. Thus, this investigation intervenes at a critical juncture to gain informed insights and strategically design future measures to minimize the vulnerability of these transportation infrastructures. For instance, proposals for damage-mitigation measures may be developed. The variations in typical superstructure construction types (cobblestone, asphalt layers, natural stone slabs, etc.) suggest that appropriate differentiations are necessary when formulating recommendations for action.

2 Method for Characterizing Vulnerability

The methodology developed in this study comprises a 4-step approach that aims to assess the vulnerability of transport infrastructure to flood events.

In a first step, the mechanisms underlying the damage caused by flood events are analyzed. The damage patterns are then classified, followed by a systematization of the different types of damage. The second step focuses on the classification of different construction types based on their susceptibility to flooding. The third step involves the assessment of the structural condition of these building types and assigning condition classes. Building on the information gained in the previous steps, susceptibility and condition class assessments are integrated in the fourth step in order to characterize the overall vulnerability of transport infrastructures.

The susceptibility, condition and vulnerability classes can ultimately be visualized on maps for all road sections. This spatial visualization enables clear and accessible identification of road sections with increased vulnerability, providing valuable insights for targeted interventions and mitigation measures.

2.1 Damage Patterns and Mechanisms

For the analysis of damage patterns, an initial step involved acquiring damage documentation and capturing damage images during relevant events. While damage patterns for analysis can be sourced from event analyses of flooding incidents, documentary literature, and press reports, they often appear in the form of highlights, showcasing particularly severe damages. However, for systematic analyses, it is necessary to obtain an overview of potential damage patterns that is as representative as possible. Therefore, we expanded this foundational set of damage patterns to encompass a broader spectrum, including less drastic extents of damage.

During our research efforts, we established direct contact with residents, enabling the extensive collection of data, photos, and press reports related to a flood event in 2002 in Schlottwitz, Germany, along the Müglitz. The 257 photographs depicting flood situations and damages to buildings, water bodies, and transportation infrastructure were digitized. Another opportunity to capture damage information arose during a heavy rainfall event in 2018 in Vogtland, Germany. This led to flash flooding and the overflow of the Weiße Elster. The subsequent day, we conducted an on-site visit to the affected area, documenting numerous flood damages. A total of 227 photographs were taken and mapped within the study area.

Table 1. Various types of damage to asphalt roads due to transverse flooding.

No.	Damage pattern	Flooding situation	Damage	Location, source
(a)		Flash flood due to heavy rainfall, localized flooding in transverse direction, road position in the cutting	Minor damage to the embankment due to erosion, deposition of silt	*)
(b)		Flash flood due to heavy rainfall, localized flooding in transverse direction, road position in the cutting	Significant damage to the verge due to receding erosion	*)
(c)		Flash flood due to heavy rainfall, localized flooding in transverse direction, road position in the cutting	Significant damage to the embankment and substructure due to receding erosion with partial collapse of the carriageway	*)
(d)		Flooding in transverse direction due to overflowing of the Weiße Elster, road in low embankment position	Superficial damage to the surface layer due to bond failure between the base and surface layers and minor damage to the embankment due to erosion	*)
(e)		Flooding in transverse direction due to overflowing of the Müglitz, road position at ground level	Complete destruction of the carriageway and substructure due to erosion	**)

*) On-site inspection in Vogtland, source F. R. Beyer/IOER Media

**) Flood event 2002 in the Müglitz valley (Schlottwitz), source D. Morgenstern/IOER Media

The analysis of damage patterns aims to identify the damage mechanisms leading to the ultimate manifestation of structural impairments. Based on these findings, correlations between specific characteristics of construction and the resulting damage mechanisms are examined. Table 1. illustrates various damage patterns on asphalt roads due to cross-flow flooding. In all five cases, the roads are constructed with asphalt, and in each instance, flooding occurred perpendicular to the roadway axis due to local conditions. However, the damage patterns vary significantly, even with comparable terrain conditions. Particularly, damage patterns (a), (b), and (c) appear as distinct stages of the same process, where differences in the extent of damage are attributed to variations in flow velocity, water level, and/or duration of exposure.

Previous analyses also suggest that the condition concerning various potential pre-damages to the structural integrity of the road network is a significant criterion for vulnerability. Comparing different construction types indicates a tendency wherein less intricate surface structures exhibit greater resistance to erosion. Naturally, cobblestone surfaces are more prone to erosive damage compared to monolithic asphalt layers. However, if monolithic layers are fragmented due to extensive cracking (considered a deterioration of condition), the intricacy, in this sense, increases, consequently raising susceptibility to erosion with the worsening of the state.

2.2 Classification of Construction Types

To address effective adaptation measures for transport infrastructures, a practical determination of existing vulnerabilities to flooding events should be established. Vulnerability fundamentally arises from the construction-specific susceptibility and the structural condition in which the transport infrastructure is situated. Impairments to the structural condition can increase vulnerability.

To achieve this, common construction types were initially compiled and classified based on their susceptibility to flooding. Paved surfaces have a small, regular or irregular structure with a high proportion of joints. The weight of the individual paving stones is generally low. In many cases, paving construction methods have a low surface bond (e.g. unbound mosaic paving). The roughness of the surface layer surface increases turbulence and hydrodynamic pressure fluctuations. These four properties consequently reduce the characteristic erosion resistance of the paving surface, so that erosion processes already start at lower flow velocities. Sufficient erosion resistance of the individual block, which is based on the verification of erosion resistance taking hydrodynamic pressure fluctuations into account, is a necessary condition for the stability of the surface layer. Loosening of the paving block bond as a result of exposed joints has a detrimental effect on erosion resistance [3]. Mortared joints have a stabilizing effect in this respect. To put it simply, the higher the proportion of joints in the surface layer, the greater the susceptibility to hydrodynamic effects. To put it simply, the higher the proportion of joints in the surface layer, the lighter the stones and the weaker the interlocking joints, the greater the susceptibility to hydrodynamic effects. Unbound base layers are even more susceptible to hydrodynamic loads due to loosening or the rearrangement of the material matrix as a result of erosion and suffusion processes.

Figure 1 illustrates the engineering assignment of construction types to susceptibility classes, which is based on this qualitative hydrodynamic assessment. The formation of the surface layer is used as the characteristic for assessing the construction type. This approach acknowledges the practical limitations in obtaining information, particularly regarding binder layers, base layers, frost protection layers, or sub-base layers during mapping activities. Additionally, it is assumed that the structural composition of the upper layer is at least tendentially related to the type of surface layer.

The construction types with monolithic surface layers (asphalt and concrete surface layers, susceptibility class 1) are considered the least susceptible. Following these are

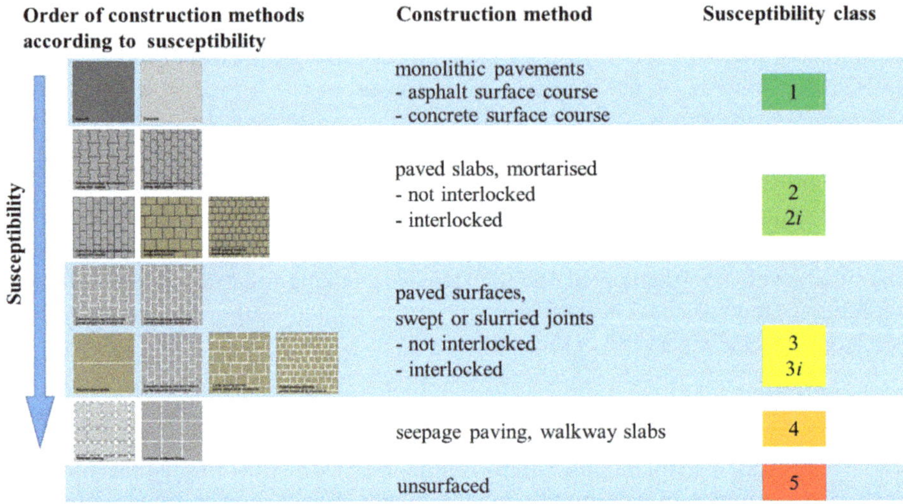

Fig. 1. Classification of construction methods (in relation to the surface layers) according to susceptibility and assignment to susceptibility classes based on hydraulic considerations in respect of the stability of the surface layer in [3].

construction types with paved surface layers featuring mortared joints, and subsequently, paved surface layers with swept or slurry-filled joints. In both cases, with form-fitting interlocking, it can be assumed that the interlocking effect, when the joint condition is diminished, leads to comparatively lower vulnerability than without it. Therefore, the aspect of interlocking (2*i* and 3*i*) is potentially considered in the respective susceptibility classes 2 and 3. Segmented surface layers with low height, such as sidewalk paving or finely textured surfaces with wide joints and thus higher susceptibility to erosion compared to susceptibility class 3, are assigned to susceptibility class 4. Unpaved traffic paths fall into the highest susceptibility class 5.

2.3 Inclusion of the Structural Condition

Identifying the structural condition is recognized as a significant influencing parameter for vulnerability, as previously mentioned. This is achieved by assigning it to a condition class. Damages such as cracks or existing openings increase turbulence and hydrodynamic pressure fluctuations; to put it simply, they also lead to a quasi-increase in the "joint" proportion, which increases susceptibility. However, the decisive criteria for this differ depending on the type of surface layer, which is why a specific designation of condition characteristics must be made separately for the different types of surface layer. The damaged areas are very localized in nature, which is why the approach of using condition classes for entire road sections seems unsuitable here. Instead, an inspection of the road section with a visual survey is more appropriate in order to detect the local types of damage and assign the appropriate condition class for this scoring approach. In Table 2, condition characteristics for assignment to condition classes are described on a scale ranging from 1 to 5.

Table 2. Construction method-specific assessment criteria. for the structural condition of road infrastructures.

Construction method	Condition class				
	1	2	3	4	5
Monolithic pavements					
- asphalt surface layer	undamaged	integrated patches	attached patches, slight cracks (localized, not extensive), separation cracks only isolated	significant cracks (covering the entire surface), frequent separating cracks, individual break-outs	extensive cracking, >20% break-outs
- concrete surface layer	undamaged	integrated patches, superficial binder discharge	attached patches, slight cracks	significant cracks, surface damage, individual break-outs	severely eroded cracks, numerous break-ups
Non-monolithic slabs (paving slabs and similar)	undamaged	low discharge of joint material	extensive discharge of joint material, uneven surface	individual missing paving stones, extensive removal of joint material	>10% missing paving stones, almost complete removal of joint material

2.4 Assessment of Vulnerability

Ultimately, the determination of vulnerability involves the integration of both the susceptibility inherent in the construction method and the prevailing structural condition. The cross table, as depicted in Fig. 2, provides a representation of the assignment to vulnerability classes, utilizing also a scale ranging from 1 to 5. While vulnerability inherently aligns with susceptibility, it is crucial to note that a compromised structural condition has the potential to negatively impact vulnerability. The scoring approach was discussed with experts as part of the joint Czech-German cross-border project STRIMA-II with regard to their valuation of applicability.

3 Map Visualization

The visualization of susceptibility, structural condition, and vulnerability is carried out using a pilot area, exemplified here by the municipality of Frýdlant (Czech Republic). Practical applications for road transportation infrastructures in pilot scenarios are primarily reliant on on-site mappings, as databases containing necessary information for classification are usually either non-existent or inaccessible.

Susceptibility class \ Condition class	1	2	3	4	5
1	1	2	2	2	3
2	2	2	3	3	5
2*i*	2	2	3	4	5
3	3	3	3	4	5
3*i*	3	3	4	5	5
4	4	4	5	5	5
5	5	5	5	5	5
	Vulnerability class				

Fig. 2. Vulnerability derived from the combination of the susceptibility of the construction method and the existing structural condition.

The municipality of Frýdlant is situated at the confluence of the Řasnice, a mid-mountain stream, and the Smědá, a mid-mountain river, thereby falling within the influence zone of two significant watercourses with regard to flood hazard. Figure 3a presents an aerial view of the Frýdlant locality overlaid with the floodplain areas of a 100-year flood, provided to us as GIS data by the Agentura Regionálního Rozvoje (ARR). The determination of the specific study area (outlined in yellow) was primarily based on these combined floodplain areas, with the immediate inclusion of adjacent roadways. Existing mapping data of the road network were intersected with the study area, partly updated and supplemented or adjusted, and segmented into study sections of approximately 100 m in length where possible. To systematically conduct on-site mapping, the study area was initially subdivided into four subareas, as indicated by colored highlights and Roman numerals in Fig. 3a.

For the determination and visualization of vulnerability, all pre-mapped road sections in the study area were physically inspected on-site. They were assigned susceptibility classes based on the visually identifiable construction type, and the respective condition class was determined according to the observed structural condition. In some of the pre-mapped segments, due to variations in construction types (for example, the transition from asphalt to cobblestone construction), a consistent classification could not be made. In these cases, the segments were appropriately subdivided during post-processing to accurately depict the actual situation.

Using color representation corresponding to Fig. 2, Fig. 3b-d visualize the determined susceptibility, condition, and vulnerability classes of the road network in the study area. For the determination and visualization of vulnerability, pre-mapped road sections in the study area were assigned susceptibility classes based on the visually identifiable construction type (Fig. 3b). Additionally, the respective condition classes were determined based on the structural condition (Fig. 3c). Segmentation occurred depending on changes in existing parameters. Vulnerability classes for all segments were determined

a. Aerial image of Frýdlant overlaid with the inundation areas of a 100-year flood, depiction of the pilot area with road network and division into sub-areas for on-site mapping.

b. Susceptibility of road infrastructure based on their construction method.

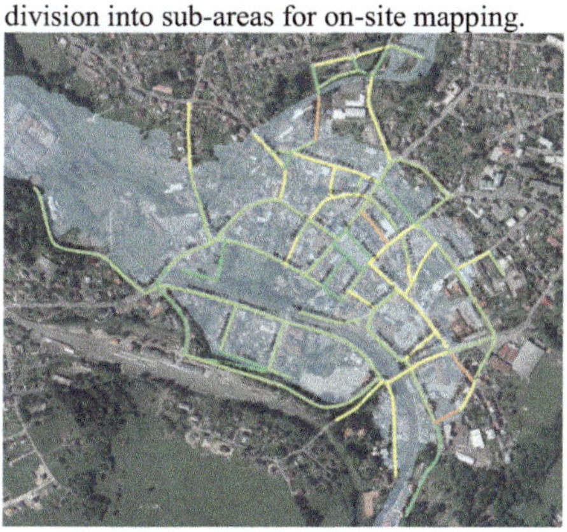

c. Determined condition classes of road infrastructure.

d. Determined vulnerability of road traffic routes based on the combination of susceptibility and structural condition.

Fig. 3. Map visualizations using the example of the pilot area in Frýdlant (Czech Republic).

from the respective combinations of susceptibility class and condition class. These were then added as attributes to the GIS data of the road network. Figure 3d visualizes the determined vulnerability classes of the road network in the study area. While many sections are generally perceived as having low vulnerability (vulnerability classes 1 and 2), some segments deviate noticeably. This is particularly concentrated in the area of the marketplace and the southern adjoining lanes with vulnerability class 3. Additionally, several other segments in the town center have vulnerability classes 4 or 5, and some roads outside or on the periphery of the 100-year flood zone exhibit vulnerability class 3.

4 Conclusions Regarding Adaptation to Reduce Vulnerability

As demonstrated by the results of the case study, the visualization of vulnerability classes within a map provides valuable insights for locating damage potential and, consequently, addressing corresponding needs for action to mitigate potential damages by reducing vulnerability.

Upon individual consideration of the two influencing factors, it is evident that impairments to the structural condition increase vulnerability, and consequently, their elimination constitutes a suitable measure for reducing vulnerability. However, the achievable extent is limited by the susceptibility of the respective construction type. In the selection of construction types for vulnerability reduction, it is fundamentally recommended to opt for those with a lower susceptibility class. Nevertheless, other aspects may be of competing significance, such as considerations related to heritage conservation or design requirements. In cases where vulnerability reduction cannot be achieved through the appropriate selection of construction type, the possible measures primarily involve preserving the building structure and rectifying impairments to the structural condition.

Our results indicate that there is often mitigation potential in avoiding impairments to the structural condition. These impairments typically result from the omission of maintenance measures of relatively low effort or improperly executed renovation or repair actions. For relevant construction types, concrete improvement proposals and recommendations for action can be provided based on this knowledge.

The proposed pre-scoring methodology can be applied across borders and helps small and medium-sized municipalities in particular to quickly assess the most vulnerable sections that require more detailed analysis.

References

1. Watson, G., Ahn, J.E.: A systematic review: to increase transportation infrastructure resilience to flooding events. Appl. Sci. **2022**(12), 12331 (2022)
2. Golz, S., Bohnenkamp, C.: Schadensanfälligkeit von Straßenverkehrsinfrastrukturen gegenüber Überflutung. In: Technische Universität Dresden, Institut für Wasserbau und technische Hydromechanik (ed.): Bemessung im Wasserbau - Klimaanpassung, Untersuchungen, Regeln, Planung, Ausführung. Dresdner Wasserbauliche Mitteilungen 58. Dresden: Technische Universität Dresden, Institut für Wasserbau und technische Hydromechanik, pp. 67–76 (2017)
3. Golz, S., Bohnenkamp, C., Heyer, T.: Überflutungsbedingte Schäden an Straßenverkehrsinfrastrukturen. Wasserwirtschaft **107**(5), 12–19 (2017)

Impact of Extreme Events on the Financial Performance of Infrastructure Projects

N. Acuña-Coll(✉) and M. Sánchez-Silva

Department of Civil and Environmental Engineering, Universidad de los Andes, Bogotá, Colombia
nm.acuna@uniandes.edu.co

Abstract. Developing and managing sustainable infrastructure in the face of climate change uncertainty is a significant challenge for governments and private entities. The increasing intensity and frequency of extreme events impact infrastructure systems, resulting in direct costs such as reconstruction expenses, as well as indirect costs such as environmental and social impacts, reduced productivity, and national/local economic consequences. This paper presents a comprehensive dynamic stochastic model to assess the financial performance of infrastructure projects subject to extreme events. The model allows for evaluating various project contractual models and diverse financial mechanisms. Likewise, it involves assessing the direct and indirect costs incurred throughout the life cycle, focusing on the impact of the increased frequency of extreme events caused by climate change. First, the paper reviews the common funding sources for infrastructure development and the traditional financial evaluations. Afterward, it explains how the model incorporates technical and financial aspects, including the mechanism to estimate direct and indirect costs. The model results show the importance of estimating indirect costs and their role in sustainability. Finally, it provides an example of the model in action with an actual infrastructure project. The results highlight the importance of considering indirect costs within a dynamic and stochastic financial evaluation. These costs are more pivotal than direct costs for evaluating the impact of extreme events on infrastructure. The presented work has important implications for evaluating sustainable infrastructure projects and improving risk allocation.

Keywords: Infrastructure Projects · Indirect Cost · Infrastructure Management · Climate Change · Extreme Events

1 Introduction

Based on the latest Intergovernmental Panel on Climate Change report [1], global warming is expected to increase from 2021 to 2040 due to rising CO_2 emissions in almost all scenarios and pathways. The report predicts that all regions worldwide will experience a higher frequency of climate hazards. The infrastructure systems of many countries are expected to face significant hazards and risks, such as extreme rainfall events that could result in flooding, increased frequency and magnitude of extreme sea-level

J. C. Matos et al. (Eds.): IPW 2024, LNCE 494, pp. 167–176, 2024.
https://doi.org/10.1007/978-3-031-60271-9_14

events, and alterations in floods, landslides, and water availability that may lead to severe consequences. Therefore, it is essential to consider the progressive risk increase when conducting financial evaluations of upcoming infrastructure projects. This consideration will affect the decision-making process and highlight the need to improve risk allocation.

It's crucial to conduct a comprehensive assessment of infrastructure projects that consider the increasing risks, the system's dynamic, and adaptability in project management and financial planning. Ulusoy and Hazır argue that new models are needed to optimize financial objectives while considering different uncertainties and risk factors [2]. Traditional financial evaluations, such as cost-benefit analysis or cost-effectiveness analysis, have limitations in evaluating system dynamics and adaptation [3]. Therefore, economic decision-making under uncertainty methods can be more realistic, though more complex. Some evaluations under uncertainty include Real Option Analysis, Robust Decision-Making (RDM), and Portfolio Analysis.

In the context of climate change, robust decision-making has been found to be the best assessment for capturing adaptability [5]. RDM identifies strategies that perform well across a range of plausible future scenarios, improving resilience against uncertainties. Yan et al. exemplify the use of Robust Decision Making (RDM) in climate change [4]. This paper proposes a stochastic financial evaluation model based on a robust decision-making method. It includes key factors that are usually studied separately in current research on infrastructure projects, such as financing methods, carbon pricing, increases in extreme events, dynamic systems, and adaptive management.

2 Infrastructure Finance: A Brief Review

2.1 Sources of Funding, Financing Methods and Carbon Pricing

Infrastructure projects' feasibility and long-term success are highly dependent on funding and financing methods. According to Grigg [6] the core pillars of financing include equity, debt through bonds, and loans from financial institutions. Equity represents investors' ownership stake in the project, but its availability can be limited. It can be challenging to attract private investors, especially for projects with uncertain revenue streams or extended gestation periods.

Debt financing is a way to inject capital into a project by obtaining bonds or loans from national, international, and development banks [7]. Although these instruments can provide significant funding, they come with interest payments and debt service requirements that can put pressure on project economics. Deciding between national and international banks requires complex negotiations, regulatory considerations, and currency risk management. Development banks may offer better terms, but accessing their funds often depends on meeting stringent criteria, affecting project viability.

The choice of financing for infrastructure projects has a significant impact on their dynamics. Some contractual options are Design-Build-Finance-Operate-Maintain (DBFOM), concessions, and fully public projects. DBFOM and concession models involve private sector participation, which shifts risks and responsibilities. While these models enhance efficiency, it is crucial to have robust regulatory frameworks to ensure public interest protection. However, fully public projects face funding constraints. It is essential to strike a balance between public and private involvement. In Public-Private

Partnerships (PPPs), Special Purpose Vehicles (SPVs) are key in ensuring efficient project management and risk allocation. Achieving a balance between private sector efficiency and public control is crucial for infrastructure projects.

Carbon pricing is a strategy that assigns a monetary value to greenhouse gas emissions and aims to reduce emissions and, in some cases, generate revenue for governments. Carbon Tax and Emission Trading System (ETS) are the two main approaches to incorporating emissions costs [8]. Carbon taxes impose a direct price on greenhouse gas emissions, incentivizing companies to transition to cleaner technologies and align financial choices with emission reduction objectives. On the other hand, ETS allocates a limited number of emission allowances to participating entities, allowing them to buy or sell allowances based on their level of emissions, creating a market-based mechanism for reducing emissions.

2.2 Traditional Financing Assessment Incorporating RDM

Cost-benefit analysis (CBA) is a widely used method for evaluating infrastructure projects, which assesses their financial viability and overall impact [3]. This analytical approach involves comparing the total costs of a project with its expected benefits, expressed in monetary terms. The project's profitability can be determined by using the Net Present Value or the Benefit-Cost ratio as an indicator. However, there are some issues with the traditional approach of CBA, which include the definition of the discount rate of cash flows and the need for more dynamic management.

Dittrich et al. argue that standard decision-making processes such as CBA are challenging to apply in the face of high uncertainty, as is the case with climate change adaptation [5]. However, it is beneficial to incorporate traditional financial assessment tools into frameworks for uncertainty, such as real option analysis, portfolio analysis, or robust decision-making (RDM). Among these methods, RDM has been identified as offering the most potential in adaptation appraisal [5]. RDM involves identifying key variables, creating diverse scenarios, and pinpointing critical factors through sensitivity testing. The approach emphasizes continuous learning and monitoring, allowing for adjustments based on real-world feedback. It offers a flexible framework to navigate complex scenarios, enhancing the likelihood of success in an ever-changing environment.

3 Comprehensive Dynamic Stochastic Model

Table 1 summarizes the main elements of the proposed dynamic stochastic financial model. in the programming language being used, each element is represented as a class. This separation is done based on the complexity of the algorithm as a whole. A main class calls other classes when required, and this class can be executed multiple times, generating a comprehensive scenario evaluation known as RDM. The following subsections provide a detailed explanation of the differentiating elements in our model.

Table 1. The main elements between the dynamic stochastic financial model for infrastructure projects

Element	Description	Main variables
Demand modeling	To generate possible demand realizations over time using probability, various growth functions, estimated growth rates, and population growth	Growth of the demand, coefficient of variation demand, coefficient of variation capacity, initial capacity
Benefits	To calculate the economic benefits that the project directly generates due to its operation	The interest rate for benefit cashflow (inflation), "Product" price for unit
Emissions	To estimate project emissions generated over the project's lifecycle and their economic value via carbon pricing	Emission factors, carbon cost, emission limit, carbon cost rate, excess emissions penalty
Extreme events	To generate extreme events with different return periods at the beginning and end of the asset's life cycle	Return period of events at time t = 0, return period of events at time t = 1, time required to go from t = 0 to t = 1, damage mean as a percentage of the system state, coefficient of variation damage, Exponent or power b
Insurance	To calculate insurance premiums and payments that cover the deductible in case of an extreme event	Deducible
Management	To simulate operation and maintenance strategies considers the evolution of system over time	Rate of maintenance occurrences (exponential), percentage invested in operation and maintenance, standard deviation maintenance sizes, standard deviation operation
Financial aspects	To estimate the cost of capital using the Capital Asset Pricing Model (CAPM)	The risk-free rate, the expected market return, and the investment's beta
Flexibility	To modify the design parameters over time based on the performance indicator evolution	Failure probability - updating criteria, forecast window for flexibility update, initial design time t = 1
Indirect costs	To estimate the indirect cost based on common stochastic indirect vs direct cost curves or direct input-output (IO) analysis	Infrastructure sector
Dynamic cash flow	To enable adaptive infrastructure management, predefined policies are implemented to continuously evaluate project financial status, including past, present, and estimated future	The time window for the analysis, time mission, total investment, investment distribution among sources, loan conditions, bond conditions, return conditions for investors, adaptive management policy

3.1 Emissions

Calculating emissions involves using an approximation of the life cycle impact assessment, which is the third phase of a life cycle analysis. Each phase of the project is assigned an emissions factor, which helps to relate the processes to the equivalent CO2 emissions generated by the project. Klöpffer et al. have highlighted the difficulty of including all the factors and indicate that specialized sciences develop the characterization models and characterization factors [9]. However, if the amount of CO2 in each stage is estimated in previous studies of the project or similar projects, it is possible to align the factors concerning cost processes. Our proposal for calculating emissions is based on this approach. Moreover, the model allows for the inclusion of carbon pricing in the project's cash flow through carbon taxes or credits. In the case of carbon taxes, it is possible to explore how the increase in the cost per ton of CO2 equivalent over time, as a consequence of regulations for decarbonization, impacts the project's return.

3.2 Extreme Events Modeling

Extreme events can be modeled in terms of their occurrence rate $\lambda(t)$, which can be constant or a function that increases due to the effects of climate change. In geophysical applications, it is expected to observe power-law interarrival times between extreme events [10]. This study modeled the occurrence rate by a power-law relationship $\lambda(t) = at^b + c\, dt$. we employ a numerical technique for generating random samples from a distribution characterized by a cumulative distribution function (CDF) derived from an integral of a rate function. First, the cumulative effect of events over time is modeled as follows:

$$\lambda(t) = \int_{t}^{t+x} \left(at^b + c\right)dt = \frac{a}{b+1} \cdot \left((t+x)^{(b+1)} - t^{(b+1)}\right) + c \cdot x \tag{1}$$

In order to ensure that the model is applicable to the specific application context, we define the parameters a and c in terms of the initial return period $T0$ and final return period $T1$, as well as the time between the beginning and the end $T01$. These parameters are defined as follows:

$$a = \frac{\left(\frac{1}{T1} - \frac{1}{T0}\right)}{T01^b}; c = \frac{1}{T0} \tag{2}$$

Including these parameters makes the model more adaptable, which is essential for simulating climate change and capturing the complexities of evolving environmental conditions. Then, the CDF is computed as:

$$F(x) = 1 - \exp(-\lambda) \tag{3}$$

The algorithm updates the lower limit of the integral in Eq. (1) with the time of discovered events, accurately capturing their temporal progression.

3.3 Indirect Costs

According to Hallegatte, research has used various economic models to estimate output losses. The Input-Output (IO) or Calculable General Equilibrium (CGE) models are the most commonly used models. These models describe the economy as a collection of economic sectors that interact through intermediate consumptions [11]. The models have demonstrated non-linearity in output losses, which can be extrapolated to the project level. If the damage to the infrastructure is minor, the repair costs (direct costs) may exceed the indirect costs. For instance, a temporary disruption to a road due to a small landslide may incur significant expenses for repairing the road itself, but indirect costs, such as lost productivity and delays in the delivery of goods, could be relatively minor. In cases where the damage is extensive, such as a major collapse in the infrastructure of a main road, it can significantly impact all sectors that rely on the connectivity of that road.

This study estimates indirect costs from an adjustment, considering the expected actual shape of the indirect vs direct cost curves (see [11]). We found that polynomial curves and a combination of linear and exponential components can be used to fit the curve. The point at which indirect costs exceed direct costs varies depending on the type of infrastructure and the redundancy of the system where it is located. For instance, in the case of a highway, if there are enough alternative routes to meet demand, a large disruption may not generate indirect costs greater than direct costs. Therefore, analyzing the data of the system and sectors involved is necessary to determine the final shape of the indirect costs versus the direct costs curve.

4 Illustrative Case

4.1 Case Description

In this study, we will utilize the available financial data of the Chad-Cameroon Pipeline case, one of Africa's largest public-private partnerships at the time, as an illustrative case to gauge the applicability of the developed model (see [12]). It is essential to highlight that the project information is openly accessible and can be used without restrictions. The project required up to \$3.7 billion in investments, and pipeline construction commenced in 2000, with completion in 2003.

The project has a time window of 28 years for analysis, which is the same as its estimated mission time. The discount rate of 9% was provided in the problem data. Negative demand growth was estimated, considering the reduced demand for petroleum products and the transition to renewable energy. The carbon emissions cost of \$20 per ton and estimated emissions factors for operation, maintenance, and construction are considered for emissions. For extreme events, there is an initial return period of 15 years and a direct cost of damage proportional to the percentage of damage. Insurance was assumed to cover up to 20% of the initial investment to cover damages caused by extreme events or unforeseen large maintenance. This insurance was not part of the original project but was included in the analysis to show its operation in the dynamic financial evaluation.

Regarding management, large unforeseen maintenance was simulated with a rate of maintenance occurrences of 0.1 (i.e., exponential). An operation according to the base cash flow of the project, including variability, was considered. The project does not illustrate flexibility, i.e., changes in its design or operational parameters, as a response to adaptability.

4.2 Financial Performance

The financial model proposed was utilized to estimate the profitability index (PI) of the private sponsors varying the impact of extreme events in the project. The impact of extreme events is defined as the total cost of the events divided by the initial investment in the project. In this model, all direct expenses caused by extreme events are assumed by the project's special purpose vehicle (SPV). The project can obtain a bank loan or issue bonds if it requires additional funds. As a last resort, the private sector may provide liquidity to the project. Figure 1 displays the probability distribution of the private sector profitability index for three different impacts.

Figure 1 illustrates how extreme events affect the returns of private investors. As the impact of these events increases, the expected return decreases, and the returns have greater uncertainty. If the cost of extreme events equals the total initial investment plus an extra 20%, the expected return from the private sector would be unacceptable. In practical terms, if only one extreme event occurs during the infrastructure's life cycle and is of that magnitude, it would be equivalent to building the infrastructure from scratch and incurring an additional 20% cost. This highlights the financial resilience required to absorb the impact of extreme events. Table 2 provides additional financial results for the impacts presented in Fig. 1.

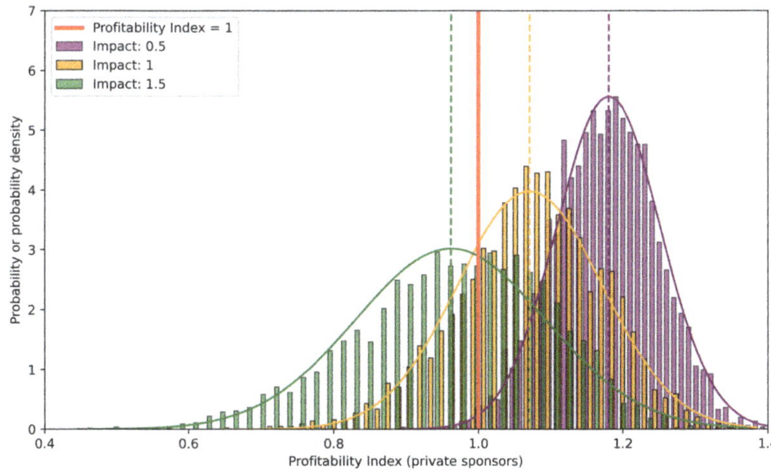

Fig. 1. Probability density of the private sponsors' profitability index for three different impacts of extreme events

Table 2 indicates that if the impact of events is 0.5 or 1, it is unlikely that a project will be refinanced through bond issuance, bank loans, or additional private-sector capital. However, if the impact level is 1.5, the probability of refinancing increases. The break-even point in the project, where total cost and total revenue are equal, is expected to increase by 40% if the impact of events goes from 0.5 to 1.5. This can affect investors' profitability by delaying their timing of returns and leading to other issues such as increased perceived risk, potential funding challenges, and the need for a reassessment of the overall feasibility of the project. Furthermore, examining the Debt Service Coverage Ratio (DSCR), we find that an increase in the impact of events from 0.5 to 1.5 would reduce the DSCR by around 4.4%. This information is essential for banking entities as it highlights the potential strain on the SPV's ability to meet its debt obligations during heightened event impacts.

Table 2 also displays the relationship between the direct and indirect costs of three reference impacts. The findings are consistent with the actual curves for indirect and direct costs. Direct costs are more significant for minor impacts, while indirect costs are more critical for large impacts. Significant events result in more extended infrastructure interruption times, leading to higher indirect costs. Remarkably, the estimated point of equality between indirect and direct costs for this project is 80% of the total initial investment, equivalent to an impact of 0.8. This basic estimate of indirect costs provides a broad perspective on the actual cost of extreme events. Usually, indirect costs are not included in the project flow, yet these costs are critical, as the results suggest. If the project is financially resilient, it could be partly accounted for by the project or the private sector.

Table 2. Expected financial results for different impacts

Impact	Profitability index (Private sponsor)	General refinancing probability [%]	Break-even point [years]	DSCR mean	Indirect cost/direct cost
0.5	1.18	0.0	9.33	2.52	0.34
1	1.07	0.0	12.84	2.47	1.57
1.5	0.96	0.4	15.07	2.41	3.57

The decision-making tool in Fig. 2 illustrates the probability of non-profitability for private sponsors, considering the expected impact of extreme events. For instance, if an investor is willing to accept a probability of losses of up to 10%, they must ensure that the expected costs of extreme events during the project's life cycle do not exceed 75% of the total initial investment cost (i.e., impact of 0.75). This tool can be applied to various projects to demonstrate their financial resilience in the face of extreme events. Note that the smaller the area under the curve, the better the project's ability to absorb the impact of events without negatively affecting the private sector. This visual representation enhances the interpretability of the analysis and provides a quick assessment of a project's overall financial resilience.

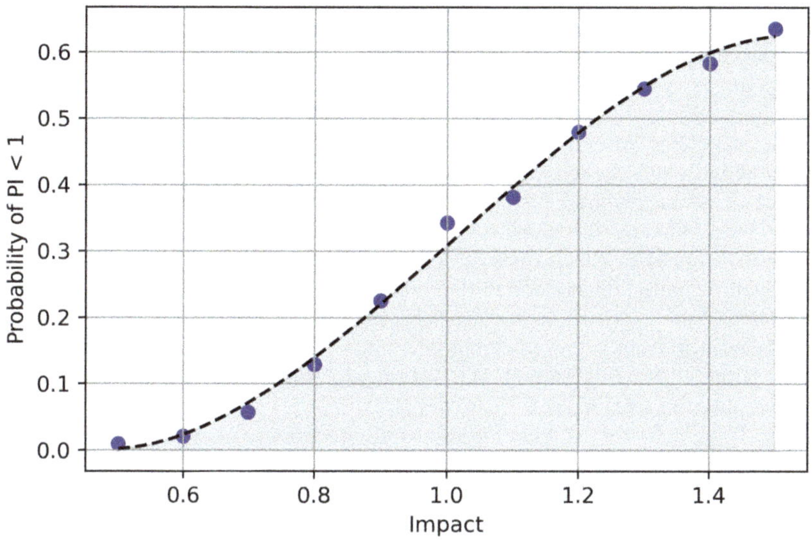

Fig. 2. Probability of private sponsors' profitability index less than 1 vs impacts of extreme events

5 Conclusion

Infrastructure systems are increasingly vulnerable to extreme events caused by climate change. However, traditional financial evaluations do not fully integrate system dynamics and adaptive management, making it difficult to evaluate the actual costs of such events. This paper aims to create a comprehensive stochastic financial model that simulates the behavior of infrastructure systems exposed to extreme events and relates it to their dynamic financial response. The model was used to determine the probability that the project will be unprofitable for the private investor, considering different levels of impact of extreme events. Additionally, the indirect costs of extreme events were estimated to provide a more complete picture of the project's financial impact. We propose to calculate a measure of financial resilience for infrastructure projects based on their results. The results showed that extreme events have an impact on the financial dynamic of a project, which is reflected in its indicators. Therefore, a conventional cost-benefit analysis may not be sufficient to reflect the actual decision-making process in infrastructure projects.

Acknowledgments. This research has received no external funding.

References

1. PCC Synthesis report - climate change 2023. An Assessment of the Intergovernmental Panel on Climate Change 335 (2023)
2. Ulusoy, G., Hazır, O.: An introduction to project modeling and planning. Springer (2021)
3. Watkiss, P., Hunt, A., Blyth, W., Dyszynski, J.: The use of new economic decision support tools for adaptation assessment: a review of methods and applications, towards guidance on applicability. Clim. Change **132** (2015). https://doi.org/10.1007/s10584-014-1250-9

4. Yan, D., Ludwig, F., Huang, H.Q., Werners, S.E.: Many-objective robust decision making for water allocation under climate change. Sci. Total Environ. **607–608**, 294–303 (2017). https://doi.org/10.1016/j.scitotenv.2017.06.265
5. Dittrich, R., Wreford, A., Moran, D.: A survey of decision-making approaches for climate change adaptation: are robust methods the way forward? Ecol. Econ. **122**, 79–89 (2016). https://doi.org/10.1016/j.ecolecon.2015.12.006
6. Grigg, N.S.: Infrastructure Finance: The Business of Infrastructure for a Sustainable Future (2011). https://doi.org/10.1002/9781118266182
7. Weber, B., Alfen, H.W., Staub-Bisang, M.: Infrastructure as an asset class: Investment strategy, sustainability, project finance and PPP, 2nd edn. Wiley (2016)
8. World Bank Carbon pricing dashboard: What is carbon pricing? (2022). https://carbonpricingdashboard.worldbank.org/what-carbon-pricing
9. Walter Klöpffer, K., Birgit Grahl, G.: Life Cycle Assessment (LCA) – A Guide to Best Practice, 1st edn.. John Wiley & Sons (2014)
10. Benson, D.A., Schumer, R., Meerschaert, M.M.: Recurrence of extreme events with power-law interarrival times. Geophys. Res. Lett. (2007). https://doi.org/10.1029/2007GL030767
11. Hallegatte, S.: The indirect cost of natural disasters and an economic definition of macroeconomic resilience, Policy Research Working Papers (2015)
12. World Bank Project Appraisal Document Chad/Cameroon Petroleum Development and Pipeline Project (2000)

Influence of Uncertainties on the Compound Rocking Failure Mechanism of Single-Nave Masonry Churches

Simon Szabó[1(✉)], Marco Francesco Funari[2], and Paulo B. Lourenço[1]

[1] Department of Civil Engineering, Institute for Sustainability and Innovation in Structural Engineering, University of Minho, 4800 - 058 Guimarães, Portugal
simon.szabo@civil.uminho.pt

[2] School of Sustainability, Civil and Environmental Engineering, University of Surrey, Guildford GU2 7XH, UK

Abstract. Façade overturning is one of the most common failure mechanisms observed in single-nave masonry churches when subjected to seismic action. The slender facades and lack of floor connections make these structures more susceptible to this mechanism than other masonry typologies. Several factors, including geometry, masonry patterns, mechanical properties, and loading conditions, influence these masonry churches' strength and displacement capacities. This study aims to assess the impact of various model parameters on the seismic assessment of single-nave masonry churches subjected to the compound rocking failure mechanism. It explores uncertainties that may arise due to incomplete surveys of the structures, and the impact of these uncertainties on the distribution of seismic behaviour is analysed through probabilistic analysis. Finally, a regression model is employed on the results, which provides valuable insight into the most influential modelling parameters and their interactions. The findings of this study highlight the importance of accurate modelling and surveying in evaluating the seismic vulnerability of single-nave masonry churches and can be useful for developing effective seismic risk mitigation strategies for these structures.

Keywords: Historic Masonry · Seismic Vulnerability · Masonry Textures Survey · Value of Information

1 Introduction

Historic masonry structures (HMS) are important assets of worldwide historical centres and are highly susceptible to experiencing localised out-of-plane (OOP) failure mechanisms when an earthquake happens, especially when lacking box-like or integral structural behaviour [1]. In this regard, single-nave masonry churches are one of the most vulnerable structural typologies and are prone to exhibit OOP failure mechanisms. The computational efficiency of advanced numerical methods (i.e. FEM and DEM) is rarely compatible with the need to have a rigorous real-time post- or pre-earthquake assessment [2]. Hence, according to the suggestions provided by national and international

J. C. Matos et al. (Eds.): IPW 2024, LNCE 494, pp. 177–186, 2024.
https://doi.org/10.1007/978-3-031-60271-9_15

standards [3], structural engineers frequently use analytical approaches based on limit analysis (LA) theorems. While the existing literature on macro-LA tools has predominantly focused on deterministic parametric studies [4]–[7], little attention has been paid to statistical investigations to capitalise on the extensive data these analytical methods can swiftly generate. Therefore, the present study aims to adopt the model originally implemented by this paper's authors [8] to perform a parametric analysis by considering different uncertainties, such as the quality of the masonry pattern and mechanical parameters. Therefore, different geometric scenarios are considered, which are affected by limited knowledge of some model parameters to assess their influence and trace conclusions regarding the direction engineers must push to achieve a reliable and not too conservative seismic assessment. Though the proposed analysis is easily extendable to any failure mechanism, it is herein referred to as the compound rocking failure of a masonry church façade (the most probable failure mode in such structures).

The paper is divided as follows. Section 2 presents the computational background of the macro-LA algorithm. In Sect. 3, the effects of incomplete knowledge on the structure are deeply discussed. In Sect. 4, a regression model is fitted on the results of Sect. 3. Finally, relevant conclusions are drawn in Sect. 5.

2 Macroblock Limit Analysis

The compound rocking failure mechanism might happen if there is a strong connection between the façade and sidewalls. In this case, the failure mechanism is identified from a section of the façade wall connected with a portion of the sidewalls (Fig. 1). The equilibrium of the green macroblock can be written by imposing the overturning equilibrium about the hinge identified at a point O, considering mass proportional horizontal loading on the structure. In this formulation, the external work contains both the overturning and the stabilising works performed by the self-weight of the façade, the part of the sidewalls involved in the failure mechanism and the overload q. In contrast, the internal work is derived from the friction forces at contact interfaces identified by the angle α_b (crack inclination upper threshold) and according to the approach developed by [9]:

$$
\begin{aligned}
W_{ext} &= \lambda \cdot \left(W_f \cdot d_{O,f} + 2 \cdot W_{ABC} \cdot d_{O,ABC} + 2 \cdot Q_{BC} \cdot d_{O,Q_{BC}}\right) - \\
&\quad -\left(W_f \cdot d_{S,f} + 2 \cdot W_{ABC} \cdot d_{S,ABC} + 2 \cdot Q_{BC} \cdot d_{O,Q_{BC}}\right) \\
W_{\text{int}} &= \left(2 \cdot F_{g,ABD} \cdot d_{S,F_{g,ABD}} + 2 \cdot F_{q,BD} \cdot d_{S,F_{q,BD}}\right) \cdot \left(1 - \frac{\alpha_c}{\alpha_b}\right)
\end{aligned}
\tag{1}
$$

with:

$$
\begin{aligned}
Q_{BC} &= q \cdot t_s \cdot (H_s - Z_O) \cdot \tan(\alpha_c) \\
F_{g,ABD} &= W_{ABD} \cdot \mu_f \\
F_{q,BD} &= q \cdot t_s \cdot (H_s - Z_O) \cdot \tan(\alpha_b) \cdot \mu_f
\end{aligned}
\tag{2}
$$

where λ is the load factor and $W\_$ are the self-weights of the façade and the sidewalls involved in the failure mechanism. Consequently, $\lambda W\_$ denote the inertial forces arising from the self-weights, $F\_$ are the frictional resistances due to the number of bed joints crossed by the crack line in the sidewalls and the overloads on them (Q_{BC}), according to the methodology reported by [9], $d_{O,\_}$ and $d_{S,\_}$ are the overturning and stabilising lever arms with respect to the pivot point O.

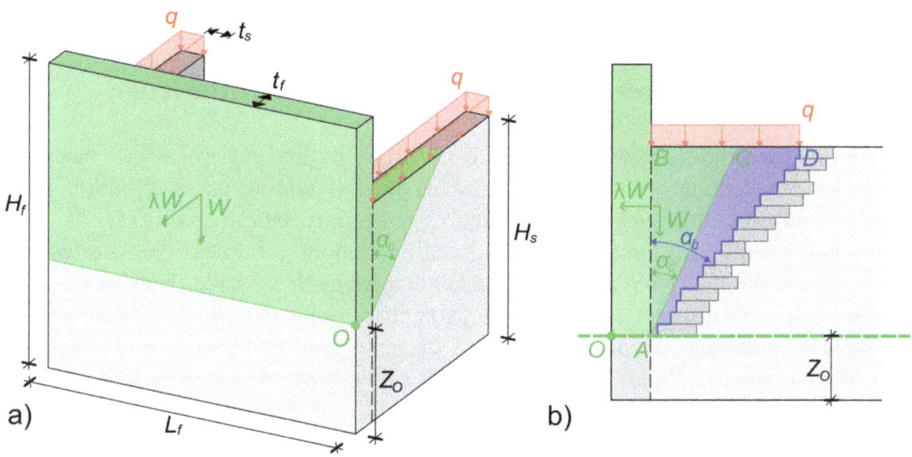

Fig. 1. Geometrical representation of the structural benchmark and the compound rocking mechanism.

Referring to coursed-rectangular irregular masonry patterns, α_b can be calculated according to the approach developed by [10]:

$$\tan(\alpha_b) = \left(M_l^{URUL} - 1\right) \tag{3}$$

where M_l^{URUL} is the structured path UP-RIGHT-UP-LEFT, defined as:

$$M_l^{URUL} = \frac{v_l^{URUL}}{H_W} \tag{4}$$

where v_l^{URUL} is the length of the polyline through the joints of the masonry pattern, going UP-RIGHT-UP-LEFT and H_w is the height of the wall panel. For more details, the reader can refer to [8].

The horizontal load multiplier can be evaluated by equating external and internal virtual works and solving for λ. According to the upper bound theorem of LA, the computation of the horizontal load multiplier requires the solution of a constrained minimisation problem, in which the parameters defining the geometry of the failure mechanism, i.e. the crack inclination α_c and pivot point height Z_O, are adopted as variables to explore all the panorama of possible solutions:

$$\begin{aligned} &\text{minimise} : \lambda \\ &\text{subject to:}\ Z_O \le H_S \\ &\alpha_c \le \alpha_b \end{aligned} \tag{5}$$

where α_c is the crack inclination for the collapse mechanism, Z_O is the height position of the pivot point and H_S is the height of the sidewalls, as can be seen in Fig. 1.

The macro-LA formulation described above has been implemented in the Java programming language, where the Nelder-Mead numerical method (NMM) is used to solve the minimisation problem of Eq. (5).

3 Treatment of Uncertainty Due to Incomplete Knowledge

This section investigates how incomplete knowledge (limited survey) affects macro-La prediction by selecting basic random variables and calculating the distribution of the response measures values. In the current analysis, masonry pattern quality (M_l^{URUL}) and friction coefficient (μ_f) have been considered as random variables. The overload (q) and the church geometry have been selected as deterministic variables because these are usually easily surveyable. Concerning the geometric configurations, 16 church geometries have been defined, according to Fig. 2, to encompass a wide spectrum of realistic cases. Each geometrical parameter corresponds to the lower or upper bounds of values observed in real masonry structures [11].

In order to accurately predict the distribution of the response and, for each combination of the two deterministic variables (church geometry and overload), the sample size has been chosen in a way the coefficient of variation of the sample mean is equal to or smaller than 5%, so the resulting response sample distribution approximates well the population distribution. The values of the random input variables have been randomly selected from a normal distribution (N) according to:

$$X = \left\{x|x \sim N\left(\mu, \sigma^2\right)\right\},\ |X| = \left(\frac{s_X}{0.05 \cdot \overline{X}}\right)^2 \tag{6}$$

where μ signifies the mean and σ the standard deviation of the distribution, $|X|$ denotes sample size, $\overline{X}$ the sample mean and s_x the sample standard deviation. As a first assumption, both variables (i.e. M_l^{URUL} and μ_f) have been selected as normally distributed. Such an assumption is supported by studies taken from the literature [12]. However, as extensively reported in the literature, the distribution choice might significantly affect the results; thus, further studies should include more emphasis on data selection to describe the population of input parameters accurately. It is worth pointing out that these distributions aim to encompass all the possible parameter values in real structures (which

Table 1. Deterministic and random parameter values

Parameter		Value
Church geometry	Deterministic	Fig. 2
q	Deterministic	$\{0.0, 0.5, 1.0, 1.5, 2.5\}$
M_l^{URUL}	Normally distributed, random	$\mu = 1.9,\ \sigma = 0.2667$
μ_f	Normally distributed, random	$\mu = 0.7,\ \sigma = 0.0667$

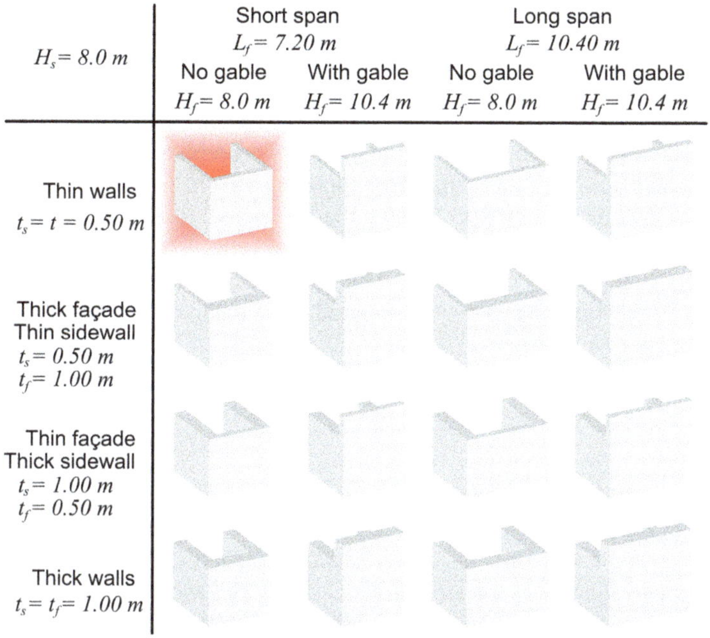

Fig. 2. Considered single-nave church prototypes.

value is unknown due to lack of survey), and not the possible random variation in a specific structure, e.g. the distribution of friction coefficients is not intended to represent the spatial variation of the parameter across a single structural benchmark.

4 Regression Model for Simplified Assessment

By representing each geometric parameter with a binary value x_i (e.g. presence ($x_i = 1$) or absence ($x_i = 0$) of the gable), any response measure (RM) can be fitted with a multiple linear regression model, where the coefficients of the model represent the presence or absence of each parameter. Furthermore, the overload is also included in the model with its magnitude expressed in kN/m^2. The model includes linear individual and interaction terms according to the following equations:

$$RM(\boldsymbol{x}) = \beta_0 + \sum_i \beta_i x_i + \sum_i \sum_j \beta_{i,j} x_i x_j \tag{7}$$

Referring to the terms of Eq. (7), the first term (β_0) defines the RM coefficient corresponding to reference configuration with all zero coefficients (i.e. specimen with thin walls, a small façade span, and no gable, as highlighted in Fig. 2). The second term considers the linear effects of parameters, while the third terms consider the two-way interactions. If the parameter is absent in the structure (0 value), the resulting product is zero, while if present (1 value), the RM is incremented (or decremented) by the corresponding coefficient β. The statistically significant predictors must be determined to construct an accurate but simple model in the form of Eq. (7). The statistical significance of a predictor is assessed through the t-Test, where the $\beta = 0$ null hypothesis' probability is calculated using the t-distribution, according to [13]. Predictors with $P(\beta = 0) \leq 5\%$ are deemed significant, having enough evidence that the coefficient is nonzero. The flowchart of the algorithm to construct the model is presented in Fig. 3.

In this work, two RM-s are considered, both calculated from the result obtained in terms of strength capacity (λ). Specifically, the five-percentile load factor $P_\lambda(5)$ defines a conservative estimate (design value) of strength capacity in case of unknown masonry pattern quality and friction coefficient. The load factor reduction $\Delta\lambda$ shows the percentile reduction of strength capacity (compared to the mean response) due to incomplete knowledge and is defined as:

$$\Delta\lambda = \left(1 - \frac{P_\lambda(5)}{\overline{\lambda}}\right) \cdot 100\% \tag{8}$$

where $\overline{\lambda}$ denotes the mean load factor. In this context, $P_\lambda(5)$ defines a lower bound of the strength capacity, providing a conservative estimate when no extensive survey was conducted on the structure. Based on its value, the structure's fragility to a loading scenario can be evaluated. If the structure's capacity is lower than the one imposed by the action, $\Delta\lambda$ can be used to assess if additional survey would lead to a substantial increase in strength capacity prediction. These equations can facilitate the survey and eventual strengthening design of the structure [14].

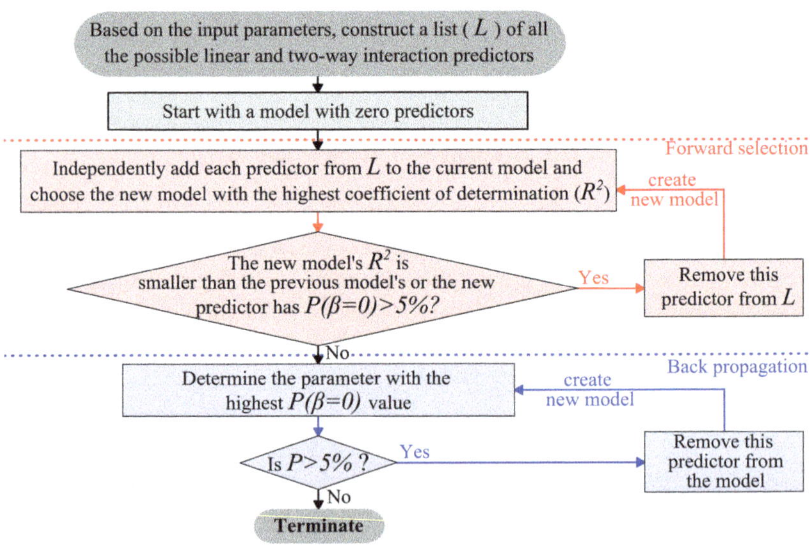

Fig. 3. Flowchart of the regression model creation algorithm

Considering $P_\lambda(5)$, Table 2 presents the coefficients (β) of the multiple linear regression model that are statistically significant ($P(\beta = 0) \leq 5\%$), notating the interaction terms by a ":" symbol, separating the interacting parameter names. The predictors' coefficients are arranged in descending order by their significance on the response measure. The model has shown a $R^2 = 0.99$. Furthermore, the fundamental assumptions of the linear regression model, such as homoscedasticity and the normality of residuals, are satisfied, as presented in Fig. 4.

Table 2. Coefficients of the multiple linear regression model of the five-percentile load factor

	Parameter	Coefficient		Parameter	Coefficient
1	Constant	0.1634	6	Gable:Sidewall	-0.016
2	With gable	-0.055	7	Gable:Span	0.0072
3	Thick sidewall	0.0497	8	Sidewall:Span	-0.005
4	Thick façade	0.0326	9	Gable:Façade	0.0045
5	Long span	-0.018	10	Façade:Sidewall	0.0045
	Overload	0.0122			

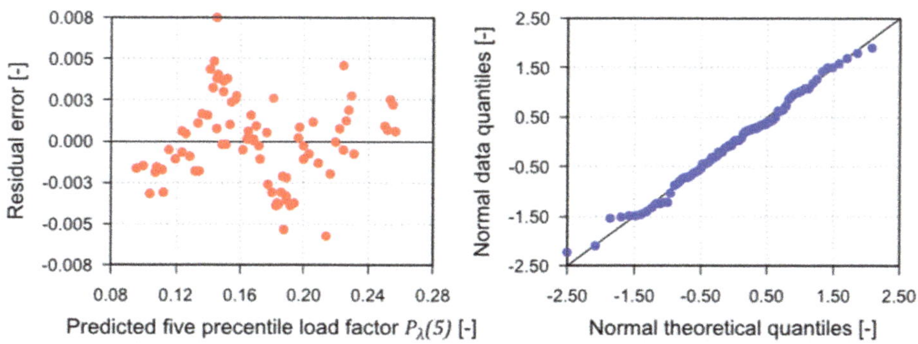

Fig. 4. Residual error and quantile-quantile plot for the five-percentile load factor model

As mentioned above, the constant term denotes the $P_\lambda(5)$ value of the specimen with thin walls, no gable, narrow span and without overload (highlighted in Fig. 2). The results show that the individual effects of the predictors are more dominant than their interactions, the gable and the wall thicknesses having the most influential effects. The presence of the gable is the most important single predictor and has significant two-way interactions with most of the other parameters. Finally, the overload has a slight positive effect, not significantly influenced by the geometric parameters.

In Fig. 3, the results underline how one- and two-way interactions of the gable and façade thickness have the most dominant effect on $\Delta\lambda$. Other significant predictors slightly increase and decrease $\Delta\lambda$. The reduction factor's value ranges between 13.49% and 44.51%, showing that depending on the church geometry, the effect of a limited survey dramatically change. Again, the model's high coefficient of determination ($R^2 =$

Table 3. Coefficients of the multiple linear regression model on the load factor reduction

	Parameter	Coefficient		Parameter	Coefficient
1	Constant	39.67%	6	Thick sidewall	4.84%
2	With gable	-24.9%	7	Gable:Span	3.37%
3	Gable:Façade	22.52%	8	Façade:Sidewall	-1.94%
4	Thick façade	-14.9%	9	Façade:Span	-1.60%
5	Gable:Sidewall	-8.50%	10	Long span	-1.30%
	Overload	1.23%			

0.98) and the satisfaction of the linear regression's fundamental assumptions (Fig. 5) show a very good fit.

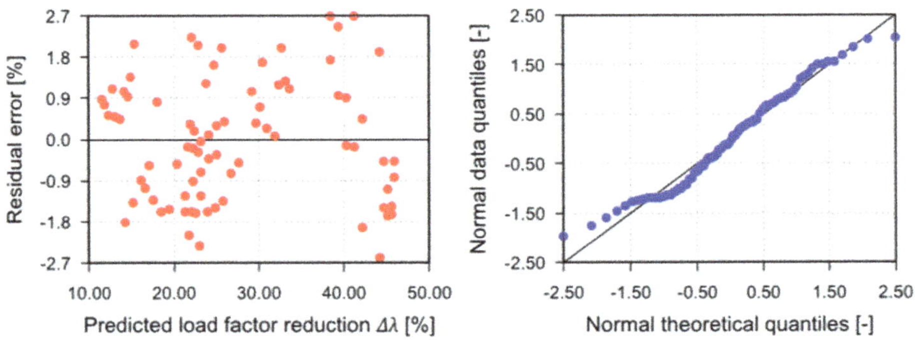

Fig. 5. Residual error and quantile-quantile plot for the load factor reduction model

5 Conclusion

A parametric study is conducted to investigate the effect of incomplete knowledge on the seismic assessment of single-nave masonry churches, with reference to the façade compound rocking failure mechanism. The church geometry and the overload on the sidewalls are systematically varied, while the friction coefficient and the sidewall masonry pattern quality are considered unknown. Macro-LA algorithm is used to compute the strength capacity. The novelty of the study is twofold: i) it assesses the uncertainties due to incomplete knowledge of the structure, and secondly, ii) it draws useful guidelines for researchers and practitioners on the use of macro-LA, underlining the most relevant parameters to survey to give more reliability to the seismic assessment of masonry buildings, using single-nave masonry churches as a demonstrator. A regression model is fitted on the results, providing a simple and efficient tool to guide such structures' survey and intervention design. It facilitates the decision-making on which buildings to focus on with respect to survey and possible retrofitting.

Acknowledgments. This work was partly financed by: (i) FCT/MCTES through national funds (PIDDAC) under the R&D Unit ISISE under reference UIDB/04029/2020, and under the Associate Laboratory Advanced Production and Intelligent Systems ARISE under reference/P/0112/2020; (ii) STAND4HERITAGE project that has received funding from the European Research Council (ERC) under the European Union's Horizon 2020 research and innovation program (Grant agreement No. 833123), as an Advanced Grant: (iii) MPP2030-FCT PhD Grants under the R&D Unit Institute for Sustainability and Innovation in Structural Engineering (ISISE), under reference PRT/BD/154348/2022.

References

1. Szabó, S., Funari, M.F., Lourenço, P.B.: Masonry patterns' influence on the damage assessment of URM walls: current and future trends. Dev. Built Environ. **13**, 100119 (2023). https://doi.org/10.1016/J.DIBE.2023.100119
2. Pulatsu, B., Funari, M.F., Malomo, D., Gonen, S., Parisi, F.: Seismic assessment of URM pier spandrel systems via efficient computational modeling strategies. Bull. Earthq. Eng., 1–24 (2023)
3. N. T. C. 2018, DM del Ministero delle Infrastrutture e dei trasporti del 17/01/2018. Aggiornamento delle Norme Tecniche per le Costruzioni (in Italian). MIT (2018)
4. Casapulla, C., Portioli, F., Maione, A., Landolfo, R.: A macro-block model for in-plane loaded masonry walls with non-associative Coulomb friction. Meccanica **48**(9), 2107–2126 (2013). https://doi.org/10.1007/s11012-013-9728-5
5. Portioli, F., Casapulla, C., Gilbert, M., Cascini, L.: Limit analysis of 3D masonry block structures with non-associative frictional joints using cone programming. Comput. Struct. **143**, 108–121 (2014). https://doi.org/10.1016/j.compstruc.2014.07.010
6. Szabó, S., Funari, M.F., Pulatsu, B., Lourenço, P.B.: Lateral capacity of URM walls: a parametric study using macro and micro limit analysis predictions. Appl. Sci. **12**(21), 10834 (2022). https://doi.org/10.3390/APP122110834
7. Colombo, C., Savalle, N., Mehrotra, A., Funari, M.F., Lourenço, P.B.: Experimental, numerical and analytical investigations of masonry corners: Influence of the horizontal pseudo-static load orientation. Constr. Build. Mater. **344** (2022). https://doi.org/10.1016/j.conbuildmat.2022.127969
8. Funari, M.F., Pulatsu, B., Szabó, S., Lourenço, P.B.: A solution for the frictional resistance in macro-block limit analysis of non-periodic masonry. Structures **43**, 847–859 (2022). https://doi.org/10.1016/j.istruc.2022.06.072
9. Casapulla, C., Argiento, L.U., Maione, A., Speranza, E.: Upgraded formulations for the onset of local mechanisms in multi-storey masonry buildings using limit analysis. Structures **31**, 380–394 (2021). https://doi.org/10.1016/J.ISTRUC.2020.11.083
10. Funari, M.F., Mehrotra, A., Lourenço, P.B.: A Tool for the Rapid Seismic Assessment of Historic Masonry Structures Based on Limit Analysis Optimisation and Rocking Dynamics. Appl. Sci. 11(3), 942 (2021). https://doi.org/10.3390/APP11030942
11. de Felice, G., Fugger, R., Gobbin, F.: Overturning of the façade in single-nave churches under seismic loading. Bull. Earthquake Eng. 20(2), 941–962 (2022). https://doi.org/10.1007/s10518-021-01243-5
12. Pulatsu, B., et al.: Probabilistic approach to assess URM walls with openings using discrete rigid block analysis (D-RBA). J. Build. Eng. 61 (2022)., https://doi.org/10.1016/j.jobe.2022.105269

13. Cannon, A.R.: STAT2 : building models for a world of data. Freeman Custom Pub (2011)
14. Funari, M.F., Greco, F., Lonetti, P.: A cohesive finite element model based ALE formulation for z-pins reinforced multilayered composite beams. Proc. Struct. Integrity (2016). https://doi.org/10.1016/j.prostr.2016.06.059

K-Means Clustering Approach for Stock Risk Assessment and Portfolio Construction: A Case Study Based on the EU-EV Risk Model

Irene Brito[1](✉) and Gaspar J. Machado[2]

[1] Center of Mathematics, Department of Mathematics, University of Minho, 4800-045 Guimarães, Portugal
ireneb@math.uminho.pt

[2] Physics Center of Minho and Porto Universities, Department of Mathematics, University of Minho, 4800-045 Guimarães, Portugal
gjm@math.uminho.pt

Abstract. The purpose of this work is to explore the approach of K-means clustering for stock risk classification and efficient portfolio construction through a case study, using data from the PSI index from January 2019 to December 2022. The classification of stock risks will be based on the expected utility, entropy and variance (EU-EV) risk model. The methodology consists in applying K-means clustering with EU-EV risk related attributes to the stocks of the Portuguese PSI index in order to obtain two classes of stocks categorized with low and with high risk. Equally weighted cluster based portfolios are built for each risk class. The performance of these portfolios will be compared with the performance of portfolios constructed by selection of the best EU-EV risk ranked stocks.

Keywords: K-means Clustering · Risk Classification · EU-EV risk · Stock Selection · Portfolio Construction

1 Introduction

In the context of financial stock and portfolio investment analysis, one important issue is the selection of stocks for an efficient portfolio construction. The mean-variance optimization model introduced by Markowitz [1] permits assessing and building up a portfolio by maximizing its expected return and minimizing its risk, expressed by variance. Other stock selection models were proposed, where risk is assessed using entropy or combinations of entropy, variance and skewness (see e.g. [2–5]). Considering machine learning methods, clustering strategies were also developed for the selection of stocks and for the design of portfolios (see e.g. [6–9]).

The EU-EV risk model, developed in [10, 11], was proposed in [12] as a portfolio stock selection model and the applications, in particular to the Portuguese PSI index, revealed that the portfolios formed with subsets containing half the number of stocks, were mean-variance efficient by comparing them with the benchmark portfolio.

J. C. Matos et al. (Eds.): IPW 2024, LNCE 494, pp. 187–196, 2024.
https://doi.org/10.1007/978-3-031-60271-9_16

A performance analysis conducted in [13], considering the PSI index, showed that the portfolios with the EU-EV risk selected stocks outperformed the benchmark portfolio, e.g. in terms of cumulative returns and Sharpe and Sortino ratios. In the previous application, the methodology consisted in ordering the stocks with respect to their EU-EV risk and picking then the half number of stocks with lowest risk to build equally weighted portfolios. In that way, a division of stocks classified with low and with high risk was achieved. Since the aim of clustering is to organize a data set into subsets, clusters, so that objects belonging to the same cluster are similar, it is interesting to investigate the ability of the clustering strategy to classify stocks with respect to their EU-EV risk and to group them into low and high risk classes. Among the various existing clustering techniques, K-means clustering constructs K clusters using the Euclidean metric to minimize the dissimilarity between elements belonging to the same cluster. This corresponds to minimize the variances in the clusters with the consequence that each element belongs to the cluster with closest mean. The aim of the present work is therefore to apply K-means clustering, with $K = 2$, to the set of stocks of the PSI index in order to obtain two subsets of stocks. Two-dimensional data with EU-EV risk related attributes are considered, namely the arithmetic mean of entropy and normalized variance and the normalized expected utility (normalized data with respect to the maximum will be used). The cluster with higher expected utility will be characterized as having lower risk and the cluster with lower expected utility will correspond to the higher risk class. Equally weighted portfolios are then constructed using the cluster stocks and the portfolios and their performances are compared with the best performing portfolios obtained by EU-EV risk selection in [12]. The clustering process will be carried out using stock return data of the PSI index in the in-sample period from January 2019 to December 2020 and the performance of the portfolios will be evaluated in the out-of-sample period from January 2021 to December 2022.

The remaining sections of the paper are arranged as follows. Section 2 describes the methodology, where in Sect. 2.1 the stock clustering process is presented and in Sect. 2.2 the performance evaluation is done. Section 3 contains the application of the proposed methodology to the PSI index. The conclusions of the work are presented in Sect. 4.

2 Methodology

The EU-EV risk model for stock selection is defined as follows (the reader is referred to [12] for further informations). Consider a set of stocks $S = \{S_1, \dots, S_I\}$ and the action space $A = \{a_1, \dots, a_I\}$, where $a_i = (x_{i1}, p_{i1}; \dots; x_{iN}, p_{iN}) \in A$ is the action of selecting stock S_i, $i = 1, \dots, I$, yielding the frequency distribution of returns, where x_{in} are the outcomes, that occur with probabilities p_{in}, $n = 1, \dots, N$, represented by the random variable X_i. The EU-EV risk for action a_i is defined by

$$R(a_i) = \frac{\lambda}{2}\left(H(X_i) + \frac{Var[X_i]}{\max\limits_{a_i \in A}\{Var[X_i]\}}\right) - (1 - \lambda)\frac{E[u(X_i)]}{\max\limits_{a_i \in A}\{E[u(X_i)]\}}, \tag{1}$$

where $0 \le \lambda \le 1$,

$$u(x) = \begin{cases} \ln(1 + x), x \ge 0 \\ -\ln(1 - x), x < 0 \end{cases} \tag{2}$$

is the utility function and

$$H(X_i) = -\sum_{n=1}^{N} p_{in} \ln p_{in} \tag{3}$$

is the entropy. The trade-off parameter λ is used to balance the uncertainty component, given by the arithmetic mean of entropy and variance, with the expected utility component. These components will be denoted by

$$EV(a_i) = \frac{1}{2}\left(H(X_i) + \frac{Var[X_i]}{\max\limits_{a_i \in A}\{Var[X_i]\}}\right) \tag{4}$$

and

$$EU(a_i) = \frac{E[u(X_i)]}{\max\limits_{a_i \in A}\{E[u(X_i)]\}}, \tag{5}$$

so that Eq. (1) can be rewritten as $R(a_i) = \lambda EV(a_i) - (1-\lambda)EU(a_i)$.

The stocks are classified with the EU-EV risk by ranking them from the lowest to the highest EU-EV risk value and using the following decision rule: given two stocks S_i and S_j, $i, j = 1, \ldots, I$, then S_i is preferred over S_j if $R(a_i) < R(a_j)$, meaning that S_i has a lower risk than S_j. From the ranked I stocks one can then choose a given number of stocks with lowest risk in order to form portfolios. This method was used in [12] to build portfolios with stocks from the PSI index, whose performance was analysed in [13], that will be considered in the present work.

2.1 Stock Clustering with EU-EV Risk Attributes

The aim of K-means stock clustering is to divide the stock set $S = \{S_1, \ldots, S_I\}$ into K clusters $C_1, \ldots, C_K$, forming a partition of S, where the objective is to minimize the variance in the clusters (or the within-cluster sum of squares).

The steps of the K-means clustering algorithm are the following [14]:

1. specify K and choose K initial centroids (means) randomly from the data;
2. assign all points to the closest cluster centroid using the squared Euclidean metric;
3. update the centroids;
4. repeat steps 2 and 3 until the centroids do not change (the clusters do not change) or a given number of iterations is attained.

In this work, we will set $K = 2$ and use two-dimensional EU-EV risk related data (attributes) $x_i = (x_{i1}, x_{i2})$, where $x_{i1} = EV(a_i)$ and $x_{i2} = EU(a_i)$, corresponding to stock S_i, $i = 1, \ldots, I$, (cf. Eq. (1)). In order to analyse the effect of entropy and variance on the clustering configuration, we will also implement the clustering process considering separately the attributes $x_{i1} = E(a_i)$ together with $x_{i2} = EU(a_i)$ and $x_{i1} = V(a_i)$ together with $x_{i2} = EU(a_i)$, were E stands for the entropy and V for the variance.

Applying the K-means stock clustering process, we will then obtain 2 clusters C_1 and C_2 containing the stocks of the whole set S, which are characterized by its centroids.

The cluster with the higher expected utility will be categorized as lower risk class and the cluster with lower expected utility as higher risk class. The stocks belonging to a given cluster will be used to construct an equally weighted portfolio. The performance of the portfolios formed with the cluster stocks can then be compared with the performance of the portfolios obtained in [13] by selection with the EU-EV risk model.

2.2 Performance Evaluation

The performance of the portfolios will be evaluated using the cumulative returns and the following performance measures (see e.g. [15, 16]): Sharpe and Sortino ratios and Alpha and Beta risk metrics. In the succeeding definitions, r_P denotes the expected return of the portfolio, r_S the expected return of the benchmark, σ_P the standard deviation of the portfolio return,σ_S the standard deviation of the benchmark return and r_f is the risk-free rate, which will be set equal to zero ($r_f = 0$).

The Sharpe ratio is defined by

$$Sharpe = \frac{r_P - r_f}{\sigma_P} \tag{6}$$

and measures the excess return - the return of the portfolio less the risk-free rate of interest - per unit of total risk of the portfolio (given by the standard deviation of the portfolio returns). A higher Sharpe ratio indicates a better performance.

The Sortino ratio, defined by

$$Sortino = \frac{r_P - r_f}{\sigma_P^-}, \tag{7}$$

is a modified version of the Sharpe ratio, where only the downside deviation, σ_P^-, is taken into account.

The risk metric Beta, given by

$$Beta = \frac{Cov(r_P, r_S)}{\sigma_S^2}, \tag{8}$$

where $Cov(r_P, r_S)$ is the covariance between the portfolio and market returns, measures the risk of a portfolio compared to the market. A portfolio with Beta > 1 can be interpreted to be riskier (or more volatile) than the benchmark. If Beta $= 1$, the portfolio has the same volatility as the market and Beta < 1 means that the portfolio is less volatile than the market.

Jensen's Alpha measures the portfolio return relative to the market return and is expressed by

$$Alpha = r_P - \left[r_f + Beta\left(r_S - r_f\right)\right]. \tag{9}$$

If Alpha > 0 the portfolio has performed better than the market, Alpha < 0 indicates that the portfolio underperformed the market and Alpha $= 0$ means that the portfolio's performance is in line with that of the market.

3 Application to the PSI Index

The EU-EV risk model was applied in [12] to the PSI index, considering daily stock return data in the in-sample period from January 2019 to December 2020. The PSI index, denoted by S, contains in the given range 18 component stocks: $S = \{S_1, \ldots, S_{18}\}$ (the names of the corresponding companies can be found in [12]). The following two subsets of S, selected with the EU-EV risk model for different ranges of the trade-off factor and consisting of nine component stocks, were shown to be mean-variance efficient [12] and also to outperform the benchmark [13] - Q_2 in the in-sample period and Q_4 in the out-of-sample period from January 2021 to December 2022 - with respect to several performance metrics presented in Sect. 2.2: $Q_2 = \{S_1, S_3, S_4, S_5, S_6, S_9, S_{11}, S_{16}, S_{18}\}$ for $\lambda \in [0.1260, 0.4685)$ and $Q_4 = \{S_3, S_5, S_6, S_9, S_{11}, S_{13}, S_{16}, S_{17}, S_{18}\}$ for $\lambda \in [0.5311, 0.7771)$.

Considering the in-sample period, where Q_2 revealed to be the best performing set, and the ordering with respect to the EU-EV risk, from the lowest to the highest risk, we can divide the 18 stocks in two risk classes, the class with low EU-EV risk is Q_2 and the remaining stocks belong to the class with high EU-EV risk, $S - Q_2$. The stocks of both risk classes are depicted graphically considering the plane defined by the EU and EV components (defined in Sect. 2), see Fig. 1.

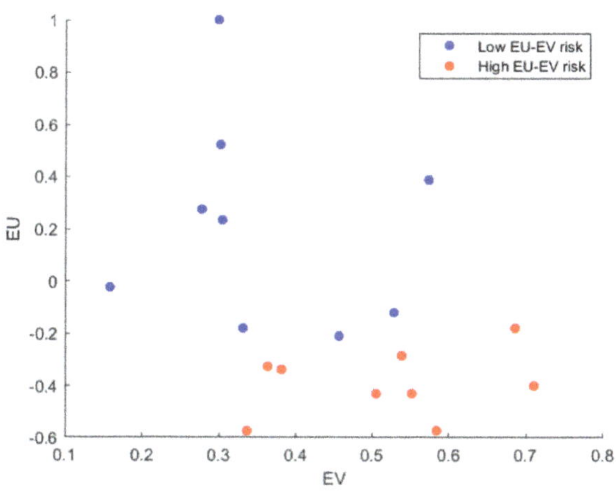

Fig. 1. EU-EV risk selected stock sets: Q_2 with low risk and $S - Q_2$ with high risk.

Since we have obtained in this way a partition of S, where the stocks were grouped according to their EU-EV risk in two subsets, we will explore if K-means clustering with $K = 2$ using the EU-EV risk attributes, is able to capture the low and high risk pattern contained in the stock data and divide S in a similar fashion. Applying the K-means algorithm, using MATLAB, to the 18 two-dimensional stock data (with components *EV* and *EU*) of the in-sample period, as explained in Sect. 2.1, leads to the partition $S = C_{1,EV,EU} \cup C_{2,EV,EU}$ presented graphically in Fig. 2.

The clusters are characterized as follows: $C_{1,EV,EU} = \{S_3, S_5, S_6, S_9, S_{11}\}$ has centroid (0.5645,0.4849) and $C_{2,EV,EU}$, containing the remaining stocks, has centroid

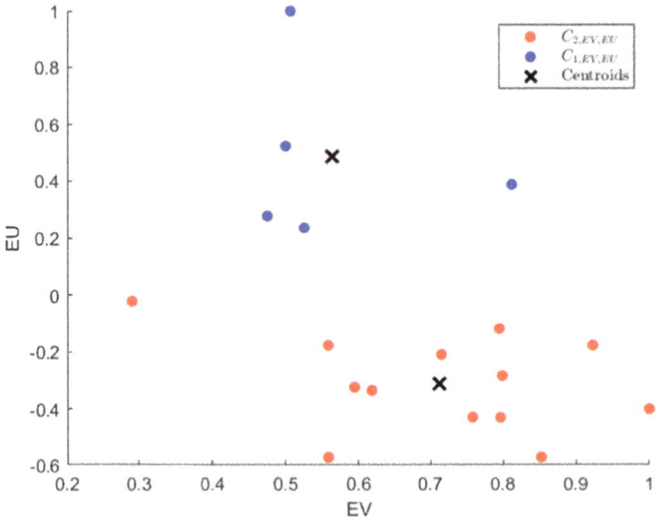

Fig. 2. Stock clusters using EU-EV risk attributes.

(0.7123, −0.3124). Since $C_{1,EV,EU}$ has lower EV centroid value and higher expected utility EU centroid value than $C_{2,EV,EU}$, it can be characterized as the lower risk cluster and $C_{2,EV,EU}$, as the higher risk cluster. Notably is that the stocks contained in $C_{1,EV,EU}$ are exactly the five stocks with lowest EU-EV risk considering the ordering according to the EU-EV risk model for $\lambda \in [0, 0.4685)$ obtained in [12]. This suggests that K-means clustering is in fact able to partition the stocks taking into account the risk pattern and ordering of the data using the related attributes (at least by empirical evidence from the present case study). One drawback may be the low number of stocks contained in the cluster (in this case in $C_{1,EV,EU}$), when forming the cluster based portfolios.

The next step consists then in forming the cluster based portfolios (equally weighted portfolios with stocks of $C_{1,EV,EU}$ and $C_{2,EV,EU}$) and analysing and comparing their performances with those of the benchmark S and of the Q_2 and Q_4 portfolios. The portfolios will be denoted using the corresponding set (or cluster) names. In addition to these portfolios, we also considered the portfolios obtained with the clustering process using the attributes V and EU, where $C_{1,V,EU} = \{S_3, S_5, S_6, S_9, S_{11}, S_{16}\}$ and $C_{2,V,EU}$ containing the remaining 12 stocks. The clustering with the attributes E and EU led to the same partition as using the attributes EV and EU.

First, the cumulative returns and the performance metrics are calculated for the in-sample period. The graphical plot of the cumulative returns in Fig. 3 shows that $C_{1,EV,EU}$ is the best performing portfolio, followed by $C_{1,V,EU}$. Oppositely, $C_{2,EV,EU}$ and $C_{2,V,EU}$ underperform all other portfolios.

Analysing the values of the performance metrics listed in Table 1, one concludes that $C_{1,EV,EU}$ is also the best performing portfolio according to these metrics, since it leads to the higher Sharpe, Sortino and Alpha values. $C_{2,EV,EU}$ is the second best performing portfolio. The Beta values indicate that these portfolios are less volatile than

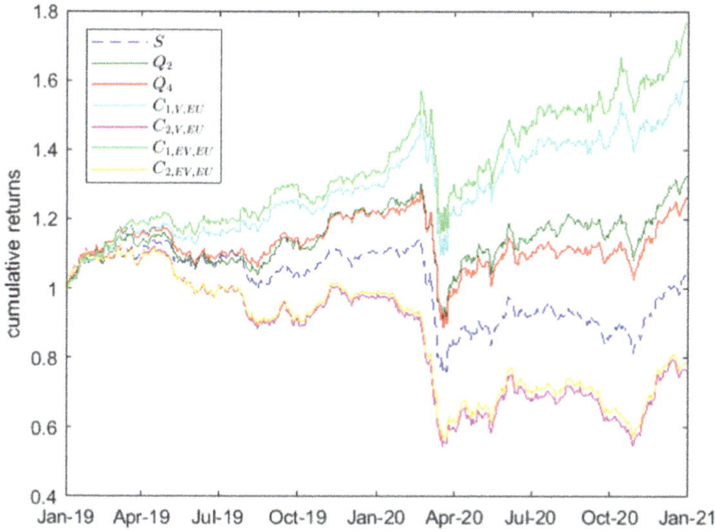

Fig. 3. Cumulative returns from January 2019 to December 2010.

the benchmark. In contrast, $C_{2,EV,EU}$ and $C_{2,V,EU}$, exhibit higher volatilities than the benchmark and have again the poorest performances.

Table 1. Performance measures from January 2019 to December 2020.

Stock set	Sharpe	Sortino	Beta	Alpha
S	0.1938	0.2289	1	0
Q_2	0.8078	0.9306	0.8650	0.1298
Q_4	0.6867	0.7924	0.8562	0.1027
$C_{1,EV,EU}$	**1.5094**	**1.8170**	0.7684	**0.3101**
$C_{2,EV,EU}$	−0.4010	−0.4900	1.0280	−0.1272
$C_{1,V,EU}$	1.3599	1.5634	0.7345	0.2478
$C_{2,V,EU}$	−0.4137	−0.5155	1.0666	−0.1354

Now, we consider the cumulative returns and the performance metrics calculated for the out-of-sample period. We can observe in Fig. 4, that the cumulative returns of $C_{2,EV,EU}$ and $C_{2,V,EU}$ surpass the other returns in the first half of 2021 and in smaller time intervals in 2022. Q_2 and Q_4 lead to higher cumulative returns in the second half of 2021 and Q_4 outperforms the other portfolios for a larger time interval in 2022.

The performance metrics in Table 2 reveal that Q_4 is the best performing portfolio in the out-of-sample period, followed by $C_{2,EV,EU}$ and then by $C_{2,V,EU}$.

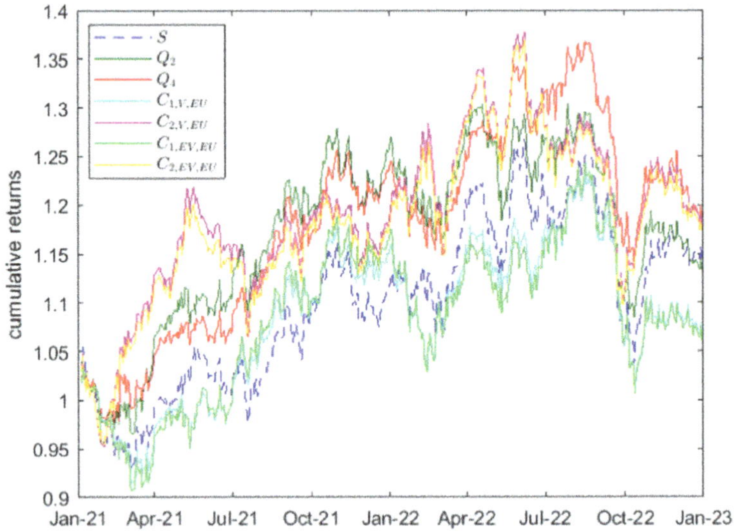

Fig. 4. Cumulative returns from January 2021 to December 2022.

Table 2. Performance measures from January 2021 to December 2022.

Stock set	Sharpe	Sortino	Beta	Alpha
S	0.4863	0.7364	1	0
Q_2	0.4790	0.6741	0.8457	0.0060
Q_4	**0.6712**	**1.0359**	0.7571	**0.0344**
$C_{1,EV,EU}$	0.2609	0.4000	0.7855	−0.0198
$C_{2,EV,EU}$	0.5747	0.7835	0.8082	0.0277
$C_{1,V,EU}$	0.2825	0.4391	0.7288	−0.0157
$C_{2,V,EU}$	0.5655	0.7666	0.8384	0.0296

4 Conclusions

The results of this research and of this case study indicate that K-means clustering, with EU-EV risk related attributes and $K = 2$, is able to classify stocks according to their risk pattern, and that the low risk cluster based portfolios perform very well in the in-sample period. However, in the out-of-sample period the portfolios constructed with stocks belonging to the higher risk cluster show to have a better performance. In that period, it is Q_4 (obtained by selection according to the EU-EV risk ordering) that outperforms the other portfolios. Since the low risk clusters contain only 5 or 6 stocks and the corresponding high risk clusters respectively 13 or 12 stocks, and Q_4 consists of 9 stocks, one can conclude that the number of stocks may play an important role for an efficient and well performing portfolio construction. A portfolio must have a minimum number of stocks in order to be sufficiently diversified. Previous studies (see e.g. [13, 17]) and the present work suggest that using 50% of stocks may be appropriate to achieve better efficiency and performance results for future investments than using approximately 25% of stocks (from an initial set of 18 stocks). In order to improve the K-means clustering for portfolio stock selection, one should introduce a constraint to impose a given number of elements in a cluster (when considering 2 risk classes each cluster should for example be formed with the half number of elements of the initial set). In the future, we will study the impact of this modification on the cluster and portfolio formation and we will apply the methodology also to larger stock sets.

Acknowledgments. IB thanks support from FCT through the Projects UIDB/00013/2020 and UIDP/00013/2020. GJM acknowledges the financial support from FCT through the Project UIDB/04650/2020.

References

1. Markowitz, H.: Mean-Variance Analysis in Portfolio Choice and Capital Markets. Wiley (2000)
2. Song, R., Chan, Y.: A new adaptive entropy portfolio selection model. Entropy **22**(9), 951 (2020)
3. Li, B., Zhang, R.: A new mean - variance-entropy model for uncertain portfolio optimization with liquidity and diversification. Chaos Solitons Fractals **146**, 110842 (2021)
4. Li, X., Qin, Z., Kar, K.: Mean - variance-skewness model for portfolio selection with fuzzy returns. Eur. J. Oper. Res. **202**(1), 239–247 (2010)
5. Mercurio, P.J., Wu, Y., Xie, H.: An entropy-based approach to portfolio optimization. Entropy **22**(3), 332 (2020)
6. Purwandari, T., Riaman, Hidayat, Y., Sukono, Ibrahim, R.A., Hidayana, R.A.: Selecting and weighting mechanisms in stock portfolio design based on clustering algorithm and price movement analysis. Mathematics **11**(19), 4151 (2023)
7. Chen, B., Zhong, J., Chen, Y.A.: Hybrid approach for portfolio selection with higher-order moments: empirical evidence from shanghai stock exchange. Expert Syst. Appl. **145**, 113104 (2020)
8. Cheong, D., Kim, Y.M., Byun, H.W., Oh, K.J., Kim, T.Y.: Using genetic algorithm to support clustering-based portfolio optimization by investor information. Appl. Soft Comput. **61**, 593–602 (2017)
9. Ren, F., Lu, Y.-N., Li, S.-P., Jiang, X.-F., Zhong, L.-X., Qiu, T.: Dynamic portfolio strategy using clustering approach. PLoS ONE **12**(1), e0169299 (2017)
10. Brito, I.: A decision model based on expected utility, entropy and variance. Appl. Math. Comput. **379**, 125285 (2020)
11. Brito, I.: The normalized expected utility - entropy and variance model for decisions under risk. Int. J. Approx. Reason. **148**, 174–201 (2022)
12. Brito, I.: A portfolio stock selection model based on expected utility, entropy and variance. Exp. Syst. Appl. **213**(A), 118896 (2023)
13. Brito, I., Machado, G.J.: Performance evaluation of portfolio stocks selected with the EU-EV risk model. In: Gervasi, O., et al.: Computational Science and Its Applications - ICCSA 2023 Workshops (ICCSA 2023). LNCS, vol. 14105, pp. 69–83. Springer, Cham (2023)
14. Everitt, B.S., Landau, S., Leese, M., Stahl, D.: Cluster Analysis, 5th edn. Wiley (2011)

15. Marhfor, A.: Portfolio performance measurement: review of literature and avenues of future research. Am. J. Ind. Bus. Manag. **6**, 432–438 (2016)
16. Samarakoon, L.P., Hasan, T.: Portfolio performance evaluation. In: Lee, C.F., Lee, A.C. (eds.) Encyclopedia of Finance. Springer, Boston (2006)
17. Marasović, B., Kalinić, T.: Is the expected utility-entropy decision model for stock selection applicable on emerging capital markets? Econ. Soc. Developm.: Book Proc. **39**, 225–235 (2019)

Modelling of Traffic Loads in a Full-Probabilistic Reassessment of Rural Bridges Based on Measurement Data – A Case Study

Christian Kainz(✉), Leo Lapidus, and Thomas Braml

University of the Bundeswehr Munich, 85577 Neubiberg, Germany
christian.kainz@unibw.de

Abstract. Bridges have been subjected to increasing traffic loads over the last decades. But not all road bridges are equally affected by traffic today - some bridges experience heavier traffic, while others do not experience intense traffic loads. Especially rural bridges are often significantly less stressed than assumed by the codes. As the preservation of infrastructure becomes important in the context of resource-efficient construction practices, new and more realistic approaches for the modelling of traffic loads are desirable. In a probabilistic calculation it is possible to use data from actual on site measurements for the assessment of bridges. If no measurement data about the traffic load is available, one can only use values from literature. Currently, there is only few specific data available about the choice for the coefficient of variation for traffic loads. Further there is no clear distinction between highway and rural road traffic. This article presents the past and current load models from the codes for bridges in Germany as well as the available recommendations for the modelling of traffic loads for probabilistic calculations from literature. On the example of the Heinrich's bridge Bamberg, where a measurement of traffic was conducted over several months recently, a dataset for the values of mean, standard deviation and coefficient of variation is given. The four span steel bridge with an orthotropic deck from 1974 carries a state road in Bavaria over a river. A contribution for the values of variability in traffic loads for road bridges with local traffic is provided.

Keywords: bridge assessment · object-specific load model · adjusted partial safety factors · traffic load · coefficient of variation

1 Introduction

The continuous growth of traffic intensity on road bridges and the increasing allowed total weight of heavy vehicles over the past decades is one major challenge for today's existing bridges. The significant increase in heavy traffic is a change in use, which effects cannot be neglected for older road bridges. Load assumptions in the code regulations aim to cover real traffic loads with load models. Since the introduction of the first German national standards with DIN 1072 in the year 1925 [1], traffic loads have been continuously adjusted to the increased traffic volume. Furthermore, specific regulations

J. C. Matos et al. (Eds.): IPW 2024, LNCE 494, pp. 197–208, 2024.
https://doi.org/10.1007/978-3-031-60271-9_17

[2, 3] for the reassessment of existing bridge structures are now available. There are approximately 130,000 existing road bridges in Germany [4, 5] and most of these bridges are not under federal management like highway bridges but managed by municipalities, states or rural districts. These bridges are typically less frequently used by heavy trucks compared to federal highways.

The growing traffic, heavier trucks, or planning and construction errors also require a reassessment of existing bridges in terms of load-bearing capacity and durability. Information about the actual effects on a structure during its service life is not available in most cases for road bridges. These must be assumed during a statical reassessment, in most times according to the specifications in code regulations. The individuality of structures regarding the actual traffic load exposure cannot be considered in many cases. However, the structures are often loaded very differently. Especially for bridges in the local road network, it is likely that they are not used as intense as assumed by the load model from the codes. The potential for a longer remaining service life, based on a more realistic assessment, is very high. Even if the direct costs of a replacement of rather small rural bridges seem to be little, the total costs including e.g. traffic routing and lost time in traffic jams must be considered as well. Against this background, enlarged use phases are preferable for already existing brides if the on-site condition of the bridge is sufficient.

2 Load Models for Bridges According to National and European Code Regulations

In Germany, the national design vehicle (German: *Schwerlastwagen SLW*) for bridges was included for the first time about 70 years ago with DIN 1072:1952-06 [6]. This model vehicle was used for the design of road bridges in Germany for the next 50 years and has an axle spacing of only 1.50 m together with a significantly higher total weight than any real vehicles available at that time. Bridges on highways were designed with bridge class 60 (this means a SLW that equals the weight of a 60 ton truck) and rural roads were designed for class 45 or 30, while municipal and economic roads were only intended to bear the smaller load of a truck (*LKW*). However, this is one big difference to today's load models in current codes, as those generally do no longer distinguish between highways and rural roads for the ultimate limit state assessment.

Since the introduction of the bridge class 60, the traffic load has significantly increased within 30 years. To accommodate the growing overtaking and oncoming traffic of approved heavy vehicles with special transports, an adjustment of the load model was necessary and performed with DIN 1072:1985-12 [7]. An additional side lane was added next to the main lane, replacing the previous classes 60 and 30 with the classes 60/30 and 30/30.

To establish a unified load model for all European countries, the different existing national load models were initially compared through computational analysis. Significant variations in design results and load effects were observed among the different national approaches [8]. Consequently, the current European traffic load model of the Eurocode 1 became largely independent of previous regulations. The calibration of the basic values was based on traffic measurements on various bridges in Europe (Belgium,

Denmark, Germany, France, Great Britan, Italy and the Netherlands) in the 1980s. Particularly the results from 1986 in Auxerre, France, were used as reference data for the traffic load model LM1 [8]. The measured traffic there most closely resembled the expected future European traffic, with a high percentage of trucks.

In November 1996, the German DIN Coordination Committee decided to transition from national regulations to European regulations. The vibration coefficients φ of DIN 1072 are already included in the new load model. The current load model LM1 for road bridges is given in DIN EN 1991-2 [9].

All new bridges in Germany shall now be designed with the load model LM1 due to the potential presence of heavy vehicles on rural or subordinate roads at any time, even if this is not expected to be on a daily basis. Subsequent adjustments to the load model were made based on recent traffic measurements, reflecting a doubling of heavy traffic on the German Brohltal Bridge within 20 years [10]. The study also revealed that in the case of an increase in the permissible total weight, using Load Model 1 according to the former DIN technical report led to computational deficiencies. For this reason, the national adjustment factors α_{Qi} and α_{qi} are increased compared to the Eurocode.

3 Probabilistic Approach for Modelling of Traffic Loads

While code regulations use the semi-probabilistic safety concept and provide characteristic values for traffic load actions and corresponding partial safety factors, there is no clear information about the actual statistical parameters behind the code values available. According to DIN EN 1991-2 [9] the returning interval of the characteristic traffic load value is 1.000 years. This equals a probability of 0,1% for the heaviest truck of one year to be greater than the characteristic value of the traffic load model. According to [11] the characteristic value of the traffic load was defined with a 10% appearance chance in 100 years and according to [12] the characteristic value should appear in 50 years with a probability smaller than 5% [13]. In the DIN technical report [14] the returning interval of the characteristic traffic load value is 50 years, but simultaneously the partial safety factor was increased from $\gamma_{LMM,EC} = 1{,}35$ to $\gamma_{LM1,DINFB} = 1{,}50$ [13]. The supplementary sheet to DIN 1072:1985 [7] states clearly, that the load models for the bridge classes are based on deterministic values and that the load assumptions cannot be defined as characteristic values according to the - at that time - new safety theory. But the former load models are still mainly used in the reassessment of existing road bridges according to the reassessment guideline [2, 3]. Probabilistic methods are also explicitly allowed in a stage 4 analysis according to this guideline.

It's important to sustain the safety margin demanded by the codes, which is not already fulfilled by reaching the necessary value for the safety factor β. The selection of the parameters for the modelling of the traffic should also be equivalent to the actual load model of the codes.

Znidaric & Znidaric [15] report about a coefficient of variation for traffic loads from 12,7 to 16,8% with a reference period of 50 years. Results from the USA from Moses & Ghosn [16] with measurements on highway bridges indicate a COV of 10%. Stewart & Rosowky [17] also suggest a COV of 10%. Eichinger [18] reports a COV from 11,9 to 16,6% for heavy good traffic in Austria in the 1990s and suggests an average COV of 10,3%. For the UDL component EICHINGER [18] suggest the use a COV of 10% and a normal distribution. NOWAK & FISCHER [19] assume for the German local and inner-city traffic a COV from 10 to 12% and a normal distribution. BRAML & WURZER suggest in [20] to use a COV = 15% and a Gumbel distribution for the probabilistic calculation of road bridges, if there is no measurement data available and the characteristic values from the code have to be converted as input parameters. In [21] the traffic load effects of LM1 for an existing two-lane state road bridge in Croatia are modelled with a Gumbel distribution and a coefficient of variation of 12,5% based on fib Bulletin 80 [22]. BOROS also uses a COV of 7,5% and a Gumbel distribution [23]. The fib bulletin 80 indicates a COV of 7,5% for traffic loads on road bridges and refers to [24], which recommends a range for the coefficient of variation from 3 to 15%. For the bridge in Croatia a Weight-In-Motion system (WIM) was installed and for the WIM-based load effect a COV of 9% was used in [21] according to the study of IVANKOVIC et al. [25]. Data from [26] indicates a COV of 6 to 7% based on measurements on a German highway in Bliesheim. For the modelling of the measured traffic load from small cars to heavy trucks, a bi- or tri-modal normal distribution is suggested in [28], which is the product of two or three single normal distributed functions and equals in one formula for the normal distributed light weight traffic and the normal distributed heavy goods traffic. STEFFENS [13] reports COVs from 9% to 13% for a prestressed box girder bridge in Germany and uses a Gumbel distribution for the modelling.

4 Case Study – Heinrich's Bridge in Bamberg

4.1 Description of the Construction

The Heinrich's Bridge (German: Heinrichsbrücke) is a steel bridge built in 1974 and located in Bamberg, Germany. The bridge carries the state street B 22 with two traffic lanes in each direction and bike lanes on both sides over the Rhine-Main-Danube Canal, a street, and the right arm of the Regnitz River. The superstructure with an orthotropic steel deck has four spans with measures of 18 m - 119 m - 54 m - 79 m, see Fig. 1.

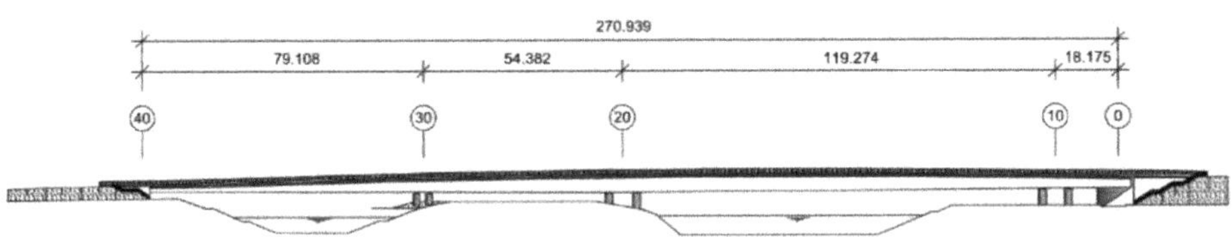

Fig. 1. View of the Heinrich's bridge in Bamberg

A structural analysis with finite-element modelling and according to the regulations of the guideline for the assessment of existing bridges [2, 3] was performed and resulted in an insufficient load bearing capacity according to the code regulations. According to [2, 3] a so called stage 3 assessment explicitly allows for the use of on-site measurement data. The bridge was then equipped with measurement instruments and the traffic loads were measured over a period of 15 months. Figure 2 shows the cross section of the superstructure.

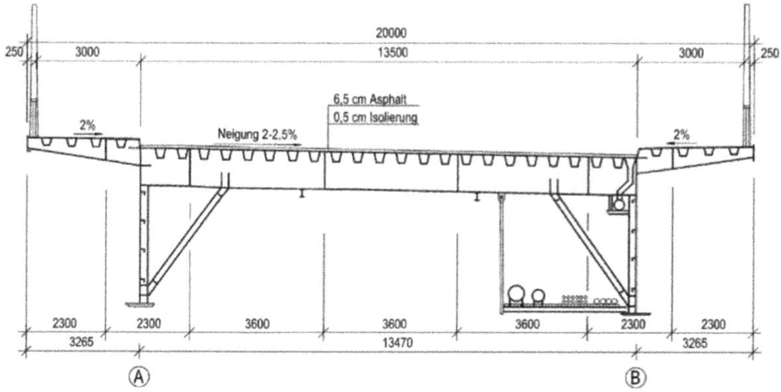

Fig. 2. Cross section of the Heinrich's bridge in Bamberg

4.2 Measurement Concept

The bridge was equipped with 13 strain gauges, 2 accelerometers and 3 temperature sensors. The sensors were placed at selected positions that are relevant locations, e.g. critical spots with a maximum stress utilization according to the finite-element calculations. For example, the strain gauge DMS1 is located on the bottom of a longitudinal girder in the middle of the main span. To measure temperature effects due to intense sunshine, the temperature sensors are distributed across the cross section. A location of one sensor each on the bottom flange of the main girder A and B and one sensor in the middle of the orthotropic deck gives sufficient information about the temperature gradient from left to right and from top to bottom with the minimal number of sensors. The accelerometers are placed in the middle of the big spans two and four and are used for information in case of unwanted resonance effects.

For a comparison of the results from the finite element model with the measurement data from the strain gauges, it is necessary to perform a calibration at first. For this purpose, a typical truck can be used. While the bridge is closed for other public traffic, the truck drives on the bridge in predefined schemes. It is necessary to close the bridge for other public traffic during this time to exclude other loads from the measurement. For a minimal disturbance of the local traffic flow it is advisable to perform this action for one night and on the weekend. All relevant data of the calibration vehicle, e.g. axle weights, geometric measurements or driving speed must be exactly known. The driving schemes for the calibration should include a variation of driving speeds and predefined lanes to drive in. A precise planning and performance are necessary for the use of the following measurement data. In this case a 7,55 m long Mercedes-Benz 3-axle truck with a total weight of 21,16 to was used.

All sensors are wired into a data-logger, which is placed on a maintenance catwalk under the bridge, see Fig. 2. The data is saved hourly in the storage format of the respective manufacturer of the measurement system, which is a.dxd-file in this case. This means each file contains the measurement values of one hour. In this project there was no data transfer via internet, e.g. mobile data UMTS or LTE, used. Ordinary external hard drives with a capacity one Terabyte (1TB) are used, which are replaced every three months due to several reasons. The first reason is the limited storage capacity of the hard drive. By changing the hard drives every few months, a sufficient data backup is ensured, the risk of a data loss is small and acceptable. The measured data can be analyzed and processed soon in case there are adjustments necessary, e.g. a replacement of faulty sensors or a change of the sampling rate.

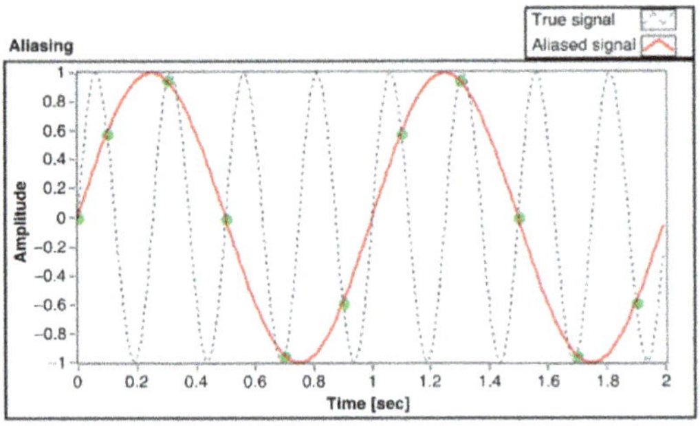

Fig. 3. Effect of an inadequate sampling rate due to aliasing [27]

After starting with a sampling rate of 500 Hz, after a few months of measuring it showed that a sampling rate of 40 Hz is sufficient enough for a correct discretization and the file size. This change reduces the file size to ca. 10% of its original size. In total there are still 5 billion measurement values documented. A too small sampling rate can lead to the known problem of *aliasing*, see Fig. 3. For a use in other projects, the authors recommend to generally start with a high sampling rate and to check the results after some time and to adjust the sampling rate afterwards.

4.3 Data Analysis

Generally, the actual weight of the traffic can not be measured directly with the used measurement setup. The strain gauges only measure a change in the electric resistance as decimal numbers at the location they are applied. They protocol the discrete measurement values with the corresponding timestamp as plain text. The strain gauges measure values based on a predefined sampling rate, they do not automatically recognize changes in the traffic flow or the position of a heavy load truck. A sampling rate of e.g. 40 Hz simply means that they save the measurement value of the strain every 25 ms.

The raw data from the sensors contain a superposition of various load effects, e.g. strains due to temperature, traffic loads, snow or wind and random noise effects. What proportion of the strains is caused by vehicle traffic is not identifiable based on the values, so the data of the other sensors have to be regarded as well to clear the raw data. This practically means, that solely based on the strain measurement it's not possible to detect if a measured strain is caused by a traffic loading or a change in temperature.

In a first step, a Fast-Fourier-Transformation (FFT) is used to determine the frequency spectrum, Fig. 4, of the measured values.

Figure 4 shows that the amplitudes are big in some areas and negligibly small in other areas. So not all frequency ranges are relevant for the analysis of the structure.

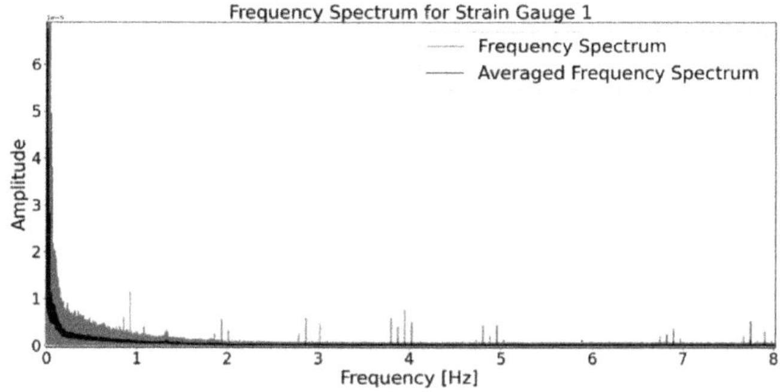

Fig. 4. Frequency spectrum for the strain gauge 1

Therefore, a filter is used to reduce the size of the data and to preserve only the data from the frequency areas of interest. A frequency of 1 Hz is selected as limit frequency because the traffic and the temperature are mainly within this range in this bridge. A Butterworth filter [28] with a limit frequency of 1 Hz is used for the filtering. Figure 5 shows the measurement data before and after the filtering over the period of 40 min on the left side. The measurement data on the left side of Fig. 5 still contains an overlay of temperature, traffic and effects due to the first natural frequency. To remove the portion of the temperature a moving average with a window of 15 min is used. The timeframe of 15 min is selected, because the temperature is assumed to be constant for this period. On the right hand side of Fig. 5 the data is cleared of the temperature.

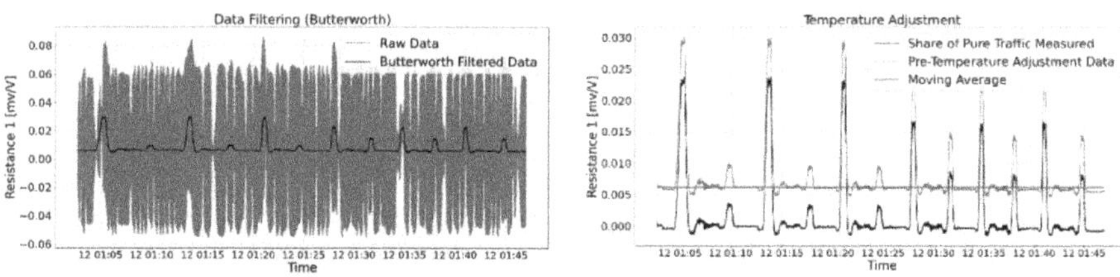

Fig. 5. Measurement data during the calibration drive for the time of 01:05 am to 01:45 am on the 12th of October 2019; left: before and after the filtering; right: with temperature adjustment

The effects of the first natural frequency could not be removed from the data but are assumed to be small due to the lower frequency of the traffic, see Fig. 4.

It is important to note that the measured data is still the electric resistance and an transformation to the traffic loads is necessary in the next step. The strain gauge technique usually uses a *Wheatstone bridge* to measure the resistances on the respective strain gauge. The manufacturers provide a product-specific factor k to transform the electric voltages V in the system to the strains ε. By further using Hooke's law of elasticity, the steel stresses σ_i [N/mm2] can then be calculated from the measurement data with Eq. (1).

$$\sigma = \frac{V_o}{V_s} \cdot \frac{4 \cdot E}{k} \tag{1}$$

Here is V_s the bridge excitation voltage and V_o the bridge output voltage. The E-modulus is set as a constant value of $E = 210.000$ N/mm2 for this steel bridge.

The actual stiffness of the bridge superstructure is regarded through a calibration factor c, which is defined as the ratio of the stress from the FEM-calculation to the stress based on measurement, see Eq. (2). While the calibration factor is theoretically slightly different for each location of a strain gauge, in a practical approach a constant value for each spot is a sufficient approximation. A non-linear response of the steel bridge is not expected. For the assessment of the bridge in this article the strain gauge DMS1 was selected as the defining spot.

$$c_i(x_i) \approx c = \sigma_{FEM} / \sigma_{Measurement} \quad (2)$$

After this transformation the stresses in the bridge construction due to the traffic are obtained. During the calibration drive, seven of the 13 strain gauges documented values that are within a 25% difference range compared to the FEM-calculation. For the further assessment three of the strain gauges were selected and in this article the values of the DMS1 are used representatively, because at this location the results from measurement and calculation fit closely together. The results of the other strain gauges are dismissed, since the numerical model does obviously not implement the actual response of the bridge correctly at certain locations or the measurement instruments did not show correct function, e.g. due to moisture or damaging. Under the load of the 21,16 to truck the measured strain on the DMS1 is 9,17 N/mm^2 while the result from the finite element model is 8,75 N/mm^2. This results in a calibration factor c = 8,75/9,17 = 0,954. The measured stresses are shown exemplarily for one month in Fig. 6.

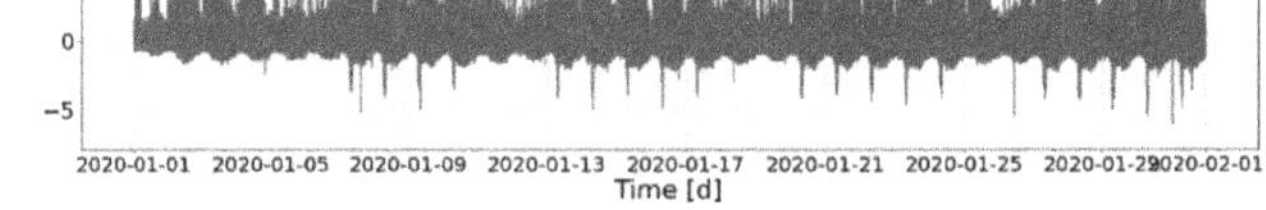

Fig. 6. Stresses due to traffic at strain gauge DMS1 in January of 2020

4.4 Results and Application

For the further statistical analysis, the equivalent total weight will be used instead of the steel stresses. Figure 7 illustrates the distribution of the equivalent total weight of the vehicles based on the results from DMS1 in a histogram.

The results can be divided into three subgroups. The first aspect is the regular traffic with an equivalent total weight of under approximately 10 tons, which represents cars and lighter trucks. The second group are the medium heavy vehicles with the peak at 16 tons. A third group of heavy vehicles can be identified with a peak at approximately 38 tons.

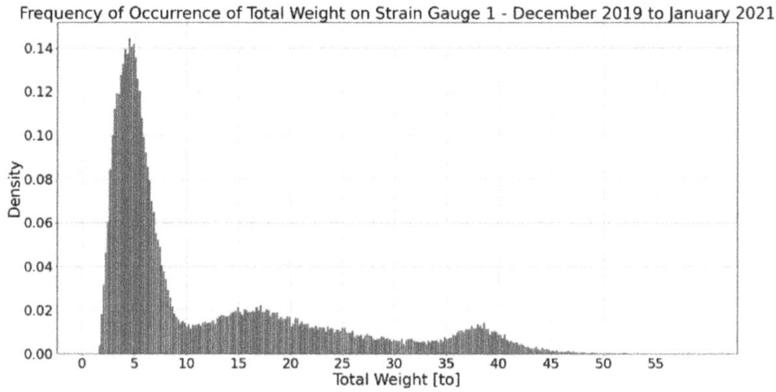

Fig. 7. Histogram for the total weight of the vehicles on the Heinrich's bridge over 14 months

Based on the method of [26] a trimodal normal distribution is used to describe the measured traffic, see Eq. (3):

$$NV(x) = \sum_{i=1}^{n} \xi_i \cdot NV_i(x, \mu_i, \sigma_i) \tag{3}$$

$$with \sum_{i=1}^{n} \xi_i = 1 \, and \, n = 3$$

Figure 8 shows the histogram with the fitted trimodal normal distribution. A big number of lighter vehicles can be observed, which is explained with the rural characteristic of the road. The normal distribution for the light traffic (red line) has a mean value μ_1 of 4.58 tons and a standard deviation σ_1 of 1.57 tons, which equals a coefficient of variation of 34%. The medium heavy traffic (green line) is approximated with a normal distribution with a mean value μ_2 of 16.20 tons and a standard deviation σ_2 of 7.77 tons, which equals a coefficient of variation of 48%. The normal distribution of the heavy traffic (yellow line) indicates a mean value μ_1 of 37.87 tons and a standard deviation σ_1 of 3.49 tons, which equals a coefficient of variation of 9% which is compliant with the literature suggestions.

For the development of a statistical parameter for the traffic load model only the heavy truck loads are considered, which are defined in this analysis as vehicles with a weight of more than 34 to. A consideration of the lighter vehicles for the characteristic value defined as 99,9% quantile according to DIN EN 1991-2 [9] underestimates the traffic loads, because it shifts the mean value from the heavy traffic to the normal traffic, which is much more frequent. It is important to note that the measurement does not indicate if the total weight of the traffic is caused by one vehicle or two separate vehicles of half the

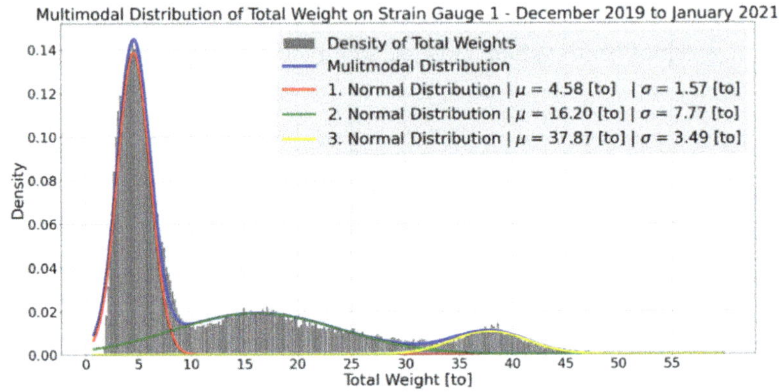

Fig. 8. Trimodal normal distribution for the total weight of the vehicles on the Heinrich's bridge

height each. Since the traffic load model LM1 is also only an imaginary single vehicle representing the actual traffic, this aspect is of subordinate importance. In literature there are basically three distribution types (Normal, log. Normal and Gumbel) suggested for modelling the traffic loads. Figure 9 shows the fitted curves for the measured data.

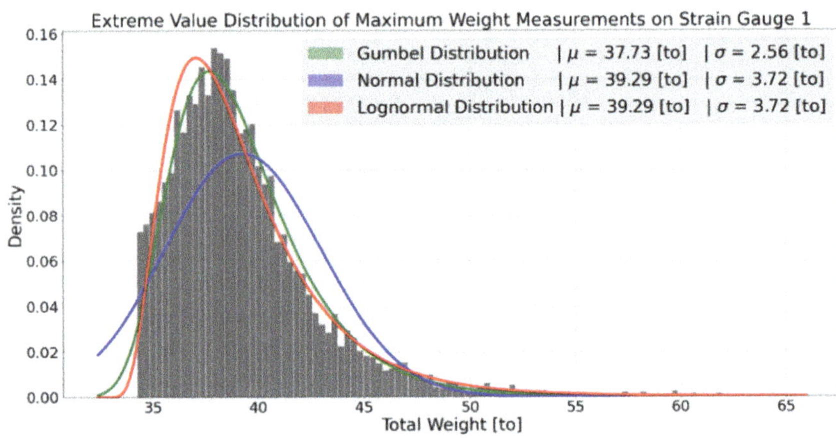

Fig. 9. Histogram for the total weight of the heavy vehicles on the Heinrich's bridge with fitted probability density functions for NV, LNV and Gumbel distributions

From visual inspection of Fig. 9 the Gumbel distribution (green line) seems to fit best and the log. Normal distribution (red line) seems also to be a good approximation. The normal distribution (blue line) does not fit to the dataset well. A QQ plot (quantile-quantile plot) is an easy graphical tool to assess whether a certain dataset follows a given theoretical distribution. The more the results tend to be on a straight line, the better the selected distribution is suited to describe the actual dataset. Figure 10 shows the results of the QQ-plots for the Heinrich's bridge. As previously assumed, a normal distribution does not fit well, especially at the tails, to the dataset. The log. Normal distribution and the Gumbel distribution both do fit the dataset better, while the log. Normal distribution is slightly more aptly in the tail areas.

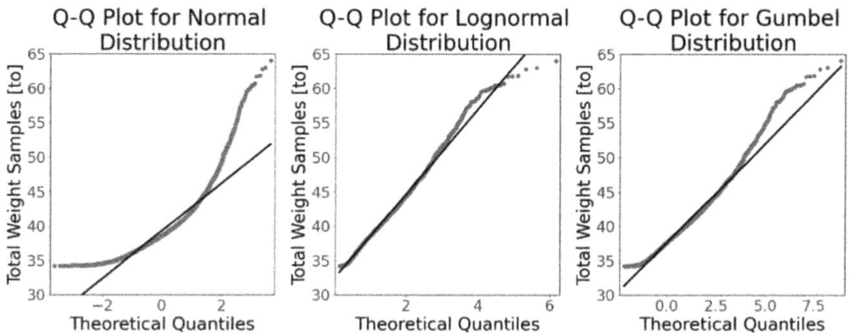

Fig. 10. Q-Q Plots of the measured traffic on the Heinrich's bridge

5 Summary and Conclusion

An on-site measurement can provide more detailed information about the actual loading of the structure. A sample rate of 40 Hz was sufficient for the strain measurement and leads to a remarkably reduced file size compared to a sample rate of 500 Hz. For a use in other projects, the authors recommend to generally start with a high sampling rate and to check the results after some time and to adjust the sampling rate afterwards.

Based on the measurement data over 15 months on the presented case study, the parameters of the probability distribution are calculated. For the distribution type a log. Normal distribution is best suited to describe the measurement data of the heavy traffic, but the choice of Gumbel distribution is also appropriate. A coefficient of variation v_x for heavy traffic load of 9,5% is derived, which is within the lower edge of the range from existing literature.

References

1. DIN 1072:1925. Straßen- und Wegbrücken. Lastannahmen, Berlin (1925)
2. Bundesministerium für Verkehr, Bau und Stadtentwicklung. Richtlinie zur Nachrechnung von Straßenbrücken im Bestand (Nachrechnungsrichtlinie) (2011)
3. Bundesministerium für Verkehr, Bau und Stadtentwicklung. Richtlinie zur Nachrechnung von Richtlinie zur Nachrechnung von Straßenbrücken im Bestand (Nachrechnungsrichtlinie) - 1. Ergänzung (2015)
4. Arndt, W.-H.: Ersatzneubau Kommunale Straßenbrücken. Endbericht, Berlin (2013)
5. Grabow, B.: Ersatzneubau kommunale Straßenbrücken - vorläufige Ergebnisse: Projekt zur Abschätzung des Ersatzbedarfes im Bereich der kommunalen Infrastruktur am Beispiel der Straßenbrücken (2013)
6. DIN 1072:1952-06. Straßen- und Wegbrücken. Lastannahmen, Berlin
7. DIN 1072:1985-12. Straßen- und Wegbrücken. Lastannahmen, Berlin

8. Merzenich, G., Sedlacek, G.: Hintergrundbericht zum Eurocode 1 – Teil 3.2: Verkehrslasten auf Straßenbrücken. Typo-Druck & Verlagsgesellschaft, Bonn (1995)
9. DIN EN 1991-2:2010-12: Eurocode 1: Einwirkungen auf Tragwerke - Teil 2: Verkehrslasten auf Brücken; Deutsche Fassung EN 1991-2:2003 + AC:2010
10. Bundesanstalt für Straßenwesen. Anpassung von DIN-Fachberichten "Brücken" an Eurocodes: Berichte der Bundesanstalt für Straßenwesen Heft B 77
11. Merzenich, G.: Entwicklung eines europäischen Verkehrslastmodells für die Bemessung von Straßenbrücken. Dissertation. RWTH Aachen (1994)
12. Bruls, A., Calgaro, J.-A., Mathieu, H., Prat, M.: Background studies to ENV 1991-3. In: IABSE Colloquium Basis of Design and Actions on Structures. Background and Application of Eurocode 1, pp. 215–227. Delft (1996)
13. Steffens, N.: Sicherheitsäquivalente Bewertung von Brücken durch Bauwerksmonitoring. Shaker Verlag, Düren (2019). ISBN: 9783844070514
14. DIN-Fachbericht 101:2009-03. Einwirkungen auf Brücken, Berlin
15. Znidaric, J., Znidaric, A.: Evaluation of the carrying capacity of existing bridges. Final Report, Institute for Testing and Research in Materials and Structures, Ljublijana (1994)
16. Moses, F., Ghosn, M.: A comprehensive study of bridge loads and reliability. Final Report No. FHWA/OH-85/005. Ohio Department of Transportation, Columbus (1985)
17. Stewart, M., Rosowsky, D.V.: Structural safety and serviceability of concrete bridges subject to corrosion. J. Infrastruct. Syst. **4**(4), 146–155 (1998)
18. Eichinger, E.M.: Beurteilung der Zuverlässigkeit bestehender Massivbrücken mit Hilfe probabilistischer Methoden, Wien (2003)
19. Nowak, M., Fischer, O., Müller, A.: Realitätsnahe Verkehrslastansätze für die Nachrechnung der Gänstorbrücke über die Donau. Beton und Stahlbetonbau 115. Heft **2**, 91–105 (2020). https://doi.org/10.1002/best.201900060.Ernst&Sohn,Berlin
20. Braml, T., Wurzer, O.: Probabilistische Berechnungsverfahren als zusätzlicher Baustein der ganzheitlichen Bewertung von Brücken im Bestand. Beton- und Stahlbetonbau 107. Heft **10**, 654–668 (2012). https://doi.org/10.1002/best.201200037.Ernst&Sohn,Berlin
21. Orcesi, A., Diamantidis, D., O'Connor, A., Palmisano, F., et al.: Investigating partial factors for the assessment of existing reinforced concrete. Bridges Struct. Eng. Int. (2023). https://doi.org/10.1080/10168664.2023.2204115
22. Fib Bulletin No. 80. Partial factor methods for existing concrete structures. In: Recommendation (2016). https://doi.org/10.35789/fib.BULL.0080
23. Boros, V.: Assessment of existing bridges – practical implementation with case studies. IABSE Webinar on 9th June 2023 (2023). https://iabse.org/elearning/webinars
24. Assessment and Rehabilitation of Central European Highway Structures, D08. Recommendations on the use of results from monitoring on bridge safety assessment and maintenance. In: Annex A: Recommendations on Bridge Traffic Load Monitoring. Final Report (2009)
25. Ivankovic, A.M., Skokandic, D., Znidaric, A., Kreslin, M.: Bridge performance indicators based on traffic load monitoring. Struct. Infrastruct. Eng. **15**(7), 899–911 (2019). https://doi.org/10.1080/15732479.2017.1415941
26. Bundesanstalt für Straßenwesen. Auswirkungen des Schwerlastverkehrs auf die Brücken der Bundesfernstraßen. Berichte der Bundesanstalt für Straßenwesen Brücken- und Ingenieurbau Heft B 68(2009). https://bast.opus.hbz-nrw.de/opus45-bast/frontdoor/deliver/index/docId/214/file/B68.pdf
27. Rainieri, C., Fabbrocino, G.: Introduction. In: Operational Modal Analysis of Civil Engineering Structures, pp. 1–21. Springer, New York (2014). https://doi.org/10.1007/978-1-4939-0767-0_1
28. Butterworth, S.: On the theory of filter amplifiers. Exp. Wirel. Wirel. Eng. **7**, 536–541 (1930)

Non-destructively Measurable Quantities Relevant to Reliability Assessment of Existing Concrete Bridges – Case Studies

Stefan Küttenbaum[1(✉)], Christian Kainz[2], and Thomas Braml[2]

[1] BAM – Federal Institute for Materials Research and Testing, Berlin, Germany
stefan.kuettenbaum@bam.de

[2] Institut für Konstruktiven Ingenieurbau, Universität der Bundeswehr Munich, Munich, Germany

Abstract. In reliability assessment of existing bridges, collected information about the structure can be utilized to modify partial safety factors and to refine characteristic values individually. In this way, uncertainties covered in initial calculation models and arising from assumptions can be reduced as well as biases identified and corrected. This allows for more realistic assessments and more appropriate decisions on actions to be taken. However, the on-site observation of information is generally associated with certain costs. For this reason, an efficient data-supported reassessment requires that only the individually relevant testing tasks are solved. Additional sensitivity analyses can further be used to specify the requirements on certain inspections, such as the testing result quality to be achieved, in order to avoid too imprecise or too expensive measurements on-site. In this contribution, the methodology for the demand-actuated definition of measurement objectives is demonstrated using the example of non-destructive concrete testing methods and findings from reassessments of two prestressed concrete bridges. Each case study includes the semi-probabilistic identification of structural weak points in ultimate and serviceability limit states, a comparison with respective probabilistic sensitivity analyses, and the subsequent derivation of parameters that are a) measurable and b) highly relevant in the individual reliability analyses.

Keywords: Reliability assessment · existing structures · concrete bridges · inspection · non-destructive testing

1 Introduction

The comprehensive replacement of structures represents a maintenance strategy for our aging and degrading transport infrastructure that is financially not feasible and would have unacceptable operational and environmental impacts. The reliability assessment of existing structures yields valuable information about structural safety, which are necessary to extend service lives and to make optimal decisions about future actions and optimal maintenance strategies. The application, however, requires manifold and individual information about the considered system. This contribution attempts to shed some

J. C. Matos et al. (Eds.): IPW 2024, LNCE 494, pp. 209–219, 2024.
https://doi.org/10.1007/978-3-031-60271-9_18

light on the effective utilization of non-destructively measurable structure parameters relevant to reliability assessment of existing concrete bridges as vulnerable elements of infrastructure networks.

Each of the following two sections deals with one actual large-scale structure as a case study including i) the semi-probabilistic identification of structural weak points in ultimate and serviceability limit states according to the Eurocodes and ii) the comparison with probabilistic sensitivity analyses. The aim is to derive the mostly non-destructively measurable structural parameters that are relevant in the individual reliability assessment to show possibilities, limits, and potentials of on-site NDT in the course of prestressed concrete bridge reassessments. Further background information can be found in a recently published state-of-the-art report on the inclusion of inspection and monitoring techniques in the assessment of concrete structures [1].

All reliability assessments are performed according to stages 1 and 2 of the German recalculation guideline [2, 3] in the following limit states: Robustness reinforcement; bending and axial force; shear and torsion; decompression; concrete, reinforcement and prestressing steel stresses; crack width limitation. The target load level has been set as load model 1 (LM1). The First and Second Order Reliability Methods acc. to [4–6] have been applied, amongst others, to compute the values of the sensitivity coefficient.

2 Case Study I – Single-Span Motorway Bridge

This prestressed concrete bridge with a length of 18,50 m and a width of 15,00 m is part of a motorway with in total four lanes. The slab-and-beam cross-section consists of six prefabricated main girders with a height of 0,71 m and an additional concrete slab with a height of 0,25 m casted on-site. The longitudinally prestressed system was constructed in 1978 and is shown in Fig. 1.

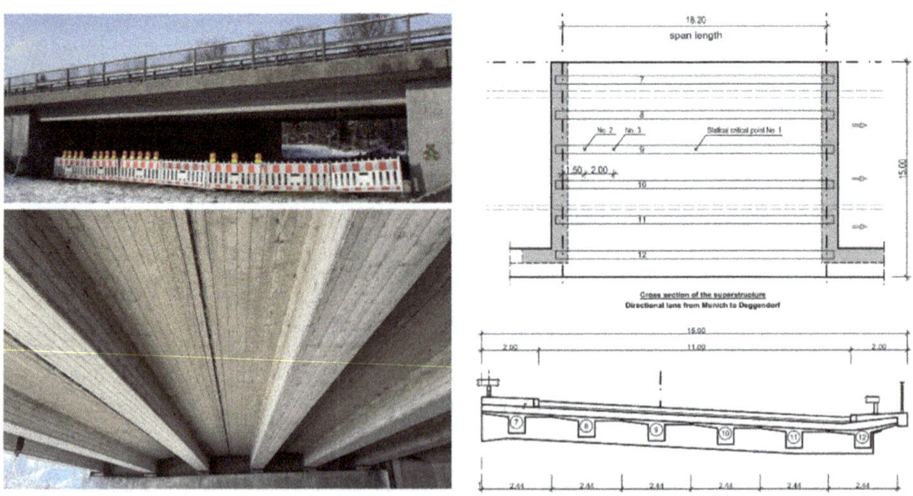

Fig. 1. Photos of the single span PC bridge (on the left), taken by *BUNG Ingenieure AG*; top-view and cross section (on the right), created by *Bundeswehr University Munich*

The following numerical weaknesses have been identified during the recalculation:

1. Longitudinal reinforcement: The amount in the t-beam is insufficient to withstand combined effects of actions due to bending and torsion as well as fatigue.
2. Transversal reinforcement: The reinforcement in the slab is insufficient to withstand the transverse bending moments.
3. Shear reinforcement: The amount in the beams is insufficient (acc. to stage 1).
4. Decompression: Evidence could only be given through the consideration of the concrete tensile strength $f_{ctk,0,05}$, so that $0 < \sigma_c < f_{ctk,0,05}$, where σ_c are the concrete stresses.

It might be obvious that the insufficient amount and spacing of reinforcement bars cannot be improved through on-site testing. However, the investigation is particularly useful, if as-built information is missing or doubts about the values assumed or stated in plans have arisen. That as-built documentations cannot always be relied upon reveals, for example, the investigation of 157 municipal bridges in Germany. No documents could be found for 42 of these structures [7]. Against this background, in addition to reducing uncertainty about the information, clarifying the internal structure and possibly identifying biases in the assumptions can also be beneficial.

The alpha values indicating the stochastic relevance of the input quantities in the ultimate limit states bending and shear are given below. The values describe the ratio of the change in reliability $\Delta\beta$ when the random variable is varied by Δu_i and are approximated using the difference quotient. For correlated quantities, the transformation relationship needs to be included in the calculation. The larger the value, the stronger the influence of the respective basic variable X_i on the results, i. e., the probability of failure p_f and the reliability index β, respectively.

The limit state equation for reinforced concrete elements from [9] has been extended by the prestressing to perform the bending proof acc. to the Eurocodes. Figure 2 shows the parameters of the decisive basic variables and the results of the sensitivity analysis. The bending moment due to the tandem system (load model 1) is the most influential basic variable besides the model uncertainties with a squared alpha-value of 26%. The plots in Fig. 3 indicate the computed reliability index of $\beta = 4{,}17$ ($T = 50$ a) and reveal, that i) a deviation between the vertical as-planned and as-built tendon duct position would affect reliability slightly and ii) the reduction of the uncertainty associated with the vertical tendon duct position through additional testing would be inefficient.

The tension strut proof is based on the limit state equation given in [9]. The resistance is the product of the steel tensile strength, the reinforcement and compression strut inclination, the inner lever arm as well as the steel cross-sectional area related to the rebar spacing of the shear reinforcement [10]. Besides the model uncertainties, the values of the sensitivity coefficients in Fig. 4 indicate, that the shear force due to the tandem system, the shear reinforcement strength and the spacing of the shear reinforcement s_w noticeably affect the computed reliability. The parameter studies in Fig. 5 show i) the computed reliability of $\beta = 3{,}79$ ($T = 50$ a) and ii) that a deviation between the assumed and the actual shear reinforcement spacing would significantly influence the results.

Basic variable (with |α| > 0,10)		Sensitivities		Parameters		
		|α|-value	α²	distr. type	mean	CoV
f_p	Yield strength of prestressed steel	0,17	3%	Gaussian	1.632,8 MN/m²	2,5%
A_P	cross-sectional area of prestressed steel	0,14	2%	Gaussian	0,0014 m²	2%
My_CS	Moment due to creep + shrinkage	0,14	2%	Gaussian	0,2550 MNm	30%
My_TS	Moment due to tandem system	0,51	26%	Gumbel	0,9293 MNm	15%
U_RM	Model uncertainty (resistance)	0,57	33%	Log.-normal	1,025	7%
U_E	Model uncertainty (effects of actions)	0,53	28%	Log.-normal	1	7%
X_i	Further quantities	—	6%	—		

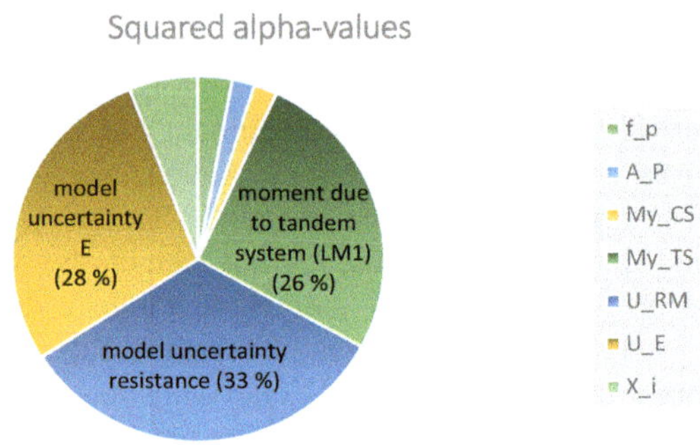

Fig. 2. Results of the sensitivity analysis (bending proof) for one position in excerpts and respective parameters of input quantities, values extracted from [8]

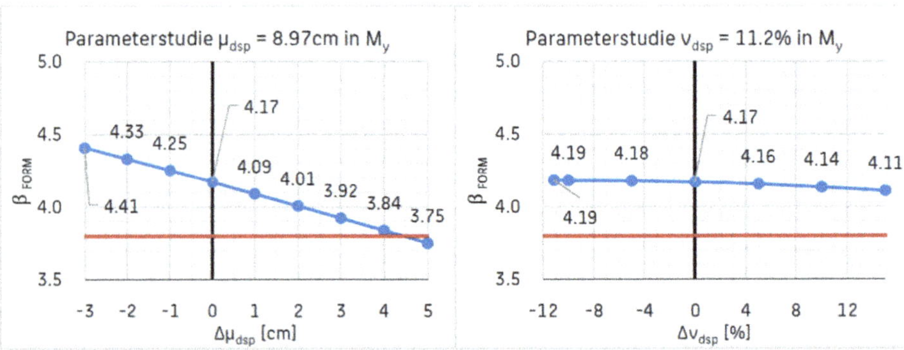

Fig. 3. Reliability index against the mean and the coefficient of variation (CoV) of the vertical tendon duct position d_{sp} for one cross-section (bending proof); extracted from [8]

The relevance of the measurable variables can be derived through a combination of engineering considerations, sensitivity analyses and recalculations. The localization of the rebars mounted in the slab may be useful to i) clarify in longitudinal direction if the consideration in the torsion proof is standard-compliant and ii) update the information about the transversal steel amount for the bending proof. Information about the amount of shear rebars is particularly relevant in the transition areas between densely and less reinforced areas in longitudinal direction, as exceedances of up to 8% were obtained in these areas while verifying the tension strut semi-probabilistically. Figure 5 additionally indicates the importance of sound knowledge about the mean of the rebar spacing.

Basic variable (with |α| ≥ 0,10)		Sensitivities |α|-value	α²	Parameters distr. type	mean	standard dev.
V_zTS	Shear force due to tandem system	0,44	19%	Gumbel	0,2201 MN	0,033 MN
s_w	Spacing of shear reinforcement	0,24	6%	Gaussian	0,1 m	0,005 m
f_yw	tensile strength shear reinforcement	0,30	9%	Log.-normal	552,9 MN/m²	33,17 MN/m²
U_Rvsyw	Model uncertainty (resistance)	0,50	25%	Log.-normal	1,1	0,11
U_EV	Model uncertainty (effects of actions)	0,60	36%	Log.-normal	1	0,12
X_i	Further quantities	—	5%	—		

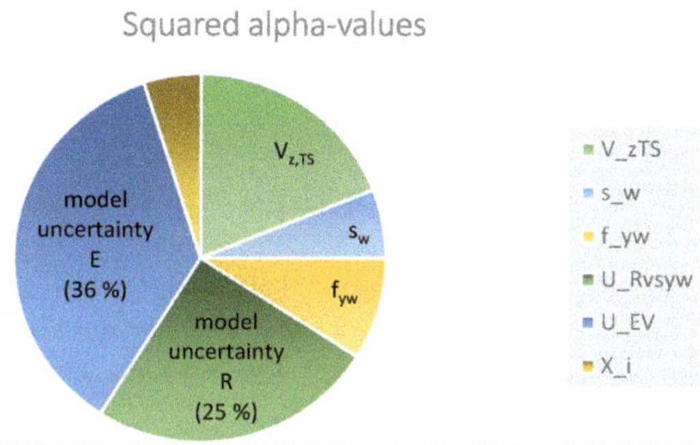

Fig. 4. Results of the sensitivity analysis (shear proof) for one position in excerpts and respective parameters of input quantities, values extracted from [8]

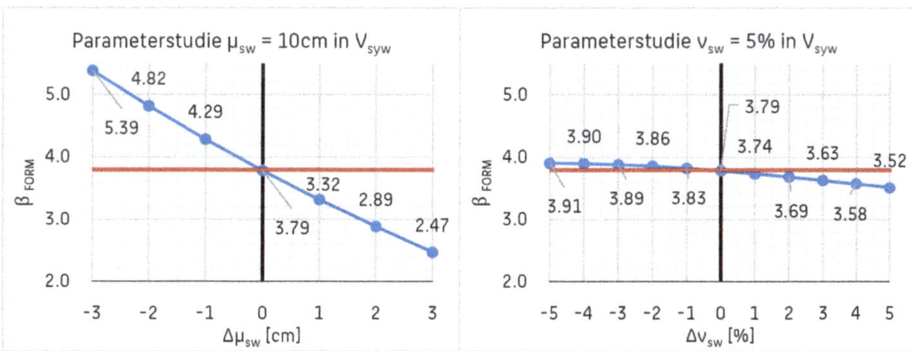

Fig. 5. Reliability index against the mean and the coefficient of variation (CoV) of the shear stirrup spacing s_W for one cross-section (shear proof); extracted from [8]

The curves of the tendons could be measured to validate the highly utilized decompression proof and FE model outputs. Testing results deviating from the assumptions would also influence the shear resistance. Collisions during core drilling could be also avoided, which is beneficial to determine the concrete quality and strength, against the background, that the tensile strength needed to be considered in decompression proof (Tables 1 and 2).

Table 1. Summary of individually identified relevant and measurable basic variables.

Basic variable	Description	Inspection technique
$A_{s,l}, \varnothing_{s,l}$	Longitudinal reinforcement (slab)	Ultrasound, GPR, x-ray
$A_{s1}, \varnothing_{s1}, d, z$	Transversal reinforcement (slab)	Ultrasound, GPR, x-ray
A_{sw}, s_w	Shear reinforcement	GPR (amount), x-ray (diameter)
A_p, d_p, z	Tendon localization	Ultrasound, GPR
f_c, f_{ct}	Concrete strength	Rebound hammer, drill cores

3 Case Study II – Three-Span Highway Bridge

This system is a longitudinally and transversely prestressed concrete bridge with a total length of 133 m and a width of 12 m as shown in Fig. 6. The height of the single-cell hollow box girder varies between 1,20 m and 1,45 m. The bridge carries a two-lane federal highway and was built in 1965.

The following numerical weaknesses have been identified during the recalculation:

1. Reinforcement: The amount is insufficient to avoid failure without prior notice, which is uncritical as only areas around the moment zero points are affected.
2. Shear proof acc. to Eurocode: Serious deficits have been identified in the tension strut and in the compression strut proof.
3. Connection between slabs and webs: The amount of reinforcement is insufficient.
4. Decompression proof and torsion analysis: Minor deficits have been identified.

The considerable variability and the (e.g., compared to the bending analysis) low level of approximation of the shear force model stated in the Eurocodes as well as the resulting recurring underestimation of the shear load-bearing capacity of existing structures motivated the verification of the main tensile stresses under predominantly static loading, which corresponds to a stage two evaluation according to the German recalculation guideline. Further information about the following limit state equation are given in [11]. The requirements can be found in section 12.4.3.3 of the recalculation guideline.

Evidence must be provided in different vertical positions *i* that the main tensile stress

$$\sigma_{I,Ed,i} \leq k_1 \cdot f_{ctd}, \tag{1}$$

where f_{ctd} is the concrete tensile strength and where the coefficient k_1 depends on the shear reinforcement amount. The main tensile stresses vary over the vertical distance to the centerline of the cross section z_i and are

$$\sigma_{I,Ed,i} = 0,5 \cdot \sigma_{cx,Ed,i} + \sqrt{0,25 \cdot \sigma_{cx,Ed,i}^2 + \left(\tau_{V,Ed,i} + \tau_{T,Ed}\right)^2} \tag{2}$$

with

$$\sigma_{cx,Ed,i} = \frac{N_{Ed}}{A_c} + \frac{M_{Ed}}{I_y} \cdot z_i, \tag{3}$$

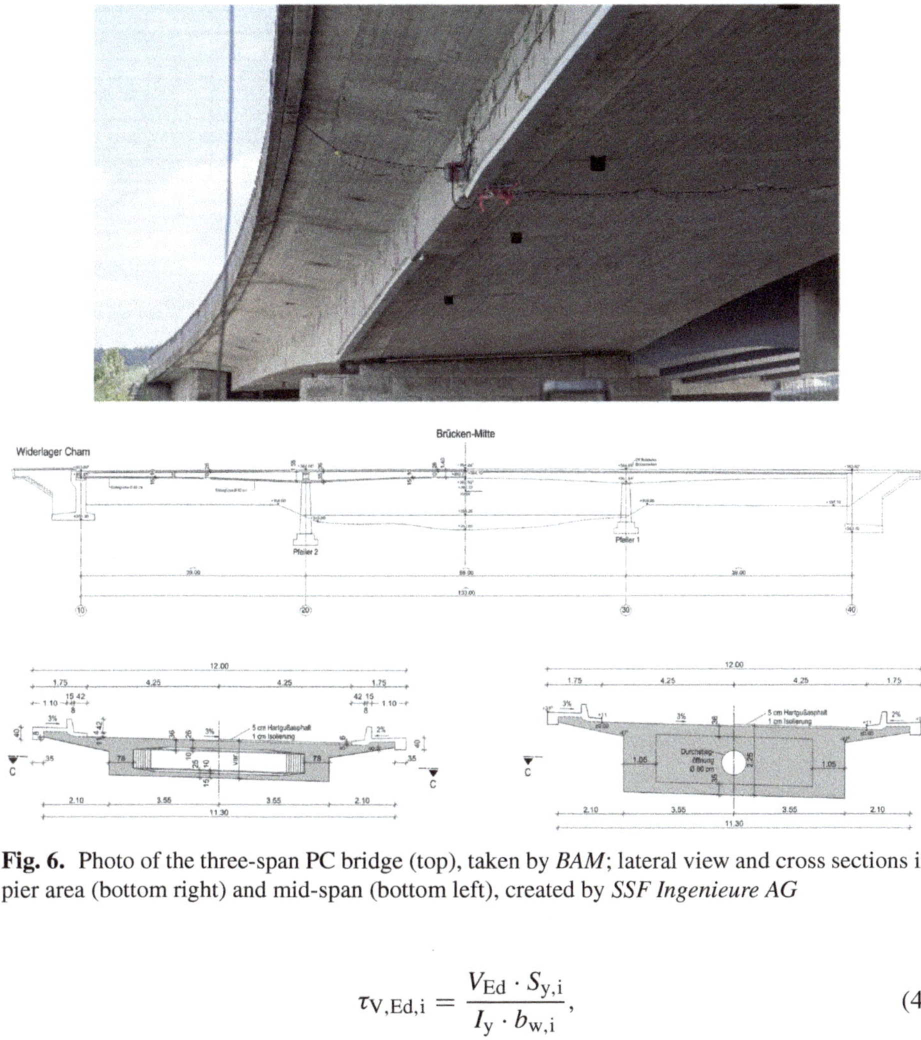

Fig. 6. Photo of the three-span PC bridge (top), taken by *BAM*; lateral view and cross sections in pier area (bottom right) and mid-span (bottom left), created by *SSF Ingenieure AG*

$$\tau_{V,Ed,i} = \frac{V_{Ed} \cdot S_{y,i}}{I_y \cdot b_{w,i}}, \tag{4}$$

$$\tau_{T,Ed} = \frac{T_{Ed}}{W_T}. \tag{5}$$

where A_c, $b_{w,i}$ (web width) and $S_{y,i}$, I_y (area moments of inertia) describe the cross-sectional characteristics and W_T is the torsion section modulus.

The distribution of the internal forces and moments are shown in Fig. 7. The traffic loads (uniformly distributed load UDL and tandem system TS) yield the major torsion effects of actions, while the prestressing represents the almost exclusive normal force action. The dead loads, prestressing and UDL contribute significantly to the bending moment and shear force.

The coefficients of variation of the relevant effects of actions were defined in a range between 10% (dead loads, prestressing) and 15% (traffic loads). The traffic loads are assumed to follow a Gumbel distribution, while the others were modelled as Gaussian

Forces and moments (with a percentage of more than 5 %)								
Load case	Torsion moment M_Z / MNm		Normal force N_X / MN		Bending moment M_Y / MNm		Shear force V_Z / MN	
Dead load	-184,5	7%			-31390,3	34%	3638,6	53%
Edge beam					-4201,2	5%	463,1	7%
Prestressing	-304,2	11%	-43214,2	99%	39573,3	43%	-1273,8	18%
UDL (LM 1)	-949,7	34%			-8077,4	9%	821,7	12%
TS (LM 1)	-811,9	29%					554,8	8%
Further quantities		19%		1%		9%		2%

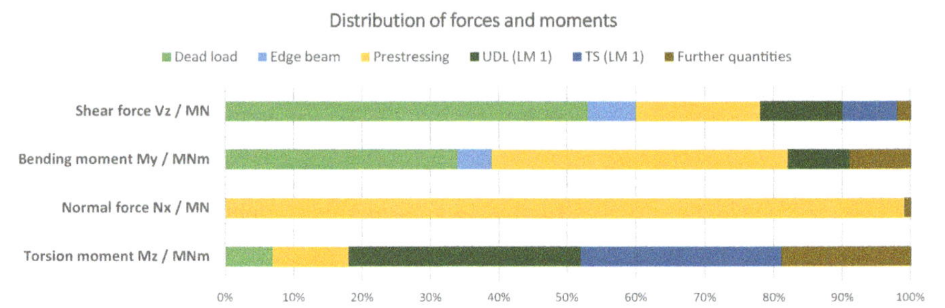

Fig. 7. Distribution of forces and moments for the main tensile proof in one exemplary position

distributions. The results of the sensitivity analyses are summarized in Fig. 8. It can be seen that the concrete tensile strength dominantly influences structural reliability in this evaluation scenario, which is consistent with the engineer's expectations when considering Eq. (1).

Basic variable		Sensitivities		Parameters		
(with |α| ≥ 0,10)		α-value	α²	distr. type	mean	standard dev.
Nx_P	Normal force due to prestressing	-0,1	1%	Gaussian	-43,21 MN	4,32 MN
Vz_1	Shear force due to dead loads	-0,16	3%	Gaussian	3,64 MN	0,36 MN
f_ctm	Concrete tensile strenth	0,94	88%	Gaussian	3,80 MN/m²	1,12 MN/m²
U_RM	Model uncertainty (resistance)	0,16	3%	Log.-normal	1,1	0,11
U_E	Model uncertainty (effects of actions)	-0,16	3%	Log.-normal	1	0,1
X_i	Further quantities	—	3%	—		

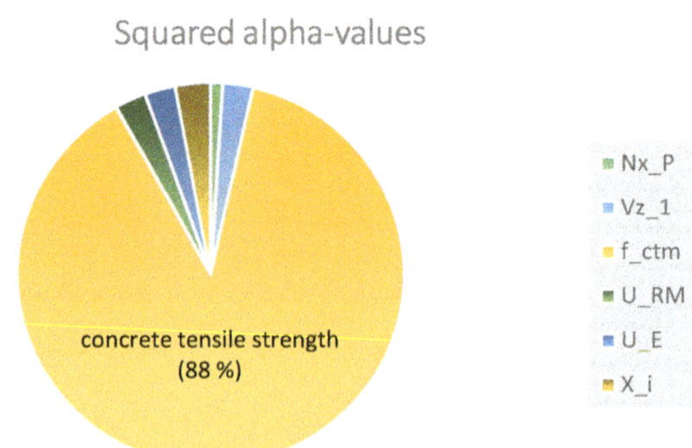

Fig. 8. Results of the sensitivity analysis (proof of main tensile stresses) for one position in excerpts and respective parameters of input quantities

The combination of engineering judgements, the semi-probabilistic recalculation outcomes, and the sensitivity analyses, from which the results of one were shown above, provides in-depth insights into the relevance of all input quantities necessary to consider for the reliability reassessment. All three approaches consistently show the great importance of the concrete tensile strength for the main tensile stress verification. Further investigations about the amount and spacing of the shear reinforcement are particularly useful if doubts exist concerning the plan specifications or if information is missing. The visual inspection of this bridge motivated further condition assessments [12].

Table 2. Summary of individually identified relevant and measurable basic variables.

Basic variable	Description	Inspection technique
f_{ct}	Concrete tensile strength	Rebound hammer, drill cores, diametric compression test
A_{sw}, s_w	Shear reinforcement	GPR (amount), x-ray (diameter)

The analysis of the relevant parameters in the reliability assessment of existing bridges and the comparison with the reliability of the available inspection and monitoring techniques also facilitates the identification of current metrological research and development needs. The key factor influencing reliability in this case study is the concrete tensile strength. However, it is well known that the measurement of this in-situ strength can be associated with comparably large uncertainties in the double-digit percentage range when performing diametric compression tests or deriving the tensile strength from the concrete strength based on hydraulic compression tests on drill cores. A more reliable, in the sense of more accurate, measuring method would be desirable.

4 Conclusion

The main findings of the case studies presented in excerpts are the following:

- Technical reasoning, engineering judgements, semi-probabilistic reassessment and sensitivity analyses can be used, also in combination, to identify the relevant testing tasks individually. Their comparison with capabilities of available testing and monitoring techniques allows for the efficient selection of a suitable measurement method and development of a reasonable measurement procedure.
- If a testing task cannot be solved with satisfactory accuracy, this results in a need for metrological research and development activities. One example in this article is the concrete tensile strength.
- The individually relevant testing tasks are: Concrete strength, area and spacing of the shear reinforcement, position and amount of reinforcement, position and cross-sectional area including the condition of tendons.
- Comparably reliably solvable testing tasks using NDT on concrete under common boundary conditions are: Localization of steel in concrete, concrete cover, concrete strength estimation using rebound hammer, component thicknesses, carbonation and

chloride ingress using laser induced breakdown spectroscopy, areas subjected to active corrosion using half-cell potential, see, e.g., [1].
- Another finding from the project work is the importance of ensuring a suitable flow of information between structural engineers and experts in inspections and probability theory, respectively. Examples include the collaborative definition of inspection objectives (such as the definition of the accuracies to be achieved and the comparison with expected achievable accuracies incl. Sample sizes) or the unambiguous expression of FE model outputs (are they characteristic values, mean values, etc.?) for the probabilistic reassessment.

A total of six prestressed and reinforced concrete road bridges are being evaluated as pilot studies within the scope of the national pre-standardization project "ZfPStatik" – with and without on-site measured data. The project aims for the development of a set of rules for the reliability assessment of existing structures on the basis of non-destructive testing results. The measured information is used to refine calculation models in order to facilitate more efficient assessments, improved service life predictions, demand-oriented maintenance strategy definitions as well as, overall, the optimization of the consumption and distribution of resources (including budgets).

Acknowledgments. The authors gratefully acknowledge the financial support of the German Federal Ministry for Economic Affairs and Climate Action under grant number 03TN0042 and the valuable activities of all project partners.

References

1. Existing Concrete Structures Life Management, Testing and Structural Health Monitoring: State-of-the-Art Report. [Lausanne]: Fédération Internationale du Béton (FIB) (2023)
2. Bundesministerium für Verkehr, Bau und Stadtentwicklung: Richtlinie zur Nachrechnung von Straßenbrücken im Bestand (Nachrechnungsrichtlinie) (2011)
3. Bundesministerium für Verkehr und Digitale Infrastruktur. Richtlinie zur Nachrechnung von Straßenbrücken im Bestand (Nachrechnungsrichtlinie) (2015)
4. Hasofer, A.M., Lind, N.C.: Exact and invariant second-moment code format. J .Eng. Mech. Div. **1974**(100), 111–121 (1974)
5. Rackwitz, R., Fiessler, B.: Structural reliability under combined random load sequences. Comput. Struct. **9**(5), 489–494 (1978). https://doi.org/10.1016/0045-7949(78)90046-9
6. Hohenbichler, M., Rackwitz, R.: Non-normal dependent vectors in structural safety. J .Eng. Mech. Div. **107**(EM6), 1227–1238 (1981)
7. Rechnungshof Rheinland-Pfalz: Bericht Nach § 111 Abs. 1 LHO über die Erhaltung und den Zustand von Brücken in kommunaler Baulast: Az.: 2-P-0057-39-1/2011. Speyer (2013)
8. Özlü C.: Probabilistischer Nachweis einer Spannbetonbrücke mit generierbaren Informationen aus der Zustandserfassung. Master-Thesis. Berlin (2023)
9. Braml, T.: Zur Beurteilung der Zuverlässigkeit von Massivbrücken auf der Grundlage der Ergebnisse von Überprüfungen am Bauwerk. Zugl.: Neubiberg, Univ. der Bundeswehr München, Diss., 2010. VDI-Verlag, Düsseldorf (2010)
10. CEN: EN 1992-2:2005-10. Eurocode 2 - Design of Concrete Structures - Concrete bridges - Design and Detailing Rules. Incl. EN 1992-2/AC:2008(1992-2) (2005)

11. Hegger, J., et al.: Zur Anwendung des Hauptzugspannungskriteriums bei der Nachrechnung bestehender Spannbetonbrücken. BUST **110**(2), 82–95 (2015). https://doi.org/10.1002/best.201400106
12. Maack, S., Torrent, R., Ebell, G., Völker, T., Küttenbaum, S.: Testing to reassess – corrosion activity assessment based on NDT using a prestressed concrete bridge as case-study. In: Pellegrino, C., Faleschini, F., Zanini, M.A., Matos, J.C., Casas, J.R., Strauss, A. (eds.) EUROSTRUCT 2021. LNCE, vol. 200, pp. 678–686. Springer, Cham (2022). https://doi.org/10.1007/978-3-030-91877-4_77

Optimization of Combustion Parameters in Heating Systems to Increase Efficiency and Environmental Protection

Diana Car-Pusic[1(✉)], Silvana Petrusheva[2], and Valentina Zileska Pancovska[2]

[1] Faculty of Civil Engineering, University of Rijeka, Rijeka, Croatia
diana.car.pusic@uniri.hr

[2] Faculty of Civil Engineering, Ss Cyril and Methodius University, Skopje, North Macedonia

Abstract. The rational consumption of energy and the consequent reduction of the emission of harmful gases into the environment are today's imperatives. We can expect for them to increase in the future. While users are still trying to reduce energy consumption to reduce costs, the idea of the need to reduce environmental pollution has not yet taken hold despite numerous appeals. Heating systems in public and private buildings offer the possibility of rationalization, especially if they are older. Practical experience and research indicate that due to wear and tear, poor maintenance and sometimes poor fuel quality in the systems, a lower degree of energy utilization (utilization coefficient ETA) is achieved and, consequently, there is higher consumption and increased emission of harmful gases CO_2, O_2 and CO into the environment.

This research aims to use predictive modelling methods that will determine the relation between the level of ETA, coefficient of energy utilization and the emission of CO_2, O_2 and CO into the environment.

For this purpose, measurement and analysis of the exhaust gases of the oil and gas heating system in 38 public and private buildings in the Primorje-Gorski Kotar County in the Republic of Croatia was carried out. Measurements were made on systems of different ages.

Appropriate measures in heating systems, even older ones, and regular high-quality maintenance can increase the ETA parameter and reduce the emission of harmful gases into the environment. The goal is to propose measures to maintain the system and increase energy utilization in heating systems.

Keywords: Heating System · Public and Private Buildings · Predictive modelling · Utilization Coefficient (ETA)

1 Introduction

The rational consumption of energy and the consequent reduction of the emission of harmful gases into the environment are today's imperatives. Heating systems in public buildings provide this possibility, especially if it is an older existing system, where in practice, it is often the case that due to wear and tear, poor maintenance and poor fuel quality, a lower level of energy consumption and increased CO_2 emissions into the

J. C. Matos et al. (Eds.): IPW 2024, LNCE 494, pp. 220–229, 2024.
https://doi.org/10.1007/978-3-031-60271-9_19

environment is achieved. The authors in the paper [1] noted that "the production of carbon dioxide is due to energy consumption in different parts of the building. Various solutions, such as the use of clean energy, the use of natural light, phase change material, heat recovery and insulation can be listed to solve these two problems."

As stated by the authors in the paper [2], "...Greenhouse gases generally include water vapour, carbon dioxide (CO_2), methane (CH_4), nitrous oxide (N_2O), tropospheric ozone (O_3) and chlorofluorocarbons (CFCs)."

The development aspect and potential of using CO_2 as a working fluid for the next generation of energy systems "due to its extraordinary thermo-physical properties and environmental friendliness" should be highlighted. Regarding CO_2 in central heating systems, fuel combustion would be complete, with as much CO_2 content as possible and as little CO and soot content as possible, it is essential to bring a sufficient amount of air into the combustion chamber [3].

The authors in [4] point out that CO_2 as natural working fluid for "the next-generation energy systems to tackle climate changes due its extraordinary thermophysical properties and environmental friendliness..." Still, there are a number of problems, especially in the production of the necessary components and the balance of costs and system efficiency. The authors in [2] point out the construction industry as a massive consumer of energy and polluter of the environment. Some authors also point to national regulation of energy efficiency as closely connected to minimizing energy consumption in buildings [5]. There, the design phase is crucial for selecting rational project solutions, both in construction and in terms of rationalizing energy consumption and maintaining environmental elements. In that relation, for minimization of life cycle cost and CO_2 emissions of buildings, authors in [6] proposed a simulation-based multi-objective optimization method.

There are other studies that are focused on different aspects of heating systems, considering technical, economic, environmental, and social factors, such as: energy-efficient building and energy indicators in European states [7]; minimization of temperature levels in district heating networks and optimization of the heating system of building [8]; proposing artificial neural network model for office building for optimizing the operation times of a heating system [9]; hot water and integrated CO_2 systems with heating, ventilation and air conditioning (HVAC) [10]; simulation model focused on energy supply systems of buildings [11], etc.

With appropriate measures in heating systems, even older ones, and with regular high-quality maintenance, it is possible to increase the ETA parameter and reduce the emission of harmful gases into the environment. In particular, the importance of regular annual service and regulation of the burner should be emphasized, along with checking the flue gases. Thus, this research is justified and aims to contribute its results to the clarification of this issue. The target is to consider the issue of improving the condition of existing central heating systems.

Research Goal

The goal of the research is to analyze the dependence of the degree of usefulness in central heating systems that use extra light fuel oil as an energy source and gas emissions into the environment by applying a neural network model to the example of public and private buildings and to propose measures that can achieve better system efficiency, i.e. reducing energy consumption and achieve less environmental pollution.

2 Research Hypotheses

1) With appropriate measures in the central heating systems, a higher utilization coefficient of the consumed energy ETA can be achieved. At the same time, this means lower energy consumption and lower emission of harmful gases into the environment.
2) Lower exhaust gas temperature at the exit from the boiler results in lower losses in the system, i.e. higher ETA.

3 Research Methodology and Description

3.1 Research Methodology

A saving device installed in front of the burner was developed by the author of the measurements made for the needs of this research. The author is an external collaborator in the research and on the project uniritehnic -18-125 of the University of Rijeka. The device was tested in the 2018 heating season. In 2019, a public facility in the Primorje-Gorski Kotar County in the Republic of Croatia showed exceptional results in fuel savings. The preparation for the application of the innovation is underway.

Furthermore, the measurement and analysis of exhaust gases in central heating systems using extra light (EL) fuel oil was carried out in 38 public and private facilities. The region covered the Republic of Croatia in the coastal area of the Primorje-Gorski Kotar and Istria counties. The recording was made in the period from 2016 to 2023. The available temperature data shown in the table were taken from the sources of the State Hydrometeorological Institute for the period from 1948 to 2021, for the city of Rijeka in the Primorje-Gorski Kotar County and the city of Pazin in the Istrian County [12] where the measuring stations are located. The altitudes of the places where the measurements were taken range from 20 to 250 m.

The coastal part of the County around Rijeka has typical Mediterranean climate, while the city of Pazin has a continental climate. The coastal part of Istria has Mediterranean climate [13]. The absolute minimum in Table 1, marked in bold, refers to the year 2021, while the other minimums are outside the data measurement period for this research, i.e. from 2016 to 2023. However, for comparison, according to the data of DHMZ [12], deviations of the mean air temperature for the measurement period from 2016 to 2023 for Pazin range, for example, for November from 0.10 °C to 2.60 °C, which is more than the average of the multi-year normal (which is the most common period from 1981–2010 or 1991–2020).

The measurement was made with a Delta 65 exhaust gas analysis device (Ser. Number 026220). The following data were measured:

- T_{dplin} – flue gas temperature in °C
- O_2 – volume fraction of oxygen in flue gases in %
- CO – share of carbon monoxide in flue gases in ppm
- CO_2 – volume fraction of carbon dioxide in dry flue gases in %

Table 1. Monthly air temperatures for the cities of Rijeka and Pazin for the period 1948–2022. [12]

Air temp. °C	Jan	Feb	Mar	Apr	Oct	Nov	Dec
Aver	5.8/2.8	6.3/3.7	9.0/6.6	12.7/10.4	14.6/11.8	10.3/7.6	7.0/3.8
Min	−11.4/−18.7	−12.8/−15.9	−7.7/14.0	−**1.2**/−7.8	−1.2/−5.6	−4.5/−10.5	−8.9/−15.5

- T_{air} – air temperature in °C
- Lambda – the ratio of the excess air coefficient (the ratio of the actual amount of air supplied per 1 kg of fuel consumed and the stoichiometric amount of air for burning 1 kg of fuel) to the content of CO_{2max} and CO_2 actually in the flue gases. [14]
- ETA – utilization coefficient (degree of usefulness) in %
- Losses – in %

The energy generating products is a fuel oil with LU EL 3% and 15.4% (CO_2 max content) characteristics. The measured data were processed using neural networks utilizing the software DTREG, where ETA is the target variable. This paper presents 3 models for predicting the target variable ETA using 7 predictors (described above).

3.2 Modelling Using Artificial Neural Networks

We used the predictive modelling software DTREG [15] for building the models, using 3 neural networks: GRNN (general regression neural network), MLP (multilayer perceptron) and RBF NN (radial basis neural network). DTREG is a professional data analysis tool that has been thoroughly tested and used by many researchers. It offers many options, and it is easy to compare model types to see which one performs best.

Neural networks (NN) are part of the *soft computing methods* that transfer the experimental data into analytical models. These methods have been used in the last several decades for solving many problems in almost every scientific area of research. Some of their advantages that make them very attractive for using are: 1) they can learn from data incorporating some human learning abilities, 2) they are robust to the noisy data, 3) their structure can be easily implemented in hardware because of their parallel structure, 4) any multivariant function can be approximated by them, they are especially good at modelling nonlinear processes, 5) many different classes of tasks can be covered by them [16].

The field of soft computing is composed of several mathematics areas, such as: statistic and probability theory, linear algebra, approximation theory, regression, classification, nonlinear optimization and multivalued logic. The models we obtained are with good accuracy, measured by the standard estimators: MAPE (mean absolute percentage error), R^2 (the coefficient of determination that indicates the proportion of variance in the target variable (the dependent variable) which can be explained by the predictors (independent variables) and the coefficient of correlation which measures the direction and strength of the linear correlation between the actual and predicted target values.

Artificial Neural Networks

The architecture and software of artificial neural networks (ANN) are inspired by the structure and functioning of the biological neurons of the human beings nervous system. Today there are many different types of neural networks for solving different types of problems, and also for different types of data. It is always recommended to try several NN models and choose the most appropriate one for the actual data and task.

ANN are composed of several computing units called neurons, organized in several layers: input, hidden and output layer. Neurons are connected with the so-called weights, which are the parameters of the network and they are subject of the learning algorithm of the NN and may have different meaning depending on the task they solve. Usually, in every layer the neurons have the same so-called activation functions, in the hidden layer the activation functions are differentiable and nonlinear, and these activation functions make the neural network universal approximator. When the input vector (with the data for the target variable and the predictors) is presented to the network in the input layer, the signals to the neurons in the hidden layer are computed and with the activations functions in the neurons in the hidden layers, the signals are processed in the similar way to the output layer, where the predicted values of the target variable are computed [16].

Some general characteristics of the NNs used in this paper will be presented below.

General Regression Neural Network

The general regression neural network (GRNN) has been proposed by D. Specht [17] and until now there has been several updates for this NN. The main and most attractive characteristic for this NN is that it is very fast one-pass NN in comparison with back-propagation NNs, which can have a large number of iterations to converge to the desired solution – and also GRNN can work with sparse data.

The idea of the algorithm of this NN is that the approximation function can be expressed as a probability density function (PDF) which can be empirically determined from the data, using Parzen window estimation [18].

Since the joint PDF for x and y is not known, it is estimated from the data using nonparametric estimators and the resulting regression equation that is implemented in this NN with parallel structure is given by Eq. (1) [17].

$$E[y/X] = \frac{\int_{-\infty}^{+\infty} yf(X, y)dy}{\int_{-\infty}^{+\infty} f(X, y)dy} \tag{1}$$

X is a particular measured value of the random variable x (vector), E[y/X] is conditional mean of y for given X (regression of y on X). Since f(x, y) is not known, it can be estimated from the data (sample of observations of x and y) by using the class of consistent estimators proposed by Parzen. The author Cacoullos [19] has shown that they can be applied to the multidimensional case. The probability estimator is given in Eq. (2), where X^i and Y^i are sample values of the random variables x and y, n is the number of samples, and p is the dimension of the vector X. This joint probability estimate given by

Eq. (2) can be substituted in Eq. (1) and the desired conditional mean of y given X can be obtained. The architecture of GRNN is given in Fig. 1 [17].

$$\hat{f}(X,Y) = \frac{1}{(2\pi)^{(p+1)/2}\sigma^{(p+1)}} \cdot \frac{1}{n}\sum_{i=1}^{n} \exp\left[-\frac{(X-X^i)^T(X-X^i)}{2\sigma^2}\right] \cdot \exp\left[-\frac{(Y-Y^i)^2}{2\sigma^2}\right] \tag{2}$$

GRNN has 4 layers. In the input layer there is the same number of neurons as the number of predictors, one neuron for a predictor. The values of the input neurons are processed to the neurons of the next hidden layer. In this hidden layer there is the same number of neurons as cases in the data set, memorizing the values of each predictor and target value of that case. This layer computes the Euclidean distance for the case from the center of the neuron and then applies the activation kernel function (RBF-radial basis function) and these values from each neuron from this layer are processed to the next summation layer. There are only 2 neurons in this layer: one is denominator summation unit which adds up the weight values which come from each neuron of the previous layer, and the numerator summation unit adds up the weight values from the neurons from the previous layer multiplied by the actual target value. The decision layer divides the value from the numerator unit by the value from the denominator and gives it as a predicted target value [15].

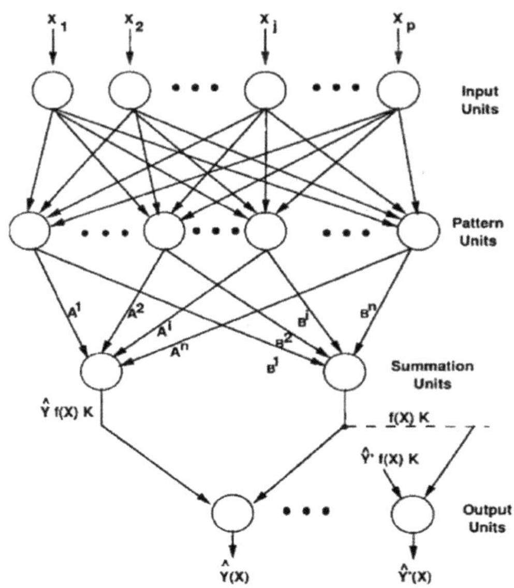

Fig. 1. Architecture of GRNN [17]

MLP (Multilayer Perceptron)

MLP is so called ‘feed forward network’ which means that the processing of the values of data moves only forward – from the input layer to the hidden layers and to the output layer. For each predictor there is one neuron in the input layer. The main goal of the algorithm of this NN is to find the optimal output from the network that will match the actual target values as accurately as possible, by finding the best set of weight values of the network. The most important issues involved in training MLP are: selecting the number of hidden layers and the number of neurons in each hidden layer, converging to an optimal solution which is the global optimum (in reasonable time), avoiding local minima; and for testing the overfitting – validating the model with some validation method.

DTREG uses special algorithm for selecting the initial weight values and afterwards the Conjugate gradient algorithm is used for optimizing them. It is recommended to allow the optimization several times with different sets of initial weight values.

In comparison with GRNN, MLP is trained slowly and is not so accurate as GRNN, but MLP requires smaller memory space than GRNN for storing the model [15].

RBF (Radial Basis Function Neural Network)
A RBF neural network, like many NNs, has 3 main layers: an input layer, a hidden layer and an output layer. Gaussian functions are used as transfer functions in the hidden layer and their output is inversely proportional to the distance from the center of the neuron. RBF with a large number of nodes is very similar to GRNN. The RBF NNs have number of neurons which is often less than the number of training points, but GRNN has one neuron for each case in the dataset in the hidden layer.

While for datasets, from small to medium size, GRNN is more accurate than RBF, for large datasets GRNN is not practical.

The algorithm of RBF NNs is similar to K-Nearest Neighbor models. The neurons of RBF NN are positioned in the space with the same dimension as the number of predictors. For computing the weight for each neuron (its influence), the Euclidean distance is computed from the center of the neuron to the point which is evaluated, and a radial basis function (kernel function) is applied to that distance. The radial basis function used most often is the Gaussian one. The best value for the predicted target is obtained by summarizing the output values of the RB functions multiplied by the weights obtained for each neuron [15].

4 Results

The dataset has 38 rows and 8 columns in the Excel file in CSV format, with the measured values of the variables used for the model: 7 of them are predictors which are explained in Sect. 3.1, and one is target variable (ETA).

The results are presented in Table 2, for three neural network models. As is visible from Table 3, among seven predictors, the most important are the losses, T_{dplin} and Lambda. GRNN gave the most accurate prediction among the three prediction models, with mean absolute percentage error (MAPE) of about 0.19% and coefficient of determination R^2 of 99.82%. The correlation coefficient between the predicted and the actual values is 0.999. The GRNN model can be updated with the new measurement data and used for new target (ETA) predictions.

For every model some adjustments for the parameters of the model should be made in the DTREG software. We shall now mention the adjustment only for the GRNN model

Table 2. Comparison of the accuracy of the three neural networks

Neural network	MAPE (mean absolute percentage error)	R^2 (coefficient of determination)	Coefficient of correlation between actual and predicted target values
GRNN	**0.1911396**	**99.822%**	**0.999135**
MLP	**0.1721962**	**98.281%**	**0.991868**
RBF NN	**0.638741**	**95.370%**	**0.977714**

Table 3. Overall importance of variables for GRNN

Variable	Importance
Losses in (%)	100.000
T_{dplin} (°C)	0.113
Lambda	0.113
O_2 (%)	0.005
CO_2 (%)	0.004
CO (ppm)	0.003
T_{air} (°C)	0.000

as most accurate for these data. The important adjustments that should be made for the GRNN model are: values for the parameter *sigma*, which controls the radius of influence for each point of the model and also the number of neurons. Also, the validation method and the type of the kernel function should be chosen. The minimal value of sigma is set to be 0.0001, maximal sigma is set to be 3. The search steps for sigma are chosen to be 20. The software offers the option for removing unnecessary neurons, which often improves the accuracy of the model. We chose this option and reduced the model to 26 neurons, improving the accuracy of the model and optimising the size of the model. The method for validation is chosen to be LOO (Leave One Out), and the type of kernel function is chosen to be the Gaussian function. [15].

It is important to mention the problem of overfitting, a frequent problem in modelling and how DTREG prevents this problem. This problem occurs when the parameters of the model are adjusted so that the training data has good accuracy, but the validation data (on which the model is tested and which are not known to the model) has bad accuracy. DTREG software has two main strategies for preventing it: 1) removing unnecessary neurons and 2) validating the model on unseen data. DTREG offers several validation methods to test the accuracy of the model on unseen data which are not used in the training phase. DTREG evaluates the model so that the parameters are being tuned and when overfitting is detected this process of tuning is stopped, known as 'early stopping' [15].

5 Measures to Improve the Efficiency of the Heating System

According to the best practice, some measures are recommended to improve the efficiency of the heating systems in terms of increasing ETA and consequently decreasing the gases emissions, as follows:

1. Regular annual general service and burner regulation and flue gas analysis
2. Measurement of fuel consumption and adequate intervention in case of determined unexpected increased consumption.
3. Use of proven higher quality fuels in practice, regardless of possible short-term savings by using cheaper types.
4. If necessary, burner regulation and exhaust gas analysis should be done more than once a year in case of increased consumption.
5. Installation of a saving device in front of the burner.

6 Conclusion

In today's situation, the need for rational consumption of energy sources and the consequent reduction of environmental pollution, improving the heating system's efficiency through optimizing system parameters, is undoubtedly justified. Scientific research supported by modern devices for measuring and analyzing exhaust gases and the application of modern scientific methods contributes to the understanding of parameter optimization, practical application and improvement of measures to be taken to increase efficiency. Neural networks are used for processing measurement data, and GRNN gives the best accuracy of 0.19% according to the standard estimator, the mean absolute percentage error (MAPE). The GRNN model can be updated with the new measurement data and used for new target (ETA) predictions.

Acknowledgments. The University of Rijeka funds this research within the project uniritehnic -18-125. Antun Pusic Graduated Mach. Ing. provided all measurements for this research and is also the saving device's constructor.

References

1. Farouk, N., El-Rahman, A.M., Sharifpur, M., Guo, W.: Assessment of CO_2 emissions associated with HVAC system in buildings equipped with phase change materials. J. Build. Eng. **51**, 104236 (2022)
2. Liu, S., Tao, R., Tam, M.C.: Optimizing cost and CO_2 emission for construction projects using particle swarm optimization. Habitat Int. **37**, 155–162 (2013)
3. Šivak, M.: Centralno grijanje, ventilacija, klimatizacija. Savez energetičara Hrvatske, Zagreb (1980). [in Croatian]
4. Li, X., Su, W., Xu, W., Dai, B., Li, J., Li, L.: Editorial: CO_2-based energy systems for cooling, heating, and power. Front. Energy Res. **10**, 993093 (2022)
5. Valančius, K., Grinevičiūtė, M., Streckienė, G.: Heating and cooling primary energy demand and CO_2 emissions: Lithuanian A+ buildings and/in different European locations. Buildings **12**(5), 1–18 (2022)

6. Xue, Q., Wang, Z., Chen, Q.: Multi-objective optimization of building design for life cycle cost and CO2 emissions: a case study of a low-energy residential building in a severe cold climate. Build. Simul. **15**, 83–98 (2022)
7. Valancius, K., Grinevičiūtė, M., Streckiene, G.: Heating and cooling primary energy demand and CO2 emissions: Lithuanian A+ buildings and/in different European locations. Buildings **12**(5), 1–18 (2022)
8. Benakopoulos, T.: Optimization of the building heating system, to minimize temperature levels in district heating networks. Ph.D. thesis, Technical University of Denmark (2021). ISBN (Electronic): 87-7877-570-1
9. Inho, Y.: Optimization of operation times of a heating system in office building. J. Asian Architect. Build. Eng. **19**(4), 400–415 (2020)
10. Smitt, S., Tolstorebrov, I., Hafner, A.: Performance improvement of integrated CO2 systems with HVAC and hot water for hotels. Thermal Sci. Eng. Progress. **23**, 100869 (2021)
11. Brennenstuhl, M., Lust, D., Pietruschka, D., Schneider, D.: Demand side management based power-to-heat and power-to-gas optimization strategies for PV and wind self-consumption in a residential building cluster. Energies (Basel) **14**(20), 6712 (2021)
12. Croatian Meteorological and Hydrological Service. https://meteo.hr/index.php. Accessed 26 Dec 2023
13. Klima u Istri. https://www.istramet.hr/klima-u-istri/. Accessed 20 Dec 2023
14. STRUNA. http://struna.ihjj.hr/search-do/?q=preti%C4%8Dak+zraka&naziv. Accessed 18 Nov 2023
15. Sherrod, P.: DTREG (Predictive Modelling Software and Tutorial) (2013). www.dtreg.com. Accessed 18 Oct 2023
16. Kecman, V.: Learning and soft computing. In: A Bradford Book. The MIT Press Cambridge, London, England (2001)
17. Specht, D.: A general regression neural network. IEEE Trans. Neural Netw. **2**(6), 568–576 (1991)
18. Parzen, E.: On estimatiojn of probability density function and mode. Ann. Matyh. Stat. **33**, 1065–1076 (1962)
19. Cacoullos, T.: Estimation of a multivariate density. Ann. Inst. Statist. Math. (Tokyo) **18**(2), 179–189 (1966)

Climate Change and Loading Uncertainties

A Stochastic Framework of Risk Assessment of Flooding on Stability and Serviceability of Bridges Under a Changing Climate

Bassel Habeeb(✉), Erwan Liberge, and Emilio Bastidas-Arteaga

Laboratory of Engineering Sciences for the Environment UMR CNRS 7356, La Rochelle University, La Rochelle, France
bassel.habeeb@univ-lr.fr

Abstract. Over time, the Earth's climate has undergone significant changes due to a variety of natural processes. During the last century, global warming has drastically increased due to human activities, which has in turn increased the greenhouse gases in the atmosphere causing an increase in global average temperature. Under a changing climate, extreme events such as floods are expected to be more severe in intensity and frequency in some regions, which in turn, could significantly impact the performance and functionality of infrastructure assets. Particularly, the built environment (bridges, roads, buildings, etc.) surrounding rivers is impacted in some regions under a changing climate, increasing the risk posed to people and assets, posing challenges in maintaining the functionality of the built environment. To address this concern, proposing a comprehensive framework under a changing climate becomes essential to evaluate the stability of a bridge due to scour and the vulnerability of the surrounding areas due to overtopping floods under a changing climate. The approach involves flood frequency analysis to determine the flood levels, their respective return periods and annual exceedance probability, in addition to determining the scour estimates corresponding to these return periods under a changing climate. Subsequently, risk assessment protocols are implemented to evaluate the bridge's performance due to scouring and the surrounding areas' vulnerability in response to overtopping floods under a changing climatic conditions. Furthermore, employing stochastic methods, particularly the Poisson process, allows for the stochastic investigation of the bridge's probability of experiencing service disruptions due to scouring in a changing climate.

Keywords: Climate Change · Stochastic process · Floods · Scouring · Risk Assessment

1 Introduction

Climate change signifies fundamental modifications in the Earth's climatic trends, resulting in global warming, rising sea levels, environmental impacts, and extreme weather events. In this context, river discharge values in some areas are expected to increase under a changing climate due to long-term shifts in weather patterns [1]. The weather

J. C. Matos et al. (Eds.): IPW 2024, LNCE 494, pp. 233–242, 2024.
https://doi.org/10.1007/978-3-031-60271-9_20

changes (temperature and humidity), materials of construction, and degree of maintenance affect the lifespan of infrastructure assets. In this context, bridges could suffer a loss in their serviceability due to progressive deterioration (corrosion, fatigue, creep, etc.) [2, 3] and they also might suffer a sudden loss in their serviceability due to shock-based deterioration (extreme events) [4].

Floods are the main cause of bridge failures due to a range of factors, e.g., scouring, hydrodynamic loads and pressures on the deck and piers, overtopping, and debris accumulation [5, 6]. In the 20th century, floods were responsible for 6.8 million deaths from natural disasters across the world [7]. In this context, scouring is the main cause of bridge failures and is influenced by river discharge values, in which the removal of sediment around the piers reduces the stiffness and the bearing capacity of the bridge foundation, resulting in structural failure.

Bridges crossing rivers are designed to maintain their serviceability based on a design flood level of a pre-determined return period of historical floods which are determined using different probability distributions [8]. In this context, flood frequency analysis estimates the frequency of occurrence of extreme events, and relates the magnitude of extreme events to their frequency of occurrence [9]. In this context, floods are expected to increase in frequency and intensity in some regions under a changing climate, inundating floodplains and adjacent lands, leading to losses in agricultural lands and infrastructure assets.

This paper proposes a framework to approach stochastically the stability and serviceability of bridges in the face of floods under a changing climate. The objectives of this paper are as follows.

- Approach the random nature of floods occurrences.
- Evaluate stochastically the stability of bridges due to local scour under a changing climate.
- Evaluate stochastically the serviceability of bridges due to the interruption of traffic circulation due to floods under a changing climate.

This paper is organized as follows. Section 2 presents the methodology. Section 2.1 illustrates the local scour which is the structural failure mode applied in this study. Section 2.2 illustrates the flood frequency analysis applied to determine the evolution of the design flood level and under a changing climate. Section 2.3 illustrates the stochastic process applied in this study to consider the random nature of floods, presents the probability of the bridge being out of service and its probability of failure, and presents the risk assessment of floods. Section 3 introduces the case study. Section 4 presents a comprehensive analysis of the results of the applied methodology.

2 Methodology

The proposed methodology investigates stochastically the impact of floods on the structural failure of bridges due to local scour and the serviceability of bridges corresponding to the flood levels under a changing climate. Section 2.1 illustrates the flood frequency analysis method applied to the case study (Gumbel distribution), which was determined to be the best fit for the case study as determined recently in [10]. Section 2.2 illustrates the structural failure due to local scour.

2.1 Flood Frequency Analysis

Gumbel distribution [11] estimates the peak discharge values and probability of exceedance for return periods of extreme hydrological events, such as floods. The probability of flood event exceedance P_T for a given return period T, writes:

$$P_T = 1 - \left[e^{-e^{ln\left[ln\left[\frac{T}{T-1}\right]\right]}} \right] = \frac{1}{T} \tag{1}$$

where Y_T is the reduced variate.

The annual peak discharge of a flood X_T for a return period T, writes:

$$X_T = \mu + \left[\sigma \times \left[\frac{Y_T - \mu_R}{\sigma_R} \right] \right] \tag{2}$$

where μ is the mean of the sample, σ is the standard deviation of the sample, μ_R is the reduced mean which signifies the location parameter, and σ_R is the reduced standard deviation which signifies the scale parameter. Detailed illustrations of the values of the reduced (mean and standard deviation) are discussed in [12].

The equation of frequency analysis $X_{T(UL,LL)}$ indicates the confidence limits CI for the annual peak discharge values of a flood X_T to account for the uncertainties of climate projections which are presented in mean and standard deviation, writes:

$$X_{T(UL,LL)} = X_T \pm [CI \times S_e] \tag{3}$$

$$S_e = \sqrt{1 + 1.3\left[\frac{Y_T - \mu_R}{\sigma_R}\right] + 1.1\left[\frac{Y_T - \mu_R}{\sigma_R}\right]^2} \left[\frac{\sigma}{\sqrt{N}}\right] \tag{4}$$

where S_e is the probable error of the distribution which depends on the dispersion (standard deviation σ) of the sample, the sample size N, and the values of the reduced mean and reduced standard deviation.

2.2 Local Scour

The HEC-18 model was developed at Colorado State University [13] and is recommended by the United States Department of Transportation Hydraulic Engineering [14]. The local scour depth D_s, writes:

$$D_s = 2HK_1K_2K_3K_4\left[\frac{b}{H}\right]^{0.65}\left[\frac{V}{[gH]^{0.5}}\right]^{0.43} \tag{5}$$

where H is the flow depth, b is the pier width, K_1 is the pier shape coefficient, K_2 is the angle of attack coefficient, K_3 is the bed condition coefficient, K_4 is the bed material coefficient, V is the mean flow velocity, and g is the gravity acceleration.

2.3 Risk Assessment

Stochastic Approach

The occurrence and magnitude of floods are random. Consequently, a stochastic Poisson process is used to simulate various scenarios of flood occurrences exceeding the design flood level under a changing climate. The number of floods $N = \{N_t : t \in [0, \infty]\}$ is defined as a Poisson process by supposing that floods are shocks of a sequence of independent, identically distributed random variables. The stochastic flood arrivals $f(t)$ simulates several flood scenarios occurrences with shock sizes ξ_i exceeding the design flood level. The stochastic flood arrivals, writes:

$$f(t) = \sum_{i=1}^{N_t} \xi_i \tag{6}$$

This approach is used to stochastically simulate and investigate the impact of climate change on the functionality and serviceability of the bridge as follows.

- The annual maximum river discharge value exceeding the magnitude of the design flood level.
- The structural failure due to local scour.

The probability of the bridge being out of service $P_s(t)$ considers the evolution of flood levels under a changing climate interrupting the bridge functionality in means of traffic circulation. The probability of bridge failure $P_f(t)$ considers the impact of local scour depth on the stability of the bridge.

$$P_{(s,f)}(t) = 1 - \sum_{t=1}^{t} \sum_{i=0}^{n} \frac{(\lambda t)^i}{i!} e^{-\lambda t}, n \in [0, t], \lambda \in \left[\lambda_s, \lambda_f\right] \tag{7}$$

where n presents the number of shocks, λ_s is the rate of flood arrivals exceeding the flood design level, and λ_f is the rate of scour depth exceeding the foundation depth under a changing climate.

Scour Risk Assessment

Scour risk assessment of bridges crossing rivers is determined by the priority rating. The scour risk assessment assesses the risk of scour based on a relation between the scour depth and foundation depth F_d, informing recommended actions to mitigate the likelihood of bridge failure due to scour.

The priority rating PR, writes:

$$PR = 15 + ln\left[\frac{D_s}{F_d}\right] + R_T + F_T \tag{8}$$

where R_T is the river type coefficient and F_T is the foundation type coefficient.

Table 1. Scour risk assessment [15].

PR	Risk	Recommended action
16–21	High	Detailed Hydraulic Assessment. The structure is included in the flood warning plan or establishing protection works
14–15.9	Medium	A repeat of assessment and underwater exams every 3 years (where necessary). The priority should be reviewed following the protection works
10–13.9	Low	A repeat of assessment every 6 years and underwater exams every 3 years (where necessary). The priority should be reviewed following the protection works

3 Application

3.1 Case Study Description

The case study is a dual-span bridge located in the United Kingdom with one pier of 2 m width and 1 m spread foundation depth. The bridge is designed to maintain its serviceability of a 200-year return period with a flood design level value of 114.19 m^3/s [16].

3.2 River Flow Projections

Optimistic (RCPs 2.6) and pessimistic (RCPs 8.5) projections scenarios from 2011 to 2095 are considered in this study to investigate the impact of climate change on the stability due to scour and serviceability due to floods causing interruption of circulation for the case study bridge. In this study, the KNMI-RACMO22E regional climate model is used to assess the impact of climate change on the stability and serviceability of the case study bridge.

4 Results and Discussions

4.1 River Flow Projections

Figure 1 presents the optimistic (RCPs 2.6) and pessimistic (RCPs 8.5) climate change scenarios. The observations indicate a higher maximum annual river discharge values under more pessimistic climate change scenarios.

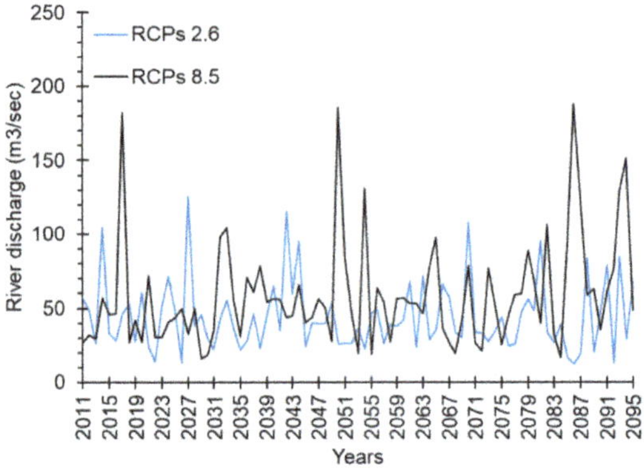

Fig. 1. River flow projections [10].

4.2 Flood Frequency Analysis

Figure 2 presents the evolution of the maximum annual river discharge values, return periods, and annual exceedance probability. The results indicate the evolution of floods under a changing climate, in which higher flood levels with higher probability of exceedance are expected in the future under a changing climate.

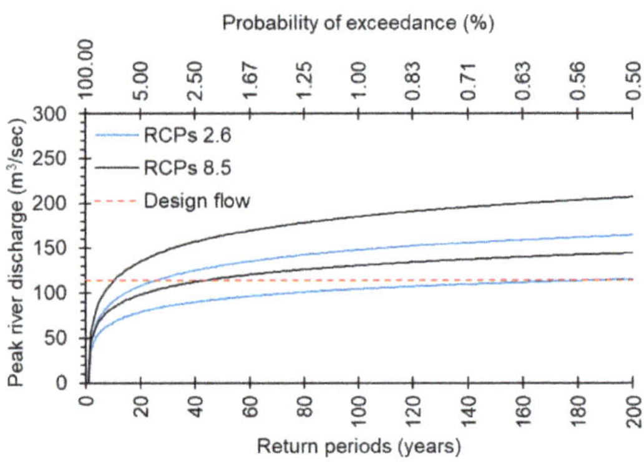

Fig. 2. Flood frequency analysis under a changing climate [10].

4.3 Scour Estimates

Figure 3 presents the scour estimates under a changing climate, in which a higher occurrence rate (Table 2) is expected for more pessimistic scenarios. In addition, considering the 200-year flood levels, the scour estimates consistently exceed the foundation depth. This indicates that exceeding the design flood level, the bridge fails.

Table 2. Occurrence rate.

Climate change scenario	Occurrence rate
RCPs 2.6	0.02
RCPs 8.5	0.08

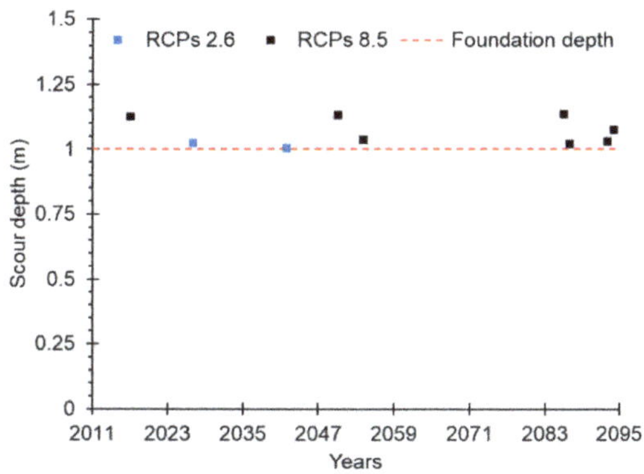

Fig. 3. Scour depth estimates.

4.4 Risk Assessment

Stochastic Results

Structural Failure

Scour estimates are stochastically simulated (Eq. (6)) based on the scour occurrences rate (Table 2) to consider the variability which is presented as Upper Limit (UL) and Lower Limit (LL) based on the quartile method [17] and to investigate the probability of failure due to scour depth exceeding the foundation depth (Eq. (7)). Table 3 indicates a higher failure rate for more pessimistic climate change scenarios.

Table 3. Failure rate.

Climate change scenario	Failure rate (λ_f)
RCPs 2.6 LL	0.01
RCPs 2.6 UL	0.02
RCPs 8.5 LL	0.05
RCPs 8.5 UL	0.09

Figure 4 presents the likelihood of failure due to local scour (Eq. (7)). The results indicate a higher likelihood of failure due to scour for more pessimistic climate change scenarios.

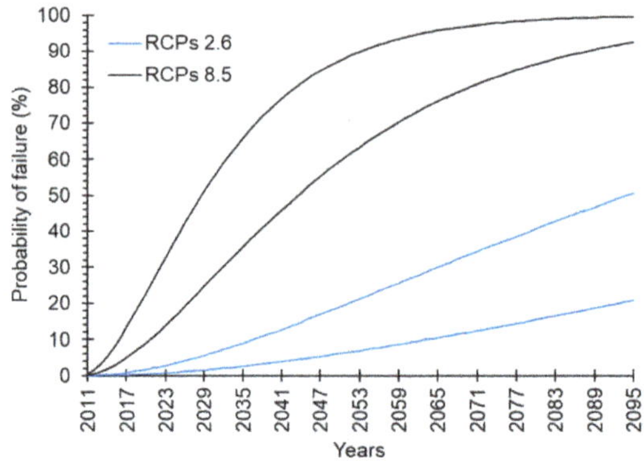

Fig. 4. Probability of failure due to stochastic scour estimates.

Serviceability

A stochastic investigation is performed to account for the random nature of floods based on their annual exceedance probability to represent their impact on the traffic crossing the bridge (Eq. (6)). Figure 2 presents the probability of exceeding the design flood level (Eq. (7)), in which the serviceability of the bridge is impacted when exceeding the design flood level under more pessimistic climate change scenarios.

The results of the stochastic simulations in Table 4 present a higher annual exceedance probability of floods exceeding the flood design level under more pessimistic climate change scenarios.

Table 4. Serviceability rate (%).

Climate change scenario	Serviceability rate (λ_S)
RCPs 2.6 LL	0.52
RCPs 2.6 UL	3.7
RCPs 8.5 LL	9.1
RCPs 8.5 UL	2.2

The probability of the bridge being out of service due to flooding events interrupting the traffic circulation due to exceeding the flood design level in Fig. 5, indicates a higher risk for more pessimistic climate change scenarios, in which the high variability in exceeding the design flood level is due to the high variability in the annual exceedance probability.

Scour Risk Assessment

Exceeding the foundation depth as presented in Fig. 3 and as determined from the evolution of the 200-year flood levels under a changing climate, both results of scour estimates present a priority rating value of a range between 16 to 21 which indicates

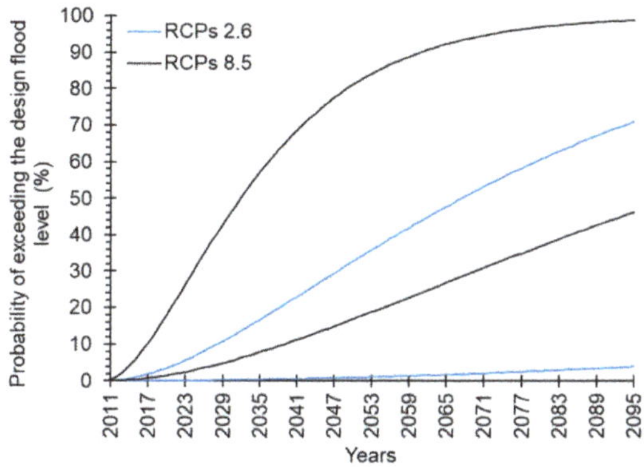

Fig. 5. Probability of exceeding the design flood level (Stochastic simulations).

a high risk due to scour. In this context, the recommended actions are mentioned in Table 1.

5 Conclusions

Under a changing climate, the annual exceedance probability and flood levels are expected to increase for the case study. Exceeding the design flood level, a higher probability of failure and a higher probability of the bridge being out of service due to interrupting the traffic circulation are found under more pessimistic climate change scenarios.

Acknowledgements. This research was supported by Project No. 101121210 "City Nature-Based Solutions Integration to Local Urban Infrastructure Protection for a Climate Resilient Society" (NBSINFRA). Funded by the European Union. Views and opinions expressed do not necessarily reflect those of the European Research Executive Agency (REA). Neither the European Union nor the granting authority can be held responsible for them.

References

1. Habeeb, B., Bastidas-Arteaga, E.: Climate change indicators dataset for coastal locations of the European Atlantic area. Data Brief **43**, 108339 (2022). https://doi.org/10.1016/j.dib.2022.108339
2. Truong, Q.C., El Soueidy, C.-P., Li, Y., Bastidas-Arteaga, E.: Probability-based maintenance modeling and planning for reinforced concrete assets subjected to chloride ingress. J. Build. Eng. **54**, 104675 (2022)
3. Bastidas-Arteaga, E., El Soueidy, C., Amiri, O., Nguyen, P.T.: Polynomial chaos expansion for lifetime assessment and sensitivity analysis of reinforced concrete structures subjected to chloride ingress and climate change. Struct. Concr. **21**, 1396–1407 (2020). https://doi.org/10.1002/suco.201900398
4. Sánchez-Silva, M., Klutke, G.-A.: Reliability and Life-Cycle Analysis of Deteriorating Systems. Springer, Cham (2016). https://doi.org/10.1007/978-3-319-20946-3

5. Davidson, M.T., Consolazio, G.R., Getter, D.J., Shah, F.D.: Probability of collapse expression for bridges subject to barge collision. J. Bridge Eng. **18**, 287–296 (2013)
6. Mondoro, A., Frangopol, D.M.: Risk-based cost-benefit analysis for the retrofit of bridges exposed to extreme hydrologic events considering multiple failure modes. Eng. Struct. **159**, 310–319 (2018)
7. Doocy, S., Daniels, A., Murray, S., Kirsch, T.D.: The human impact of floods: a historical review of events 1980–2009 and systematic literature review. PLoS Curr. (2013). https://doi.org/10.1371/currents.dis.f4deb457904936b07c09daa98ee8171a
8. Singh, V.P., Yadava, R.N.: Watershed Hydrology: Proceedings of the International Conference on Water and Environment. Allied Publishers, New Delhi (2003)
9. Te Chow, V., Maidment, D.R., Mays, L.W.: Applied Hydrology (1988)
10. Habeeb, B., Bastidas-Arteaga, E.: Assessment of the impact of climate change and flooding on bridges and surrounding area. Front. Built Environ. **9**, 1268304 (2023). https://doi.org/10.3389/fbuil.2023.1268304
11. Gumbel, E.J.: The return period of flood flows. Ann. Math. Stat. **12**, 163–190 (1941)
12. Gumbel, E.J.: Statistics of Extremes, Columbia University Press, N.Y. (1958)
13. Arneson, L.A., Zevenbergen, L.W., Lagasse, P.F., Clopper, P.E.: Evaluating Scour at Bridges, Fifth Edition, U.S. Department of Transportation Federal Highway Administration (2012)
14. Annad, M., Lefkir, A., Mammar-kouadri, M., Bettahar, I.: Development of a local scour prediction model clustered by soil class. Water Pract. Technol. **16**, 1159–1172 (2021). https://doi.org/10.2166/wpt.2021.065
15. Wallingford, H.: Hydraulic Aspects of Bridges: Assessment of the Risk of Scour (EX 2502). Hydraulics Research Ltd., Oxfordshire, United Kingdom (1992). http://eprints.hrwallingford.com/id/eprint/315
16. Matthew, L.: Western Route Scour Assessment Programme, Network Rail – Western Route Western House, JBA Consulting, North Yorkshire, United kingdom (2013)
17. Hyndman, R.J., Fan, Y.: Sample quantiles in statistical packages. Am. Stat. **50**, 361–365 (1996). https://doi.org/10.1080/00031305.1996.10473566

Some Limitations of Statistical and Probabilistic Models and Current Developments

Dirk Proske(✉)

Bern University of Applied Sciences, 3400 Burgdorf, Switzerland
dirk.proske@bhf.ch

Abstract. Nuclear power plants and dams must demonstrate their safety against events with a return period of 10,000 years. However, determining such rare events from observations alone is difficult because time series often cover only years or decades, and in some very rare cases, centuries. Various statistical techniques attempt to overcome the limitations of extrapolating such extreme return periods. In some cases, pre-instrumental data can be used, such as paleo-seismic or paleo-hydraulic data. In recent years, a new approach has emerged that promises to cover phenomena that cannot be detected in the data because of the limitations of the data series. These are model-based methods. Especially in Switzerland, this approach has been used for example in the EXAR project for precipitation and floods, but also in other recent studies on extreme precipitation and temperatures. This article aims to highlight the opportunities, but also the risks, of this approach. The importance of such considerations is that this approach is not only found in the field of climatology, but also in the field of structural safety or earthquake engineering.

Keywords: Extreme Value · Climate Change

1 Introduction

The safety of some critical infrastructure elements must be demonstrated for extreme actions and loads. For example, the functionality and safety of nuclear power plants and dams must be demonstrated for earthquakes with a return period of 10,000 years or an annual exceedance probability of 10^{-4} (UVEK 2009, ENSI 2012). For nuclear power plants, the actions are not limited to earthquakes, but also include natural hazards such as floods, precipitation, wind, and snow loads.

The calculation of such extreme loads is mainly based on the application of statistical methods, such as extreme value theory. However, data availability is often very limited, and extrapolation is required. This leads to significant epistemic uncertainties. In recent years, model-based simulations have been established and widely used as a competing method to compute such extreme values. Both methods will be briefly reviewed, and their advantages and disadvantages discussed and related to approaches in structural reliability.

J. C. Matos et al. (Eds.): IPW 2024, LNCE 494, pp. 243–249, 2024.
https://doi.org/10.1007/978-3-031-60271-9_21

2 State of the Art

2.1 Introduction

The consideration of rare and extreme values is of utmost importance for engineering and insurance. The Roman builders already took extreme floods into account, as almost two hundred Roman bridges still in existence today prove (Mehlhorn and Curbach 2014). Especially in the last hundred years and decades, scientists and engineers have worked intensively to solve the problem of calculating such rare event using extreme value statistics.

2.2 Extreme Value Statistics

In general, extreme value statistics discusses the development of statistical distributions based on extreme values over a period (e.g., block maxima method) instead of using actual values. There are different types of distributions, e.g., the Weibull or Gumbel distributions, which cover different conditions, such as minimum or maximum values. In recent decades, the general extreme value distribution has been used very successfully.

Unfortunately, using only extreme values results in a limited sample size. Therefore, alternative approaches can be used, such as peaks over threshold or r largest order statistics. Peaks over threshold considers not only the maximum values over a given period, but all samples above a certain threshold. In this case, however, the choice of threshold is of utmost importance. The use of r largest order statistics extends the use of maximum values to include the second or third largest values. On the other hand, the independence of these values must be carefully considered.

The extreme value methods can be adapted to seasonal and annual trends. Daily or monthly samples can be used instead of annual values to extend the samples. However, different granularities can lead to different results. As shown for precipitation, this can have a significant impact. For example, 10,000-year hourly precipitation events can mathematically be greater than 10,000-year daily precipitation events, which is theoretically not possible. In addition, the data often must be homogenized, meaning measurements must be converted to different heights, different measuring devices or different building conditions in the surrounding area.

In summary, the data must meet requirements such as homogeneity, stationarity, independence, and representativeness. Statistical artifacts such as outliers, censored data, and multimodal data must be considered. However, most of these tests, in turn, require a sufficient number of samples. Recommendations for extrapolating data periods range from one to four (Pugh 2004,). In some cases, however, extrapolation factors of up to 100 are used. Swan et al. (2005) gives exceedance probabilities based on the data sources (Table 1).

A general introduction to the use of extreme values in engineering can be found in Castillo et al. (2008), Davidson and Huser (2015). Successful applications in various fields are shown in Davidson (2014), Davidson et al. (2013), Murensee (2020). The block maxima method and peaks over threshold have been explicitly mentioned by the Swiss regulator as possible analyses for the calculation of 10,000-year return period values for various natural hazards such as temperature, precipitation, snow, and wind loads (ENSI 2012).

Table 1. Estimation of extreme low probability events for floods based on different data basis (Swain et al. 2004).

Type of data used for the flood frequency analysis	Limit of credible extrapolation for the annual exceedance probability
Local runoff data	$10^{-2} - 5 \times 10^{-3}$
Regional runoff data	$2 \times 10^{-3} - 10^{-3}$
Local runoff data and paleo-hydraulic data	$2.5 \times 10^{-4} - 10^{-4}$
Regional precipitation data	$5 \times 10^{-4} - 10^{-4}$
Regional runoff data and paleo-hydraulic data	$6.7 \times 10^{-5} - 2.5 \times 10^{-5}$

2.3 Model Based Simulation

Model-based simulation calculates extreme values with the required return periods by simulating the underlying processes. For example, extreme floods can be simulated by extreme precipitation using rain generators and consideration of hydrological and hydraulic conditions. One application is EXAR for Switzerland (Staudinger and Viviroli 2020, Andres et al. 2021). Extreme temperatures for Switzerland were modeled in Fischer et al. (2023) and extreme precipitation for Central Europe in Gessner et al. (2023).

3 Discussion and Conclusion

3.1 Natural Hazards

The Swiss regulator has recently broadened the acceptance of methods from pure statistical methods to simulation methods, so that for example for precipitation either statistical analysis or the EXAR results can be used. At the same time, the Swiss regulator has downgraded the importance of historical data, e.g., for precipitation, which was a major concern a decade ago (ENSI 2012).

Many researchers have addressed the problem of using historical data for investigations, such as paleo-seismic (Schwarz-Zanetti and Fäh 2011a, b, Proske 2022), paleo-hydraulic (Mudelsee 2017) and paleo-climatic data (Proske 2015, Proske 2016). The errors introduced by using historical data are just one part of many different types of uncertainty that affect the calculation, see Table 2. However, neglecting such data does not seem to be a solution, as such data are an important resource for testing the developed models. It must be concluded that it is currently not possible, or only possible to a limited extent, to test the results of both methods by observation. The current solution is the application of hybrid models which have been used for some natural hazards (see Table 3). The weighting of different methods in the seismic hazard estimation process has been formalized (Ake et al. 2018) and can be applied in a broader approach.

Table 2. Uncertainty in the statistical analysis of floods (Mudelsee 2017).

Random Fluctuations	Systematic Fluctuations
Reading Error	Zero Point Setting
Measurement error discharge (calibration water level - discharge)	Functional form of calibration curve
	Non-updated calibration curve (e.g., after flood events)
Short-term fluctuations (precipitation)	Spatial extrapolation (e.g., tributaries)
Spatial fluctuations (soil runoff conditions)	Over/underestimations in historical climate documents
Indicator variable of the natural climate archives (pre-instrumental period)	Systematic external influences (i.e., influences of factors other than floods) on indicator variables of the pre-instrumental period
Limited data availability (statistical models)	Spatial extrapolation (e.g., tributaries)
Limited computing capacity (statistical models)	Over/underestimations in historical climate documents

Table 3. Examples for hybrid approaches.

	Statistical analysis	Model based analysis
Air Temperatures	×	×
Precipitation	×	×
Snow		×
Floods	×	×

3.2 Structural Safety

The same situation can be found in the field of structural safety. Safety proofs, such as static proofs or failure probability calculations, are increasingly based on sophisticated simulations, such as finite element models or stochastic finite element models. These models and simulations are mainly based on real-world test series with limited laboratory sample sizes, such as bending tests, but rarely at the structural level.

However, structural collapse and damage data on structural level should be used to verify the simulations. Such checks have been carried out intensively in the field of seismic engineering (Bachmann 1997, 2002, Wyss & Trendafiloski 2009, Basöz et al. 1999, Wang & Lee 2009).

3.3 Critics

Science relies heavily on the development and testing of theories (Goodman 1978, Golomb 1970). However, in most cases today, theories cannot be directly observed. The question then arises as to how such theories can be confirmed if their truth cannot be directly established by observation. A theory is then confirmed or refuted by its observable consequences. If the observable consequences materialize, the theory is partially confirmed. However, such an observation also confirms many other theories, though not an infinite number. A general problem in various fields of science is that:

- Many theories and the measures developed within them are not directly observable in real world, or not observable over a reasonable period, such as risk measures. There are many variations of risk measures (Proske 2022), none of which satisfies all requirements.
- Interactions are often unknown, unobservable, or uncontrollable due to permanent and highly dynamic changes in boundary conditions.
- The competition of methods is based on criteria such as applicability, efficiency, and research mode rather than their performance (availability bias) (Kuhn 2012).

Currently, large parts of academic research are focused on the development of models, such as probabilistic models of structures, bridges, e.g., and their application. However, an equal amount of effort in science should be spent on testing the models and methods with real-world observations. This testing is usually done intensively during the development of the methods themselves, see for example tests of FORM/SORM and some variance reduction methods of Monte Carlo simulation (Proske 2003). Also, general purpose finite element packages such as ANSYS have large verification manuals (ANSYS 2013). The Riskman software can only be used if a verification run is performed after installation (Wakefield 2004). However, this is only one part of testing the methods.

The question remains, what are the observable effects of the application of such methods in real world. This concern also applies to simulation-based approaches to natural extreme events. From the point of view of the author, such models cannot stand alone - they are partly based on data, at least to develop and validate the models. Therefore, different models compete, but also provide additional information. For example, some can be as considered bottom-up and others as top-down methods. One possibility is to extend the methods and explicitly state the epistemic uncertainty (see e.g., MeteoSwiss 2023).

Unfortunately, this may lead to the return of at least partially arbitrarily chosen deterministic values, as the regulator has done for the $10'000$ yearn values of seismic and climatological hazards (ENSI 2016, 2022). In other words, statistical and probabilistic methods will lose their relevance if they do not strive to capture all uncertainties (Mudelsee 2017), such as human error, measurement uncertainties, etc., and move toward testing real-world data.

References

Ake, J., Munson, C., Stamatakos, J., Juckett, M., Coppersmith, K., Bommer, J.: Updated Implementation Guidelines for SSHAC Hazard Studies (NUREG-2213), U.S. NRC, Washington (2018)

Andres, N., Steeb, N., Badoux, A., Hegg, C.H. (eds.): Extremhochwasser an der Aare. Hauptbericht Projekt EXAR. Methodik und Resultate. Heft 104 Eidg. Forschungsanstalt für Wald, Schnee und Landschaft WSL, Birmensdorf (2021)

ANSYS: ANSYS Mechanical APDL Verification Manual, Ed. 15, Canonsburg, USA (2013)

Bachmann, H.: Erdbebensicherung der Bauwerke. In: Mehlhorn, G. (ed.) Der Ingenieurbau: Grundwissen, Teil 8: Tragwerkszuverlässigkeit, Einwirkungen. Verlag Wilhelm Ernst & Sohn, Berlin (1997)

Bachmann, H.: Erdbebensicherung von Bauwerken, 2. Aufl. Birkhäuser, Basel (2002)

Basöz, N.I., Kiremidjian, A.S., King, S.A., Law, K.H.: Statistical analysis of bridge damage data from the 1994 Northridge, CA, earthquake. Earthq. Spectra **15**(1), 25–54 (1999)

Bayliss, A.C., Reed, D.W.: The use of historical data in flood frequency estimation, Centre for Ecology and Hydrology, Natural Environment Research Council, Oxfordshire, March 2001

Burton, H.: Causal inference on observational data: opportunities and challenges in earthquake engineering. Earthq. Spectra **39**(1), 54–76 (2023)

Castillo, E., Castillo, C., Minguez, R.: Use of extreme value theory in engineering design. In: Martorell, C.S., et al. (eds.) Safety, Reliability and Risk Analysis: Theory, Methods and Applications. Taylor & Francis, London (2008)

Davidson, A.C.: Extremal Analysis: Methods and Specific Considerations, Report, January 27, 2014. EPFL, Lausanne (2014)

Davidson, A.C., Huser, R.: Statistics of extremes. Annu. Rev. Stat. Appl. **2**, 203–235 (2015). https://doi.org/10.1146/annurev-statistics-010814-020133

Davidson, A.C., Huser, R., Thibaud, E.: Geostatistics of dependent and asymptotically independent extremes. Math. Geosci. **45**, 511–529 (2013). https://doi.org/10.1007/s11004-013-9469-y

Embrechts, P., Klüppelberg, C., Mikosch, T.: Modelling Extremal Events for Insurance and Finance, pp. 152–168. Springer, Berlin (1997). https://doi.org/10.1007/978-3-642-33483-2

ENSI: Fukushima-Aktionsplan. Extreme Wetterbedingungen. 14/12/039. Brugg, 4th Juli 2012 (2012)

ENSI: Verfügung betreffend Erdbebengefährdungsannahmen ENSI-2015, rechtliches Gehör: Stellungnahme des ENSI zu den Anhörungskommentaren der Kernkraftwerksbetreiber, 30th Mai 2016, Brugg (2016)

ENSI: ENSI legt Gefährdungsannahmen für Extremwetter-Ereignisse neu fest, 14 Juli 2022. https://www.ensi.ch/de/2022/07/14/ensi-legt-gefaehrdungsannahmen-fuer-extremwetter-ereignisse-neu-fest/. Accessed 28 Aug 2023

Fischer, E.M., et al.: Storylines for unprecedented heatwaves based on ensemble boosting. Nat. Commun. **14**, 4643 (2023)

Gessner, C., Fischer, E.M., Beyerle, U., Knutti, R.: Developing low-likelihood climate storylines for extreme precipitation over central Europe. Earth's Future **11**, e2023EF003628 (2023). https://doi.org/10.1029/2023EF003628

Golomb, S.W.: Mathematical models – uses and limitations. SIMULATION **4**(14), 197–198 (1970)

Goodman, N.: Ways of Worldmaking, Hackett Publishing Company (1978)

Hennemuth, B., et al.: Statistical methods for the analysis of simulated and observed climate data, applied in projects and institutions dealing with climate change impact and adaptation. CSC Report 13, Climate Service Center, Germany (2013)

Kuhn, T.: The Structure of Scientific Revolutions, 4th edn. University of Chicago Press, Chicago (2012)

Mehlhorn, G., Curbach, M.: Brückenbau auf dem Weg vom Altertum zum modernen Brückenbau. Aktuelle Entwicklungen. In: Mehlhorn, G., Curbach, M. (eds.) Handbuch Brücken. Springer Fachmedien, Wiesbaden (2014). https://doi.org/10.1007/978-3-658-03342-2_1

MeteoSchweiz: Fachinformationen und Interpretation der Extremwertanalysen und des Intensitätsdiagramms anhand der Beispiele von Genève-Cointrin und Lugano (2023). https://www.meteoschweiz.admin.ch/klima/klima-der-schweiz/rekorde-und-extreme/extremwertanalysen/hintergrundinformationen/fachinformationen.html. Accessed 2 Aug 2023

Mudelsee, M.: Exkurs: Unsicherheiten bei der Analyse und Attribution von Hochwasserereignissen. In: Brasseur, G., Jacob, D., Schuck-Zöller, S. (eds.) Klimawandel in Deutschland, pp. 103–109. Springer, Heidelberg (2017). https://doi.org/10.1007/978-3-662-50397-3_11

Mudelsee, M.: Statistical Analysis of Climate Extremes. Cambridge University Press (2020)

Proske, D.: Beitrag zur Risikobeurteilung von alten Brücken unter Schiffsanprall. Ph.D. thesis, University of Technology Dresden, Dresden (2003)

Proske, D.: Zur Unsicherheit der Abschätzung von natürlichen Einwirkungen, Bautechnik. Ernst und Sohn, Berlin (2015)

Proske, D.: Do new findings in science of history influence risk analysis? ESRA-Newsletter, December 2015, p. 2. February 2016

Proske, D.: Katalog der Risiken, 2nd edn. Springer, Wiesbaden (2022). https://doi.org/10.1007/978-3-658-37083-1

Pugh, C.: Changing Sea Levels – Effects of Tides, Weather and Climate. Cambridge University Press, Cambridge (2004)

Schwarz-Zanetti, G., Fäh, D.: Grundlagen des Makroseismischen Erdbebenkatalogs der Schweiz, Band 2 1681–1878, Schweizerischer Erdbebendienst (Hrsg). vdf Hochschulverlag AG an der ETH, Zürich (2011a)

Schwarz-Zanetti, G., Fäh, D.: Grundlagen des Makroseismischen Erdbebenkatalogs der Schweiz, mit Beiträgen von Virgilio Masciadri und Philipp Kästli, Band 1 1000–1680, Schweizerischer Erdbebendienst (Hrsg). vdf Hochschulverlag AG an der ETH, Zürich (2011b)

Staudinger, M., Viviroli, D.: Extremhochwasser an der Aare. Detailbericht A Projekt EXAR. Hydrometeorologische Grundlagen. Universität Zürich, Zürich (2020)

Swain, R.E., England, J.F. Jr., Bullard, K.L., Raff, D.A.: Hydrologic hazard curve estimating procedures. DSO-04-08, U.S. Department of the Interior (2004)

UVEK: Verordnung des UVEK über die Gefährdungsannahmen und die Bewertung des Schutzes gegen Störfälle in Kernanlagen UVEK. Eidgenössisches Departement für Umwelt, Verkehr, Energie und Kommunikation, Version from 01.02.2019 (2009)

Wakefield, D.: Example Application of Riskman, Global Aviation Information network, 10 p. ABS Group, Irvine, Ca, USA (2004)

Wang, Z., Lee, G.C.: A comparative study of bridge damage due to the Wenchuan, Northridge, Loma Prieta and San Fernando earthquakes, June 2009, Earthquake Engineering and Engineering Vibration, vol. 8, Heft 2, pp. 251–261 (2009)

Wyss, M., Trendafiloski, G.: Trends in the casualty ration of injured to fatalities in earthquakes. In: Second International Workshop on Disaster Casualties, 15–16 June 2009, University of Cambridge, UK, pp. 1–6 (2009)

The Role of Sustainability in Target Reliability Assessment

Lloyd Chaitezvi(✉) and Christina McLeod

University of KwaZulu-Natal, Durban, South Africa
213566488@stu.ukzn.ac.za, mcleodc@ukzn.ac.za

Abstract. This research focusses on the effects of including sustainability factors in determining the target reliability of a structure. Target reliability is a measure of structural performance, and whose values depend on the costs of increasing safety together with economic, societal, and environmental consequences of failure. With the advent of climate change, structural performance must be complemented by appropriate sustainability levels. Structural standards, particularly in South Africa, have received minor to no updates with regards to sustainability. Sustainability is measured in terms of economic, social, and environmental aspects further classified by the Life Cycle Cost Analysis, the Life Quality Index, and Life Cycle Assessment respectively. The research methodology is based on the *fib* Model Code where indicators are proposed to assess performance criteria such as sustainability. However, as no values for these indicators have been determined, this research aims to determine benchmark values for the evaluation of sustainability parameters. Three methods which entail the Environmental Priority Strategy, the Panel and No Effect Level methods, will be considered in the quantification of indicators. This study considers the serviceability limit state (SLS), complementing a wider research project aiming to determine suitable SLS target reliabilities. Reliability models for two cases where serviceability requirements govern the design, namely, a bridge and a water retaining structure, will be developed. The sustainability framework will be integrated into the reliability model through probabilistic analysis. This will enable the impact of sustainability indicators on determining the target reliability of a structure to be modelled in a developing country's context.

Keywords: Reliability · Sustainability · Serviceability · Developing Country

1 Introduction

This study focusses on the inclusion of sustainability factors in determining the target reliability of a structure. Reliability is a measure of structural performance, forming the basis for the design of new and the assessment of existing structures for current standards such as SANS 10160-1 (SANS 2018), Eurocode (EN 1990, 2002), ISO 2394 [1], the *fib* Model Code 2010 [2], and draft Model Code 2020 [3]. Target reliability is a minimum reliability corresponding to a certain risk acceptance. This is an aspect of structural safety, which entails limiting the risk posed to human life through death or

J. C. Matos et al. (Eds.): IPW 2024, LNCE 494, pp. 250–258, 2024.
https://doi.org/10.1007/978-3-031-60271-9_22

injury in the event of failure. Limit states govern the minimal structural performance that is required or regarded as acceptable. They are defined in terms of the Ultimate Limit State (ULS) and the Serviceability Limit State (SLS). ULS governs the overall safety of the structure and people, while SLS focusses on general everyday use of the structure. This ensures that a structure functions optimally while considering safety, aesthetics, and user comfort [4]. Durability and robustness are other aspects considered in the performance of a structure. Durability focusses on meeting serviceability targets of the structure throughout its design life. Robustness focusses on safety and is the ability of a structure to resist exceptional actions [5]. These concepts are extensively considered in current structural design standards.

With the need to move towards protecting the environment and limiting the use of natural resources, it is imperative that structural design methods consider sustainability measures. Sustainability measures consist of economic, social, and environmental aspects. Economic and social aspects are considered and quantified through cost optimisation and the Life Quality Index (LQI) respectively. However, structural standards mostly consider the environmental aspects qualitatively [6]. The *fib* Model Code 2010 and the draft Model Code 2020 outline a framework to determine sustainability which includes the assessment of environmental impacts. However, SANS standards have not been fully updated to consider environmental aspects of sustainability. ISO 2394, adapted by South Africa as SANS 2394, only mentions sustainability in terms of human health, social property, and biodiversity. ISO 2394 states, however, that the concept of environmental sustainability is still under development, whose constraints still need to be determined by committees of standards [7]. Presently, target reliability has been mostly assessed for the Ultimate Limit State. However, research by Way, et al. [8] states that the determination of target reliability with regards to SLS has not been fully considered in current design standards. This study, therefore, aims to introduce a framework which includes serviceability and sustainability measures in determining target reliability for current structural design standards, particularly for concrete structures.

2 Structural Performance

Structural performance outlines how a structure/structural component behaves in response to an action or a combination of actions during its life cycle. Requirements for optimum structural performance entail meeting stakeholder needs through safety, sustainability, durability, and robustness. The randomness of structural properties and actions means that performance requirements cannot be outlined in an exact form but can be prescribed to meet performance criteria at a certain reliability [5].

2.1 Performance Evaluation

Evaluation of structural performance is conducted through verification of limit states, which separate desirable from undesirable conditions. Ultimate Limit State (ULS) corresponds to user and structural safety, while Serviceability Limit State (SLS) focusses on the general function and the fitness of use of a structure. The limiting values for SLS are defined depending on the consequences of exceedance and costs of repair or malfunction. Exceedance of SLS may be reversible, but in the case where irreversible damage has occurred, repair may be necessary to restore the structure to a fit for use state. ULS is irreversible, and failure of the structure occurs upon the first instance of exceedance [4]. Reliability is a measure of structural performance, which is the ability of a structure/structural component to function effectively over a period of which it was designed, under certain conditions, in an economic way [5]. Target reliability is a measure of reliability, which refers to the assurance of structural performance above a certain probability of failure corresponding to a specific reference period [1].

2.2 Target Reliability

Target reliability represents a balance between the cost and safety of a structure. It is derived through cost optimisation and is dependent on the costs of increasing safety together with economic, societal, and environmental consequences of failure [9]. Target reliability with regards to ULS is widely considered in design codes, with a quantitative standard of measure with respect to failure costs. However, the derivation of target reliability with regards to SLS is ambiguous and has only been considered qualitatively by design standards. This ambiguity presents challenges for structures whose designs are governed by SLS. Serviceability, by definition, is also subjective. The distinction between reversible and irreversible limit states further complicates the derivation of SLS target reliabilities. Serviceability limits between failure and non-failure are not as well defined as those for the Ultimate Limit State [8]. Equation 1 shows the function to determine target reliability through cost optimisation:

$$z(d) = C_0 + C_1.(d) + A(d) + D(d) \tag{1}$$

where C_0 are the initial construction costs without the decision parameter, C_1 are the costs associated with changes in the decision parameter (d), $A(d)$ represents obsolescence costs, and $D(d)$ represents failure costs.

The advent of climate change and controlling resource usage requires structures to be designed in a sustainable manner [3]. Progress in the incorporation of sustainability methods in determining target reliability has been slow, with most structural standards having received minor to no updates in this regard [10].

There is a need for structural design standards to include sustainability models in the design of new and the assessment of existing structures. The effects of environmental factors on target reliability must also be investigated, particularly in the context of developing countries through the derivation of sustainability indicators suitable to such conditions.

3 Sustainability

The concept of sustainability covers a wide range of aspects and cannot be defined unanimously [10]. The *fib* Model Code 2010 [2] defines sustainability as "the ability of a structure to function in a way that serves current environmental and resource needs without compromising the needs of future generations." This definition is in line with the

Sustainable Development Goals (SDGs), which aim to reduce greenhouse gas emissions, limit excessive usage of natural resources, and reduce waste production. Resilient urban development requires infrastructure [11], which bears directly on the construction industry, with concrete as the most widely used material. Production of cement products to make concrete contributes approximately 5% of global greenhouse gas emissions, with the construction sector contributing to more than 20% of global CO_2 emissions [12]. Frameworks and guidelines have been developed to encourage the design and management of more sustainable infrastructure. Lepech, et al. [6] classified the frameworks into knowledge based, rating based and performance based frameworks, which range from national to regional and to being project specific.

Knowledge based frameworks rely on how sustainability is defined, consisting of manuals, guidelines, and relevant design recommendations such as the Green Infrastructure Planning Guide (EU). The major challenge in knowledge-based frameworks is their dependency on how sustainability is defined, which is usually vague, resulting in difficulties in determining sustainability parameters. Rating-based frameworks use checklists and calculators used to determine a sustainability profile, such as LEED (US). The main challenge of rating frameworks is due to their discrete nature which does not allow compromise, which limits optimisation of resources and cannot be easily calibrated for most assessment methods. Performance-based tools consist of Life Cycle Assessment (LCA) methods and material flow analyses to determine environmental impacts. The use of continuous variables to determine sustainability overcomes the challenges posed by knowledge and rating-based frameworks. Examples of performance based tools entail programs such as SimaPro [6] and OneClick LCA.

Performance based frameworks consider economic, social, and environmental impacts holistically in the life cycle of a structure. A framework of quantifiable sustainability indicators can be derived to address the ambiguity in defining sustainability. The continuous characteristics of these frameworks allow for the use of probabilistic approaches towards sustainability in the same way they are used in structural reliability. Limit States can be applied to compare environmental impacts and environmental performance. Currently, no tools or guidelines in structural standards exist to set performance requirements or to evaluate the limit state function of sustainability. It is essential that the design, operation, maintenance, rehabilitation, and retirement of concrete infrastructure be conducted in a way that limits harm to humans and the environment alike. The performance of the structure throughout its life cycle must be in line with the three pillars of sustainability namely economic, social, and environmental aspects.

3.1 Sustainability Performance

As with structural performance, the sustainability performance of a structure is dictated by the relevant stakeholders, who influence how the structure functions to meet their basic needs [5]. This entails achieving a balance between economic efficiency, social responsibility, and environmental quality [12], while complying with statutory responsibilities and requirements. Structural performance with regards to SLS must be in line with the sustainability approach [3].

Economic Performance

Economic performance is assessed through a Life Cycle Cost Analysis (LCCA) to determine the feasibility of projects. This allows designers to develop different economic alternatives optimised to meet the requirements of the clients. Costs accrued in a project are categorised into direct and indirect costs. Direct costs are those borne by the owner during the life cycle of the structure and consist of design and construction costs, overheads, costs of maintenance and operation, and end of life costs. Indirect costs are those borne by society as a result of the project. The most common indirect costs entail user costs, societal costs, and environmental costs. Environmental costs are costs incurred due to impact or damage to surrounding areas and contribution to global challenges such as climate change. These can be translated in terms of energy and resources during the life cycle of a structure, together with accompanying environmental emissions [3].

Social Performance

Social performance focusses on the satisfaction of stakeholders during the life cycle of the structure, fostering a connection between structural engineering and the society. The first aspect of social performance is the health and quality of the built environment. This influences how people perceive, function, and behave as a result of the structure. Performance qualities entail acoustic comfort, thermal comfort, maintainability, flexibility, and healthiness. The second aspect is safety and security, where a structure should address security risks which are a function of threats, vulnerabilities, potential impacts and expected losses. The third aspect is concerned with aesthetics and cultural heritage. This is subjective, with performance requirements such as visual appearance, textures, integration into heritage/culture, and how they affect the aesthetics of the surrounding landscape. The fourth aspect entails the impact on the local communities, considering negative consequences such as noise pollution, traffic disturbance, time delays and positive influences such as improved infrastructure [3].

Environmental Performance

Environmental performance considers the effects of the structure on the environment and addressing challenges such as global warming, resource usage and waste management. These effects cannot be fully eliminated but can be managed to meet the standard global requirements. The draft Model Code 2020 [3] presented two basic principles to determine environmental impacts. The first is the Life Cycle Concept which considers the total environmental effect of a structure, from raw materials to disposal. The second principle entails the evaluation of Environmental Impact, which is a weighted sum of values representing various criteria. The Environmental Impact related to a particular step in the life cycle of a structure must incorporate all environmental damages and is represented by Eq. 2:

$$E_i = \sum W_j . Q_j \tag{2}$$

where $[w_j] = (w_1 \ldots, w_m)^{\mathrm{T}}$ is the vector weighting representing importance of individual criteria.

m = number of essential environmental criteria.

$[Qj] = (Q_1 \ldots, Q_m)^{\mathrm{T}}$ is the vector of embodied values of environmental criteria.

The determination of weighting factors is complex and subjective due variety in criteria associated with different features. These factors must cover specific conditions and boundaries, and must be chosen with sensitivity and transparency to ensure accountability for all stakeholders [13]. The Model Code proposed three approaches in determining the weighting of these factors. The first approach is the Environmental Priority Strategies (EPS) which highlights the price to be paid by society to prevent environmental impacts. The second approach is Panel Method, which includes the determination of weighting factors by a panel of experts. The third method is the No Effect Level (NEL), which relates the current level of an environmental aspect with a zero effect, with the difference being called a Sustainability Indicator [3]. Sustainability Indicators are mainly used for three purposes which are as explanatory tools, for conducting pilot projects or for assessing the performance of a city [14]. Three basic characteristics of indicators include the ease of interpretation, being well defined, and easy to collect. Data used to formulate these indicators must be easily understood by the public and policy makers. Proper implementation of sustainability indicators will result in more communication between different stakeholders, thereby holding relevant authorities accountable in terms of progress. Their reliability is often affected by lack of data from municipalities and government institutions, particularly in developing countries [15].

4 Methodology

The main aim of the methodology is to integrate a structural performance model with a sustainable design framework, developed from the derivation of sustainability indicators tailored to a developing country. Two representative structures, namely a bridge and a water retaining structure, will be chosen as their designs are governed by SLS. The bridge and WRS will be analysed at the operation and maintenance point of their life cycles. The first step is to determine the annual target reliabilities of the representative structures from an SLS perspective. This will be conducted through cost optimisation and FORM analysis as highlighted in Eq. 1. The structural model will be analysed at a point where it fails to meet the minimum performance criteria (falls below the required target reliability). Annual target reliabilities to restore performance will be derived from economic, social, and environmental aspects. Target reliability from an economic perspective will be determined through cost optimisation, while target reliability from a social perspective will be derived through the Life Quality Index, with requirements on human safety and individual risk.

The environmental aspects will be considered by first determining and quantifying sustainability indicators tailored to a developing country. The *fib* Model Code 2010 and Draft Model Code 2020 outlined some indicators which include CO_2, SO_x, and NO_x emissions, resource usage and the amount of waste produced. However, more elementary indicators will need to be derived, which must be precise, explanatory, and easily understood. These elementary indicators will be derived theoretically using various literature sources and previous research. Structural models of the bridge and water retaining structure will be imported into OneClick LCA to determine basic numerical values of the indicators. Their weighting will be conducted through a combination of the No Effect Method and the Panel Method. The No Effect Method will be firstly applied using available data on each indicator. This will be confirmed by a panel of experts before an indicator can be included in the sustainability model. The values from the sustainability model will be used to calculate the target reliability using environmental factors, which will be compared to target reliabilities from cost optimisation and the Life Quality Index. Figure 1 shows a flowchart of the methodological approach.

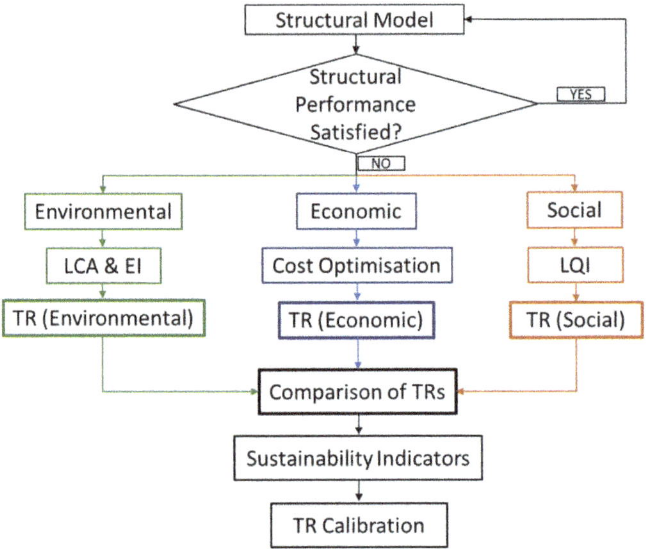

Fig. 1. Methodological Approach

5 Expected Results and Discussion

The results of this research will entail an integrated structural model incorporating a sustainability framework. This framework will be used to calibrate the target reliability of a bridge and a Water Retaining Structure, whose target reliabilities will have been derived considering the Serviceability Limit State. Research by Way, et al. proposed SLS target reliabilities for minor, moderate and great cases of failure costs. For minor failure costs ($0.01 < C_F/C_0 \leq 0.05$), the annual SLS target reliabilities for Low, Medium, and High costs of safety were 3.5, 2.9, and 2.2 respectively. For moderate failure costs, where ($0.05 < C_F/C_0 \leq 0.20$), the annual SLS target reliabilities for Low, Medium, and High costs of safety were 3.9, 2.3, and 2.6 respectively. For High failure costs, where ($0.20 < C_F/C_0 \leq 1.0$), the annual SLS target reliabilities for Low, Medium, and High costs of safety were 4.2, 3.6, and 3.0 respectively [8].

Inclusion of environmental factors ensures that the three main pillars of sustainability are considered holistically. Milan and Holicky [10] considered the effect of target reliability on sustainability indicators through the Building Material Sustainability Potential (BMSP), citing the RC slab on the deck of a bridge that has been service for 30 years. BMSP links the performance of the structure/structural member, service life and the

Environmental Impact with a focus on ULS. It was concluded that probabilistic and risk-based approaches were the most effective in determining sustainability, tallying with the methods employed in analysing target reliability. However, focus was on the Ultimate Limit State, and future research will investigate how BMSP can be incorporated in target reliability with regards to SLS. This requires the quantification of sustainability indicators in an unambiguous manner, which will be derived in further research. The Draft Model Code 2020 recommends using the Panel methods for weighing indicators. This is still subjective and complex, depending on the background knowledge and the interpretation of sustainability. An indicative relationship between the SLS target reliability and the effects of sustainability indicators will be determined. The combined NEL Panel method is expected to improve the subjectivity of the definition of sustainability. This will allow the development towards a sustainability limit state. Treating sustainability using probabilistic methods will allow for uncertainties in the methods used to evaluate the environmental impact.

6 Conclusion

Target reliability is a major aspect of structural design and is well documented in structural design standards, mostly for the Ultimate Limit State. However, current structural standards give little attention with a less clear basis in the determination of target reliabilities with regards to the Serviceability Limit State. This paper highlights the need to include quantitative bases for determining annual SLS target reliabilities, particularly for structures whose designs are governed by serviceability limits. A 1-year reference period will be used for structures governed by SLS, which differs as compared to ULS, which uses a standard reference period of 50 years. New structures, whose design is governed by SLS, will be considered as case studies in determining the framework for sustainability in terms of serviceability. This is due to the fact that South Africa is a developing country, with a lot of infrastructural development tailored towards the construction of new structures. The risk based approached proposed by [10], will be used as a basis in determining and quantifying sustainability indicators to determine the environmental effects on the target reliability with regards to Serviceability. Further research will extend the application of these indicators to structures whose design is governed by the Ultimate Limit State.

References

1. Sykora, M., Diamantidis, D., Holicky, M., Jung, K.: Target reliability for existing structures considering economic and societal aspects. Struct. Infrastruct. Eng. **13**(1), 181–194 (2016). https://doi.org/10.1080/15732479.2016.1198394
2. fib: fib Model Code for Concrete Structures 2010. Ernst & Sohn, Lausanne, Switzerland (2013)
3. fib: 2nd Draft Model Code 2020. Ernst & Sohn, Lausanne, Switzerland (2022)
4. JCSS: Probabilistic Model Code Part 1 - Basis of Design (2001)
5. Bigaj-van Vliet, A., Vrouwenvelder, T.: Reliability in the performance-based concept of fib model code 2010. Struct. Concr. **14**(4), 309–319 (2013). https://doi.org/10.1002/suco.201300053

6. Lepech, M.D., Geiker, M., Stang, H.: Probabilistic design and management of environmentally sustainable repair and rehabilitation of reinforced concrete structures. Cem. Concr. Compos. **47**, 19–31 (2014). https://doi.org/10.1016/j.cemconcomp.2013.10.009
7. General principles of reliability on structures, ISO 2394 (2014)
8. Way, A.C., McLeod, C.H., Viljoen, C.: Basis of serviceability limit state target reliability: fact or fiction? Presented at the 14th International Conference on Applications of Statistics and Probability in Civil Engineering ICASP 14, Dublin, Ireland, 9–13 July 2023 (2023)
9. Way, A.C., de Koker, N., Viljoen, C.: Target reliability for new road bridges in South Africa. J. South Afr. Inst. Civil Eng. **64**(3), 10–19 (2022). https://doi.org/10.17159/2309-8775/2022/v64n3a2
10. Holický, M., Sýkora, M.: Reliability approaches affecting the sustainability of concrete structures. Sustainability **13**(5) (2021). https://doi.org/10.3390/su13052627
11. UN-Habitat: Habitat III Paper 18 - Urban Infrastructure and Basic Services including Energy. In: Habitat III Issue Papers (2016)
12. Hingorani, R., Köhler, J.: Towards optimised decisions for resource and carbon-efficient structural design. Civ. Eng. Environ. Syst. **40**(1–2), 1–31 (2023). https://doi.org/10.1080/10286608.2023.2198767
13. Carraro, C., Cruciani, C., Ciampalini, F., Giove, S., Lanzi, E.: Aggregation and projection of sustainability indicators: a new approach (2009)
14. EU: Indicators for Sustainable Cities. In: Science for Environment Policy (2015)
15. Aschkenazi, K., et al.: Indicators for sustainability: how cities are monitoring and evaluating their success. In: Sustainable Cities International, Vancouver, Canada (2012)

Novel Decision-Making Frameworks

Effect of Seismic Source Assumption on the Value of Information from a Network of Ground Motion Sensors

Nilgün Merve Çağlar(✉) and Maria Pina Limongelli

Politecnico di Milano, Milan, Italy
nilgunmerve.caglar@polimi.it

Abstract. The seismic emergency management of the network of bridges is one of the most complex topics in the field of structural engineering due to the high level of uncertainty associated with the demand posed by the future possible seismic events, and the capacity of the bridges. Moreover, the selection of the optimal emergency actions must be performed considering the financial constraints, which increases the computational burden considerably. Additional information about the demand provided by the ground motion sensors improves the knowledge about the managed system and thereby the decision-making process. The Value of Information (VoI) from Bayesian decision analysis is a tool to quantify the benefits gained from additional information. The quantification of such benefits requires the modeling of the seismic demand which affects the estimated Value of Information. In this study, the impact of the seismic source assumption and modelling on the Value of Information provided by ground motion seismic sensors is investigated.

Keywords: Value of Information · Seismic Source Modelling · Probabilistic Seismic Hazard Assessment · Network of Bridges

1 Introduction

The Value of Information (VoI) from Bayesian decision analysis emerges as a tool that can be used to quantify the benefits to be gained using monitoring to support decision-making. The selection of the optimal sensor layout to support seismic emergency management based on the maximization of the VoI is pursued by modelling the future seismic demand and structural performance. In previous studies, the VoI acquired from the sensing schemes is evaluated for specific types of seismic sources and relevant models. The seismic source model extends beyond geographical locations and encompasses variations in both minimum and maximum magnitudes, as well as the frequency of events at specific magnitudes. The area source assumption, which describes regions of homogenous seismicity, is the most common seismogenic source representation and is adopted for example in reference [1]. The fault source assumption is used to model the seismicity of regions where well-identified and characterized faults exist and is employed for example in reference [2]. The estimated seismic demand changes with the adopted seismic

J. C. Matos et al. (Eds.): IPW 2024, LNCE 494, pp. 261–270, 2024.
https://doi.org/10.1007/978-3-031-60271-9_23

source model and thereby it is expected to affect the quantification of the VoI provided by a set of ground motion sensors. To the best of the author's knowledge, there is a gap in the literature regarding the impact of the seismic source model on the VoI. In this paper, the VoI acquired through ground motion monitoring is quantified using two different seismic source modeling approaches, namely area, and fault, to assess their effect on the VoI from a set of ground motion sensors. An exemplary case study focusing on a bridge network of three bridges connecting three cities is considered for the application. Based on the considered measurement schemes, the effect of the source modeling assumption on the optimum sensor layout for the seismic stations is also investigated.

2 Seismic Source Modelling

VoI quantification for seismic emergency management entails a Probabilistic Seismic Hazard Assessment (PSHA) of the region under consideration to estimate the expected cost (e.g. the risk) associated to the considered emergency management actions. The PSHA provides a quantification of the seismic demand that a site may encounter in terms of Intensity Measures (IMs) and corresponding rate of exceedance. The PSHA procedure may employ diverse seismic source models and Ground Motion Models (GMMs) to represent both epistemic and aleatory uncertainties associated with the seismic sources and earthquakes originating from them. A streamlined approach employs simplified seismic source assumptions to alleviate computational burdens. For example, the recent European Seismic Hazard Model 2020 (ESHM20) [3] incorporates a set of source models that considers the effect of the uncertainty associated with the maximum magnitude bound on the magnitude-frequency distribution. In contrast, its companion, the European Seismic Risk Model 2020 (ESRM20) [4] employs a collapsed version of the source models used in ESHM20 to reduce the computational complexity and correlation of the epistemic uncertainty. Event-based methods (Monte Carlo Simulation (MCS) technique and its variants) are generally employed to assess the seismic risk of infrastructures [5] due to the difficulties in describing the ground-shaking level over a region [6].

3 Methodology

The primary objective of this research is to investigate how the source model assumption employed in seismic demand modeling affects the optimal number and layout of seismic ground motion sensors at the locations of bridges in a network. Therefore, the benefit to be gained from the considered ground motion sensors' scheme, amongst the candidate configurations, is quantified using the VoI theory. The measurement layout that yields the greatest benefit in terms of monetary units is selected as the optimal one.

The initial stage in the methodology involves modeling the seismic demand on the bridges due to potential future seismic events. An event-based technique is utilized to calculate the seismic demand. Therefore the magnitude of each event is sampled from the associated magnitude frequency distribution of the considered seismic source using MCS. The seismic sources and associated magnitude frequency distributions can be extracted from an existing database if available (e.g., [4]). The magnitude frequency distribution can also be developed based on the earthquake catologs, which provide information about earthquakes that have occurred in a specific region over a certain period of time. For each scenario earthquake the associated rupture surface is determined using the magnitude-area scaling relationship [7]. The distance parameters required by the GMM are calculated as per [8]. Using the GMM applicable to the region, the seismic demand acting on each bridge is derived in terms of IMs. The appropriate GMM can be selected by referencing previous studies conducted to establish PSHA for the specific region. Alternatively, one can utilize existing methodologies to evaluate the goodness of fit of the GMM to the region based on the recorded ground motions. Since the paper compares the VoI from ground motion monitoring based on different seismic source assumptions, namely fault and area sources, this step is conducted by individually considering either area sources or fault sources.

The outputs of the first step are utilized to derive the VoI from ground motion monitoring, which is presented in detail in the following section.

The effects of the source modelling assumption on the selection of the optimal layout for the ground motion sensors and on the calculated VoI are investigated by comparing the analysis results carried out by considering each of the two seismic sources. An exemplary case study provides insights into the effects of seismic source assumptions on the VoI and optimal ground motion monitoring layout.

3.1 Value of Information Analysis

The VoI for each considered sensor layout is calculated as the difference between the expected costs calculated not accounting for the monitoring information (prior analysis) and accounting for the monitoring information. If the computation of VoI is carried out before collecting the information but assuming that they are available, thereby based on the probabilistic modelling of the information that will be collected, the analysis is called pre-posterior.

After an earthquake, in a network consisting of N bridges, where each bridge may be in one of L damage states, E different emergency actions can be applied to each bridge. Therefore, $C = E^N$ combinations of actions and $K = L^N$ combinations of damage states must be considered in the analysis. Each of the C-th alternative actions on the bridge network is the combination of the actions performed on the individual bridges and is denoted as $\boldsymbol{A^c}$. The k-th damage state of the network is the combination of the damage states of the individual bridges and is denoted as $\boldsymbol{DS_k}$. Considering the damage state of the n-th bridge in the k-th network damage state, $DS_{k,n}$, the probability of each network damage state is calculated as $P(\boldsymbol{DS}_k) = \prod_{n=1}^{N} P(DS_{k,n})$. The expected cost of the network action $\boldsymbol{A^c}$ in the network damage state $\boldsymbol{DS}_k$ is the sum of the expected direct costs, $E[Cost_D(\boldsymbol{A}^c)|\boldsymbol{DS}_k]$, and indirect costs, $E[Cost_I(\boldsymbol{A}^c)|\boldsymbol{DS}_k]$, which are calculated as given in Eqs. (1) and (2), respectively. The expected direct cost associated with each combination of action $\boldsymbol{A^c}$ given the k-th damage state of the network $\boldsymbol{DS_k}$ is the sum of the direct costs associated with each bridge. This is calculated by considering the action to be applied to n-th bridge given the A_n^c and the damage state of the n-th bridge in the $\boldsymbol{DS}_k$.

$$E[Cost_D(\boldsymbol{A}^c)|\boldsymbol{DS}_k] = \sum_{n=1}^{N} \left[P(F_n|A_n^c, DS_{k,n})c_{Fn}(A_n^c) + P(S_n|A_n^c, DS_{k,n})c_{Sn}(A_n^c)\right] \tag{1}$$

$$E\left[Cost_I\left(\boldsymbol{A}^c\right)|\boldsymbol{DS}_k\right] = f\left(\Delta\overline{\eta}\left(\boldsymbol{A}^c\right)\right) \tag{2}$$

where, $c_{Fn}(\boldsymbol{A}^c)$ and $c_{Sn}(\boldsymbol{A}^c)$ are the costs associated with the failure and survival of the n-th bridge in the network under action A_n^c. $P\left(F_n|\boldsymbol{A}^c, DS_{k,n}\right)$ and $P\left(S_n|\boldsymbol{A}^c, DS_{k,n}\right)$ are respectively the failure and survival probabilities of the n-th bridge, conditional on the action in combination A_n^c on this component, and the damage state $DS_{k,n}$. $\Delta\overline{\eta}$ is the loss of connectivity in the network [9].

Prior Analysis. The seismic demand resulting from an earthquake, denoted as e, at N bridge locations, is modeled using the Ground Motion Model as a Gaussian Multivariate Distribution $\boldsymbol{D}(e) \sim N\left(\boldsymbol{\mu}_{\boldsymbol{D}}(e), \sum_{\boldsymbol{D}}(e)\right)$. $\boldsymbol{\mu}_{\boldsymbol{D}}(e)$ is the vector of median estimates of the GMM at N bridge locations specific to the considered earthquake e. $\sum_{\boldsymbol{D}}(e)$ is the covariance matrix, which accounts for the spatial correlation of within-event residuals at N bridge locations. The vector of $\boldsymbol{D}$(e) is populated with the IM s corresponding to N bridge locations. For simplicity of notation, the dependency of the demand on the earthquake will not be explicitly reported in the rest of the section. The capacity, $\boldsymbol{C}_{\boldsymbol{DS}_k}$, of N bridges in network damage state $\boldsymbol{DS}_k$ is modeled as a Gaussian Multivariate Distribution $\boldsymbol{C}_{\boldsymbol{DS}_k} \sim N\left(\boldsymbol{\mu}_{\boldsymbol{C}_{\boldsymbol{DS}_k}}, \sum_{\boldsymbol{C}_{\boldsymbol{DS}_k}}\right)$ [9]. $\boldsymbol{\mu}_{\boldsymbol{C}_{\boldsymbol{DS}_k}}$ and $\sum_{\boldsymbol{C}_{\boldsymbol{DS}_k}}$ are the mean vector and covariance matrix of $\boldsymbol{C}_{\boldsymbol{DS}_k}$, respectively.

The probability of failure associated with the k-th network damage state is derived based on the reliability analysis methods as presented in [9]. In this context, the limit state functions are employed as an intermediate step to derive the reliability index linked to each $\boldsymbol{DS}_k$. The limit state functions are modeled as Gaussian Multivariate Distribution $\boldsymbol{G}_{\boldsymbol{DS}_k} \sim N\left(\boldsymbol{\mu}_{\boldsymbol{G}_{\boldsymbol{DS}_k}}, \sum_{\boldsymbol{G}_{\boldsymbol{DS}_k}}\right)$ where: $\boldsymbol{\mu}_{\boldsymbol{G}_{\boldsymbol{DS}_k}} = \boldsymbol{\mu}_{\boldsymbol{C}_{\boldsymbol{DS}_k}} - \boldsymbol{\mu}_{\boldsymbol{D}}$ and $\boldsymbol{\Sigma}_{\boldsymbol{G}_{\boldsymbol{DS}^k}} = \boldsymbol{\Sigma}_{\boldsymbol{C}_{\boldsymbol{DS}_k}} + \boldsymbol{\Sigma}_{\boldsymbol{D}}$. The reliability index, $\boldsymbol{\beta}_{\boldsymbol{DS}_k}$, associated with each $\boldsymbol{DS}_k$ is calculated as given in Eq. (3).

$$\boldsymbol{\beta}_{\boldsymbol{DS}_k} = \boldsymbol{\mu}_{\boldsymbol{G}_{\boldsymbol{DS}_k}} / \sqrt{diag\left(\boldsymbol{\Sigma}_{\boldsymbol{G}_{\boldsymbol{DS}_k}}\right)} \tag{3}$$

For the n-th bridge in a given l-th damage state $DS_{n,l} = DS_{k,n}$, the probability that the seismic demand D_n exceeds the associated capacity is calculated as given in Eq. (4).

$$P\left[D_n \geq C_{DS_{n,l}}\right] = \Phi\left(-\beta_{DS_{n,l}}\right) \tag{4}$$

The probability of the l-th damage state for the n-th bridge $DS_{n,l}$, $P\left(DS_{n,l}\right)$, is assigned as given in Eq. (5).

$$P\left(DS_{n,l}\right) = \begin{matrix} P\left[D_n \geq C_{DS_{n,l}}\right] - P\left[D_n \geq C_{DS_{n,l+1}}\right] & if\ l \neq L \\ P\left[D_n \geq C_{DS_{n,L}}\right] & if\ l = L \end{matrix} \tag{5}$$

The probability of each of network damage state, $\boldsymbol{DS}_k$, is calculated as described earlier as the product of the $P\left(DS_{k,n}\right)$. The total expected cost of the actions $\boldsymbol{A^c}$ is calculated as given in Eq. (6).

$$E\left[Cost\left(\boldsymbol{A}^c\right)\right] = \sum_{k=1}^{K} E\left[Cost\left(\boldsymbol{A}^c\right)|\boldsymbol{DS}_k\right]P(\boldsymbol{DS}_k) \tag{6}$$

The optimal network action $\hat{\boldsymbol{A}}$, resulting in the minimum expected cost is calculated in Eq. (7).

$$\hat{\boldsymbol{A}} = \underset{\boldsymbol{A}^c}{\operatorname{argmin}}\, E\left[Cost\left(\boldsymbol{A}^c\right)\right] \tag{7}$$

The prior expected cost is given in Eq. (8). It is worth noting that, this expected cost is estimated based on the demand $\boldsymbol{D}(e)$ defined for a given earthquake, e.

$$E'\left[Cost\left(\hat{\boldsymbol{A}}\right)\right] = \sum_{k=1}^{K} E\left[Cost\left(\hat{\boldsymbol{A}}\right)|\boldsymbol{DS}_k\right]P(\boldsymbol{DS}_k) \tag{8}$$

Pre-posterior Decision Analysis. It is assumed that strong ground motion sensors provide information (measurements) that can be used to update the demand $\boldsymbol{D}(e)$ posed by an earthquake e. Therefore, the probability of damage states, and thereby the expected costs of the considered management actions change accordingly. The m-th measurements of the demand $\boldsymbol{D_m}(e)$ can be modeled as given in Eq. (9).

$$\boldsymbol{D_m}(e) = \boldsymbol{\Omega_m D}(e) + \boldsymbol{\varepsilon} \tag{9}$$

where $\boldsymbol{\Omega_m}$ is the measurement matrix indicating the location of the M measurements. The error $\boldsymbol{\varepsilon}$, represents the difference between the $\boldsymbol{D}(e)$ at measurement locations and the $\boldsymbol{D_m}(e)$ can be modeled as Gaussian Multivariate Distribution $\boldsymbol{\varepsilon} \sim N\left(\mu_{\varepsilon}, \sum_{\varepsilon}\right)$. The functional dependence on the earthquake e is again dropped in the following for simplicity of notation. Using the Gaussian Multivariate Distribution, the measurements are represented as $\boldsymbol{D_m} \sim N\left(\mu_{D_m}, \sum_{D_m}\right)$ where, $\mu_{D_m} = \boldsymbol{\Omega_m \mu_D} + \boldsymbol{\mu_\varepsilon}$ and $\boldsymbol{\Sigma_{D_m}} = \boldsymbol{\Omega_m \Sigma_D \Omega_m^T} + \boldsymbol{\Sigma_\varepsilon}$. The distribution of the demand parameters can be updated conditional on the measurements, $\boldsymbol{D}|\boldsymbol{D_m} \sim \mathcal{N}(\boldsymbol{\mu_D}|\boldsymbol{D_m}, \boldsymbol{\Sigma_D}|\boldsymbol{D_m})$, as given in Eqs. (10) and (11)

$$\boldsymbol{\mu_D}|\boldsymbol{D_m} = \boldsymbol{\mu_D} + \boldsymbol{\Sigma_D \Omega_m^T \Sigma_{D_m}^{-1}}\left(\boldsymbol{D_m} - \boldsymbol{\mu_{D_m}}\right) \tag{10}$$

$$\boldsymbol{\Sigma_D}|\boldsymbol{D_m} = \boldsymbol{\Sigma_D} - \boldsymbol{\Sigma_D \Omega_m^T \Sigma_{D_m}^{-1} \Omega_m \Sigma_D} \tag{11}$$

The limit state functions can be updated by replacing the prior with the posterior distribution parameters as follows: $\boldsymbol{\mu_{G_{DS_k}}}\Big|\boldsymbol{D_m} = \boldsymbol{\mu_{C_{DS_k}}} - \boldsymbol{\mu_D}\Big|\boldsymbol{D_m}$ and $\boldsymbol{\Sigma_{G_{DS_k}}}\Big|\boldsymbol{D_m} = \boldsymbol{\Sigma_{C_{DS_k}}} + \boldsymbol{\Sigma_D}\Big|\boldsymbol{D_m}$. The updated reliability indices and can be calculated using Eq. (3) by incorporating the parameters of the updated limit state functions. Following the same procedure as in the prior analysis, updated damage states can be derived. The expected cost of each network action $\boldsymbol{A}^c$ given the measurements m is calculated as given in Eq. (12) using the updated damage states given the measurement, $\boldsymbol{DS}_k|\boldsymbol{D_m}$.

$$E\left[Cost\left(\boldsymbol{A^c}\right)|\boldsymbol{D_m}\right] = \sum_{k=1}^{K} E\left[Cost\left(\boldsymbol{A^c}\right)|\boldsymbol{DS}_k\right]P(\boldsymbol{DS}_k|\boldsymbol{D_m}) \tag{12}$$

The optimal action $\tilde{\boldsymbol{A}}$ resulting in the minimum expected cost for the measurement set is the optimal pre-posterior action given the measurement as given in Eq. (13).

$$\tilde{\boldsymbol{A}}|\boldsymbol{D_m} = \underset{\boldsymbol{A^c}}{\operatorname{argmin}}\, E\left[Cost\left(\boldsymbol{A^c}\right)|\boldsymbol{D_m}\right] \tag{13}$$

The pre-posterior expected cost for each sensor layout is obtained considering all the possible outcomes of the monitoring system and their probability of occurrence as given in Eq. (14).

$$E''\left[Cost(\tilde{\mathbf{A}})\right] = \sum\nolimits_{m=1}^{M} E\left[Cost(\tilde{\mathbf{A}}|\mathbf{D_m})\right]P(\mathbf{D_m}) \tag{14}$$

Quantification of Value of Information. A vast number of earthquakes characterized by their epicentral locations, p, and magnitudes, m, are simulated, $e(m, p)$. For each earthquake, $e(m, p)$, the *VoI* for a given measurement scheme is given by Eq. (15).

$$VoI\,(\mathbf{D_m}|e(m,p)) = E'\left[Cost(\hat{\mathbf{A}})\right] - E''\left[Cost(\tilde{\mathbf{A}})\right]] \tag{15}$$

The VoI derived from each measurement scheme is calculated by integrating the VoI obtained from each earthquake over the probabilities of the magnitudes, f_m, and epicentral locations f_p as given in Eq. (16).

$$VoI\,(\mathbf{D_m}) = \int_m \int_p VoI\,(\mathbf{D_m}|e(m,p))f_m\, f_p dm\; dp \tag{16}$$

Since a Poisson process is adopted to model the occurrence of the earthquakes, the net VoI for a given reference time of t, is calculated as given in Eq. (17) [10].

$$VoI_t = \frac{\lambda}{r} VoI\,(\mathbf{D_m})\left(1 - e^{rt}\right) \tag{17}$$

where, λ is the rate of earthquakes, extracted from the magnitude-frequency distribution defined for the adopted source assumption and r is the nominal discount rate. For the reference time of $t_{max} = t \to \infty$ Eq. (17) can be re-written as given in Eq. (18)

$$VoI_{t=t_{max}} = \frac{\lambda}{r} VoI\,(\mathbf{D_m}) \tag{18}$$

The net VoI of the measurement scheme is calculated as the difference between the VoI calculated for the reference time and the cost of the monitoring system.

4 Case Study

The seismic emergency management exemplary problem addressed in this paper considers a network of three identical bridges with natural period T = 1 s connecting three cities located in Sicily, Italy. The region has experienced several devastating earthquakes such as the 1693 Mw 7.4 and the 1542 Mw 6.7 Sicily Earthquakes. The goal of the study is to investigate the impact of the seismic source model adopted in the analysis, on the quantification of the Value of Information (VoI) from ground motion monitoring. The VoI analysis aims to identify the benefit of installing such sensors to manage the seismic emergency of the bridge network. In the context of the seismic demand modelling the area and fault source assumptions utilized in the ESRM20 [2] are employed. To reduce the computational burden, seismic sources are extracted from the complete set in ESRM20 based on whether they entirely or partially fall within the specified 200 km distance limit. Figure 1 demonstrates the area sources (red polygons) and fault sources (black lines) considered in the analysis. The location of the bridges and cities are also demonstrated in Fig. 1.

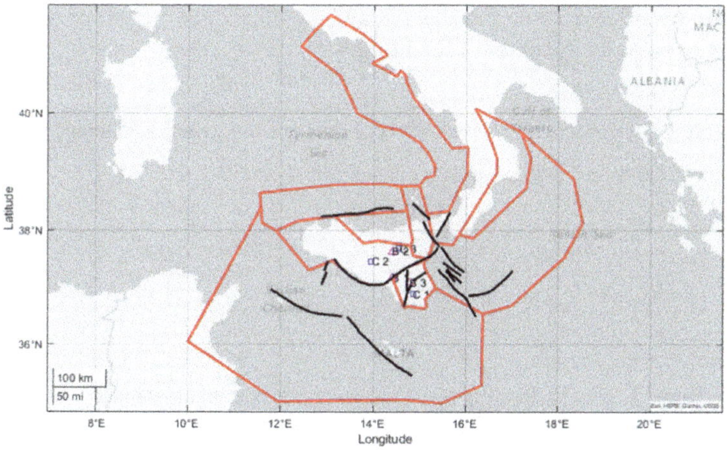

Fig. 1. Location of the bridges (magenta triangles), cities (blue squares), area sources (red polygons) and fault sources (black lines).

The magnitude of the scenario earthquakes are sampled from the respective magnitude frequency distribution provided in ESRM20 using MCS technique. The epicentral location of the seismic events is assumed to be uniformly distributed over and within the considered seismic sources. The dimension of the ruptured zone associated with each scenario earthquake is derived using the magnitude-area scaling relationship given in [7]. The distance metrics that are required to evaluate ground motion intensity measures corresponding to each bridge location are calculated following the procedure provided in [8]. The spectral acceleration at the first natural period of the bridges, ($S_a(T = 1s)$), is calculated using the GMM provided in [11]. A total number of approximately 1450000 and 50000 scenario earthquakes are generated based on area and fault source assumptions to model the seismic demand for each bridge. Scenarios exceeding a distance of 200 km are eliminated due to the distance limitations of the GMM. Since the aim of this study is to identify the impact of the source modelling assumption on the VoI, the site conditions have been considered identical. Two alternative seismic management action were considered in the analysis namely: keeping the bridge open to traffic (*Open*) or closing the bridge (*Close*), after the earthquake. Therefore, for the 3 bridges, 2^3 sets of possible combinations of actions are considered. The direct costs associated with the failure of each component are assumed to be 10M and 0 € for actions open and close, respectively. The direct costs associated with the survival of a bridge are assumed to be 0 €. The indirect costs are calculated as a function of loss of connectivity ($1000000\Delta\overline{\eta}(\boldsymbol{A}^c)$) and the cost of each measurement is assumed equal to 1000 €. For each bridge three damage states, light, DS_1, medium, DS_2, and severe damage, DS_3, states which correspond to 27 combinations of damage states of the bridge network are considered. The distribution parameters of the capacity associated with each damage state are presented in Table 1. The probability of failure given the damage states is given in Table 2. The measurement error is modeled as a lognormal variable with zero mean and a standard deviation of 0.2. The nominal discount rate is assumed to be 5%. Several measurement schemes are considered assuming the availability of one, two or three ground motion sensors deployed at the location of the bridges.

Table 1. The distribution parameters of the capacity given the damage states.

Damage State	$\mu_{C_{DS_l}}$	$\sigma_{C_{DS_l}}$
DS_2	0.26 g	0.6
DS_3	0.65 g	0.6

Table 2. The probability of failure given the action and damage states.

	DS_1	DS_2	DS_3
$P(F\|Open, DS_l)$	1E−03	1E−02	9.99E−01
$P(F\|Close, DS_l)$	0	0	0

5 Results and Discussion

The results of the VoI analysis for area and fault sources are compared with each other in this section. It is worth noting that the VoI results presented in this section are calculated using Eq. (18). Therefore they reflect the net value of information for each measurement layout considering the lifepan of $t_{max} = t \rightarrow \infty$. Although the substantial difference in the number of earthquake scenarios required for presenting the seismic hazard associated with area and fault sources results in a considerable variation in the execution time of the analysis, while calculating the net VoI only the cost of sensors are assumed and the cost associated with the computation is not considered. The left panel of Fig. 2 illustrates the hazard curves obtained for each bridge, considering each source assumption separately. As depicted in Fig. 2, the estimated seismic demand on Bridge 3 (B3) is higher with respect to the other bridges, regardless of whether area or fault sources are taken into account whereas the lower demand is computed for Bridge 2 (B2). The right panel of Fig. 2 illustrates the calculated net VoI for each measurement scheme based on the fault and area source assumptions. The horizontal axis indicates the location(s) of the sensor(s) while the vertical axis shows the net VoI in €. It is important to emphasize that negative values found for the net VoI indicates that the cost of the sensor placed at location 2 exceeds the benefits gained from monitoring the seismic demand at that locations. The net VoI attains its lowest value for Bridge 2. This observation is significant, as Bridge 2 undergoes the least seismic demand among the others (as shown in Fig. 2, left panel), highlighting the influence of seismic demand levels on the VoI. A similar trend is apparent when comparing hazard curves derived from area and fault sources, along with their respective VoI analyses. Notably, assuming the area source model, a higher seismic demand is estimated for all bridges and this leads to a net VoI calculated for area sources substantially higher than the corresponding value estimated assuming fault sources. In the case of the fault source assumption, measurement schemes 1-3 and 1-2-3 result in approximately similar values. However, when adopting the area source assumption, the measurement scheme 1-2-3 emerges as the optimal choice with scheme 1-3 as the second best choice.

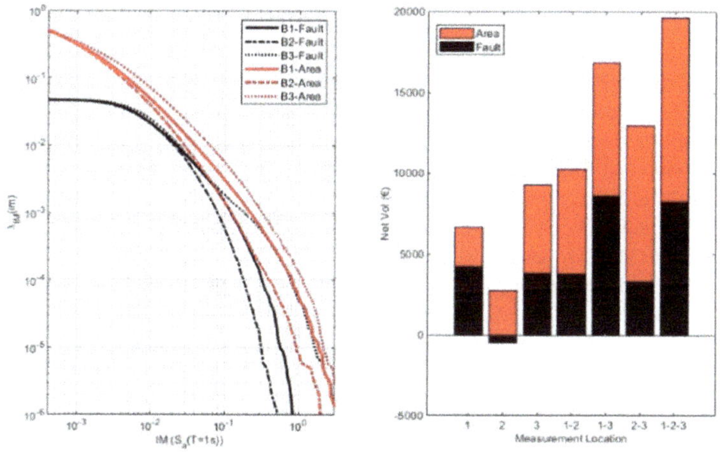

Fig. 2. Hazard curves associated with each bridge considering fault (black) and area (red) (left panel), the net value of information obtained for each measurement scheme (right panel).

These results confirm the findings of paper [12] that monitoring bridges facing the highest seismic demand offers a more substantial benefit in monetary terms. Furthermore, the highlight the importance of the seismic source modeling on the selection of the optimal sensors scheme, assumed as the one leading to the highest VoI.

6 Conclusion

This study investigates the effect of seismic source modelling assumptions on the Value of Information from ground motion monitoring. The analysis is carried out comparing the seismic demand modeled using fault and area source assumptions provided in ESRM20 and the respective computed VoI. A network of three identical bridges located in Sicily, Italy is employed and several schemes of sensors, deployed at bridge locations are considered. The analysis results show that the seismic source assumption considered in the seismic hazard assessment may significantly change the Value of Information and therefore, the selection of the optimal seismic sensing scheme. Specifically, the level of the intensity measures and their respective exceedance rates play a crucial role on the benefits to be gained from the sensing schemes. The results derived from the conducted analysis show that as the seismic demand increases, the VoI also increases. Although those results are specific to the study region, source models (earthquake rupture forecast models), and ground motion model employed in the analysis, it highlights the importance of the adopted source modeling technique, and therefore of the seismic demand modeling, in VoI analysis.

Acknowledgments. This project has received funding from the European Union's Horizon 2020 research and innovation programme under the Marie Skłodowska-Curie grant agreement No 10110717.

References

1. Giordano, P.F., Limongelli, M.P.: The value of structural health monitoring in seismic emergency management of bridges. Struct. Infrastruct. Eng. **18**(4), 537–553 (2022)
2. Malings, C.A.: Optimal sensor placement for infrastructure system monitoring using probabilistic graphical models and value of information. Doctoral dissertation, Department of Civil and Environmental Engineering, Carnegie Mellon University, Pittsburgh, PA (2017)
3. Danciu, L.: The 2020 update of the European seismic hazard model: model overview. EFEHR Technical Report 001, v1.0.0 (2021). https://doi.org/10.12686/a15
4. Crowley, H., et al.: European Seismic Risk Model (ESRM20). EFEHR Technical Report 002, V1.0.0 (2021). https://doi.org/10.7414/EUC-EFEHR-TR002-ESRM20
5. Miller, M., Baker, J.: Ground-motion intensity and damage map selection for probabilistic infrastructure network risk assessment using optimization. Earthquake Eng. Struct. Dynam. **44**(7), 1139–1156 (2015)
6. Jayaram, N., Baker, J.W.: Efficient sampling and data reduction techniques for probabilistic seismic lifeline risk assessment. Earthquake Eng. Struct. Dynam. **39**(10), 1109–1131 (2010)
7. Leonard, M.: Earthquake fault scaling: self-consistent relating of rupture length, width, average displacement, and moment release. Bull. Seismol. Soc. Am. **100**(5A), 1971–1988 (2010)
8. Kaklamanos, J., Baise, L.G., Boore, D.M.: Estimating unknown input parameters when implementing the NGA ground-motion prediction equations in engineering practice. Earthq. Spectra **27**(4), 1219–1235 (2011)
9. Giordano, P.F., Limongelli, M.P., Iannacone, L., Gardoni, P.: Value of information analysis for degrading engineering systems. In: 7th International Symposium on Life-Cycle Civil Engineering, IALCCE 2020, pp. 671–678. CRC Press/Balkema (2020)
10. Yeo, G.L., Cornell, C.A.: Building life-cycle cost analysis due to mainshock and aftershock occurrences. Struct. Saf. **31**(5), 396–408 (2009)
11. Lanzano, G., et al.: A revised ground-motion prediction model for shallow crustal earthquakes in Italy. Bull. Seismol. Soc. Am. **109**(2), 525–540 (2019)
12. Giordano, P.F., Limongelli, M.P.: The value of seismic structural health monitoring for road bridges. In: SHMII-10, Porto, Portugal, 30 June–2 July 2021 (2021)

Physics-Informed Neural Networks for Nonlinear Analysis of Reinforced Concrete Beams

Vera M. Balmer(✉), Walter Kaufmann, and Michael A. Kraus

Institute of Structural Engineering (IBK), ETH Zürich, Design++ and ETH AI Center
Stefano-Franscini-Platz 5, 8093 Zürich, Switzerland
vera.balmer@ibk.baug.ethz.ch

Abstract. Reinforced concrete (RC) beams are a common structural component in buildings and bridges and play a pivotal role in design and verification processes. Current structural computations often rely on linear elastic analyses which potentially yield overly conservative representations of the load-deformation behaviour of RC. Despite the superior accuracy offered by nonlinear analyses, they are limited in engineering practise due to computational inefficiency and intricate dependencies on uncertain material parameters. The scarcity of nonlinear calculations hinders exploration of diverse design, verification, and uncertainty quantification scenarios, necessitating a computational tool for improved assessment.

This paper proposes physics-informed neural networks (PINNs) as an innovative differentiable computational approach for analysis and design of RC beams. We investigate the training and predictive quality of PINNs for a single-span simply supported beam structure, where we start with a linear elastic material model and from this deduce a PINN for the commonly used multi-linear (uncracked-cracked-plastic) material law idealisation of RC beams. To achieve this PINN representation, we propose a novel mixed-approach reformulation of PINNs. A hyperparameter tuning study reveals medium-sized networks of 8 layers with tanh or SiLU activation to be most suitable. By comparing the PINN predictions to analytical as well as Finite-Element solutions we demonstrate its accuracy in forecasting displacements and rotations, however no increase in computational time can be identified. The outlook highlights the enhancement of physics-informed deep operator networks to allow fast and accurate analysis and design of parametric beam structures.

Keywords: Physics-Informed Neural Networks · Mixed-Approach Formulation · Multi-linear Reinforced Concrete Material Model

1 Introduction

In light of the increasing demand in urbanisation as well as associated environmental concerns, reinforced concrete (RC) plays an ever-so important role: On one hand its beneficial mechanical properties allow for durable structures, on the other hand its detrimental environmental impact needs to be considered by utilising the material responsibly. RC

J. C. Matos et al. (Eds.): IPW 2024, LNCE 494, pp. 271–280, 2024.
https://doi.org/10.1007/978-3-031-60271-9_24

structures are typically segmented into their constituent elements for numerical analyses, where RC beams are a common structural component in buildings and bridges and hence play a pivotal role in design and verification processes.

Current design and verification tools for RC structures often rely on linear elastic analyses. These simulation models approximate the strongly nonlinear material behaviour of RC with linear models and may thus be overly conservative [1]. Such an approximation can hence lead to economically as well as environmentally inefficient designs. Nonlinear analyses (e.g. finite element analyses (FEA)) on the other hand are able to reproduce the real behaviour of RC structures more accurately and can be applied universally. However, they lack computational efficiency and mostly depend on multiple uncertain material parameters hindering exploration of diverse designs, verification and uncertainty quantification.

Novel digital processes such as machine and deep learning (ML/DL) are the current state-of-the-art for differentiable modelling and gradient-based optimisation. Recent advances of physics-based machine learning have additionally shown the potential of ML methods in combination with conventional computational physics models across various domains [2]. This paper proposes the utilisation of physics-informed neural networks (PINNs) [3] as computational approach for analysis and design of RC beams. PINNs have been leveraged in multiple applications ([4, 5]) to make solution approximations of PDEs more understandable by incorporating the PDE directly in the loss terms of the NN. This is a crucial point for the application of ML methodologies in the civil engineering sector, as it allows for embedding domain knowledge within the ML system. Additionally, PINNs do not require any data to be trained on and hence circumvent the problem of data scarcity in the construction sector.

Therefore, this study aims at (i) providing insights into modelling RC structures with a novel mixed-approach for PINNs and (ii) examining the feasibility and effectiveness

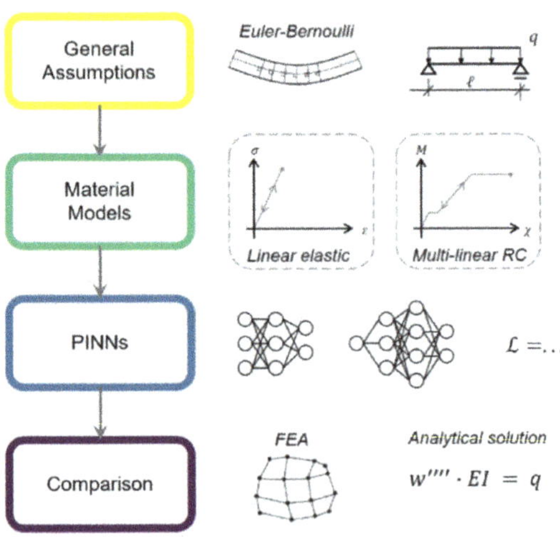

Fig. 1. Workflow of the research carried out in this study

of using this novel methodology in RC structural analysis. To this end, a workflow comprising four steps was defined (see Fig. 1): Defining the general assumptions, identifying material models, fine-tuning the PINN method and finally comparing it to state-of-the-art computational methodologies such as FEA or the analytical solution. Overall, these steps shall provide an overview of how PINNs can be utilised as a suitable alternative in terms of modelling RC structures accurately, even in case of material nonlinearities.

2 Methodology

This section presents the research methodology according to the steps shown in Fig. 1.

2.1 General Assumptions

A simply supported, single-span beam of length $\ell = 10$ m under a uniform line load q is analysed in this study (see Fig. 1). The bending stiffness EI is determined depending on the material law (see Sect. 2.2) and assumed to be piecewise constant. Additionally, the Euler-Bernoulli assumption of cross-sections remaining plane and orthogonal to the axis of the beam is considered. Hence, the Euler-Bernoulli beam differential equation (DE) is valid:

$$EI^{(i)}\, w''''(x) = q \tag{1}$$

2.2 Material Models

Two material models with increasing complexity are tested, each with its corresponding PINN architecture (see Fig. 1). In the following, more detailed descriptions of both material models are provided.

Linear Elastic: The linear elastic behaviour is schematically shown in Fig. 1. The stiffness EI is chosen as the uncracked stiffness of the RC cross-section, as explained below for the multi-linear RC material. As the linear elastic model does not depend on the loading level, only a load of $q = 5$ kN/m is tested.

Multi-linear RC Material: The multi-linear material model for RC is shown in Figure 2a) as moment-curvature relationship. It consists of three regimes: (I) The uncracked elastic regime, within which the material behaves linearly elastic with bending stiffness EI^{I}, (II) the cracked elastic region, within which RC again behaves linearly elastic but with a reduced bending stiffness EI^{II}, and (III) the perfectly plastic regime, where the bending resistance of the cross-section is reached. A rectangular cross-section with the characteristics shown in Fig. 2b) is adopted, from which the stiffnesses EI^{I} and EI^{II} can be deduced in closed form. The plastic range of the multi-linear RC material model is not incorporated in the modelling with the PINN. Three load levels are tested, which correspond to different locations in the χ-M diagram: $q_1 = 5$ kN/m (uncracked elastic), $q_2 = 7.5$ kN/m (cracked elastic, right after crack formation), $q_3 = 40$ kN/m (cracked elastic, just before yielding).

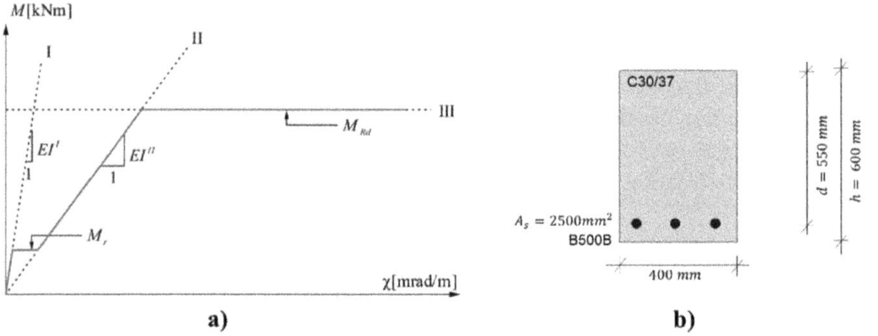

Fig. 2. a) Multi-linear material model for reinforced concrete, after [6] (I: uncracked elastic, II: cracked elastic, III plastic); b) geometrical and material parameters of the RC cross-section

2.3 Finite Element Modelling

To be able to compare the predictions of the PINNs to a numerical solution, a simulation is carried out in the FEA software RFEM 5.33.01 [7]. The software enables the implementation of a nonlinear concrete material law for the analysis of a RC beam. Compared to the analytical solution, the FEA simulation relies on the method of Quast [8], which reduces the stiffness incrementally with increased moment, considers tension stiffening up to f_{ctm}. The target load q is applied in three incremental steps with equilibrium iterations each.

2.4 Physics-Informed Neural Networks (PINNs)

PINNs are a type of unsupervised ML, which by definition only require input data for training. In this study, the input data is sampled uniformly along the beam axis (spatial coordinate x), where the number of segments (i.e. number of sampling points) is a hyperparameter (see Table 2).

Network Architecture. Depending on the type of material model, different PINN architectures need to be defined. While a single NN with x as input and w as output suffices to describe the behaviour with the linear elastic material model, two stacked NNs are required to describe the multi-linear RC behaviour (see Fig. 3). The stacked PINNs version will subsequently be refered to as "mixed-PINNs", analogously to mixed FEA formulations [9]. Mixed-PINNs allow decoupling of M- and w-related terms and hence enable the incorporation of nonlinear material models. Thus, the constraints of the Euler-Bernoulli equation can be applied to the piecewise formulation of the RC multi-linear material while maintaining mechanical consistency. The mechanical formulae (i.e. "physics") are included as loss functions within the training of the PINNs architecture and can be interpreted as follows:

- $\mathcal{L}_\Omega$ represents the loss corresponding to the force equilibrium
- $\mathcal{L}_\Gamma$ penalises PINNs such that they adhere to the boundary conditions
- $\mathcal{L}_\Upsilon$ enforces the coupling of the material model with the acting moment, hence allowing for a separate introduction of EI (decoupled from the DE)

Additionally, potential data from experiments could be added in the loss terms (with $\mathcal{L}_\Psi$) to enhance the performance of the model. All loss terms are individually multiplied with hyperparameters λ_i and summed up to create a single scalar loss which is used to optimise the weights and biases (ϑ and μ) of the networks predicting M and w. The activation function α is required to be at least twice differentiable to be able to predict all desired quantities of the simply supported beam. It is empirically found that alternating optimisation of the M- and w-net outperforms simultaneous optimisation of both nets at the same time.

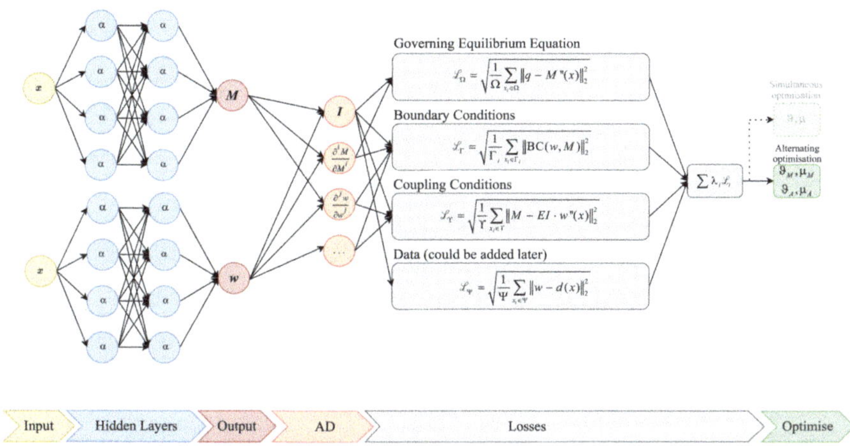

Fig. 3. PINN architecture for multi-linear RC material law (AD: automated differentiation)

Table 1 presents an overview of the PINNs architectures for the two different material models which were inspected.

Table 1. Overview of modified PINN architectures for different material models

Material Model	No. of stacked NNs	Direct predictions	Loss functions
1 – Linear Elastic	1	w	$\mathcal{L}_\Gamma = \Gamma^{-1}\sum_i \lvert\lvert BC(w,M)\rvert\rvert_2$
2 – Multi-linear RC	2	w, M	$\mathcal{L}_\Omega = \sqrt{\Omega^{-1}\sum_i \lvert\lvert q - M''(x)\rvert\rvert^2}$ $\mathcal{L}_\Gamma = \sqrt{\Gamma^{-1}\Sigma_i \lvert\lvert BC(w,M)\rvert\rvert^2}$ $\mathcal{L}_\Upsilon =$ $\sqrt{\Upsilon^{-1}\Sigma_i \lvert\lvert M(x) - EI(i)w''(x)\rvert\rvert^2}$

Hyperparameter Tuning. Hyperparameter tuning was carried out using a manual as well as the ReLoBraLo [10] approach. The manual setting lead to an optimal fit of the results and reduced the MSE or RMSE respectively to a minimum. An overview of the most important hyperparameters in this study is given in Table 2.

Table 2. Hyperparameters and their tuning range as well as chosen parameters for different material models (i: Linear elastic, ii: Multi-linear RC)

Hyperparameter	Range	Final Choice	
Learning rate l_r	$0.1 \div 0.001$	*i*	0.01
		ii	0.001
Network architecture (constructed as FFNN[1])	$8 \times [32 \div 256]$	*i*	$1 + 7 \times 32 + 1$
		ii	w-net: $1 + 8 \times 64 + 1$ M-net: $1 + 5 \times 64 + 1$
Loss parameters of boundary conditions w, M $\lambda_{BC,w,L}, \lambda_{BC,w,R}, \lambda_{BC,M,L}, \lambda_{BC,M,R}$	$1 \div 1e4$	*i*	1e2, 1e2, 1e2, 1e2
		ii	1e3, 1e3, 1e3, 1e3
Other loss parameters $\lambda_{DE}, \lambda_{coupling,w}\lambda_{coupling,M}$	$1 \div 1e5$	*i*	1e3, –, –
		ii	1e4, 1e4, 0
Loss measure	RMSE, MSE, MLSE	*i*	MSE
		ii	RMSE
Amount of sampling points x used as input for PINN		*i*	50
		ii	100
Activation function α	ReLU, SiLU, tanh	*i*	tanh
		ii	w-net: SiLU M-net: SiLU

[1]Feed Forward NN

3 Results

The following subsections provide an overview of the performance of the PINNs compared to the analytical as well as to FEA simulations. The results are structured according to the material models, where the full details of all results are provided together with interactive visualisations on the project homepage.

3.1 Linear Elastic

Figure 4 shows the results of the trained PINN for the linear elastic material law. The figure can be read in a clockwise manner, starting from top left, where every graph shows the subsequent level of differentiation in the Euler-Bernoulli beam equation. Hence, the deflections, rotations and curvatures are shown in the top row of Fig. 4, the moment, shear force and load in the bottom row, where the PINN solution is represented with a continuous line, the analytical solution with a dashed line and the FEA solution with a dotted line. The PINN prediction fits the analytical solution very well, with the highest mean absolute error in the prediction of the load (12.16%), followed by the rotation

(5.88%). The FEA results show a perfect alignment with the shear force, bending moment and curvature of PINN and analytical solution, but exhibit slight numerical resolution errors in the deflections and rotations (limited floating points).

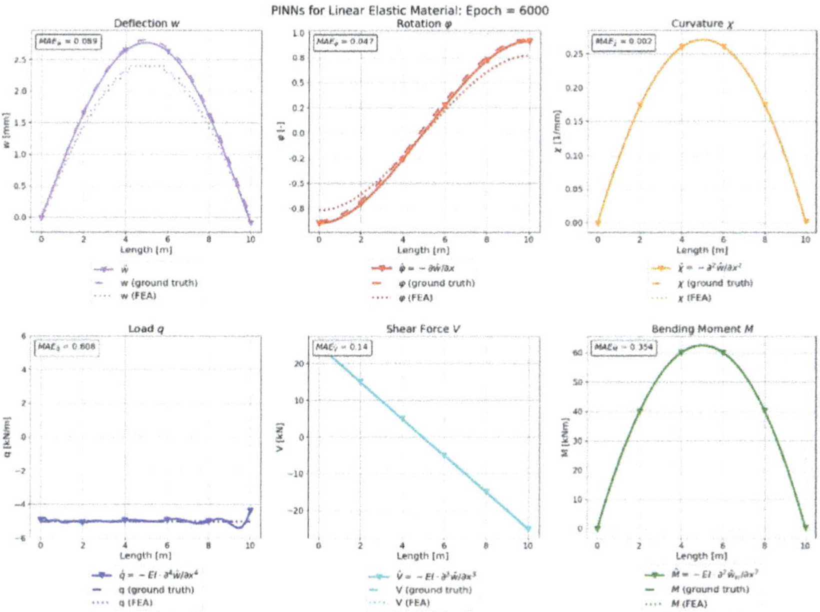

Fig. 4. Results of the PINN for a linear elastic Euler-Bernoulli beam

3.2 Multi-linear RC Materials

The mixed-PINN results of loading level q_1 can be found on the project homepage, as they resemble the results previously reported for the linear elastic material model. Figure 5 and Fig. 6 show the results of the loading levels q_2 and q_3. They correspond well with the analytical and the FEA solution, except for the discontinuous regions, i.e., where the stiffness changes abruptly from uncracked to cracked. The mixed-PINN exhibits difficulties in predicting the discontinuity, but otherwise provides better results than the FEA at q_2.. Given the chosen analysis setting of the FEA (see Sect. 2.3), there are small deviations within the magnitudes of the two bending stiffnesses EI_I and EI_{II} computed by the analytical consideration compared to RFEM (see project homepage for a graphical comparison), resulting in some differences in the computed fields in Figs. 5 and 6. Finally, the load prediction via the second derivative of the M-net is non-smooth with distinct peaks for the q_3 load case, which leads to non-vanishing errors for q_3. Similarly, the w-net cannot reproduce the moment function as accurately and smoothly as the M-net.

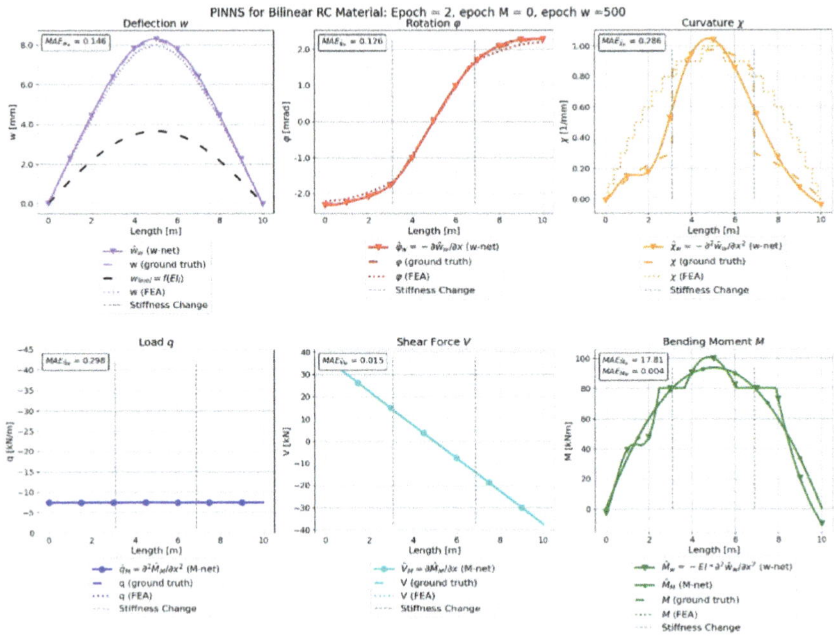

Fig. 5. Results of PINN for multi-linear RC material law, with $q_2 = 7.5$ kN/m

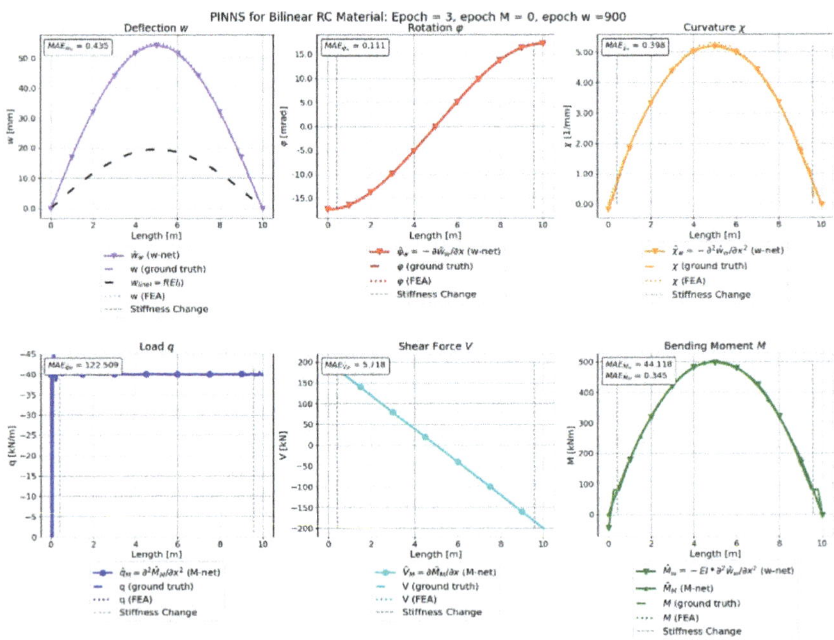

Fig. 6. Results of PINN for multi-linear RC material law, with $q_3 = 40$ kN/m

4 Discussion

Overall, the developed novel methodology of a mixed-approach reformulation of PINNs presented in this paper performs well for the two investigated material models. Hence, a PINN for nonlinear material models was successfully developed, representing a direct mapping between the inputs x and outputs w and M. This allows replacing the iterative task of manually selecting the correct bending stiffness $EI^{(i)}$ at each loading level.

The PINN for the linear elastic material model (with just one NN) performs equally well as the mixed-PINN in the linear elastic setting (at $q_1 = 5$kN/m), showing consistency in the novel decoupled approach. The most challenging part for the mixed-PINN to predict, is the discontinuity of the moment-curvature relationship at the cracking moment. Thus, if predictions close to this point are made, uncertainties remain compared to the analytical solution. Further, simulation results or experimental data could be added to the PINN training to improve the overall predictive power of the mixed-PINN. The PINN also struggles with predicting a smooth load function, because the third derivative of the moment is not regularised, which could be added in a later stage. Lastly, this study excludes a modelling of the horizontal (plastic) branch in the multi-linear RC material model, hence the mixed-PINN also cannot capture the ultimate load. A suitable extension should be explored in the future.

For the structural system of the simply supported beam analysed in this study, a FEA solution can easily be found with little to no computational overhead. However, if the mesh is not chosen densely enough, our PINN models can outperform the FEA, as highlighted by Fig. 5 e.g. for the curvature. To examine the computational efficiency stemming from the use of the mixed-PINN instead of the FEA, an analysis on a larger-scale structure would be necessary. Based solely on the results of this study, an improved computational time cannot be claimed.

All three mixed-PINN models for the different load levels were trained with the same hyperparameters. Hence, additional transfer learning capabilities between different load cases could be explored to reduce training times. The method developed in this paper hence provides a prototype for further development in the direction of ML-FEA hybrid methodologies.

5 Conclusion and Outlook

In conclusion, this study successfully introduces the novel concept of mixed-PINNs and implements this for a simply supported RC beam structure with uniform loading. In doing so, the application boundaries of PINNs are extended and limitations for multi-linear discontinuous material laws are overcome. Hence, the mixed-PINNs formulation can be further applied to nonlinear material laws beyond the scope of this paper by changing the coupling loss terms according to the material model.

Future research is concerned with generalisation of the physics-informed deep learning for parametric problems, employing different types of deep operator networks. This allows for generalisation of parametrised simulations for beams with different span lengths, different forces and materials. Furthermore, the concept of finding a surrogate mapping to replace the nonlinear dependency between parameters should be investigated for shell structures, which carry an even larger potential for computational time savings.

Acknowledgments. The authors thankfully acknowledge the facilities of the ETH Zürich AI Center and Design++ as well as the contributions of Prof. Stelian Coros and Dr. Bernhard Thomaszewski for their fruitful feedback during discussions.

References

1. Thoma, K., Roos, P., Weber, M.: Finite-Elemente-Analyse von Stahlbetonbauteilen im ebenen Spannungszustand: Scheiben- und Plattenberechnungen auf der Grundlage des gerissenen Scheibenmodells. Beton- Stahlbetonbau **109**(4), 275–283 (2014). https://doi.org/10.1002/best.201300087
2. Shen, C., et al.: Differentiable modelling to unify machine learning and physical models for geosciences. Nat. Rev. Earth Environ. (2023). https://doi.org/10.1038/s43017-023-00450-9
3. Raissi, M., Perdikaris, P., Karniadakis, G.E.: Physics-informed neural networks: a deep learning framework for solving forward and inverse problems involving nonlinear partial differential equations. J. Comput. Phys. **378**, 686–707 (2019). https://doi.org/10.1016/j.jcp.2018.10.045
4. Bastek, J.-H., Kochmann, D.M.: Physics-informed neural networks for shell structures. Eur. J. Mech. – A Solids **97**, 104849 (2023). https://doi.org/10.1016/j.euromech-sol.2022.104849
5. Zhou, M., Mei, G.: Transfer learning-based coupling of smoothed finite element method and physics-informed neural network for solving elastoplastic inverse problems. Mathematics **11**(11), 2529 (2023). https://doi.org/10.3390/math11112529
6. Marti, P.: Baustatik - Grundlagen, Stabtragwerke, Flächentragwerke, 2., Korrigierte Auflage. Wilhelm Ernst & Sohn (2014)
7. Dlubal, RFEM - Dlubal. https://www.dlubal.com/en
8. Dlubal Software Inc., RF-Concrete Members, User Manual. Philadelphia, USA: Dlubal Software GmbH (2020). https://shorturl.at/xJPR5
9. Taylor, R.L., Filippou, F.C., Saritas, A., Auricchio, F.: A mixed finite element method for beam and frame problems. Comput. Mech. **31**(1–2), 192–203 (2003). https://doi.org/10.1007/s00466-003-0410-y
10. Bischof, R., Kraus, M.A.: Multi-Objective Loss Balancing for Physics-Informed Deep Learning. https://arxiv.org/abs/2110.09813

Probabilistic Forecast of Concrete Compressive Strength Using ML

Asma Yahiaoui[1](✉), Jose C. Matos[2], and Saida Dorbani[3]

[1] Department of Structures and Materials, Built Environmental Research Laboratory, Civil Engineering Faculty, University of Sciences and Technology Houari Boumediene (USTHB), BP 32 El Alia, Bab Ezzouar, 16111 Algiers, Algeria
ayahiaoui1@usthb.dz, yahiaouiasma14@gmail.com

[2] University of Minho, ISISE, ARISE, Department of Civil Engineering, Guimarães, Portugal

[3] University of Sciences and Technology Houari Boumediene (USTHB), Bab Ezzouar, Algeria

Abstract. Concrete compressive strength is one of the critical indices of concrete characteristics. Researchers have developed several models to predict the concrete compressive strength using various machine learning techniques, which are considered deterministic methods. This paper aims to forecast the concrete compressive strength using a probabilistic prediction method. For this purpose, two boosting algorithms were selected mainly: LightGBM and XGBoost, where their combination with quantile regression has the ability to predict intervals and quantile forecasts. Quantile XGboost outperformed quantile lightGBM in all metrics with a lower quantile score, indicating a better quantile prediction and lower error values for both the RMSE and MAE for all the quantiles except the 0.05 quantile, where quantile lightGBM performed better. In the prediction of 95% intervals, XGboost achieved better performance with a narrow average width equal to 0.4375 and higher coverage of data with a value of the prediction interval coverage probability equal to 0.9369.

Keywords: Concrete compressive strength · Quantile · Prediction interval · LightGBMorth · XGBoost

1 Introduction

Reinforced concrete (RC) structures are widely used in construction worldwide [1], due to their benefits in terms of durability, low cost, and strength [2]. Structural performance and safety depend mainly on the quality of the concrete used in construction, and one of the most critical concrete properties is the compressive strength [3], especially the one at 28 days, since it is important to consider it in the design of RC buildings [4]. As concrete strength is a key factor in construction, many researchers are interested in developing reliable models that have the ability to accurately predict it. For this purpose, various datasets were created to compare experimental and numerical results.

One of the most used datasets by researchers to forecast concrete compressive strength is the one developed by Yeh [5], where the most common method to predict it was

J. C. Matos et al. (Eds.): IPW 2024, LNCE 494, pp. 281–286, 2024.
https://doi.org/10.1007/978-3-031-60271-9_25

the use of machine learning techniques, using different techniques. Aalimahmoody et al. conducted a comparative study between genetic algorithm, teaching-learning-based-optimization, multi-linear regression, as well as BAT- optimized ANN, and concluded that the latter method is more precise [6]. Liu and Sun [4], in order to go deeper into the findings of the black box models, the explainable boosting machine was applied to different ML methods, and the obtained results revealed that age and cement are the most influential parameters. Akpinar and Khashman [7], using neural networks, proposed a classification system for concrete into low, normal, and high strength. Muliauwan et al. [8] demonstrated that artificial intelligence tools have the ability to predict reliable concrete compressive strength without the need to conduct experiments. Although the aforementioned literature achieved high accuracy in the estimation of concrete strength, the predictive models are based on a point estimate and cannot take account of the uncertainty. To this end, quantiles and prediction intervals are the common methods used in probabilistic forecasting [9, 10].

This research aims to develop a probabilistic prediction model of the concrete compressive strength using quantile XGBoost and quantile lightGBM, as well as the 95% prediction intervals for both XGBoost and lightGBM.

2 Dataset

The data used in this paper is the Yeh dataset [5], an open-source database available at the UCI Machine Learning Repository [11]. The database consists of 1030 cases to predict the concrete compressive strength, with eight main features: cement (kg/m^3), blast furnace slag (kg/m^3), fly ash (kg/m^3), water (kg/m^3), superplasticizer (kg/m^3), coarse aggregate (kg/m^3), fine aggregate (kg/m^3), and age (day). The dataset is divided randomly into 80% for training and 20% for testing the performance of the predictive model, and the statistical properties for both the training and test are given in Table 1.

The present study was conducted using Python [12] as a programming language on the Kaggle platform [13].

Table 1. Statistical properties of the training and test dataset for the prediction of concrete compressive strength

Features	Training				Test			
	Mean	SD	Min	max	Mean	SD	Min	max
Cement (kg/m^3)	283.36	107.54	102	540	272.4	91.1	116	540
Blast furnace slag (kg/m^3)	74.37	86.98	0	359.4	71.99	83.61	0	342.1
Fly ash (kg/m^3)	53.16	64	0	195	58.3	63.97	0	200.1
Water (kg/m^3)	181.80	21.32	121.8	247	180.65	21.51	126.6	228
Superplasticizer (kg/m^3)	6.16	5.97	0	32.2	6.37	6.01	0	32.2
Coarse aggregate (kg/m^3)	973.55	78.69	801	1145	970.4	74	801.1	1125
Fine aggregate (kg/m^3)	772.11	80.98	594	992.6	779.47	76.76	594	992.6
Age (day)	44.66	60.48	1	365	49.67	72.99	3	365

3 Methodology

The probabilistic prediction of the concrete compressive strength was achieved using the combination of two boosting methods, with quantile regression mainly: lightGBM and XGBoost. The following quantiles were applied: 0.05, 0.25, 0.5, 0.75, and 0.95. The advantage of using these approaches is that the computation time is much faster than that of Bayesian neural networks.

The metrics used to evaluate the performance of quantile lightGBM and quantile XGBoost are: quantile score (QS), which is a probabilistic forecast error, it is defined as the average of pinball losses across the prediction horizon and across all targeted quantiles [10], lower values indicate better quantile prediction, root mean squared error (RMSE), and mean absolute error (MAE).

The evaluation of the 95% prediction interval is based on the prediction interval coverage probability (PICP), which measures the data that falls within the upper and lower bounds of the 95% prediction interval, and the prediction interval normalized average width (PINAW), which is a value between 0 and 1, it computes the average width of the prediction interval [14].

4 Results and Discussion

In order to achieve optimum model performance, hyperparameter tuning was performed, which is required in machine learning to minimize errors and increase the accuracy of the predictive models as described in [15, 16]. Hyperparameter tuning was carried out using the Optuna framework [17], and Table 2 illustrates the tuned hyperparameters for both lightGBM and XGboost used in this study.

Table 2. Tuned hyperparameters for both lightGBM and XGboost

Models	Hyperparameter								
	boosting_type	n_estimators	learning_rate	subsample	num_leaves	max_depth	reg_alpha	reg_lambda	colsample_bytree
LightGBM	dart	1715	0.176	0.843	48	4	0.892	0.475	0.681
XGboost	/	4253	0.01	0.18	/	19	0.956	0.847	0.865

Table 3 illustrates the performance of lightGBM and XGboost for the following quantiles: 0.05, 0.25, 0.5, 0.75, and 0.95. LightGBM performed better for the low quantile of 0.05, and as the quantile increased, the RMSE and MAE increased. In contrast, for XGboost, as the quantile increased, the RMSE and MAE decreased, and the lowest error is achieved at the 0.5 quantile; from this, the error metrics increased. It is also important to note that, as quantile XGBoost has the lowest RMSE and MAE, it indicates that the prediction interval will be narrow, which will be demonstrated in Table 4. XGboost outperformed lightGBM in all error metrics except the 0.05 quantile. In terms of QS, which is a metric for probabilistic prediction error, XGboost performed better, achieving the lowest QS in all quantiles, emphasizing that the quantile XGboost achieved better quantile prediction than the quantile lightGBM. It should be noted that as the quantiles

Table 3. Metrics for each quantile of both lightGBM and XGboost

Metrics		Quantiles				
		0.05	0.25	0.5	0.75	0.95
LightGBM	QS	1.0497	1.2012	1.3906	1.5799	1.7314
	RMSE	**0.66**	0.79	0.75	0.93	1.78
	MAE	**0.50**	0.59	0.60	0.68	1.22
XGboost	QS	**1.0314**	**1.1474**	**1.2924**	**1.4373**	**1.5533**
	RMSE	0.76	0.70	**0.64**	0.78	1.47
	MAE	0.58	0.52	**0.49**	0.56	0.99

increase, the difference between quantiles XGBoost and lightGBM also increases, illustrating that the uncertainty is more pronounced for the latter methods as well as for high quantiles for both methods.

Figure 1 illustrates the 95% prediction interval for both lightGBM and XGboost; nearly the same plot is displayed for the two methods. The 95% prediction interval covers the majority of the data with a few outliers, highlighting low uncertainty in the prediction of concrete compressive strength for both models. As it is not apparent from Fig. 1, the best prediction interval for this purpose is Table 4, which evaluates the quality of 95% prediction intervals in terms of PICP and PINAW for the test sets. XGBoost has the largest PICP with a value equal to 0.9369, which is 1.02 times higher than the 95% prediction interval for lighGBM, demonstrating that it covers more data, and the smallest PINAW with a value equal to 0.4374, which is lower by 0.972 times than the second adopted method for this paper. Although there is a small difference between the two methods in the evaluation of the 95% prediction, it underlines that lighGBM incorporates more uncertainty in the probabilistic forecast of concrete strength.

Overall, XGBoost outperformed lightGBM in all metrics, with a narrow interval width and large coverage of the data.

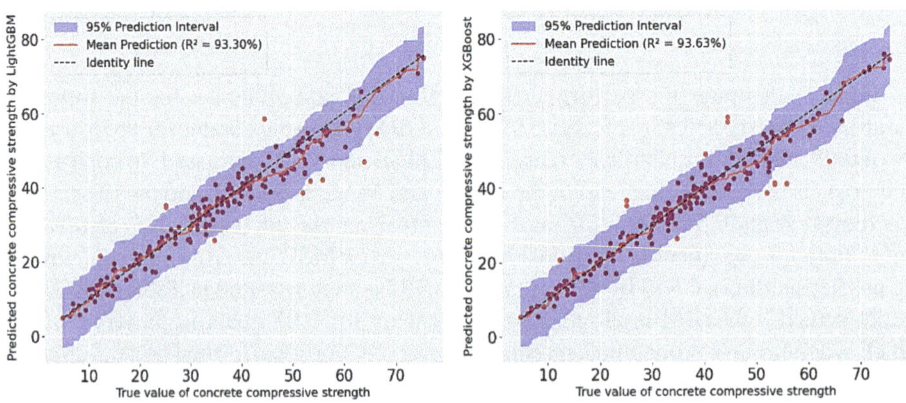

Fig. 1. 95% prediction interval for lightGBM and XGboost respectively

Table 4. Prediction interval results for the forecast of concrete compressive strength

Models	PICP	PINAW
LightGBM	0.9223	0.4498
XGBoost	0.9369	0.4374

5 Conclusion

In this paper, a probabilistic forecast of concrete compressive strength was conducted using the combination of two boosting methods, namely lightGBM and XGBoost, with quantile regression. The results can be summarized as follows:

1) Quantile XGBoost achieved the lowest error values in all quantiles except the 0.05 quantile, indicating a better prediction interval with low uncertainty.
2) The findings from the probabilistic forecast error 'quantile score' revealed that the best quantile prediction was achieved by XGBoost for all quantiles except the 0.05 quantile.
3) As the quantiles increased, the quantile score for both methods also increased, revealing more uncertainty for high quantiles as well as less accuracy.
4) The evaluation of the 95% prediction interval showed that XGBoost has a narrow width interval and a larger coverage of the data than lightGBM.
5) The benefit of using quantile tree methods and 95% prediction intervals is that, in addition to the uncertainty incorporation in the models, the resulted models give information on the reliability of the results.

References

1. Khelfi, M., Bourahla, N., Remki, M.: Performance evaluation of masonry infilled RC frame structures under lateral loads. Građevinar **73**(3), 219–234 (2021). https://doi.org/10.14256/JCE.2647.2019
2. Dabiri, H., Kioumarsi, M., Kheyroddin, A., Kandiri, A., Sartipi, F.: Compressive strength of concrete with recycled aggregate; a machine learning-based evaluation. Clean. Mater. **3**, 10004 (2022). https://doi.org/10.1016/j.clema.2022.100044
3. Rathakrishnan, V., Beddu, S.B., Ahmed, A.N.: Predicting compressive strength of high - performance concrete with high volume ground granulated blast - furnace slag replacement using boosting machine learning algorithms. Sci. Rep. 1–17 (2022). https://doi.org/10.1038/s41598-022-12890-2
4. Liu, G., Sun, B.: Concrete compressive strength prediction using an explainable boosting machine model. Case Stud. Constr. Mater. **18**, e01845 (2023). https://doi.org/10.1016/j.cscm.2023.e01845
5. Yeh, I.-C.: Modeling of strength of high-performance concrete using artificial neural networks. Cem. Concr. Res. **28**(12), 1797–1808 (1998)
6. Aalimahmoody, N., Bedon, C., Hasanzadeh-Inanlou, N., Hasanzade-Inallu, A., Nikoo, M.: Bat algorithm-based ann to predict the compressive strength of concrete—a comparative study. Infrastructures **6**(6) (2021). https://doi.org/10.3390/infrastructures6060080

7. Akpinar, P., Khashman, A.: Intelligent classification system for concrete compressive strength. In: 9th International Conference on Theory and Application of Soft Computing, Computing with Words and Perception, vol. 120, pp. 712–718 (2017). https://doi.org/10.1016/j.procs.2017.11.300
8. Muliauwan, H.N., Prayogo, D., Gaby, G., Harsono, K.: Prediction of concrete compressive strength using artificial intelligence. J. Phys. Conf. Ser. **1625**(1), 012018 (2020). https://doi.org/10.1088/1742-6596/1625/1/012018
9. Xie, J., Hong, T., Laing, T., Kang, C.: On normality assumption in residual simulation for probabilistic load forecasting. IEEE Trans. Smart Grid **8**(3), 1046–1053 (2017). https://doi.org/10.1109/TSG.2015.2447007
10. Zhang, W., Quan, H., Srinivasan, D.: An improved quantile regression neural network for probabilistic load forecasting. IEEE Trans. Smart Grid **10**(4), 4425–4434 (2019). https://doi.org/10.1109/TSG.2018.2859749
11. Concrete Compressive Strength - UCI Machine Learning Repository. https://archive.ics.uci.edu/dataset/165/concrete+compressive+strength. Accessed 25 Dec 2023
12. Welcome to Python.org. https://www.python.org/. Accessed 28 Aug 2022
13. Kaggle: Your Home for Data Science. https://www.kaggle.com/. Accessed 25 Dec 2023
14. Yin, X., Fallah-Shorshani, M., McConnell, R., Fruin, S., Chiang, Y.Y., Franklin, M.: Quantile Extreme Gradient Boosting for Uncertainty Quantification. arXiv Prepr. arXiv2304.11732 (2023). http://arxiv.org/abs/2304.11732
15. Yahiaoui, A., Dorbani, S., Yahiaoui, L.: Machine learning techniques to predict the fundamental period of infilled reinforced concrete frame buildings. Structures **54**, 918–927 (2023). https://doi.org/10.1016/j.istruc.2023.05.052
16. Markou, G., Bakas, N.P., Chatzichristofis, S.A., Papadrakakis, M.: A general framework of high-performance machine learning algorithms: application in structural mechanics. Comput. Mech. Press (2023). https://doi.org/10.13140/RG.2.2.14962.89287
17. Akiba, T., Sano, S., Yanase, T., Ohta, T., Koyama, M.: Optuna: a next-generation hyperparameter optimization framework. In: Proceedings of the 25th ACM SIGKDD International Conference on Knowledge Discovery & Data Mining, pp. 2623–2631 (2019). https://doi.org/10.1145/3292500.3330701

Railway Inspection and Information Model (RIIM): An Intelligent Decision-Making Tool for Enhanced Infrastructure Management

E. Aldao(✉), E. Ríos-Otero, F. Veiga-López, H. González-Jorge, and E. Balvís

Research Institute of Physics and Aerospace Sciences, University of Vigo, Campus As Lagoas, 32004 Ourense, Galicia, Spain
enrique.aldao.pensado@uvigo.gal

Abstract. High-speed railway tracks represent critical infrastructures exposed to cyclic loads of high intensity, arising from the operation of high-speed or heavy-haul trains. Consequently, regular inspections are required to guarantee the safety and efficiency of the transportation system. Traditionally, track inspections were carried out by human operators that manually checked the state of the different track elements. However, this method is not only time-consuming but also incurs significant costs for infrastructure management companies. To meet the increasing demand for high-speed operations and ensure sustainable development in the railway sector, there is a pressing need to shift towards intelligent inspection systems. In this context, the University of Vigo and COPASA are collaborating on the RIIM project (Railway Inspection and Information Model), dedicated to the development of an efficient inspection solution for enhancing the safety of railway infrastructure. This system utilizes vehicle-embedded sensors for autonomous and georeferenced data collection. With this information, statistical data analysis and Deep Learning techniques are applied to systematically assess the health condition of the infrastructure. The outcome is a decision-making tool seamlessly integrated into a Geographic Information System (GIS) for efficient railway maintenance task management and planning. In this work, the main results of this project are outlined as well as the implications for future research and improvements.

Keywords: Intelligent decision-making · Geographic Information System (GIS) · Deep Learning · Statistical Analysis · Building Information Model (BIM)

1 Introduction

Railway maintenance and inspection programs are vital for ensuring the safety and efficiency of transportation. These programs involve periodic inspections to assess the condition of various infrastructure elements, including sleepers, rail fastening elements, and the ballast bed morphology [1, 2]. Traditionally, trained operators manually conducted these inspections, incurring significant economic and resource costs for infrastructure management companies. In fact, European Union countries spent 20.8 billion US dollars on the maintenance of their railway infrastructure, as reported by the European Commission [3].

J. C. Matos et al. (Eds.): IPW 2024, LNCE 494, pp. 287–295, 2024.
https://doi.org/10.1007/978-3-031-60271-9_26

According to the International Energy Agency, passenger and freight activity is expected to more than double by 2050 [4]. This surge will be accompanied by electrification and the implementation of high-speed lines, offering sustainable and ecological alternatives to traditional transportation means. However, the increased high-speed traffic will intensify wear on existing infrastructure, leading to higher inspection and maintenance requirements [5]. To ensure a sustainable and economically viable development of railway industry, future inspection trends need to shift towards the automation and the development of autonomous inspection systems based on sensors embedded in trains or track vehicles. This transition has the potential to reduce maintenance costs and promote the development of Structural Health Monitoring programs [6, 7]. The implementation of these systems, along with the use of Deep Learning techniques and statistical data analysis, will improve the operational efficiency of the railway system by preventing and detecting possible failures or wear in the infrastructure at an early stage [8, 9].

In this context, the University of Vigo and COPASA are collaborating on the RIIM project (Railway Inspection and Information Model), dedicated to the development of autonomous railway inspection solutions based on embedded systems [10]. In this initiative, optical sensors such as global-shutter cameras are being used for the visual inspection of track elements, as well as LiDAR technology for assessing the morphology of ballast shoulders. All this information is referenced in a Geographical Information System (GIS), thus, facilitating its integration into a Building Information Model (BIM) of the infrastructure.

Another key aspect of RIIM is the development of intelligent information management systems. Using georeferenced data from sensors, statistical data analysis and Deep Learning techniques, are employed to automatically characterize the state of the infrastructure. The final output of this project is a decision-making tool for infrastructure managers to more efficiently manage and plan maintenance tasks. This work presents the main results of RIIM, as well as an analysis of how these systems can be integrated into Structural Health Monitoring programs of railway infrastructures.

The remainder of the manuscript is structured as follows: Sect. 2 introduces the data acquisition system, presenting the data acquisition hardware as well as the integration of information within a GIS. Section 3 outlines the data analysis tools and the development of a decision-making framework. Finally, Sect. 4 presents the conclusions and future developments.

2 Autonomous Railway Inspection and Monitoring

2.1 Data Acquisition Hardware

For the development of the autonomous data acquisition system, the rail draisine shown in Fig. 1 is used in the RIIM initiative. This vehicle is equipped with a combustion engine and is designed for transporting operators during maintenance tasks. In this project it is being employed as a testing platform for the development of hardware and the implementation of autonomous inspection and data analysis algorithms.

Figure 2a presents an image of the developed inspection system prototype and its integration onto the rail draisine. In the depicted configuration, four cameras and two LiDAR sensors are positioned at the front. The main function of the cameras is to capture

Fig. 1. COPASA's rail draisine.

images of the concrete sleepers and the rail fastening elements. Conversely, the LiDAR sensors capture 3D point clouds, which are employed to analyse the ballast bed geometry. All these data are transmitted to the central processing unit depicted in Fig. 2b, where they are synchronised with the navigation system and stored for subsequent analysis (Fig. 3).

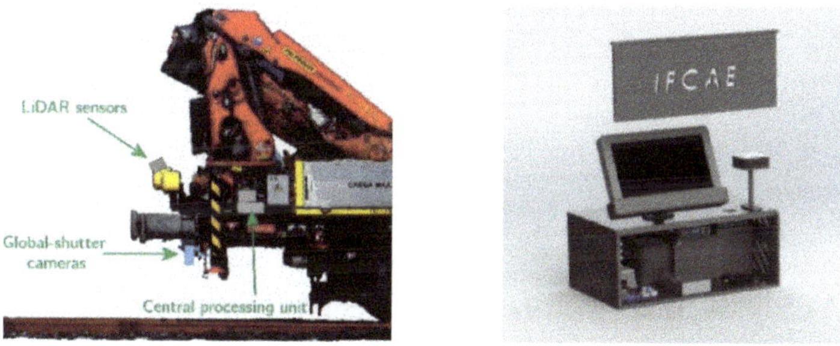

a) Onboard inspection sensors. b) Central processing unit.

Fig. 2. Autonomous inspection prototype.

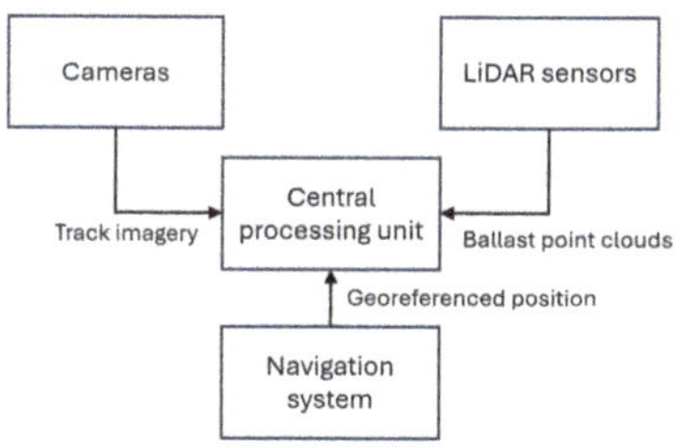

Fig. 3. Inspection system architecture.

2.2 Navigation System and Integration Within a GIS Tool

During the inspections, it is necessary to accurately calculate the position of the inspection system to georeference the data in a GIS. For this purpose, the central processing unit is equipped with a navigation system based on RTK GPS (Real Time Kinematic Global Positioning System) and an Inertial Measurement Unit (IMU). The RTK GPS provides position measurements with relatively high precision, but its measurements have a certain amount of noise on the order of several centimetres. On the other hand, the IMU measures linear accelerations and angular velocities, allowing for a continuous estimation of motion. The information of both sensors is combined through a Kalman filter, which improves the system's state estimates [11]. This information is used to correct the displacements in the LiDAR data and generate georeferenced orthophotos. All the data is stored in a GIS for subsequent processing and analysis (Fig. 4).

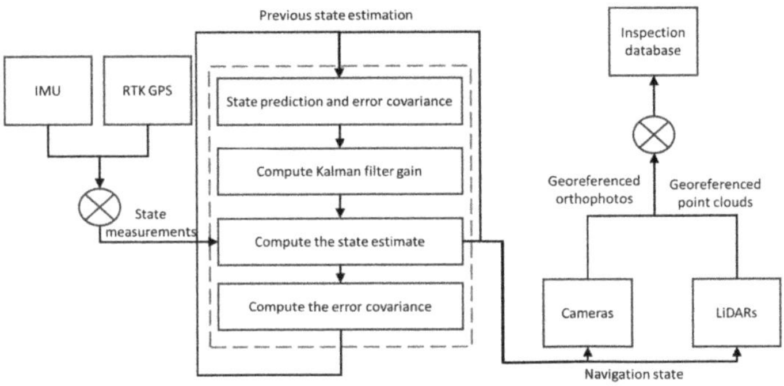

Fig. 4. Navigation system implementation scheme.

3 Data Analysis and Maintenance Decision-Making

The georeferenced data from the inspection system are processed to automatically characterize the health status of different sections of the infrastructure. This analysis process is carried out in the following two steps:

3.1 Automatic pathology detection: Using Computer Vision algorithms, raw sensor data is processed to identify and quantify the magnitude of pathologies.

3.2 Damage characterization: Subsequently, a systematic evaluation takes place, considering the severity of the detected pathologies. The objective of this process is to provide infrastructure managers with detailed information and assistance for the planning and allocation of maintenance resources.

3.1 Automatic Pathology Detection

The first step in the detection of pathologies is to identify the different elements of the railway within the inspection images. For this purpose, downscaled images of inspection data are employed (Fig. 5) and a DeepLab V3 + semanic classification model is utilised [12]. This model was trained using hundreds of manually labelled track images. For training, image augmentation techniques were employed, introducing synthetic variations in lighting conditions, as well as in white balance. In this way, with this state-of-the-art model, a high classification accuracy was achieved with a mean Intersection over Union of 94% [13].

Fig. 5. Track elements semantic segmentation.

Using this pixel-wise classification, the system verifies the positioning of bolts and fastening clips. Various alternatives based on Image Processing and Deep Learning were considered in the project, which were analysed and compared in the following work [13]. The most effective approach utilized a custom Histogram-based network, enabling detection of alignment errors in bolts within a millimetre range and identifying rotated fasteners with a one-degree precision. This system's perception capabilities are comparable to or slightly exceed those of a human operator. Further information about the system can be found in the provided reference.

To analyse the crack propagation in concrete sleepers, high-resolution pictures are utilized to achieve greater definition. If all the image pixels were to be analysed, computing costs would increase considerably due to the dimensionality of the problem. Therefore, only the snippets outlined in Fig. 6a are analysed. These are the sections most prone to crack proliferation, owing to the high shear stresses generated by the rail fastening elements. YOLO-V8 object detectors are subsequently applied to each of these sections to characterize the damage magnitude, as illustrated in Fig. 6b [14].

The training of these models is challenging since it is difficult to obtain large databases of cracked sleepers. Therefore, synthetic images of cracks were generated using a customized version of the Stable Diffusion generative network. The results are published in the following article [15], demonstrating the suitability of this approach for training object detectors, achieving detection performance comparable to human perception.

For the analysis of ballast shoulder geometry, LiDAR point clouds are employed, as the one shown in Fig. 7a. This geometry is compared with a BIM model of the infrastructure to detect the detachment or movement of layers in the ballast bed. The point clouds are registered with the geometry of the BIM model, and deformation maps are obtained, as shown in Fig. 7b. The main advantage of LiDAR technology is its high measurement accuracy, on the order of 2 cm, which allows for a quantitative assessment of the health status of the ballast bed, eliminating the subjective component of inspection through human operators.

a) Slicing of areas of interest.

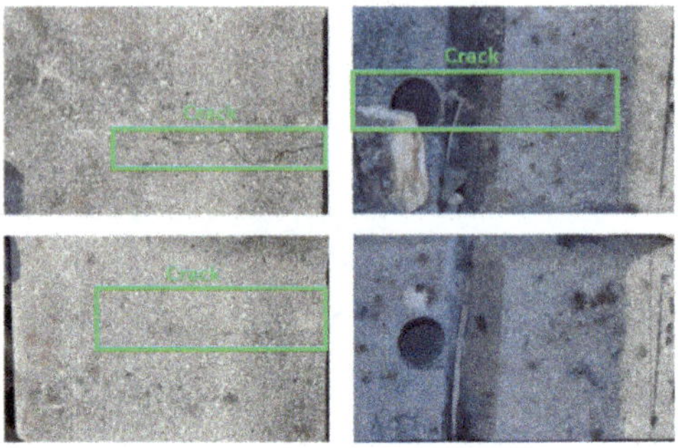

b) Autonomous crack detection.

Fig. 6. Concrete surface inspection [15].

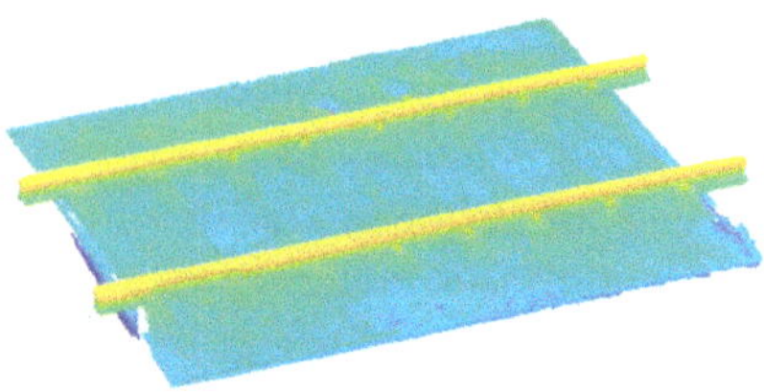

a) Ballast point cloud.

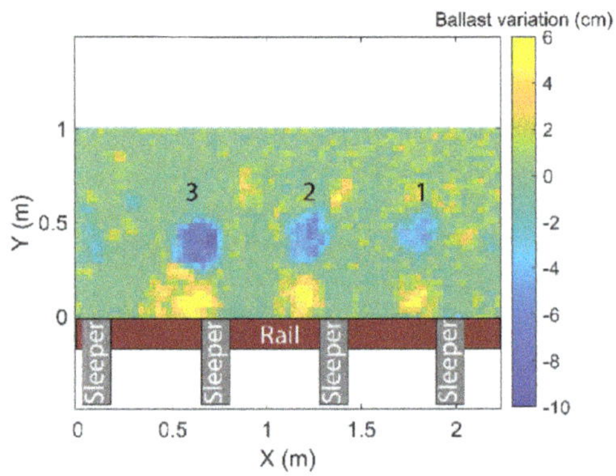

b) Ballast deformation map.

Fig. 7. Ballast bed autonomous inspection [16].

3.2 Damage Characterization

The detection of pathologies presented in Sect. 3.1 quantifies the extent of damages in terms of parameters such as ballast deformation or crack size but does not establish any objective criteria for the actions to be taken. To address this issue, the University of Vigo and COPASA are currently developing a decision-making framework to systematically characterize damages in the following categories based on their severity:

- Healthy condition: The infrastructure shows no damage during inspection.
- Minor damage: Small imperfections exist in some of the elements, but they do not compromise the safety of operations.
- Moderate damage: Refers to damage of greater magnitude that may worsen in the future and become critical.
- Critical damage: Immediate repair is required.

For this classification, professionals in the field are actively helping to create a database of railway damages characterized in the aforementioned categories. Based on

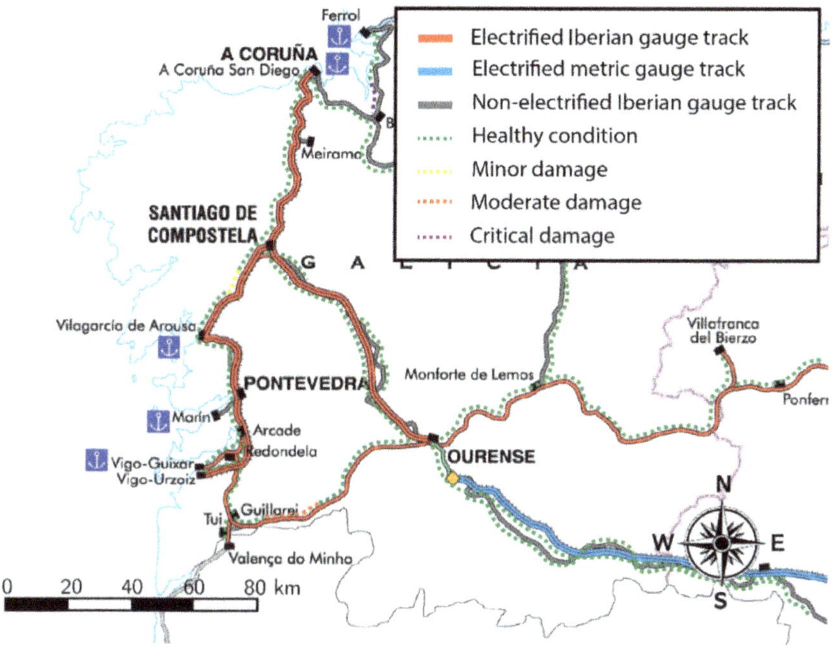

Fig. 8. Georeferenced health state of railway network.

this classification and representative magnitudes of each pathology (i.e., crack size, degree of misalignment of fastening elements…), a Support Vector Machine (SVM) is being developed [17]. This SVM characterizes the health status of each section of the infrastructure systematically considering the damage representative magnitudes. In this way, infrastructure managers will have a georeferenced decision-making tool, as illustrated in Fig. 8, that facilitates improved decision-making and resource allocation in maintenance operations.

4 Conclusions

In this work, we presented the main results and lessons learned in the RIIM project, jointly developed by the University of Vigo and COPASA. In it, an autonomous inspection system was developed, embedded in a rail vehicle for the acquisition of infrastructure state data. This information is processed through Deep Learning algorithms and statistical analysis, providing the infrastructure manager with a systematized characterization of the health status of different georeferenced track sections.

In future work, the use of inspection data will not be limited solely to assessing current infrastructure damage but is also expected to play a crucial role in the creation of Structural Health Monitoring programs. These programs will enable continuous and detailed monitoring of the infrastructure's condition over time. Additionally, the use of Deep Learning algorithms for pathology prediction and the development of Predictive Maintenance systems will be explored.

Acknowledgments. COPASA participated in the development of this work by providing support and resources to obtain the experimental data. They also collaborated in the definition of inspection system requirements and their experience was essential for the development of data analysis algorithms and decision-making tools.

Funding. E.A. is funded by the Spanish Ministry of Universities (grant ref: FPU21/01176)

Grant PLEC2021–007940 funded by MCIN/AEI/ https://doi.org/10.13039/501100011033 and by the European Union NextGenerationEU/PRTR

Grant CPP2022–009679 funded by MCIN/AEI/ https://doi.org/10.13039/501100011033 and by European Union NextGenerationEU/PRTR

The presented work was also performed in context of the Horizon Europe project SARIL which is funded by the European Union under grant agreement ID 101103978. Views and opinions expressed are however those of the author(s) only and do not necessarily reflect those of the European Union or the European Climate, Infrastructure and Environment Executive Agency. Neither the European Union nor the granting authority can be held responsible for them.

References

1. Sedghi, M., Kauppila, O., Bergquist, B., Vanhatalo, E., Kulahci, M.: A taxonomy of railway track maintenance planning and scheduling: a review and research trends. Reliab. Eng. Syst. Saf. **215**, 107827 (2021). https://doi.org/10.1016/J.RESS.2021.107827
2. Fernandez-Bobadilla, H.A., Martin, U.: Modern tendencies in vehicle-based condition monitoring of the railway track. IEEE Trans. Instrum. Meas. **72**, 1–44 (2023). https://doi.org/10.1109/TIM.2023.3243673
3. European Commission. Seventh monitoring report on the development of the rail market under Article 15 (2019). https://eur-lex.europa.eu/legal-content/EN/TXT/?uri=CELEX%3A52021DC0005
4. International Energy Agency. The future of rail (2019). https://www.iea.org/reports/the-future-of-rail. Accessed 25 Nov 2023
5. Gattuso, D., Restuccia, A.: A tool for railway transport cost evaluation. Procedia – Soc. Behav. Sci. **111**, 549–558 (2014). https://doi.org/10.1016/j.sbspro.2014.01.088
6. Wang, Y.-W., Ni, Y.-Q., Wang, S.-M.: Structural health monitoring of railway bridges using innovative sensing technologies and machine learning algorithms: a concise review. Intell. Transp. Infrastruct. **1**, liac009 (2022). https://doi.org/10.1093/iti/liac009
7. Iyer, S., Velmurugan, T., Gandomi, A.H., Noor Mohammed, V., Saravanan, K., Nandakumar, S.: Structural health monitoring of railway tracks using IoT-based multi-robot system. Neural Comput. Appl. **33**(11) (2021). https://doi.org/10.1007/s00521-020-05366-9
8. Tang, R., et al.: A literature review of Artificial Intelligence applications in railway systems. Transp. Res. Part C Emerg. Technol. **140** (2022). https://doi.org/10.1016/j.trc.2022.103679
9. Gibert, X., Patel, V.M., Chellappa, R.: Deep multitask learning for railway track inspection. IEEE Trans. Intell. Transp. Syst. **18**(1), 153–164 (2017). https://doi.org/10.1109/TITS.2016.2568758
10. COPASA Group (2022). Railway Inspection and Information Model research project. https://www.copasagroup.com/es/riim-proyecto-de-idi/. Accessed 25 Nov 2023
11. Gu, S., Dai, C., Mao, F., Fang, W.: Integration of Multi-GNSS PPP-RTK/INS/vision with a cascading kalman filter for vehicle navigation in urban areas. Remote Sens. **14**(17) (2022). https://doi.org/10.3390/rs14174337

12. Chen, L.-C., Zhu, Y., Papandreou, G., Schroff, F., Adam, H.: Encoder-Decoder with Atrous Separable Convolution for Semantic Image Segmentation, pp. 833–851 (2018). https://doi.org/10.1007/978-3-030-01234-2_49
13. Aldao, E., Fernández-Pardo, L., González-deSantos, L.M., González-Jorge, H.: Comparison of deep learning and analytic image processing methods for autonomous inspection of railway bolts and clips. Constr. Build. Mater. **384**, 131472 (2023). https://doi.org/10.1016/j.conbuildmat.2023.131472
14. Home - Ultralytics YOLOv8 Docs. (2024). (https://docs.ultralytics.com/). Accessed 21 Feb 2024
15. Aldao, E., Fernández-Pardo, L., Veiga-López, F., González-deSantos, L.M., González-Jorge, H.: Assessment of synthetic data generation techniques for enhancing crack detection in railway concrete sleepers. Submitted to Automation in Construction (Under review)
16. Aldao, E., González-Jorge, H., González-deSantos, L.M., Fontenla-Carrera, G., Martínez-Sánchez, J.: Validation of solid-state LiDAR measurement system for ballast geometry monitoring in rail tracks. Infrastructures **8**(4), 63 (2023). https://doi.org/10.3390/infrastructures8040063
17. Abedi, M., Shayanfar, J., Al-Jabri, K.: Infrastructure damage assessment via machine learning approaches: a systematic review. Asian J. Civ. Eng. **24**(8), 3823–3852 (2023). https://doi.org/10.1007/s42107-023-00748-5

Significance of Measurement Data for the Probabilistic Assessment of New and Existing Infrastructure on the Example of the Pilot Project Bridge Schwindegg

Thomas Braml(✉), Christian Kainz, and Johannes Wimmer

University of the Bundeswehr Munich, 85577 Neubiberg, Germany
thomas.braml@unibw.de

Abstract. Bridge maintenance is an essential task to establish a strong and efficient infrastructure. Due to the even more increasing shortage of qualified personnel in the future the development of concepts for a semi-automated bridge inspection is necessary. Especially for the large number of small and medium span bridges in the municipal road network, there is often no specialized personnel available for the tasks of structural inspection and maintenance. An individual on-site condition assessment in defined and regular time intervals is the current state of the art. This assessment includes a close inspection of the bridge but usually not an end-to-end documentation or continuous measurement since its time of construction. Measurement data contains important information about the actions and resistance of the structures. This information gives a proposition about the structure's condition. Digital twins can be a concept for the use of this information. While full probabilistic methods deliver exact information about the reliability of the examined structure, they require representative input data. Measurement data is best suited to provide this input, otherwise the stochastic parameters from literature may not necessarily match for the specific structure. In the future, the acquisition of measurement data can be performed from digital twins or the actual structure. This article presents the concept of digital twins for medium span concrete bridges. The large amount of collected data over the entire lifespan of the structure will significantly enhance knowledge and calculation models. Especially in the construction of new bridge structures, where the installation of the measurement technology can be performed simultaneously with the actual construction, the advancement of digital twins provides many advantages for the quality of a reliability assessment.

Keywords: digital twin · bridge construction · big data · data measurement

1 Introduction

The load bearing capacity and the structural safety of a building or bridge cannot be directly measured. The fact that a structure does not collapse, is free of damages or a determination of material properties on the object only indicates the actual stability.

J. C. Matos et al. (Eds.): IPW 2024, LNCE 494, pp. 296–302, 2024.
https://doi.org/10.1007/978-3-031-60271-9_27

Stability, in this context, implies a sufficiently low probability of failure. The static calculations serve as a computational proof of this stability. They are typically conducted before the construction of the structure in most cases within the field of civil engineering. During the design of a new structure, construction is still in the future and associated with many uncertainties. In contrast, an existing structure has already been completed, and results are documented. The actual parameters may not necessarily match the assumptions from the code, emphasizing the need for a more realistic assessment and predictive maintenance.

In recent years, the concept of a "Digital Twin" has gained attention in civil engineering as a tool for digitization. While digital twins are well-established in many other industries, the necessary implementations for the transportation infrastructure on a technical and national level, such as the mandatory semantics of the ASB-ING [1] (German: *Anweisung Straßeninformationsbank Segment Bauwerksdaten*), are still unavailable. A data-driven digital twin, utilizing an Asset Administration Shell (AAS) and sensors with data transfer between the real and digital object, could be one solution for the utilization of existing parameters and measurement data. The goal is to create an overall platform for condition assessment and management as a tool for a strategy and financial needs determination for the maintenance of the structures.

While many approaches use the BIM (Building Information Modeling) method as a central element for a digital twin, it may not be optimally suited for real-time data integration [2], as the focus of BIM was traditionally on the data integration and exchange during different stationary planning stages. This concept was initially designed for a static versioning, while real-time measurement data are constantly changing.

2 Bridge Maintenance According to DIN 1076

The assessment of bridges in Germany is carried out at regular intervals according to the code regulations of DIN 1076 [3]. During this process, damages to the structure are recorded and evaluated in terms of their impact on stability, durability, and traffic safety. The structural inspector conducts the classification of the damage intensity individually, while being supported by a guiding damage catalog with example damages and the national code RI-EBW-PRÜF [4]. Visual inspections are conducted at regular intervals and the results are usually documented in an inspection report. Typically, the results and information of one structure are archived for each specific structure, but not used across other structures.

This inspection is associated with certain costs and requires personnel resources. Only technically qualified engineers are suited for the inspections of the structures. A completed university degree in structural engineering and several years of professional experience in the field of bridge design and construction are necessary. Also, extensive experience with structures of similar scale and complexity is required.

Facing the increasing shortage of highly trained personal in general and in the construction industry especially as well as the aging infrastructure, it is an open challenge to ensure that all necessary inspections can be conducted by qualified personnel in the future. In many cases, external engineering companies conduct the inspection for the structure owners. These services need to be tendered, awarded, supervised, and billed.

In contrast, measurements with monitoring systems can be carried out regularly or even permanently.

3 Concept of Digital Twins on the Example of the Schwindegg Bridge

The scope of the pilot project is to collect a variety of measurement data from different locations on a bridge structure starting directly from its first day opening to traffic. The focus was on gaining experience in the fields of planning, implementing the measurement techniques both on the construction site and in the precast concrete plant, data storage, and data analysis. The bridge is a single-span frame bridge with a superstructure made of precast prestressed concrete elements supplemented with in-situ concrete. It is located in Schwindegg, which is a rural community in Bavaria. The span is approximately 19.8 m, which is a typical design for small and medium spans. The foundations of the abutments are constructed using Ø120 cm bored piles. Figure 1 shows a view of the pilot bridge.

A real bridge structure as a full-size laboratory for implementing a digital twin from the first day of opening to traffic, was realized for the first time. With a total of approximately 170 independent measurements from approximately 140 installed sensors, extensive data on the effects on the structure and its resistances can be obtained. This pilot project forms the basis for further research into intelligent structural management and was presented in [5–12].

Fig. 1. View of the Schwindegg bridge over the Isen lake

In the first step, the realization of the structure as a digital bridge required the planning and installation of the sensors. The suitability of different sensor types for implementation in digital twins and the associated installation efforts were examined. Since there are currently no standardizations available for intelligent bridge maintenance management, a wide variety of parameters were measured at different locations on the structure. These measurements should form the basis for initial concepts.

The sensors are integrated into the cross-section, i.e., embedded in concrete, as well as applied externally. The sensors for the measurement of earth pressure become inaccessible due to backfilling. Accessible systems are suitable for measuring acceleration, inclination, and deformation. Embeddable strain sensors are located in the field center, at the abutment area of two of the four precast beams, at the corner of the in-situ concrete frame, and at the pile head. To perform modal analysis investigations, three-dimensional accelerometers are placed at the deflection points of the first three eigenmodes, which were previously calculated in a simulation using a FEM (finite element method) model. In addition to the accelerations, the inclinations are measured in two directions. Environmental conditions (air temperature, pressure, humidity, wind speed and direction, sun intensity, and precipitation) are externally recorded by a weather station.

The planning of the monitoring system was carried out in parallel with the construction planning. Another essential task is the transport of the on-site data to the analysis office. Data loggers initially collect the measurement data. Since only 4G network connectivity was available at the bridge location, a request was made to the local network operator for a fiber optic connection at the technology block. The measurement data is transmitted redundantly via mobile internet (currently) and a fiber optic connection to the AAS using two IoT routers. Local storage serves as a data buffer in case of a data connection failure.

The AAS configurator (called *Bridge box*, short *BBox*) contains information about geometric properties, damages, or used construction materials. The sensors are integrated, which provides the advantage of an established library for users. A sensor has an active sensor configuration as a sub-model with a unique ID, a temporal validity range, and basic data such as sampling frequency or unit. The sensor configuration is connected to the measurement point in the BBox and receives a unique qualification to stream real-time measurement data using the MQTT protocol communication technique. There is a bidirectional connection between the AAS and the sensor. Figure 2 illustrates the concept of data management.

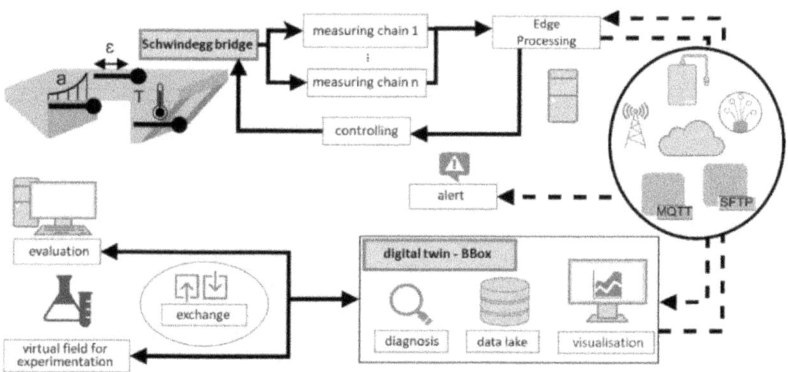

Fig. 2. Concept of data management from the real object to its digital twin (Source: Johannes Wimmer, translated from [12] and based on [7])

4 Use of Measurement Data from Digital Twins for Probabilistic Calculations

The possibility for a direct implementation of measurement data is a big advantage of full probabilistic calculations [13]. While full probabilistic methods provide precise information about the reliability of the examined structure, their results depend strongly in the input data. For this reason, any information gained from the actual structure is very valuable for an assessment, because it is directly related to the on-site conditions. Since the AAS of the bridge does not only communicate with the AAS of the sensors but also interacts with the AAS of other bridges, the information gained from sensor data on one structure can be utilized for other structures as well. Figure 3 illustrates the configuration of the measuring sensors, the connection of the AAS of the sensors to the AAS of the bridge and its various submodels with categorized information.

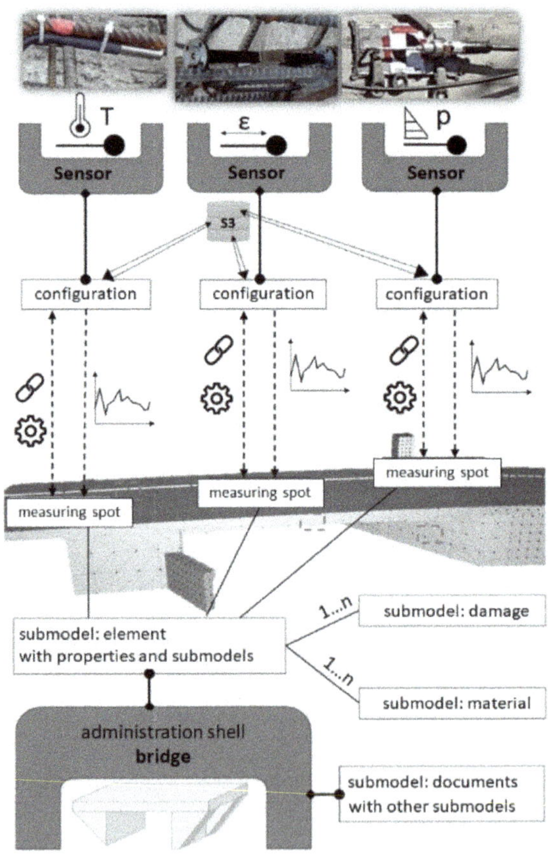

Fig. 3. Connection of the administration shells of the sensors to the administration shell of the bridge (Source: Johannes Wimmer, translated from [12])

Over time, an individual traffic load model for the bridge or certain road segments can be developed and updated as needed. For example, the current traffic load model LM1 from DIN EN 1991–2 [14] was calibrated for highway bridges in Europe, especially in Auxerre, France [15] and does not match the actual loads on the structure as it is a model. Monitoring the actual traffic on a structure over its entire lifecycle provides more reliable data due to the longer measurement period than a short term measurement. This aspect supports a more realistic choice of statistical parameters for the distribution functions of the probabilistic calculation. While the calibration of load assumptions can lead to more realistic models, one must always critically ensure, that a compliance with the necessary reliability levels according to the codes is still present.

The measurement data serves as basis for developing statistical models to describe variation properties. Typically, the code regulations refer to the semi-probabilistic safety concept, providing characteristic calculation values. In contrast, the mean values and variation parameters for full-probabilistic reliability assessment are often based on literature references, which are not officially legitimated as national standards. In Germany this also includes the JCSS Probabilistic model code [16]. However, data from continuous measurements are much more robust than those from short-term measurements. Measurement data explicitly obtained from the structure also aligns much better with the actual conditions.

Continuous measurements represent an ongoing structural condition monitoring and allow for the early identification of anomalies, which can be investigated by engineers on-site as part of an object-specific examination.

5 Summary and Conclusion

Regulations and standards are primarily focused on the construction of new structures. Only few codes focus on the assessment of existing structures. Measurements on existing structures are often limited for various reasons (e.g. costs or traffic disruption) and are typically only conducted for a few months or years. During the construction phase, the installation of sensors is easier since the construction site setup (machines, traffic control) are already in place and there is no traffic passing over the bridge. To determine the potential scope of a monitoring concept, the existing boundary conditions must be considered. Bridges are often located in remote areas without reliable power sources. Furthermore, it may be necessary for the collected data to be available online, requiring a data connection (e.g., via mobile networks).

With the pilot project of the Schwindegg bridge a real-size laboratory was realized for the first time, demonstrating its feasibility. Extensive data on the effects on the structure and its resistance can be obtained and used for full-probabilistic calculations. In the long term, measurement data can serve as a basis for updating building regulations and standards. The possibility to collect a large amount of information was not available in the past but can lead to new insights about the structural safety of infrastructure. The insights gained from sensor data on one structure can also be utilized for other structures or entire sections of infrastructure. For these reasons the use of digital twins in the construction industry should be further established in other projects.

References

1. Bundesanstalt für Straßenwesen Anweisung Straßeninformationsbank für Ingenieurbauten, Teilsystem Bauwerksdaten (ASB-ING), Bergisch Gladbach (2013)
2. Naraniecki, H., et al.: Concept for a digital twin of railway bridges on the example of the new Filstal bridges. In: ce/papers 6, Nr. 5, pp. 711–717 (2023). https://doi.org/10.1002/cepa.2050
3. DIN 1076:1999–11: Ingenieurbauwerke im Zuge von Straßen und Wegen - Überwachung und Prüfung. Beuth Verlag, Berlin
4. Bundesanstalt für Straßenwesen Richtlinie zur einheitlichen Erfassung, Bewertung, Aufzeichnung und Auswertung von Ergebnissen der Bauwerksprüfungen nach DIN 1076 (RI-EBW-PRÜF), Bergisch Gladbach (2017)
5. Wimmer, J., Braml, T.: Digital bridge Schwindegg: Ab initio bridge monitoring for a sustainable lifecycle management. In: IABSE Congress. New Delhi (2023)
6. Braml, T., Haslbeck, M., Wimmer, J.: Digitalisierung in der Versuchsführung und Monitoring von Bauwerken. In: Betonkalender 2024, pp. 574–603. Ernst & Sohn, Berlin (2024)
7. Braml, T., et al.: Standardisierung im structural health monitoring (SHM) – Konzeptvorschlag. In: 2. Fachkongress Digitale Transformation der Verkehrsinfrastruktur, pp. 71–81. Expert Verlag, Tübingen (2023)
8. Braml, T., Wimmer, J., Seitz, F., Spannaus, M.: Standardisation in structural health monitoring (SHM) – a concept proposal. In: 13th - German-Japanese Bridge Symposium. Osaka, Japan (2023)
9. Wimmer, J., Braml, T.: Geo-referenced localisation of SHM sensors on new bridge construction based on the example of the digital bridge Schwindegg (Germany). In: 13th - German-Japanese Bridge Symposium. Osaka, Japan (2023)
10. Wimmer, J., Braml, T.: Permanent structural health monitoring of a new prestressed concrete bridge. In: 2nd Conference of the European Associa-tion on Quality Control of Bridges and Structures – EUROSTRUCT. Ernst & Sohn, Berlin(2023)
11. Wimmer, J., Braml, T., Martinez, R.: Digitale Zwillinge für Brücken mittlerer Stützweite – Pilotprojekt Brücke Schwindegg – Teil 1: Sensorik. In: Beton- und Stahlbetonbau 118, Heft 12. Ernst & Sohn, Berlin (2023). https://doi.org/10.1002/best.202300062
12. Wimmer, J., Braml, T., Kaiser, M.: Digitale Zwillinge für Brücken mittlerer Stützweite – Pilotprojekt Brücke Schwindegg – Teil 2: Verwaltungsschale. In: Beton- und Stahlbetonbau 119. Ernst & Sohn, Berlin (2024). https://doi.org/10.1002/best.202300096
13. Braml, T., Kainz C.: Practical concepts for the use of probabilistic methods in the structural analysis and reassessment of existing bridges -presentation of latest research and implementation. In: 19th International Prob-abilistic Workshop. Acta Polytechnica CTU Proceedings, Stellenbosch (2022). https://doi.org/10.14311/APP.2022.36.0047
14. DIN EN 1991–2:2010–12: Eurocode 1: Einwirkungen auf Tragwerke - Teil 2: Verkehrslasten auf Brücken; Deutsche Fassung EN 1991–2:2003 + AC:2010. Beuth, Berlin
15. Merzenich, G., Sedlacek, G.: Hintergrundbericht zum Eurocode 1 – Teil 3.2: Verkehrslasten auf Straßenbrücken. Typo-Druck & Verlagsgesellschaft, Bonn (1995)
16. Joint Committee on Structural Safety JCSS Probabilistic Model Code (2001). ISBN 978-3-909386-79-6

The Role of Uncertainty Propagation for Digital Twins

Matthias Haslbeck(✉) and Thomas Braml

Institute for Structural Engineering, University of the Bundeswehr Munich, Werner-Heisenberg-Weg 39, Neubiberg, Germany
matthias.haslbeck@unibw.de

Abstract. The great demand for digital maintenance concepts for bridges has given an enormous boost to research in the field of digital twins.

The many areas of application of the digital twin include the planning of maintenance measures as well as the prediction of future damage and life cycle management. The correct quantification of uncertainties in the underlying model plays an enormous role both in decision-making and in the acceptance of the concept by practitioners. The article provides an overview of current developments in digitalization in civil engineering and shows the application of the Monte Carlo method to determine uncertainties in the predictions using the digital twin. The theoretical explanations are underpinned using the example of a mid-span slab bridge. A non-parametric method is presented to monitor the convergence of the resulting probability distributions in order to determine the accuracy of the estimation and to minimize the computational effort.

Keywords: Uncertainty Propagation · Uncertainty Quantification · Digital Twin · Laser scanning · Bootstrapping

1 Motivation and Scope

The generation of digital twins through measurement data plays a central role for digitization in the construction industry.

Especially simulation based digital twins are very promising for the prediction of an engineering structure's behaviour. However, to ensure an acceptable level of uncertainty in the derived model, the impact of remaining uncertainties in the system's parameters needs further investigation. Additionally, the (a-priori) model can be validated concerning its predictive power.

This can be done using a probabilistic framework, where parameter uncertainty is propagated through the model to observe the uncertainty in the remaining output prediction. The approach most commonly adopted for this task is to use sampling-based Monte-Carlo.

In this method, a probability distribution function is determined that reflects the probability of occurrence for each parameter. The set of parameter values is then propagated as input through the finite element model to observe the frequency of the output value(s).

J. C. Matos et al. (Eds.): IPW 2024, LNCE 494, pp. 303–312, 2024.
https://doi.org/10.1007/978-3-031-60271-9_28

Sampling-based approaches for this task are easy to use and are free from assumptions on the structure of the input-output-relationship. Convergence monitoring can be challenging as the type of distribution function cannot be predicted beforehand and the assumption of a gaussian normal distribution does not hold in most cases. However, it is absolutely required for computationally expensive models.

In this paper, the importance of uncertainties in the predictive power of digital twins is discussed and a method is proposed to evaluate the uncertainty in the estimation of the standard deviation in the prediction.

The paper is organized as follows: After a review of current developments in the application of digital twins for civil engineering structures, the uncertainties attributed to finite element modelling is discussed. Subsequent to a short introduction to sampling-based uncertainty propagation, a framework is presented to confirm convergence of the results using the non-parametric technique of bootstrapping. The proposed framework is then presented using the example of a large-scale structure.

2 Digital Twins

2.1 Definition of Digital Twins and Implementation

Performance as well as Structural Health Management concepts require the collection of data and decision making from (simplified) models of structures during the lifecycle of bridges [1].

The most promising concept for linking our models to the physical structure is a digital twin that represents the behaviour of the structure by means of a digital presentation of the knowledge of a structure.

First implementations of digital twins are reported from NASA to support the Apollo program in space by a physical counterpart on earth and the from the University of Michigan as a first sketch of the future of product design [2]. Even though there is currently no consistent definition of the Digital Twin, most references describe the concept as 'a virtual representation of a physical system (and its associated environment and processes) that is updated through the exchange of information between the physical and virtual systems' [3]. To reach that goal, three types of representations have evolved. The simplest representation focuses on geometry as the digital representation of a structure, while data and simulation based digital Twins rely on sensing to draw conclusions on the condition or behaviour [4, 5].

Geometry is often represented by point clouds. These can be gained by methods of laser scanning or photogrammetry.

When referring to digital twins as a true representation of the underlying physical relationship between the input, i.e. the parameters of a system and the system's response to loads, uncertainty quantification comes into play.

The first step in uncertainty quantification is commonly the definition of an a-priori-model and the propagation of the uncertainty of the parameters through the model to quantify the uncertainty in the assumed system's response. Most common in this approach is the statistical modelling of the parameters via probability distributions to gain frequencies of the responses via Monte Carlo simulation. As an add-on to the propagation of uncertainties, the evaluation of the share the parameters contribute to the uncertainty results, sensitivity analysis is a means to investigate on a model [6] and to possibly overthink model building [7]. Only then a reverse engineering is advisable that updates the model parameters using in situ data.

2.2 Application of DT in Civil Engineering

Applications of the digital twin concept are applied the mechanical engineering for more than 20 years [8]. However, the subject has only attracted interest very recently in the construction industry. Projects that may pave the way to a broader application of digital twins in the fields of monitoring, maintenance and recalculation are very promising concerning their acceptance by practitioners. Among them are:

- **smartBRIDGE Hamburg:** Using the Example of a large-scale bridge in the center of the port of Hamburg, a digital-twin-based maintenance concept is implemented that is based on a large amount of heterogeneous data in order to optimize the life-cycle-management and condition assessment [9–11]
- **Brücke Schwindegg:** A mid-size frame bridge with span length 22 m was equipped with a large number of sensors transferring its data to a digital repository where data is stored, processed and visualized [12, 13].

In addition to questions of data assimilation and data management, the interpretation of the data and the adaptation of the digital twin to measured responses represents a major challenge for the application of digital twins in science and practice [14].

Especially when referring to simulation-based digital twins, 3D high fidelity models are the core of this new tool [2]. The benefits of model-based approaches lie in particular in their quantitative results. These can be used, for example, to create 'what if' simulations, to derive condition indicators or to trigger automated recalculations that can result in automatic blocking in the event of critical conditions.

This directly leads the way to method of uncertainty quantification and system identification to create a back loop from measurement to the digital representation in a digital twin concept.

2.3 Uncertainty Propagation for Digital Twins

Finite element models, just like all other models in science and technology, can only represent a section of reality. This requires special treatment in the application of the models. Methods of uncertainty propagation are used for this purpose. In this approach, the effects that scatter in the model parameters has onto the output variables of the model relationship is examined [15]. For a comparison of the different methods, the reader is referred to [16].

In the context of digital twins, methods of uncertainty propagation are required to quantify the predictive power of an a-priori-model as well as those uncertainties that remain in the model after updating. In this way, a level of uncertainty that is considered to be 'acceptable' can be compared to the remaining ones. Additionally, in the phase before installing a monitoring system to feed the digital twin with data, uncertainty propagation is a mighty tool to pic relevant and uncertain system reactions, even in the presence of measurement uncertainty. Previous work on this subject for applications in civil engineering has been accomplished e.g. in [17].

3 Uncertainties in Finite-Element Models of Bridges

3.1 Sources of Uncertainties and Statistical Modelling

Models applied in structural analysis are often considered to be a perfect representation of the truth. However, numerical and experimental studies repeatedly show that the accuracy of these models, even FEM-based calculations have lower accuracies than are often attributed to them and show considerable uncertainties and biases [18]. Therefore, the modelling of engineering structures must be rethought to produce the overall goal of a digital representation. Based on a literature review, the uncertainties in modelling can be traced to the following sources:

- simplification of the model and its underlying assumptions
- human error in the construction phase
- degradation in the lifetime of the structure
- uncertain material properties
- unconsidered participation of non-structural elements

To take the different sources of uncertainties into account, the a-priori-FEM is commonly parametrized in such a way that the influences can be mapped to certain (physically interpretable) variables or reduction factors.

3.2 Reduction of Uncertainties and Measurement Error

Reducing uncertainties in the model is the main task of system identification. In order to reduce parameter uncertainty, direct investigations can be carried out in some cases, such as material tests or indirect identification methods like Bayesian updating.

However, direct measurements are often infeasible or lack accuracy concerning the identification of parameters. Measurement uncertainties in material testing and local variations in material testing lead to uncertainties in the estimation that can only be reduced by a large sample size. A survey on the amount of data needed for certain levels of uncertainties in the parameter estimation based on Bayesian Updating can be found in [19]. Most common is the identification of system parameters using indirect measurements. Especially in Bayesian updating, special attention should be given to the attribution of uncertainties to the data used for updating. This can be done using the GUM-framework [20]. Practical applications of the two concepts can be found in [21] for deformation measurements and in [22] for the application of strain gages.

4 Theory of Uncertainty Propagation

4.1 Monte Carlo Simulation

Even though some few methods to analytically determine the uncertainty in a model exist, most uncertainty quantification methods rely on a numerical study to propagate the input uncertainty through the model. This is especially beneficial because of the restrictive assumptions made in most analytic derivations of the output uncertainties. These methods are commonly free from overly restrictive assumptions, but can be demanding for computationally expensive models.

Monte Carlo sampling from the output distribution is done implementing a probabilistic solver that uses samples of the parameters values that are propagated through the deterministic model function. Basis of each Monte Carlo technique is the sampling methodology. In this paper, the random sample generator MT19937ar developed by Matsumoto and Nishimura in [23]. This sampling strategy generates independent samples which allows for constant monitoring of convergence and gives a good representation of the underlying population. Even though variance reduction techniques lead to faster convergence, monitoring is in some cases demanding and has difficulty to show extreme output values.

4.2 Evaluation of Convergence

One of the major tasks in numerical simulation of probabilities is the evaluation of the uncertainty contained in the results. Therefore, special attentions should be given to this task.

As most Monte Carlo simulations aim to evaluate some kind of mean value, this is most often done using the standard error or standard intervals. The underlying assumption of a normally distributed approximation error can be rectified by the central limit theorem. However, this is no longer necessarily true when variance or standard deviation is estimated. When the kind of distribution function is unknown beforehand, this can no longer be assumed.

In this case non-parametric approximations come into play. These remain correct under very mild conditions and make it possible to estimate convergence even if the resulting distribution is unknown. The task in convergence monitoring is to evaluate an interval where the true value of the parameter lies in between with a certain probability $1 - \alpha$. The lower and upper interval boundaries are monitored to ensure convergence. This is formulated in Eq. (1) for the mean value and in Eq. (2) for the estimated standard deviation.

$$P(\widehat{\mu}_{low} \leq \widehat{\mu} \leq \widehat{\mu}_{up}) \geq 1 - \alpha \tag{1}$$

$$P(\widehat{\sigma}_{low} \leq \widehat{\sigma} \leq \widehat{\sigma}_{up}) \geq 1 - \alpha \tag{2}$$

The non-parametric method of bootstrapping uses resamples of the output sample to evaluate the distribution of the estimator [24]. From the output sample artificial resamples are derived to generate a statistic on the observed distribution parameter. In this way, the upper and lower boundaries can be estimated regardless of the estimator's distribution [25, 26].

An alternative to the derivation of percentiles from the gained sample of the estimations are so called 'bias-corrected and accelerated' (BC_a) – Intervals. These are superior to the derivation of percentiles and converge faster [27]. Due to reasons of brevity, the reader is referred to [27] for further details and the construction of these intervals.

In order to confirm convergence, the derived interval for a certain parameter is commonly compared to a threshold value and the simulation is stopped as soon as the requirement is met. In case of heterogeneous data, however, convergence cannot be confirmed on a common level for all responses. Therefore, we propose the threshold to be of relative nature. Convergence is met, when for all system responses the ratio of the length of the $1-\alpha$ - confidence interval to the estimation is equal or less than a certain threshold value.

5 Example

5.1 Example Bridge and Simulation Settings

The structure under investigation is a two-span, prestressed slab bridge. It was modelled using shell elements where the bearings are modelled as springs. The calculations were conducted using SOFISTIK 2024. The wheel loads of a heavy goods vehicle were applied as load for the simulation of the uncertainty in the system's response.

The simulation has been implemented in MATLAB, where sampling and translation to input files is done that are executed by SOFISTIK until the results are returned to MATLAB for postprocessing and convergence monitoring. For a more detailed explanation of the implementation and a comparable simulation, please refer to [28].

To describe the uncertainty in the model, the following parameters were used:

- The Young's modulus of the concrete slab, separately for each span
- The stiffnesses of the caps in span 1 and span 2 respectively
- The stiffnesses of the springs to represent the bearing stiffnesses and the foundation soil at the abutments and the pier respectively
- The weight of the structural concrete in span 1 and span 2 respectively
- The slab thickness in span 1 and 2
- The transverse strain coefficients of each of the two spans

Figure 1 shows a side view of the bridge, which was generated from the point cloud.

For the study in this paper, the deformations as well as strain on the bottom side of the bridge are considered.

Fig. 1. Side view of the bridge considered for the simulations in Sect. 5.

5.2 Example of Simulation Results

In this section, the resulting histograms and convergence charts are given. For reasons of brevity, only a few results can be shown here, but they can underpin the importance of uncertainty propagation in digital twin concepts.

The simulation has been conducted using 95% intervals for the estimation of the parameter. Convergence has been assumed when the length of the interval for value of the standard deviation is less than 10% of the parameter estimation. This resulted in a number of 75.000 samples.

Figure 2(a) shows one of the results of the uncertainty propagation as a histogram. Here, one of the studied deflections is given when taking parameter uncertainty into account. The mean value for the output-sample is 5.13 mm while the standard deviation is at 0.66 mm. It can be observed that the system contains relevant uncertainties which can be seen from the coefficient of deviation $\nu = 0.13$. It can also be seen that the assumption of a normal distribution does not hold and the distribution is skewed.

The strain values from Fig. 2(b) shows negatively skewed distribution with a mean value of $-6.2 \cdot 10^{-6}$ and standard deviation of $0.77 \cdot 10^{-6}$. The coefficient of variation is $\nu = 0.12$.

The method for convergence monitoring presented in Sect. 4.2 can be visualized in convergence diagrams that show the lower and the upper boundaries of the BC_a-intervals as well as the estimation depending on the number of model evaluations. Examples of such diagrams are shown in Fig. 3. After a number of 75.000 simulation runs, the observed ration of length of the 95%-BC_a-interval and the estimation is at 1.2% for the standard deviation and 0.19% for the mean value.

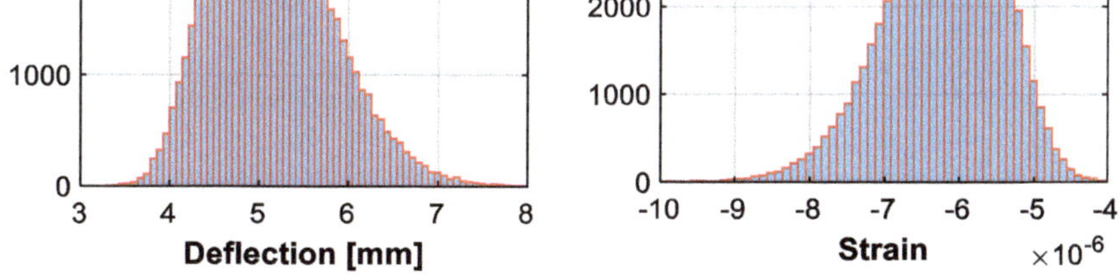

Fig. 2. Examples of histograms of the output for (a) the deformation of the slab and (b) for strain at two evaluation positions

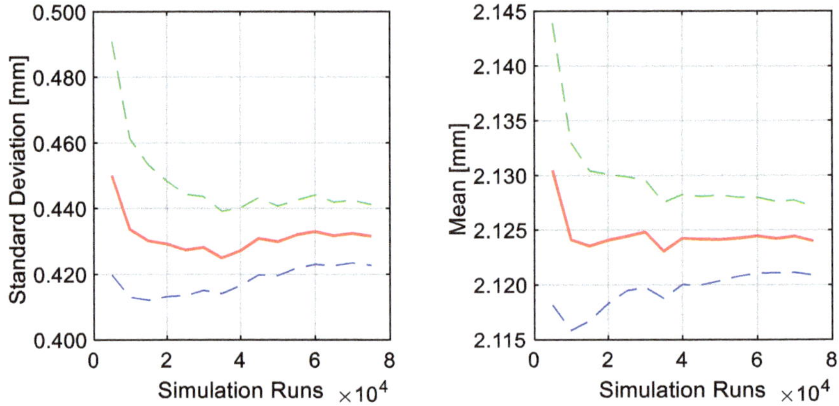

Fig. 3. Examples of Convergence diagrams for (a) the estimation of the standard deviation an (b) for the mean value.

6 Summary and Outlook

The paper presents the advantages and necessity of uncertainty propagation for designing digital twins and the evaluation of their predictive power. After a review on the definition and examples of the application of digital twins, the basics of uncertainty quantification are explained. A brief example for such an evaluation is given using a mid-span prestressed concrete bridge.

The summary of the state of the art as well as the theoretical explanations on Monte Carlo simulation show that models always need to be treated with great care, as uncertainty is always present when models are used in civil engineering. Future work on this subject may concentrate to the backpropagation of uncertainties to lower uncertainty in the description of the digital twin. For this task the updating of the parameters using Bayesian inversion via MCMC is a mighty tool that will become standard for connecting the physical object to the digital twin.

References

1. Zimmert, F., Braml, T., Keuser, M.: Decision-making for the efficient life cycle management of structures. Struct. Concr. **21**(4), 1217–1226 (2020)
2. Opoku, D.-G.J., et al.: Digital twin application in the construction industry: a literature review. J. Building Eng. **40** (2021)
3. VanDerHorn, E., Mahadevan, S.: Digital twin: generalization, characterization and implementation. Decision Support Syst. 145 (2021)
4. Bertola, N., et al.: Methodology for selecting measurement points that optimize information gain for model updating. J. Civil Struct. Health Monit. **13**(6–7), 1351–1367 (2023)
5. Arcones, D.A., et al.: A Bayesian framework for simulation-based digital twins of bridges. ce Papers **6**(5), 734–740 (2023)
6. Saltelli, A., et al.: Global Sensitivity Analysis – The Primer. Wiley, Chichester, West Sussex (2008)
7. Saltelli, A., Tarantola, S., Campolongo, F.: Sensitivity analysis as an ingredient of modeling. Stat. Sci. **15**(4), 377–395 (2000)

8. Marx, S., et al.: Digitaler Zwilling – Strategie für den Bestandserhalt (DBV-Heft 51). Eigenverlag Deutscher Beton-Verein e.V, Berlin (2021)
9. Wenner, M., et al.: smartBRIDGE hamburg: a digital twin to optimise infrastructure maintenance. In: Casas, J.R., Frangopol, D.M., Turmo, J. [Hrsg.] Bridge Safety, Maintenance, Management, Life-Cycle, Resilience and Sustainability. London: CRC Press, pp. 964–970 (2022)
10. Grabe, M., et al.: smartBridge Hamburg – prototypische Pilotierung eines digitalen Zwillings. In: Bautechnik, vol. 97, no. 2, pp. 118–125 (2020)
11. Wedel, F., et al.: Das 3-D-Modell als Grundlage des digitalen Zwillings. Bautechnik **99**(2), 104–113 (2022)
12. Braml, T., Haslbeck, M., Wimmer, J.: Digitalisierung in der Versuchsführung und Monitoring von Bauwerken. In: Bergmeister, K., Fingerloos, F., Wörner, J.D. [Hrsg.] 2024 BetonKalender. Wiley, pp. 573–603 (2023)
13. Wimmer, J., Braml, T., Martinez, R.: Digitale Zwillinge für Brücken mittlerer Stützweite – Pilotprojekt Brücke Schwindegg – Teil 1: Sensorik. Beton- und Stahlbetonbau **118**(12), 889–896 (2023)
14. Hu, W., Lim, K.Y.H., Cai, Y.: Digital twin and industry 4.0 enablers in building and construction: a survey. Buildings **12**(11), 2004 (2022)
15. Ghanem, R., Higdon, D., Owhadi, H., [Hrsg.].: Handbook of Uncertainty Quantification. Springer International Publishing, Cham (2017)
16. Lee, S.H., Chen, W.: A comparative study of uncertainty propagation methods for black-box-type problems. Struct. Multi. Optim. **37**(3), 239–253 (2009)
17. Santos, I.C.E., Brito, J.L.V., de; Caetano, E.D.S.: Uncertainty quantification: data assimilation, propagation and validation of the numerical model of the Arade river cable-stayed bridge. Struct. Infrastruct. Eng. **18**(10–11), 1410–1427 (2022)
18. Structural identification of constructed systems – Approaches, methods, and technologies for effective practice of St-Id. Reston, Va.: American Society of Civil Engineers (2013)
19. Haslbeck, M., et al.: Uncertainty assessment for the Bayesian updating process of concrete strength properties. In: Acta Polytechnica CTU Proceedings, vol. 36, pp. 76–83 (2022) . https://doi.org/10.14311/APP.2022.36.0076
20. BIPM et al.: Evaluation of measurement data – Guide to the expression of uncertainty in measurement (GUM) – JCGM 100:2008. https://www.bipm.org/documents/20126/2071204/JCGM_100_2008_E.pdf/cb0ef43f-baa5-11cf-3f85-4dcd86f77bd6
21. Haslbeck, M., Strübing, T., Braml, T.: Quantification of uncertainties for geodetic observations in the context of bridge surveillance. In: Pellegrino, C. et al. [Hrsg.] Proceedings of the 1st Conference of the European Association on Quality Control of Bridges and Structures. Cham, Springer International Publishing, pp. 1–13 (2022). https://doi.org/10.1007/978-3-030-91877-4_1
22. Haslbeck, M., Böttcher, J., Braml, T.: An uncertainty model for strain gages using Monte Carlo methodology. Sensors (Basel, Switzerland) **23**, 21 (2023). https://doi.org/10.3390/s23218965
23. Matsumoto, M., Nishimura, T.: Mersenne Twister: A 623-dimensionally equidistributed uniform pseudo-random number generator. ACM Trans. Model. Comput. Simul. **8**(1), 3–30 (1998)
24. Chernick, M.R.: Bootstrap Methods – A Guide for Practitioners and Researchers, 2nd edn. Wiley-Interscience, Hoboken, NJ (2008)
25. Efron, B., Tibshirani, R.: An Introduction to the Bootstrap. Chapman & Hall, Boca Raton, Fla. (1998)
26. Efron, B.: The Jackknife, the Bootstrap, and Other Resampling Plans. Society for Industrial and Applied Mathematics, Philadelphia, Pa (1982)

27. Martinez, W.L., Martinez, A.R.: Computational Statistics Handbook with MATLAB. Chapman & Hall/CRC, Boca Raton (2021)
28. Haslbeck, M., Kraus, M.A., Braml, T.: Bayesian reliability assessment and system identification for existing concrete bridge structures – elaboration of a template for existing civil engineering structures. In: Proceedings of the 17th International Probabilistic Workshop.

Fundamentals and Application of Probability Theories

A Guideline for Probability-Based Structural Design: Cornerstones and Principles

Tânia Feiri[1(✉)], Jan Philip Schulze-Ardey[2], Udo Wiens[3], Marcus Ricker[1], and Josef Hegger[2]

[1] Chair of Structural Concrete, TU Dortmund University, 44227 Dortmund, Germany
tania.feiri@tu-dortmund.de
[2] Institute of Structural Concrete, RWTH Aachen University, 52074 Aachen, Germany
[3] German Committee for Structural Concrete, 10787 Berlin, Germany

Abstract. It is widely acknowledged that probability-based concepts are the logical basis for dealing with uncertainty and the basis for structural safety – the prime requirement of structural engineering. In structural engineering, design codes and standards provide a foundation for good engineering practice. The use of probabilistic methods in such codes and standards, which have been continuously evolving, is the result of several decades of structural reliability research alongside remarkable advances in computational capabilities. Albeit such progress and developments, the widespread adoption of probability-based approaches among structural engineering practitioners does not seem to have occurred yet. To bridge such gap and promote a wider acceptance of these approaches in structural design, the German Committee for Structural Concrete (DAfStb), together with multiple research partners, has been developing a guideline entitled *„Procedure for the derivation of safety factors in concrete structures using probabilistic methods"*. This two-part document covers the basics for time-invariant design considerations and guides the use of methods for deriving PSF when decoupling the load and resistance sides. This paper introduces the cornerstones and principles of this guideline and discusses to which extent this document can contribute to a broader acceptance of probability-based approaches. The adoption and dissemination of this guideline and its principles calls for a broader participation of the research and practising engineering communities. This discussion opens new avenues for future initiatives and developments in the field of structural reliability design.

Keywords: Guideline · Probability theory · Structural safety · Structural reliability

1 Introduction

In all the lifecycle phases of structures – and particularly, during early design stages – uncertainties arise from multiple sources [1]. For many decades, it has been established that probability-based concepts are the logical basis for dealing with uncertainties and the foundation for structural safety – the prime requirement of structural engineering

J. C. Matos et al. (Eds.): IPW 2024, LNCE 494, pp. 315–323, 2024.
https://doi.org/10.1007/978-3-031-60271-9_29

(e.g., [1, 2]). Nowadays, due to the urgency to reduce material and resource consumption – and, consequently, generate sustainable and economic benefits – the potential offered by probability-based concepts has been explored (e.g., [3–5]). A specific application of probability-based concepts is to reduce partial safety factors (also known as *safety factors* or *partial factors*) (PSF) on the resistance side related to concrete γ_C and reinforcing or prestressing steel γ_S. Alternative design solutions can be proposed by exploiting safety reserves (or margins) due to more comprehensive design models, while still complying with specified target safety levels [6]. There are two ways to achieve this goal: With the support of reliability-based analyses, the safety margin can be expoited for the existing design models. Also, more comprehensive design models can be adopted for a better description of the bearing behaviour. Ideally, both approaches can be utilised and combined.

In this context, in the past few years, notable research developments have motivated the introduction of more detailed probability-based provisions in codes and standards. For example, the recently approved Annex A (informative annex) of FprEN 1992–1-1:2023 (D) [7] – part of the second generation of Eurocodes – includes now a set of revised provisions to modify PSF on the resistance side alongside more detailed stochastic models for a relative wide range of basic variables. For practitioners, reducing PSF while complying with structural safety levels requires the use of advanced reliability-based methods (e.g., Level II methods, as First or Second Order Reliability methods and Level III methods, as Crude Monte Carlo Simulations), which can be supported with modern computational capabilities (e.g., [8–10]).

It is widely acknowledged that technological advances of computational capabilities have also occurred during the last few decades. For example, more modern reliability-based methods such as Monte-Carlo with Importance Sampling or Monte-Carlo with Subset Sampling and other even more advanced approaches have emerged or even become more robust (e.g., [9, 10]). Simultaneously, multiple computational platforms have been introduced and made publicly available to support structural engineering practitioners and scientific research communities. For example, the *TesiproV* software package [11] – an open-source package for structural reliability assessments developed by the Hochschule Biberach and RWTH Aachen (both in Germany) in the free software environment *R* [12] – was released in 2022.

Attempts to support structural engineering practitioners to adopt probability-based approaches have been made before. For example, one of the first roadmaps for codified probabilistic approaches was proposed by Marek et al. [13]. Albeit such progress and developments, the widespread adoption of probability-based approaches among structural engineering practitioners does not seem to have occurred yet. A recent study from Feiri et al. [6] conducted among manufacturers of construction products suggests that their familiarity with probability-based safety concepts and techniques and their interest to statistically evaluate experimental data to assess structural safety remain somewhat limited. In this context, the German Committee for Structural Concrete (in German: *Deutscher Ausschuss für Stahlbeton e.V.*, DAfStb), together with multiple research partners, has been developing a draft guideline (hereafter denoted as DAfStb-guideline) entitled *„Procedure for the derivation of safety factors in concrete structures using* *probabilistic methods"*. The goal of the DAfStb-guideline is to support the safety assessment of structural components and systems and the derivation of PSF through the use of probability-based safety concepts and methods. An additional goal is to promote a wider acceptance of these concepts and methods by communities of practice and encourage the development of new construction methods and solutions (e.g., FRP reinforced concrete) by avoiding overly conservative structural safety – and, ultimately, increasing economic and sustainable efficiency. This paper introduces the cornerstones and principles of the new DAfStb-guideline.

2 General Considerations, Regulatory Context and Fundamentals

2.1 General Considerations

The DAfStb-guideline was established in the context of the *TesiproV* project, a research project comissioned by the Federal Ministry for Economic Affairs and Climate Action (in German: *Bundesministerium für Wirtschaft und Klimaschutz*, BMWK) to a German-based consortium composed by academic partners (RWTH Aachen University and Hochschule Biberach), a specialised committee association (DAfStb), a consultant partner (*Deutsches Institut für Bautechnik*, DIBt), and industry partners (solidian GmbH and H + P Ingenieure GmbH). The project took place between December 2019 and March 2022 and its results have enabled the development of the DAfStb-guideline. As the name implies, the guideline was produced under the leadership and coordination of DAfStb. Currently, it is being discussed in the context of an expertise committee and it is expected to be formally concluded by the end of 2024. Note that the DAfStb is a German technical and scientific committee that aims to promote concrete construction as a safe, durable, economical and environmentally friendly method while being in line with national and international references and best practices. This body is also engaged in the development of other guidelines and standards (e.g., DIN standard in Germany) and coordinates standardisation work.

2.2 Regulatory Context

The DAfStb-guideline sits within the regulatory context of existing structural codes, norms and guidelines. On the international level, the guideline considers that overarching principles of FprEN 1990:2022 [14] and Probabilistic Model Code postulated by the Joint Committee on Structural Safety [15]. Additionally, the guideline is in line with the provisions of ISO 2394:2015 [16], the document that governs the general principles for structural reliability. Also, the principles of the first order reliability concepts for design codes prescribed in CEB *Bulletin* 112 [17] are considered. Other references, such as the *fib Bulletin* 80 [18] and *fib Bulletin* 86 [19] are taken into account despite of targeting existing structures. On the national level, the fundamentals of *GruSiBau* (DIBt) [20] are considered, a document that sets the principles for the definition of safety requirements for structural systems and is currently under revision.

3 Structure of the Guideline

The newly developed DAfStb-guideline is composed by two parts, whose content is listed in Table 1. *Part I* sets the fundamentals for time-invariant design considerations and *Part II* describes a procedure for the derivation of PSF with decoupling of load and resistance side.

Table 1. Table of content of *Parts I* and *II* of the DAfStb guideline draft

Part I: Fundamentals for time-invariant considerations 1 Scope and limits of application 2 References to standards and other technical regulations 3 Terms, definitions, symbols 3.1 Terms and definitions 3.2 Symbols, abbreviations and formula symbols 4 Determining the statistical parameters of the basic variables 4.1 General information 4.2 Determining a suitable distribution function 4.2.1 General information 4.2.2 Graphical methods 4.2.3 Goodness-of-fit tests 4.3 Detection of outliers 4.4 Parameter estimators 5 Calculation of the failure probability 5.1 General 5.2 Level II methods 5.3 Level III methods 5.4 Determining the probability of system failure 6 Procedure for determining safety factors 6.1 General 6.2 Procedure for assessing the safety level and calculating structural safety 7 References
Part II: Procedure for the derivation of PSF with decoupling of load and resistance side 1 Scope and limits of application 2 Normative references 3 Terms, definitions, symbols 4 Procedure for calculating PSF 5 References

4 *Part I*: Fundamentals for Time-Invariant Considerations

4.1 Procedures to Derive the Statistical Parameters of Basic Variables

Reliability-based assessments are not limited to the calculation of failure probabilities and reliability indices. The evaluation of various statistical properties, such as probability distribution functions for the description of basic variables and confidence intervals affecting structural responses also play an important role [21]. For example, the misrepresentation of the stochastic description of a variable property (e.g., concrete compressive strength or effective depth) due to an inaccurate choice of the distribution type may greatly influence the calculated statistics and lead to biased estimates of structural safety.

Also, the stochastic description of model uncertainties is not always properly derived since existing knowledge of model uncertainties seems to be mostly based on intuitive judgements or suffer from imprecise definitions [21]. In other cases, experimental data for robust statistical inferences seems to be lacking. If sufficient experimental data exists, innacuracies can emerge from incorrect statistical treatment of such data.

It is precisely the statistical assessment of data samples and the consequent derivation of suitable stochastic models that are one of the cornerstones of *Part I* in the DAfStb-guideline. The goal of this part is to propose a standard procedure to derive stochastic models for basic variables (e.g., Lognormal distribution for the characterisation of concrete compressive strength). The proposed procedure is divided in two parts (Fig. 1): *Part A* concerning the procedures for the statistical treatment of data samples (i.e., mostly concerning experimental data) and *Part B* refering to the procedures for the parameter estimation of the stochastic models resulting from *Part A*.

Part A entails three steps: *Steps 1, 2* and *3*. *Step 1* proposes a preliminary investigation of a data sample, where "rule-of-thumb" tools (i.e., visual/graphical and numerical-based tools) are recommended with the goal to narrow down the scope of possible stochastic models. *Step 2* suggests the use of goodness-of-fit tests. Goodness-of-fit tests (GOF) describe how well a stochastic model represents a set of observations (e.g., test values of concrete compressive strength). GOF tests can be used to decide if the available data sufficiently supports the hyptothesis that the set of observations follows a certain distribution with a specific level of significance (e.g., whether the set of observations follows a normal distribution). This is done with provisions for the selection of specific statistical tests (e.g., Kolmogorov-Smirnov [22] tests or Anderson-Darling [23] tests) on the basis of the characteristics of the dataset. Finally, *Step 3* suggests means to detect outliers (or extreme observations). Outliers can stem out from errors in test execution, records and computations, non-standard test conditions, failure mode beyond the scope of a model, specimens significantly different from other in the sample space, among others [21]. Their presence may increase the variability of a data sample and decrease the sensitivity of subsequent statistical tests. Examples of these tests are the Dean-Dixon [24] tests (for the detection of a single outlier) but only for small size datasets ($n < 30$ values) that follow a normal distribution or the Walsh [25] tests for larger datasets ($n >$ 60 values) that are non-parameteric. *Steps 2* and *3* can be repeated iteratively until some confidence is reached regarding the statistical model adopted for the basic variables being investigated.

Part B provides guidance on the estimation of population parameters from a limited dataset (with size n). Based on appropriate sampling distributions, two types of estimates are considered: interval estimates (for $n \leq 100$) and point estimates (for $n > 100$). *Interval estimates* are obtained as the intervals of a relevant sampling distribution that cover the population parameters with a given probability (i.e., confidence level) [26]. Information about the precision of the estimation is conveyed by the length of the interval (e.g., a short interval implies precise estimation). Note that the confidence interval shall be built in a way that analysts have high confidence that the sample contains the unknown population parameter. *Point estimates,* in their turn, refer to a single number based on the sample data that is the most plausible value for the unknown population parameter that is being investigated.

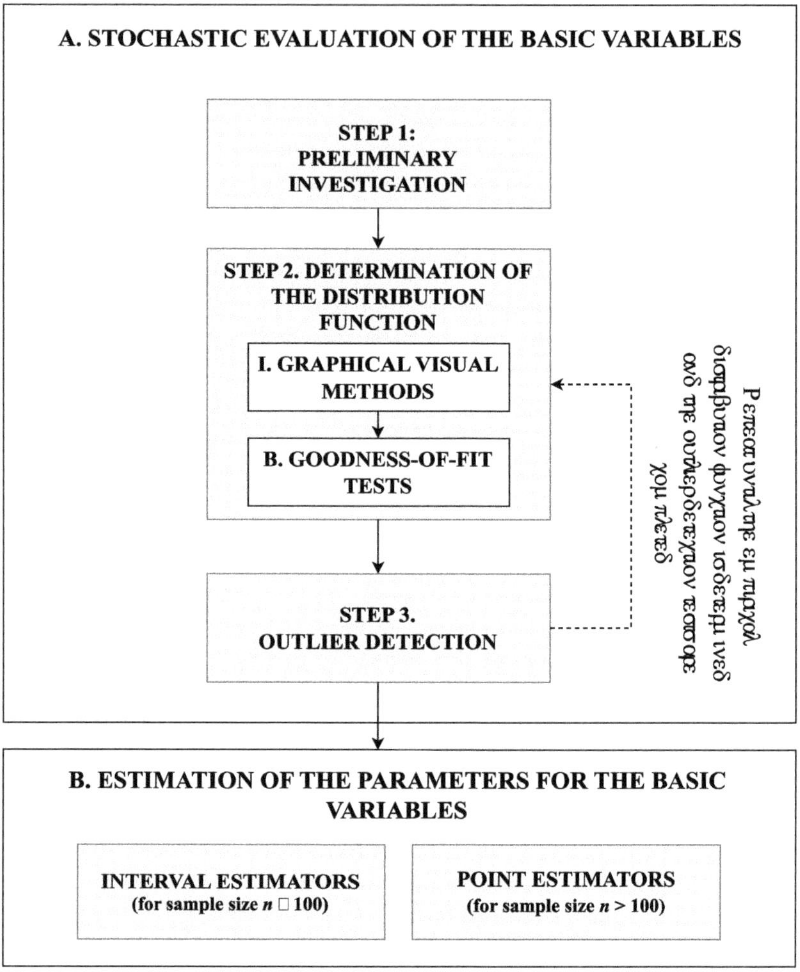

Fig. 1. Standard procedure for deriving suitable stochastic models for basic variables.

For each of these parts, the DAfStb-guideline offers provisions as well as flowcharts that support the selection of the prescribed statistical approaches or techniques.

4.2 Procedure to Calculate Failure Probabilities

In addition to the procedure described in Sect. 4.1, the DAfStb-guideline also provides guidance on the selection of suitable reliability-based methods for the resolution of practical structural reliability problems. Provisions are given for individual structural components and are divided in recommendations for the use of Level II methods (e.g., first order reliability methods) or Level III methods (e.g., Crude Monte Carlo simulations or Monte Carlo with variance reduction as Importance Sampling). Structural systems (i.e., parallel, series or hybrid systems) are also contemplated. Also, flowcharts are given; yet, they are not presented here due to space limitations.

4.3 Procedure to Determine Partial Safety Factors

In this section a general procedure is given to derive PSF, global safety factors and additional safety elements (Δa in FprEN 1990:2022 [14] refering to changes made to nominal geometrical data for particular design purposes, e.g., assessment of effects of imperfections) while specifying a target reliability value. It is a six-step procedure for assessing the safety level of designs solutions or for calculating PSF. Also, provisions and flowcharts are given; yet, they are not presented here due to space constraints.

5 *Part II*: Procedure for the Derivation of PSF with Decoupling of Load and Resistance Side

Part II of the guideline proposes a seven-step procedure to derive PSF on the resistance side or to assess the structural safety level of a component or system (Fig. 2). Here, it is assumed that the target reliability level is achieved through PSF, which can be determined with the use of Level II and/ or Level III methods. The action and resistance sides are considered separated by using constant weighting factors viewed in a decoupled manner.

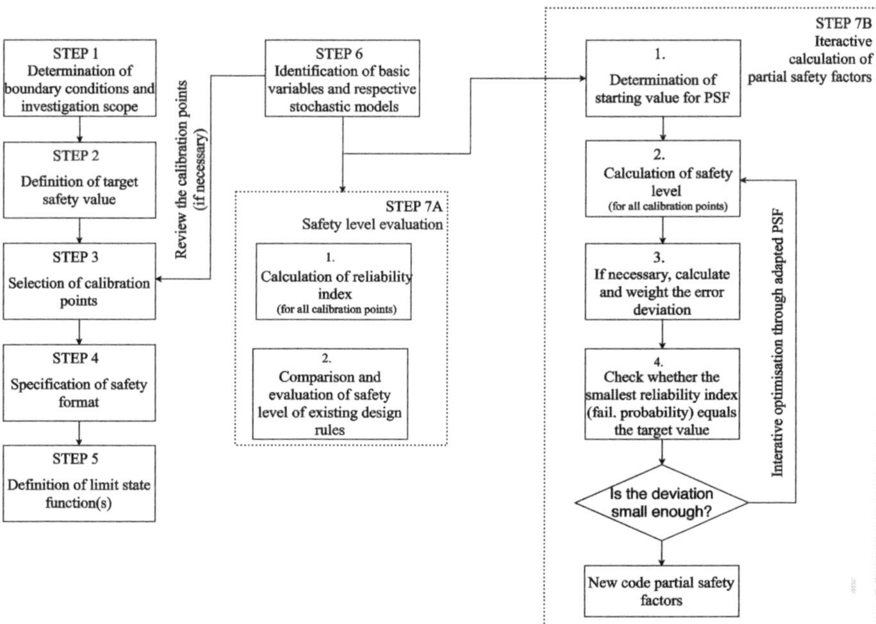

Fig. 2. Standard procedure for the derivation of PSF load and resistance side decoupled.

6 Conclusions

This manuscript has led to the following conclusions:

- The DAfStb-guideline is a vehicle to conduct statistical treatement of data samples (including experimental datasets), to estimate failure probabilities and to evaluate structural safety as well as to derive PSF.
- Such guidance is needed by the communities of practice since the misrepresentation of the stochastic description of a variable property (e.g., concrete compressive strength or effective depth) due to an inaccurate choice of the distribution type may greatly influence the calculated statistics and lead to biased estimates of structural safety.
- The guideline presents a wide range of probability-based techniques to treat data and assess the safety level of a structural component. The choices of the techniques and approaches made during such assessments are vital to the results and their interpretation. Biased results may not only compromise the structural safety but can also lead to overconservative and, consequently, anti-economic and anti-ecological structural solutions.
- Thus, it is expected that a wider dissemination of the guideline promotes more robust and reliable structural reliability assessments.

Future work shall include methods for the derivation of PSF considering the load and resistance side (combined), the assessment of existing structures and as well as the consideration of the effects of quality control measures for the derivation of PSF or for structural safety assessments.

Acknowledgments. The financial support from the Bundesministerium für Wirtschaft und Klimaschutz, BMWK (German Federal Ministry for Economic Affairs and Climate Action, reference number 03TNK003E) is gratefully acknowledged.

References

1. Ellingwood, B.: Probability-based codified design: past accomplishments and future challenges. Struct. Saf. **13**(3), 159–176 (1994)
2. Sexsmith, R.: Probability-based safety analysis – value and drawbacks. Struct. Saf. **21**(4), 303–310 (1999)
3. Holický, M., Sýkora, M.: Reliability approaches affecting the sustainability of concrete structures. Sustainability **13**(5), 2627 (2021)
4. Poutanen, T.: Code calibration of the eurocodes. Appl. Sci. **11**(12), 5474 (2021)
5. Feiri, T., Schulze-Ardey, J., Ricker, M., Hegger, J.: Using system reliability concepts to derive PSF for punching shear with shear reinforcement: an explorative analysis. Appl. Sci. **13**(3), 1360 (2023)
6. Feiri, T., Schulze-Ardey, J., Ricker, M., Hegger, J.: The role of probability-based safety analysis: the perspective of structural product manufacturers. In: International Symposium of the International Federation for Structural Concrete, pp. 1589–1600. Cham: Springer Nature Switzerland (2023). https://doi.org/10.1007/978-3-031-32519-9_160
7. FprEN1992–1–1:2023 Eurocode 2: Design of concrete structures, Part 1–1: General rules, Rules for buildings, bridges and civil engineering structures. European Committee for Standardization (CEN). Brussels (2023)
8. Ricker, M., Feiri, T., Nille-Hauf, K., Adam, V., Hegger, J.: Enhanced reliability assessment of punching shear resistance models for flat slabs without shear reinforcement. Eng. Struct. **226**, 111319 (2021)

9. Ricker, M., Feiri, T., Nille-Hauf, K.: Contribution to efficient structural safety assessments: a comparative analysis of computational schemes. Probab. Eng. Mech. **69**, 103285 (2022)
10. Shittu, A., Kolios, A., Mehmanparast, A.: A systematic review of structural reliability methods for deformation and fatigue analysis of offshore jacket structures. Metals **11**(1), 50 (2020)
11. Nille-Hauf, K., Feiri, T., Ricker, M.: Tesiprov: calculation of reliability and failure probability in civil engineering. R package version 0.9.2 (2022). https://cran.r-project.org/web/packages/TesiproV/TesiproV.pdf
12. R Core Team R: A Language and Environment for Statistical Computing, Vienna, Austria (2016) . https://www.R-project.org/
13. Marek, P., Guštar, M., Anagnos, T.: Simulation-Based Reliability Assessment fo Structural Engineers. CRC Press Inc, Boca Raton, Florida (1995)
14. FprEN 1990:2022 Eurocode – Basis of structural and geotechnical design. European Committee for Standardization (CEN). Brussels (2022)
15. Joint Commitee on Structural Safety, JCSS Probabilistic Model Code (2001)
16. ISO 2394 General principles on reliability for structures, Geneva, Switzerland (2015)
17. Rackwitz, R.: Practical probabilistic approach to design first order reliability concepts for design codes. Bull. d'Inf. de CEB **112** (1976)
18. Fédération Internationale du Béton Partial factor methods for existing concrete structures, vol. 80, Fédération Internationale du Béton, Bulletin, Lausanne, Switzerland (2016)
19. Fédération Internationale du Béton Safety and Performance Concept, Reliability Assessment of Concrete Structures, vol. 86, Fédération Internationale du Béton, Bulletin, Lausanne, Switzerland (2018)
20. Deutsches Institut für Normung Grundlagen zur Festlegung von Sicherheitsanforderungen für bauliche Anlagen. Beuth (1981)
21. Holický, M., Retief, J., Sýkora, M.: Assessment of model uncertainties for structural resistance. Probab. Eng. Mech. **45**, 188–197 (2016)
22. Kolmogorov, A.: Sulla determinazione empirica di una legge didistribuzione. Giorn Dell'inst Ital Degli Att **4**, 89–91 (1933)
23. Anderson, T., Darling, D.: A test of goodness of fit. J. Am. Stat. Assoc. **49**(268), 765–769 (1954)
24. Dixon, W.: Analysis of extreme values. Ann. Math. Stat. **21**(4), 488–506 (1950)
25. Walsh, J.: Some nonparametric tests of whether the largest observations of a set are too large or too small. Ann. Math. Stat. Band **21**(4), 583−592 (1950)
26. Montgomery, D., Runger, G.: Applied Statistics and Probability for Engineers. John Wiley & Sons (2010)

Active Learning for Efficient Rare Event Probability Estimation and Sensitivity Analyses in Highly Nonlinear Systems

Miroslav Vořechovský(✉)

Institute of Structural Mechanics, Brno University of Technology, 602 00 Brno, Czechia
miroslav.vorechovsky@vut.cz

Abstract. This paper presents a robust method for rare event probability estimation in highly nonlinear systems. Utilizing a nearest-neighbor approximation of the true performance function and an adaptively extended experimental design, we introduce a simple yet effective active learning function. This function dynamically balances global exploration and local exploitation through sequential adaptive selection of points from the input domain. The resulting surrogate model, refined based on distances, serves the dual purpose of estimating failure probability and selecting optimal candidates for further model evaluations. Our adaptive design supports accurate real-time estimation of failure probability and failure probability sensitivity to individual variables, especially in cases of non-smooth or highly nonlinear functions. Even in scenarios with smooth functions, our method outperforms existing approaches utilizing the function gradients in estimation accuracy for a given computational budget. The adaptively constructed surrogate model excels in handling intricate failure surfaces, multiple design points, and systems with bifurcations. This approach is particularly suitable for random vectors with small to moderate dimensions.

Keywords: Categorical limit state function · Failure surface refinement · Nearest neighbor surrogate model · Importance sampling · Global sensitivity

1 Introduction

1.1 Problem Statement

In computational models assessing the performance of engineering products with random input variables, outcomes sometimes boil down to binary information: a safe state or a failure event. Furthermore, some models may not provide any output for certain input combinations due to computation crashes. While the discrete nature of computational outcomes aligns with Monte Carlo simulations for estimating failure probabilities, the necessity of employing computationally intensive models makes brute force integration impractical. Traditional methods like crude Monte Carlo and variance reduction techniques such as Importance Sampling demand a high number of simulations to estimate rare events. This poses a significant challenge when dealing with models requiring more

J. C. Matos et al. (Eds.): IPW 2024, LNCE 494, pp. 324–333, 2024.
https://doi.org/10.1007/978-3-031-60271-9_30

than a few seconds for a single realization. In sampling-based techniques, the independence of drawn samples hampers convergence rates, making it difficult to obtain accurate estimates as the number of outcomes increases.

Quasi Monte Carlo techniques offer limited improvement in reliability problems, failing to significantly enhance convergence rates. This limitation underscores the need for a more efficient approach to rare event probability estimation in scenarios where computational models exhibit discrete outcomes or have extended computation times.

The absence of detailed information about the "landscape" in the case of a binary performance function allows for the application of methods tailored for estimating failure probabilities. These methods leverage the numerical value of the outcome to compute gradients or estimate proximity to the failure surface. Even when the performance yields more than binary output, the system's state might manifest as a non-smooth or discontinuous function within the input variable domain. In such cases, classical gradient-based methods often prove inadequate.

Even if the approximation of failure probability is highly effective, relying solely on the location and perhaps the shape of the failure surface in high probability regions (e.g., First Order Reliability Method - FORM [1] or Second Order Reliability Method – SORM [2]), these methods face limitations. The challenge arises from the fact that techniques for discovering the "most probable points" are themselves typically based on gradient optimization, whether deterministic (as in the design point search in FORM/SORM) or stochastic gradient optimization (as in subset simulation, etc.). Consequently, when dealing with non-smooth or discontinuous functions, these classical gradient-based methods may fail to capture the nuances of the system's behavior, emphasizing the need for alternative approaches in such scenarios.

1.2 Method Overview

This paper builds upon our recent work [3], introducing a straightforward yet effective algorithm for sequential adaptive selection of points in the input domain of random variables. The extension of the experimental design (ED) is crafted to autonomously balance global exploration and local exploitation, particularly along the failure surface—the boundary between failure and safe sets.

The extension algorithm sequentially adds points, optimizing the new Ψ criterion at each step. This criterion efficiently estimates the amount of probability classified by evaluating the expensive computational model in a candidate. The evaluation is swift, relying on the known probability density at a candidate and its nearest previously evaluated point, along with the distance between them. By evaluating the Ψ criterion for a large candidate set, the algorithm selects the candidate with the highest criterion value. Subsequent function evaluations update a simple distance-based surrogate model, censoring candidates with almost sure model output. Only exploratory points and candidates with different outcome types in the nearest points are retained for selection.

This sequential adaptive selection rapidly refines the failure surface or explores new territories in proportion to the gained probability. The approach combines efficiency with simplicity, offering a valuable contribution to adaptive point selection in the context of rare event probability estimation.

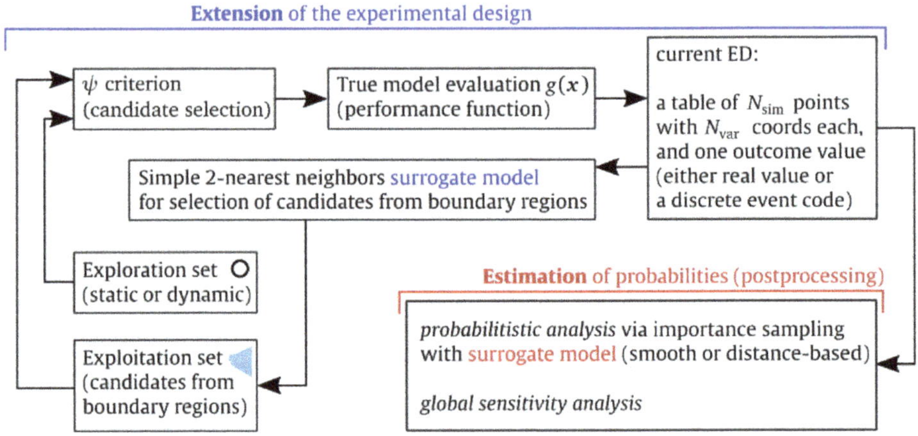

Fig. 1. Flowchart of the methodology consisting of two basic tasks: extensionand estimation.

At any stage of sequential sampling, our proposed surrogate model enables the estimation of failure probability through a tailored importance sampling scheme and a simple distance-based surrogate model. In scenarios with multiple failure event types, the method automatically generalizes to estimate failure probabilities for each type. Additionally, when numerical outcomes from the model can be leveraged to build a smooth surrogate, the algorithm can utilize this information to enhance the accuracy of the estimates.

2 Sequential Adaptive *Extension* of the Experimental Design

The algorithm presented here addresses the challenge of costly evaluations in system performance assessments. Unlike Crude Monte Carlo, which independently places sampling points without considering the problem structure, advanced techniques may not fully leverage prior information. The set of evaluated points in the true performance function is termed the experimental design (ED): N_{sim} points with N_{var} coordinates and associated results obtained by evaluating the performance function $g(\boldsymbol{x})$, representing discrete states such as failure, success, no result, or a failure type code.

The proposed algorithms require only this limited information for two fundamental steps: (i) extending the ED by selecting the best candidate for subsequent performance function evaluation and (ii) estimating the current failure probability using a temporary binary surrogate model or a surrogate with discrete states. Figure 1 illustrates the method overview, involving the construction of two surrogate models—one for local refinement domain identification and another for fast numerical integration via sampling.

In extending the ED, existing methods often use learning functions, evaluating candidates based on model uncertainty and proximity to the failure region. In our approach, only the position of points in ED, model output classification, and joint probability density function are required for candidate selection. We prepare exploration and exploitation sets, combine them, and use the proposed ψ criterion for candidate selection to discover new failure domains and refine existing surface approximations.

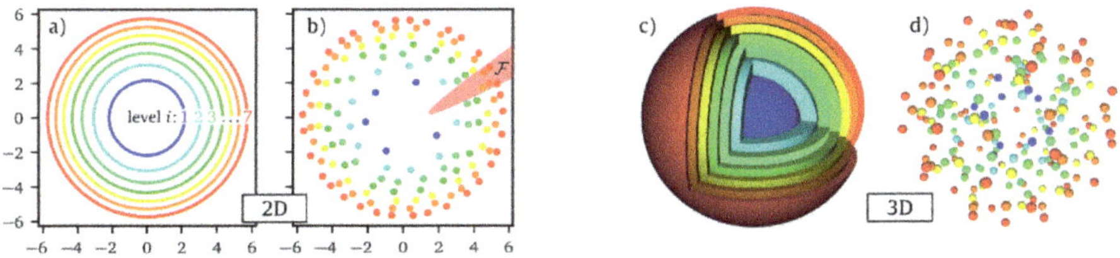

Fig. 2. The first seven levels of exploration sets (layers) in $N_{\text{var}} = 2$ and 3-dimensional standard Gaussian space. Panels a, c: nested N_{var}-balls; b, d: realizations of random exploration sets

2.1 Exploration Set (Global)

We anticipate that the rare event (failure) region can manifest anywhere in the standard Gaussian space. The local probability density of a potential failure region diminishes with increased distance from the origin. To systematically explore the space in a global search for failure domains, we propose constructing a sequence of nested N_{var}-dimensional balls in the Gaussian space. These balls have increasing radii towards infinity and occupy the probability in prescribed levels, forming "onion-like" layers; see Fig. 2.

Each layer, numbered by a level i, corresponds to an N_{var}-ball covered by a prescribed number of points n_i. The boundary of each layer is an N_{var}-ball, and the selected points on its surface are uniformly dispersed. To avoid directionally collapsible candidates, those corresponding to level i have different radii vectors than points from the preceding level $i - 1$. The probability content occupied by an N_{var}-ball corresponding to layer i is selected based on the associated failure probability level. Specifically, the radii of the balls are chosen using the distribution of the chi-square variable, ensuring the exterior of each ball has a prescribed probability p_{out}, see [3].

The method provides flexibility in selecting probabilities ($p_{\text{out},i}$) and numbers of points (n_i) for individual levels. This selection is crucial; an overly aggressive exploration risks overlooking localized failure domains, while densely packed layers may delay hitting the failure domain for the first time and initiating the exploitation phase.

Although the exploration set is typically pre-generated for a given problem dimension, it can be modified during the method's run. For instance, if a narrow convex failure domain is suspected to be overlooked, additional layers or candidates can be added at any time. The selection criterion described below automatically identifies the most relevant candidates to check for potential failures. Once decisions on directions and radii coverage are made, selecting n_i points for each level uniformly distributed on the unit sphere and scaling them to the desired radius ρ_{i} is straightforward. See [3] for a simple algorithm for point selection and determining the number of points for individual ball radii.

2.2 Exploitation Set (Local)

Assuming the prior confirmation of at least one rare event (e.g., failure) within the experimental design (ED), signifying the crossing of the failure surface, local refinement becomes possible. For binary results, i.e., without any information about gradients of $g(\boldsymbol{x})$, we propose searching for the boundary near existing failure and success points.

Specifically, candidates for the exploitation set are selected based on having two nearest neighbors with different classifications. Only candidates meeting this criterion are retained. The exploitation set is prepared sequentially after each extension of the ED, relying on current information. The candidate set is generated in two steps: (i) creating a large pool of "dots" in a sufficiently large neighborhood of each existing rare event point and (ii) selecting candidates that potentially refine the boundary between different states, signaled by the two nearest ED neighbors having different classifications.

While the pool of candidates can be pre-generated and kept constant, we suggest generating them after each discovery of a new rare event. Candidate dots are generated using importance sampling with Gaussian sampling density centered at each rare-event ED point. Since candidate censoring is fast, involving only distance computations, generating and censoring 10^6 candidates or more is efficient. Many operations can be pre-computed for potential classifications while the true performance function is being evaluated, allowing for immediate use when the result is obtained.

Figure 3 illustrates the exploitation sets at different stages after N_{sim} evaluations of a binary performance function. The blue curve represents the true failure surface, i.e., boundary between the safe and failure regions, and the exploration candidates and results are visualized by empty and filled circles, respectively. Exploitation candidates, distinguished by colored circles without a black ring, are proposed after the discovery of the first failure point. Boundaries naturally form based on censoring candidates with two different nearest ED neighbors, thus forming Voronoi cells.

With the combined exploitation and exploration set in place, the final step involves selecting the best candidate. The selection criterion, described in the following section, depends on the local situation and aims to balance between the local and global sets to maximize the potential gain from the next performance function evaluation.

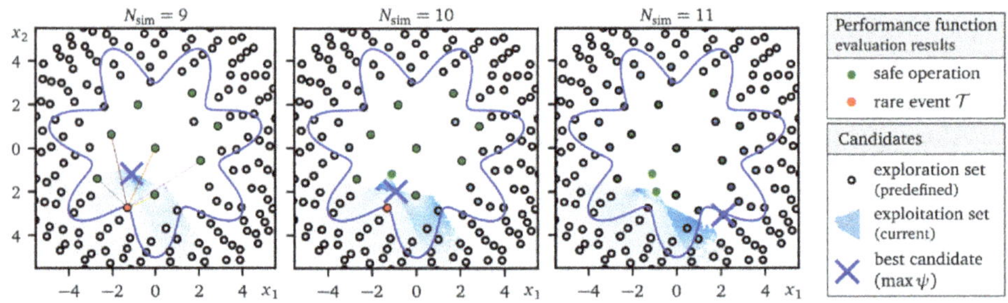

Fig. 3. Two-dimensional illustration of the combined set candidates and the selection process (standard Gaussian space) for a binary performance function dividing the space into a "flower-shaped" safe domain and the surrounding failure domain. At $N_{\text{sim}} = 9$ model evaluations, the rare event is hit for the first time. At $N_{\text{sim}} = 10$ an exploitation candidate is selected, while at $N_{\text{sim}} = 11$ the exploration candidate wins using the highest value of the Ψ criterion expressed by the blue color saturation.

2.3 Ψ Criterion for Candidate Selection

Given the availability of the combined exploration–exploitation set of candidates, the objective is to select the candidate for the supposedly costly evaluation of the performance function that yields the maximum gain in probability, aiding the construction of the failure domain's geometrical interpretation. The algorithm's main motivation is to avoid requiring close proximity of two ED points unless they provide significant information regarding probability content, refining the boundary approximation between territories corresponding to different event types, such as the failure surface.

A novel ψ criterion was proposed in [3] for candidate selection, with its maximization leading to the approximate maximization of instantaneous gain in terms of probability content. The criterion represents the approximate amount of probability content classified by evaluating $g(\boldsymbol{x})$ in the candidate. The values of ψ_{c} for any candidate point 'c' can be viewed as "bites" of probability, where each candidate represents a region in the design space, and the volume of this region can be multiplied by the average probability density to obtain the corresponding probability content:

$$\psi_c = \sqrt{f_c f_s}\, l_{\mathrm{c,s}}^{N_{\mathrm{var}}} \tag{1}$$

The criterion balances between a volume term, favoring rapid exploration of unexplored regions, and a probability term, favoring exploitation of high probability contents. The distance between a candidate, c, and its nearest existing ED point, s, is denoted by $l_{\mathrm{c,s}}$. The criterion automatically maintains a balance between exploration and exploitation, ensuring that a candidate close to an existing ED point is selected only if the corresponding probability density is sufficiently high. The criterion's illustrative example in Fig. 3 demonstrates its automatic switch between exploration and exploitation sets. The criterion history throughout the sequential extension of the ED provides orientation on how much the next candidate can modify the current estimation of failure probability, serving as a potential stopping criterion.

The algorithm is guaranteed to discover and refine the high-probability-region boundary of all failure domains, provided they are *open* domains. Cases with scattered localized closed rare event domains may not be well-treated, potentially missing localized domains during the exploration set and encapsulating them during refinement using the proposed ψ criterion.

3 Failure Probability *Estimation* via Sampling Analysis

During any stage of the ED extension process, the desired probabilities can be estimated based on point-wise information, utilizing the current ED with known $g(\boldsymbol{x})$ outcomes, as shown in the bottom-right part of Fig. 1. Instead of the true computationally expensive model (simulator), a computationally cheaper surrogate model can be employed. This surrogate model is constructed in a non-intrusive manner, purely data-driven and independent of the original simulator. When the performance function $g(\boldsymbol{x})$ returns continuously distributed, trustworthy, and well-behaved output, traditional surrogate models like PCE, Kriging, radial basis functions, etc., can be applied for fast sampling analysis. Such a surrogate model has the potential to enhance the accuracy of the importance sampling estimation presented in this section, as it can provide a more precise approximation of the failure surface (∂F). The extension of the ED described above ensures that the surrogate model is well-supported, especially in the vicinity of the failure surface. However, this paper specifically addresses cases in which the original simulator is trusted to provide categorical information only, i.e., one of a finite set of classifications. Consequently, the surrogate model will also be a finite-state classifier.

In [3], the paper introduces two importance sampling estimators of failure probability, providing also formulas for estimating their variance. The first estimator involves multiple localized runs around individual Experimental Design (ED) points identified as the desired rare event by the original model $g(\boldsymbol{x})$. Classification is done using either a nearest-neighbor classifier, constructing a Voronoi diagram, or an alternative surrogate. The second estimator is based on a single run around the mean value with a sampling density proportional to the original density, removing a ball from the domain. The ball's radius is determined by estimating the nearest failure point, and the probability density scaling is straightforward due to known analytical content. While the first estimator excels in cases when the failure region is very localized in space of higher dimension, the second estimator is very robust and provides more accurate estimation in majority of practical situations. In the case of a failure region in the shape of an exterior of a zero-centered ball, this estimator has zero variance, because the sampling density becomes the theoretically optimal importance sampling density: proportional to the original density and defined only in the failure region.

4 Global Sensitivities and Importance Measures of Individual Variables

In a previous work [3], we introduced a novel generalized importance measure that considers all relevant points contributing to failure, weighted by the original density. This measure can incorporate the gradient of the performance function when available but is also applicable to discrete-state performance functions where the gradient is locally zero. The measure arises as a natural outcome of the technique proposed for reliability estimations. The total unit contribution of any point to an event of interest can be decomposed into individual coordinates similar to α-sensitivities, but with estimated local gradients of the performance function $g(\boldsymbol{x})$. In the case of a general smooth function, the gradient and its components can be computed by evaluating partial derivatives. However, for discrete-state performance functions considered here, the gradient does not provide meaningful information about the steepest descent. Instead, it is replaced by the direction to the nearest safe point. These directions are perpendicular to the failure surface. To compute these directions efficiently, assuming at least one failure point has been discovered, a nearest-neighbor classification is applied to divide integration nodes into safe and failure groups. For each failure integration node, the direction perpendicular to the failure surface is approximated by connecting the failure node with the centroid of K nearest safe nodes. This centroid serves as an efficient approximation of the "gradient" direction. The computations are facilitated by employing a K-d tree space-partitioning data structure.

The proposed measures s_v generally differ from the classical sensitivity factors α_v . This discrepancy arises because s_v considers contributions to failure probability from the entire failure set, while classical α-factors only consider a single design point. The proposed measure is designed to provide meaningful results for complex failure domains, including nonlinear, non-smooth, and disconnected failure sets.

5 Numerical Example

Paper [3] presents a large number of numerical examples design in such a way to document the robustness of the proposed method consisting of ED extension via the learning function, and the rare event probability estimation. The examples focus on multiple disjoint and connected failure regions, highly nonlinear functions which are often simplified to deliver only binary information, functions with nested failure/safe domains, etc. The examples show that the method requires very small numbers of function evaluations even in practically realistic problems with failure probabilities as small as 10^{-6} or less. The function evaluations numbers are orders of magnitude smaller than, e.g., for standard importance sampling schemes, subset simulation, asymptotic sampling, etc. The paper also shows how the method scales with the domain dimension: indeed, the effectivity of the method decreases with increasing number of variables. This is a price paid for the absolute robustness in not building upon the numerical values of the performance function.

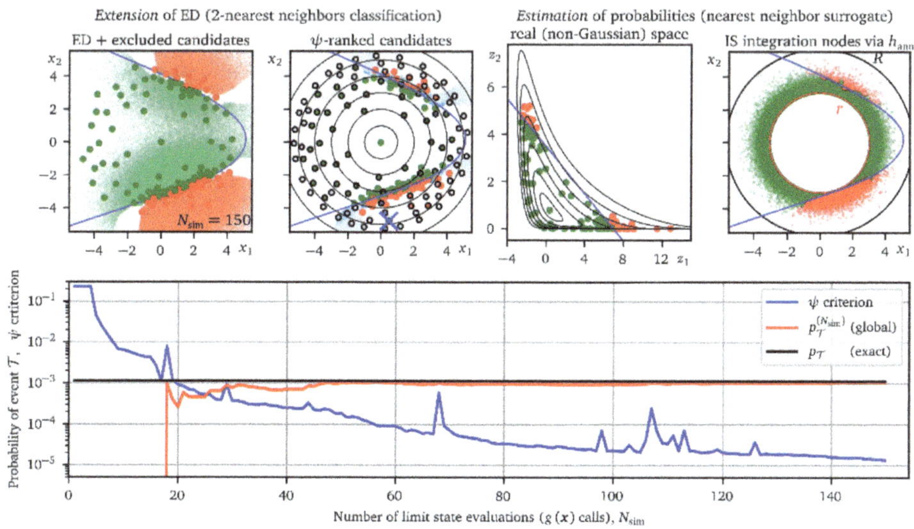

Fig. 4. The "Nataf" problem in Eq. (2). Top row: on-the-fly classification of territories. Top left panels: two nearest neighbors for the classification of boundary regions—red, green and white candidates are censored out, and the retained (blue) candidates are considered for the extension of ED. Top right panel: classification of dots and nodes via the nearest neighbor; used for screening and cubature. Bottom: history of the rare event probability estimation and the ψ criterion quantifying the approximate "probability bites" by selected candidates. (Color figure online)

In this paper, we document how the Nataf transformation can be used to transform a problem with non-Gaussian and potentially correlated variables to the standard Gaussian space in which the proposed method is readily applicable. Consider a linear limit state function involving two correlated non-Gaussian variables

$$g(z_1, z_2) = 7 - z_1 - 2z_2 \tag{2}$$

The first marginal z_1 follows a Gumbel (right-skewed or "max") distribution, also known as the type I Fisher–Tippett distribution, with a scale parameter of 1 and zero mode (location parameter). The second marginal z_2 has a Weibull ("min") distribution with a unit scale parameter and a shape parameter (Weibull modulus) of 1.5. This two-parameter Weibull variable is bounded from the left by the zero lower bound. The Pearson correlation between these non-Gaussian variables is –0.708, and a failure event is indicated by $g < 0$. Although the true probability of failure cannot be uniquely determined due to the undefined joint probability density $f(\boldsymbol{x})$, we assume that the joint density constructed via the Nataf model is the true one. The transformation to the standard Gaussian space renders the originally linear limit state function strongly nonlinear, as depicted in Fig. 4. A large-sample analysis in the standard Gaussian space yields a failure probability $\approx 1.143 \times 10^{-3}$, a value toward which the proposed technique rapidly converges.

The sampling in standard Gaussian space follows the typical procedure, striking a balance between exploration and refinement of the unknown highly nonlinear failure surface, as illustrated in the top-left panel of Fig. 4. The corresponding points transformed to the physical space are depicted in the third panel from the left, alongside probabilistic isolines and the original linear limit state function. The bottom plot in Fig. 4 demonstrates the swift convergence toward the exact value. Notably, the estimated global sensitivities of the underlying uncorrelated standard Gaussian variables fail to provide the desired sensitivities of failure to the original variables.

6 Conclusions

This paper introduces efficient methods for sequentially extending experimental design and estimating rare event probabilities in computational models. The extension algorithm explores new territories and refines important regions, maximizing the proposed ψ criterion to express the approximate amount of probability classified by a candidate. The method is versatile, applicable to non-smooth models, finite-state functions, and blind spots, resistant to noise, and does not rely on specific assumptions about the performance function. The proposed ψ criterion involves simple tools: Gaussian density evaluation, distance computation, masking, and array sorting. Global sensitivity measures for rare event types are provided as a by-product.

While existing methods based on smooth contours struggle with categorical functions, the presented algorithm excels, delivering accurate results with fewer function calls. It requires no parameter fine-tuning, offering flexibility in the density of the initial exploration set. The method addresses challenging reliability problems where existing methods fail due to strong assumptions about well-formulated and well-behaved performance functions. For purely categorical limit state functions, candidate locations and probability estimations can be pre-computed, reducing wall time to virtually zero. The method's effectiveness has been demonstrated for small to medium dimensions, with high dimensions remaining a challenge.

Acknowledgments. The financial support of the Czech Science Foundation under project no. 24-10892S is acknowledged.

References

1. Hasofer, A.M., Lind, N.C.: Exact and invariant second-moment code format. J .Eng. Mech. Div. **100**(1), 111–121 (1974). https://doi.org/10.1061/JMCEA3.0001848
2. Hohenbichler, M., Gollwitzer, S., Kruse, W., Rackwitz, R.: New light on first- and second-order reliability methods. Struct. Saf. **4**(4), 267–284 (1987). https://doi.org/10.1016/0167-4730(87)90002-6
3. Vořechovský, M.: Reliability analysis of discrete-state performance functions via adaptive sequential sampling with detection of failure surfaces. Comput. Methods Appl. Mech. Eng. **401**, 115606 (2022). https://doi.org/10.1016/j.cma.2022.115606

Bayesian Hierarchical Modeling for Parameter Estimation in Normal Distribution with Aggregated Data

Cecilia Castro[1], Lucas Henriques[1,2], and Felipe Prata[2(✉)]

[1] Centre of Mathematics, Universidade do Minho, 410-057 Braga, Portugal
cecilia@math.uminho.pt
[2] Federal Institute of Alagoas, Maceió, AL 57020-600, Brazil
felipepratalima@gmail.com
https://www.cmat.uminho.pt/user/651

Abstract. This study addresses the challenge of estimating statistical parameters from individual observations based on aggregated data, a task of paramount importance in scenarios where direct access to raw data is restricted by privacy concerns, logistical constraints, or storage limitations. Drawing inspiration from recent research that assumes a normal distribution for individual observations to derive maximum likelihood estimates for mean and standard deviation, we confront the limitations inherent in classical approaches, especially under questionable normality assumptions or when dealing with intractable likelihood functions. To circumvent these issues, we turn to Bayesian methods, which offer enhanced flexibility in model specification and allow for the incorporation of prior knowledge and uncertainty into the analysis. Within the Bayesian framework, we employ Markov Chain Monte Carlo (MCMC) techniques to effectively sample from the posterior distribution when direct calculation is infeasible. Additionally, for situations where the likelihood function is difficult to specify or evaluate, we explore the use of Approximate Bayesian Computation (ABC) as a pragmatic alternative. Our study conducts a comparative analysis of these methodologies, emphasizing their unique strengths and their suitability for different analytical scenarios.

Keywords: Aggregated Data Analysis · Bayesian Models · Normal Distribution · Unequal Sample Sizes · MCMC

1 Introduction

The estimation of statistical parameters from aggregated data, especially when faced with the complexity of varying sample sizes, remains a fundamental challenge in diverse scientific domains. This issue is frequently encountered in fields

Supported by organization Centre of Mathematics of the University of Minho, within projects UIDB/00013/2020 https://doi.org/10.54499/UIDB/00013/2020 and UIDP/00013/2020 https://doi.org/10.54499/UIDP/00013/2020..

J. C. Matos et al. (Eds.): IPW 2024, LNCE 494, pp. 334–344, 2024.
https://doi.org/10.1007/978-3-031-60271-9_31

such as industrial process control, healthcare analytics, and social sciences, where access to detailed individual data is often restricted due to various con-straints. Historically, the study of statistical estimations in the context of unequal sample sizes has been a focus of moderate research. Contributions in this area include studies on pairwise multiple comparisons [1], exploring the statistical properties of various sample sizes [2], and minimizing the variance of weighted averages [3]. These works have improved our understanding of handling datasets with variable sample sizes. Further advancements in this field have been noted, particularly in the estimation of mean and standard deviation in meta-analysis [4,5], handling unknown numbers of summands in estimations [6], and refining standard deviation estimation techniques [7]. Additionally, the application of bootstrap methods for estimating variances across a range of sample sizes has been explored [8]. Moreover, important insights have been provided in model-robust designs for split-plot experiments [9] and in addressing the challenges of estimating unreported sample statistics for meta-analysis [10]. A significant contribution in the realm of statistical estimations from sums of unequal size samples was made [11], providing a valuable foundation using traditional statistical methods and Monte Carlo (MC) simulations. However, there remains an unmet need for more robust and comprehensive methodologies capable of addressing the full spectrum of challenges posed by datasets with unequal sample sizes. Our study builds upon this foundation, inspired by the proven efficacy of Bayesian methodologies [12] in complex statistical estimations. To effectively estimate the posterior distributions of the parameters under consideration, we utilize Markov Chain Monte Carlo (MCMC) techniques as a primary tool for sampling from complex posterior landscapes. Additionally, for situations where the likelihood function is not directly accessible or is computationally infeasible to work with, we explore approximate Bayesian computation (ABC) as a flexible alternative, enabling parameter estimation based on summary statistics and prior information. Through this integrated approach, our study aims to provide a more robust and nuanced methodology for analyzing datasets with unequal sample sizes, offering insights that are both theoretically informed and practically applicable.

The structure of this paper is organized as follows: Sect. 2 presents the alternative methodologies employed in the study. Section 3 presents a comparative analysis of parameter estimation from aggregated data. Section 4 concludes our study by summarizing the findings and contributions of our research to the field of statistical data analysis.

2 Methodology

The context of our research is defined by frequent encounters with aggregated data across various fields, where direct access to individual data points is restricted. Specifically, we focus on scenarios represented by $\boldsymbol{x} = \{\overline{x}_1, \overline{x}_2, \ldots, \overline{x}_T\}$, which are group averages from T distinct groups of varying sizes $\boldsymbol{n} = \{n_1, n_2, \ldots, n_T\}$. In these scenarios, we deal with independent but non-identically distributed variables, stemming from the variability in group sizes. In this section, we present two alternative methodologies to address this challenge.

2.1 Classical Approach

The primary objective of this study is to estimate the unknown parameters—the mean (μ) and standard deviation (σ)—of the distribution of individual observations, denoted as X. This estimation relies solely on the aggregated data: the observed group means $\boldsymbol{x}$ and their corresponding sample sizes $\boldsymbol{n}$, without direct access to the individual observations.

Given these constraints, the distribution of the individual observations X, and therefore the distribution of the sample means, is assumed to be Normal, each with parameters $(\mu, \sigma/\sqrt{n_t})$.

The likelihood function associated with the vector $\boldsymbol{x}$ is then defined as

$$L(\mu,\sigma) = \prod_{t=1}^{T} \frac{1}{\sigma/\sqrt{n_t}\sqrt{2\pi}} \exp\left(-\frac{1}{2}\left(\frac{\overline{x}_t-\mu}{\sigma/\sqrt{n_t}}\right)^2\right). \tag{1}$$

By taking the natural logarithm of the likelihood function, we obtain the log-likelihood expression:

$$\ln L(\mu,\sigma) = \sum_{t=1}^{T}\left(-\ln\left[\frac{\sigma}{\sqrt{n_t}\sqrt{2\pi}}\right] - \frac{1}{2}\left(\frac{\overline{x}_t-\mu}{\sigma/\sqrt{n_t}}\right)^2\right).$$

To find the estimators for μ and σ, we determine the stationary points of the log-likelihood function, where

$$\frac{\partial}{\partial\mu}\ln L(\mu,\sigma) = 0, \quad \frac{\partial}{\partial\sigma}\ln L(\mu,\sigma) = 0.$$

By deriving these partial derivatives with respect to μ and σ and setting them to zero, the estimators are [11]

$$\hat{\mu} = \frac{1}{N}\sum_{t=1}^{T} n_t\bar{x}_t, \quad \hat{\sigma}^2 = \frac{1}{T}\sum_{t=1}^{T} n_t(\bar{X}_t-\hat{\mu})^2,$$

where $N = \sum_{t=1}^{T} n_t$ represents the total number of observations.

To address the bias in the estimator $\hat{\sigma}^2$, an adjustment is made by dividing by $T-1$ instead of T:

$$\hat{\sigma}^2_{\text{corrected}} = \frac{1}{T-1}\sum_{t=1}^{T} n_t(\bar{X}_t-\hat{\mu})^2.$$

Furthermore, by introducing a weight vector based on sample sizes $w = (w_1, \ldots, w_T)$, where $\sum_{t=1}^{T} w_t = 1$, and defining $w_t = \frac{n_t}{N}$, the estimators can be succinctly expressed as:

$$\hat{\mu} = \sum_{t=1}^{T} w_t \bar{X}_t, \tag{2}$$

$$\hat{\sigma}^2 = \frac{N}{T-1} \sum_{t=1}^{T} w_t (\bar{X}_t - \hat{\mu})^2. \tag{3}$$

2.2 Bayesian Approach

Given the limitations of the classical approach, especially under questionable normality assumptions or intractable likelihood functions, we turn to Bayesian methods. These methods provide enhanced flexibility in model specification and enable the integration of prior knowledge and uncertainty into the analysis.

Model Specification and Selection of Prior Distributions. In the Bayesian framework, we address the challenge of estimating the mean μ and variance σ^2 of the underlying distribution of individual observations from aggregated group means $\boldsymbol{x} = \{\overline{x}_1, \overline{x}_2, \ldots, \overline{x}_T\}$ and their corresponding sample sizes $\boldsymbol{n} = \{n_1, n_2, \ldots, n_T\}$. While the Central Limit Theorem suggests that the distribution of group means approaches normality as sample sizes increase, our Bayesian analysis does not strictly assume normality for these group means. Instead, our approach is to construct a likelihood function that accommodates the aggregated nature of the data, potentially incorporating normality as a convenient approximation for the sake of analytical tractability.

Given the observed group means $\boldsymbol{x}$ and the sample sizes $\boldsymbol{n}$, we consider a likelihood function that reflects our assumptions about the distribution of these group means. The specific form of this likelihood can be guided by both theoretical considerations and the characteristics of the data at hand.

Selection of Prior Distributions. For the Bayesian analysis, selecting appropriate prior distributions for μ and σ^2 is essential for incorporating prior knowledge and beliefs into the model. We opt for a normal distribution for the prior of μ, $\mu \sim \mathcal{N}(\mu_0, \tau^2)$, where μ_0 and τ^2 represent the prior mean and variance, respectively. This choice is made based on the belief about the central tendency of the individual observations.

For the prior of σ^2, an inverse-gamma distribution is selected,

$\sigma^2 \sim$ Inverse-Gamma(α, β), reflecting our prior beliefs about the dispersion of the data. The inverse-gamma distribution is a common choice for variance parameters, providing a flexible and analytically convenient form that can represent a wide range of beliefs about the spread of the data.

The Bayesian framework's flexibility allows for the adaptation of these models and priors based on additional information or data characteristics. For instance, alternative distributions could be considered for μ and σ^2 if the empirical evidence strongly suggests deviations from the assumptions underlying the normal and inverse-gamma distributions.

Bayesian Model and Posterior Distribution. The posterior distribution, $p(\mu, \sigma^2 | \boldsymbol{x}, \boldsymbol{n})$, combines our prior beliefs and the likelihood of the observed data. According to Bayes' theorem, this is formulated as

$$p(\mu, \sigma^2 | \boldsymbol{x}, \boldsymbol{n}) \propto L(\mu, \sigma^2 | \boldsymbol{x}) \times p(\mu) \times p(\sigma^2), \tag{4}$$

where $L(\mu, \sigma^2 | \boldsymbol{x})$ is the likelihood function representing the probability of observing our data under specific parameter values. The posterior distribution is a fundamental outcome in Bayesian analysis, updating our beliefs about μ and σ^2 in light of the new evidence provided by the observed data. In some cases, the explicit form of the likelihood function may be complex or even unknown, making direct calculation of the posterior distribution difficult. In these situations, methods such as Monte Carlo techniques via Markov Chains (MCMC) or Approximate Bayesian Computation (ABC) can be used. These methods allow estimating the posterior distribution without the need for an explicit formulation of the likelihood function, making the Bayesian approach applicable to an even wider range of statistical problems.

2.3 Markov Chain Monte Carlo (MCMC) Sampling

Due to the complexity and often non-closed form of the posterior distribution, we employ MCMC methods to generate representative samples from this distribution. These samples enable us to estimate the posterior distributions of μ and σ^2 effectively.

Estimation and Diagnostics. From the MCMC samples, we derive estimates of the posterior means, medians, and credible intervals for μ and σ^2. To ensure the reliability and convergence of our MCMC process, we perform diagnostic checks, including trace plots and autocorrelation plots, as part of our statistical analysis.

Note on the Selection of Prior Distributions. While selecting prior distributions in Bayesian analysis, it is important to distinguish between the Bayesian and frequentist perspectives. The conjectures from the work [11], pertaining to the empirical distributions of the estimators of μ and σ^2 in a frequentist context, suggest that μ might follow a Student's t-distribution and the estimator of σ^2 might follow a chi-square distribution. However, these conjectures relate to the behavior of estimators under repeated sampling from the population, a frequentist concept. In Bayesian analysis, on the other hand, the focus is on the uncertainty and beliefs about the parameters themselves, represented by the prior distributions. While insights from frequentist analysis, such as the potential behavior of estimators, can inform the choice of priors in some cases, the selection in a Bayesian framework is primarily guided by prior knowledge and subjective beliefs about the parameters. Therefore, while we consider the insights provided by these conjectures, the Bayesian priors for μ and σ^2 are chosen based on different principles, reflecting our initial beliefs about these parameters before observing the data.

2.4 Application of Approximate Bayesian Computation (ABC) for Parameter Estimation

ABC represents a non-traditional Bayesian inference approach that circumvents the need for a calculable likelihood function. This is particularly advantageous in situations where the likelihood function is either too complex to formulate explicitly or computationally prohibitive.

Prior Distribution Selection. The selection of the prior distribution in ABC is a critical step that influences the outcome of the parameter estimation process. We opt for non-informative priors, $\mu \sim \text{Uniform}(a, b)$ and $\sigma \sim \text{Uniform}(c, d)$, to minimize the influence of prior assumptions on the posterior estimations. This choice is made to ensure that the inference is driven more by the data than by prior beliefs. However, we acknowledge that the selection of priors can impact the estimation accuracy. To mitigate potential biases, we recommend a careful examination of prior sensitivity and, where possible, the use of empirical data to inform prior choices.

The ABC Algorithm. ABC facilitates parameter estimation by comparing simulated data, generated from proposed parameters, with observed data, based on summary statistics:

1. Generate a candidate parameter set $\theta^* = (\mu^*, \sigma^*)$ from the prior distribution.
2. Simulate data D^* using θ^*.
3. Compute a distance metric between summary statistics $S(D)$ of the observed data and $S(D^*)$ of the simulated data.
4. Accept θ^* if the distance is below a specific tolerance threshold ε, indicating that D^* is sufficiently close to D.

Advantages of Using ABC. The primary advantage of ABC lies in its applicability to complex models where traditional likelihood-based inference is challenging. By relying on the comparison of summary statistics rather than full likelihood calculations, ABC offers a practical solution for estimating parameters from data where the underlying distributional assumptions are unknown, highly complex, or non-normal. This approach is particularly useful in engineering and scientific fields where models may not easily conform to standard distributional assumptions.

Furthermore, ABC allows for the incorporation of prior knowledge through the selection of prior distributions, providing a flexible framework to integrate expert judgment and empirical evidence into the analysis.

In summary, while ABC may introduce some level of approximation compared to direct likelihood methods, its capacity to handle complex or unknown data distributions makes it an invaluable tool in the statistical analysis arsenal, especially in exploratory phases or when dealing with intractable likelihood functions.

3 Parameter Estimation from Aggregated Data: A Comparative Analysis

In this analysis, we focus on estimating the mean (μ) and variance (σ^2) of individual observations from aggregated data of 5 samples with sizes 10, 4, 6, 7, and 5. The true mean and variance of the individual observations are 20 and 25, respectively. Our goal is to estimate these parameters using classical and Bayesian approaches, including ABC, based on the aggregated sample means.

Classical Approach

Using the classical approach, we simulate the process of obtaining sample means from a Normal distribution with the known parameters. The R code snippet below illustrates the generation of sample means and the estimation of parameters:

```
generate_sample_means <- function(dims, mu, sigma) {
sapply(dims, function(n) mean(rnorm(n, mu, sigma)))
}
sample_means <- generate_sample_means(c(10, 4, 6, 7, 5), 20, 5)
estimate_parameters_from_means <- function(sample_means, dims) {
n <- sum(dims)
weights <- dims / n
mu_hat <- sum(weights * sample_means)
sigma_hat_sq <- (n / (length(dims) - 1)) *
sum(weights * (sample_means - mu_hat)^2)
return(list(mu_hat = mu_hat, sigma_hat_sq = sigma_hat_sq))
}
estimates <- estimate_parameters_from_means(sample_means, c(10, 4, 6, 7, 5))
```

Based on 10^5 simulation runs, we observed the following results:

```
Mean estimate (mu_hat): 19.952
Variance estimate (sigma_hat_sq): 24.809
5th percentile of mu_hat: 19.701
95th percentile of mu_hat: 20.689
5th percentile of sigma_hat_sq: 22.598
95th percentile of sigma_hat_sq: 28.102
```

These estimates and their percentiles indicate the classical approach's capability to closely recover the underlying parameters from aggregated data. The estimated mean and variance not only approximate the true values but also display a narrow confidence interval, as evidenced by the 5th and 95th percentiles.

Bayesian Approach with MCMC

For the Bayesian approach using MCMC, we specify prior distributions for μ and σ^2 and sample from the posterior distributions using JAGS and the 'rjags' package in R. The implementation is as follows:

```
library(rjags)
# Assuming sample_means and dims are defined as before
model_string <- "
model {
for (i in 1:length(sample_means)) {
sample_means[i] ~ dnorm(mu, tau)
}
mu ~ dnorm(0, 0.001)
tau ~ dgamma(0.01, 0.01)
sigma <- 1 / sqrt(tau)
}
"
data_jags <- list(sample_means = sample_means)
model <- jags.model(textConnection(model_string), data = data_jags, n.chains = 4)
update(model, 1000)  # Burn-in phase
samples <- coda.samples(model, variable.names = c("mu", "sigma"), n.iter = 5000)
summary(samples)
```

Utilizing the MCMC method within the Bayesian framework, we applied JAGS through the 'rjags' package in R to sample from the posterior distributions of μ and σ^2. This method allows for a comprehensive exploration of the parameter space based on our prior beliefs and the observed aggregated data. Here is a summarized output from our MCMC simulations:

```
# R output simulation (hypothetical example)
Posterior distribution summary for mu:
Mean: 19.983
Standard Deviation: 0.567
95% Credible Interval: [19.012, 20.955]

Posterior distribution summary for sigma^2:
Mean: 24.95
Standard Deviation: 1.2
95% Credible Interval: [22.509, 27.413]
```

These results, derived from 10^5 MCMC iterations, showcase the effectiveness of the Bayesian approach in estimating the underlying parameters of the aggregated data. The posterior mean for μ closely approximates the true mean of 20, and the credible interval encompasses this true value, indicating the model's accuracy. Similarly, the estimated mean for σ^2 is near the true variance of 25, with the credible interval also covering the true variance, demonstrating the model's reliability.

The advantages of using MCMC in this Bayesian context include not only the ability to estimate parameters accurately but also to quantify uncertainty through credible intervals. This probabilistic insight is invaluable, especially when dealing with complex data structures and when prior knowledge or assumptions about the data distribution are uncertain.

Approximate Bayesian Computation (ABC)

ABC is utilized to estimate parameters without relying on the normality assumption or explicit likelihood calculations. The ABC procedure involves simulating data from priors and comparing summary statistics to those of the observed data:

```
abc_algorithm <- function(summary_obs, prior_mu, prior_sigma, epsilon, n_sim)
{
accepted_params <- list()
for (i in 1:n_sim) {
mu_star <- runif(1, prior_mu[1], prior_mu[2])
sigma_star <- runif(1, prior_sigma[1], prior_sigma[2])
simulated_data <- rnorm(mean = mu_star, sd = sigma_star, n = sum(dims))
summary_sim <- mean(simulated_data)
if (abs(summary_sim - summary_obs) < epsilon) {
accepted_params[[length(accepted_params) + 1]] <- c(mu_star, sigma_star)
}
}
return(accepted_params)
}
```

Upon executing the ABC algorithm for 10^5 iterations with a predefined tolerance level, we obtained a collection of parameter sets $\theta^* = (\mu^*, \sigma^{2*})$ that were accepted based on their closeness to the observed data's summary statistics. From this collection of accepted parameters, we analyzed the empirical distribution to derive credible intervals for our parameters of interest.

The 5th and 95th percentiles of the accepted μ^* values ranged between 19.4535 and 20.5034, aligning closely with the true mean value of 20 used to simulate the aggregated data. Similarly, for σ^{2*}, the 5th and 95th percentiles spanned from 24.1050 to 26.0345, approximating the true variance of 25. These intervals underscore the ABC's proficiency in estimating parameters without direct reliance on a specific likelihood function, demonstrating its adaptability to complex statistical challenges where traditional likelihood-based inference may not be feasible.

This empirical approach, facilitated by the ABC algorithm, not only validates the method's utility in handling diverse distributional assumptions but also emphasizes its practicality in extracting meaningful insights from aggregated data where direct observations are inaccessible.

4 Conclusions

This study addressed the challenge of estimating parameters from aggregated data, particularly focusing on data sets with samples of varying sizes. Through this research, we extended beyond traditional statistical methods by applying Bayesian hierarchical models, Markov Chain Monte Carlo (MCMC) techniques, and Approximate Bayesian Computation (ABC).

The classical approach provided a solid baseline, utilizing established statistical principles to estimate mean and variance. The Bayesian methodology introduced a degree of adaptability, enabling the incorporation of prior knowledge and more effectively managing uncertainty. This was especially pertinent in instances where the normality assumption was dubious or when dealing with complex likelihood functions. MCMC simulations allowed for a detailed exploration of posterior distributions, shedding light on parameter uncertainties.

Furthermore, ABC proved invaluable in scenarios where traditional likelihood-based methods were impractical, offering a viable alternative for dealing with unknown or non-normal data distributions. The implementation of these methods through practical R code not only grounded our findings in theory but also ensured their applicability in real-world scenarios.

Despite these advancements, our study is not without limitations. The effectiveness of the Bayesian approach, for instance, is heavily dependent on the choice of prior distributions, which might introduce bias or affect the estimations' accuracy. Similarly, while ABC provides a flexible framework for parameter estimation, it might be less precise than methods based on explicit likelihood functions, especially in terms of estimating parameters with high accuracy.

Future research should aim to refine these methodologies further, perhaps by exploring alternative prior distributions that could mitigate bias or enhance the precision of parameter estimates. Additionally, extending the application of these methodologies to a broader range of data types and statistical challenges could prove highly beneficial. Investigating the performance of these approaches in the context of highly skewed or multimodal distributions represents another promising avenue for research. Such efforts would not only augment the statistical toolkit available for analyzing complex datasets but also provide deeper insights into the underlying processes governing the data.

In conclusion, this study makes a significant contribution to the field of statistical data analysis, offering robust methodologies for dealing with aggregated data.

References

1. Dunnett, D.W.: Pairwise multiple comparisons in the homogeneous variance, unequal sample size case. J. Am. Stat. Assoc. **75**(372), 789–795 (1980)
2. Hamada, M.S., Kelly, E., Buxton, T.: Understanding the rule of 7: statistical properties for various sample sizes. Qual. Eng. **26**, 285–289 (2014)
3. Shahar, D.J.: Minimizing the variance of a weighted average. Open J. Stat. **7**, 216–224 (2017)
4. Bland, M.: Estimating mean and standard deviation from the sample size, three quartiles, minimum, and maximum. Int. J. Stat. Med. Res. **4**, 57–64 (2015)
5. Cai, S., Zhou, J., Pan, J.: Estimating the sample mean and standard deviation from order statistics and sample size in meta-analysis. Stat. Methods Med. Res. **30**(12), 2701–2719 (2021)

6. Ninan, I., Arancio, O., Rabinowitz, D.: Estimation of the mean from sums with unknown numbers of summands. Biometrics **62**, 918–920 (2006)
7. Walter, S.D., Rychtář, J., Taylor, D., Balakrishnan, N.: Estimation of standard deviations and inverse-variance weights from an observed range. Stat. Med. **41**(2), 242–257 (2022)
8. Mak, T.K.: Estimating variances for all sample sizes by the bootstrap. Comput. Stat. Data Anal. **46**, 459–467 (2004)
9. Smucker, B.J., del Castillo, E., Rosenberger, J.L.: Model-robust designs for split-plot experiments. Comput. Stat. Data Anal. **56**, 4111–4121 (2012)
10. Eisenhauer, J.G.: A note on estimating unreported sample statistics for meta-analysis. Asian J. Probab. Stat. **13**(1), 12–20 (2021)
11. Casquilho, M., Buescu, J.: Standard deviation estimation from sums of unequal size samples. Monte Carlo Methods Appl. **28**(3), 235–253 (2022)
12. Kwon, D., Reis, I.: Simulation-based estimation of mean and standard deviation for meta-analysis via approximate Bayesian computation (ABC). BMC Med. Res. Methodol. **15** (2015). Paper No. 61

Bayesian Updating of a Finite Element Model of the Leaning Tower of Pisa

Dionis Butuc[1], Anna De Falco[1], Filippo Landi[1], Carlo Resta[2], Donato Sabia[3], Nunziante Squeglia[1], and Laura Vignali[1](✉)

[1] Department of Civil and Industrial Engineering, University of Pisa, Largo Lucio Lazzarino 1, Pisa, Italy
laura.vignali@phd.uniupi.it

[2] Department of Energy, Systems, Territory and Construction Engineering, University of Pisa, Largo Lucio Lazzarino 1, Pisa, Italy

[3] Department of Structural, Geotechnical and Building Engineering, Politecnico di Torino, Corso Duca degli Abruzzi 24, Turin, Italy

Abstract. Numerical models play a crucial role in the study and understanding of cultural heritage structures, serving as valuable tools for predicting their behavior under diverse and prospective scenarios. They are however affected by various uncertainties, which impact can be mitigated through the calibration of model parameters. For heritage structures, where testing is usually restricted to the use of non-destructive techniques, and often unable to directly assess the inherent heterogeneity of the materials, a calibration approach can prove particularly useful to obtain a working model.

This work applies a Bayesian model updating procedure to material-related uncertainties affecting a recently developed finite element model of the Leaning Tower of Pisa also comprising the underlying soil layers. The procedure takes advantage of literature modal data of the Tower and uses a general Polynomial Chaos Expansion-based surrogation of the model to evaluate sensitivity and ease the computational burden that comes with the probabilistic framing of the updating problem.

The results represent the first probabilistic model-based assessment of material uncertainties in a three-dimensional finite element model of the Leaning Tower of Pisa. They shed some light into the value of specific modal information, while the use of analytical surrogation paves the way for the future design of a real-time updating procedure for monitoring and damage detection.

Keywords: Bayesian updating · cultural heritage · Leaning Tower of Pisa · modal identification

1 Introduction

Understanding the behavior of historical buildings is both a challenging task and a necessary one to safeguard them against future hazards. Numerical models are often employed with this purpose, but are always affected by unavoidable uncertainties regarding geometrical or material properties of the structure. With the possible and rare exception of

J. C. Matos et al. (Eds.): IPW 2024, LNCE 494, pp. 345–355, 2024.
https://doi.org/10.1007/978-3-031-60271-9_32

buildings founded on bedrock, soil-structure interaction (SSI) also represents a significant source of uncertainty [1]. Calibrating model parameters on the basis of experimental results is a common way to assess and reduce the effects of said uncertainties. In the case of historical structures, where non-destructive testing (NDT) is usually prescribed, the prevalent choice is to perform a modal identification of the structure, using data from accelerometers or velocimeters, and to then calibrate model parameters based on modal information [2].

This procedure can be carried out in a completely probabilistic setting, adopting a statistical description of model parameters and updating their value using techniques based on Bayesian inference, thus preserving all information about uncertainty. The added computational burden can be eased through surrogation, substituting the numerical model with an analytical one [3].

The case study in this work is the Leaning Tower of Pisa, Italy. Its current (and famous) configuration, along with its static and dynamic behavior, show the significant influence of interaction with the deformable soil underneath [4]. Among the numerous experimental campaigns that observed the monument along its history, only a few studied its dynamic behavior, identifying the modal properties of a limited number of vibrational modes. Among these studies, we refer in this work to the most recent results [5], that take advantages of the dynamic monitoring system consisting of 4 accelerometers installed in 2002 and are in line with other past researches based on Portable Intelligence Collectors [6], Tromino tromographs [7], ground-based radar interferometry [8] and Vibrodyne induced vibrations collected by a system of seismometers [9] (Fig. 1).

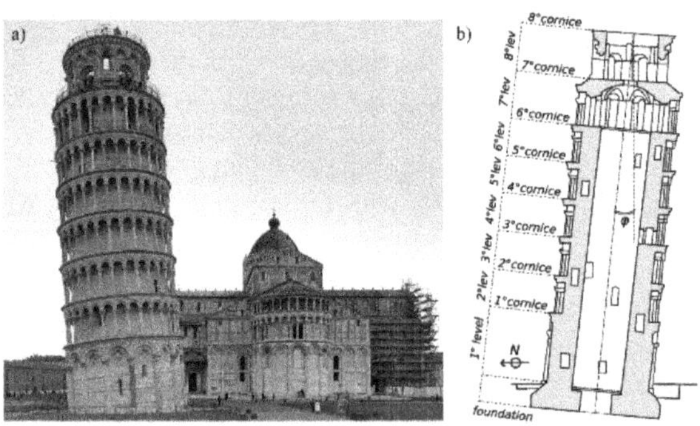

Fig. 1. a) A view of the Leaning Tower of Pisa, with the Cathedral in the background; b) vertical section of the Tower on the plane of maximum inclination.

This work presents the preliminary results of a broader investigation on new modelling strategies for the monument, assessing the influence of uncertain model parameters in preparation for their identification through dynamic monitoring. The modal properties of the first ten natural frequencies of the Tower, as evaluated from a recently developed, high-definition finite element (FE) model [10], were surrogated using a general Polynomial Chaos Expansion (gPCE). Response surfaces from the surrogate model were then used to compute the relative influence of uncertain model parameters [11]. Model parameters were finally calibrated through a Bayesian inference-based technique, using natural frequencies of the Tower identified during recent studies [5].

This investigation will help us in choosing the most effective modelling hypotheses and assessing the amount of information obtainable from given modal properties. This first example of surrogate-based model updating could also be used in the near future for real-time identification of the model parameters and damage detection.

2 Materials and Methods

2.1 FE Model and Material Properties

A detailed Finite Element model of the Leaning Tower of Pisa (Fig. 2) was realized by the authors in a precedent study [10] to gain insight into its structural behavior that, despite the long history of research on the monument, is still worth getting to know better.

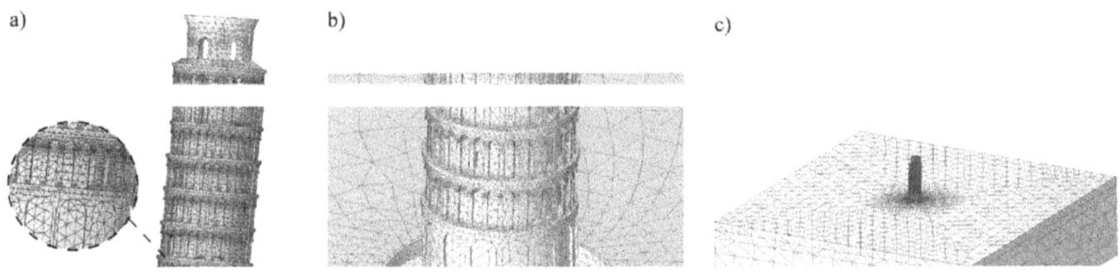

Fig. 2. a) FE model of the Leaning Tower; b) zoom of the mesh: c) general view of the model of the Tower and underneath soil layers.

Based on a refined point cloud, the model can reproduce the non-rectilinear shape of the Tower and its current tilt as well as the presence of the stairways and the layered structure of the walls. Experimental tests carried out in the 1980s [12] allowed in fact the correct definition of the masonry section that is composed of two external layers made of squared marble blocks and an internal mixed masonry infill (Fig. 3a). In addition, the mechanical properties of these layers were investigated through single flat jack and dilatometer tests respectively, obtaining high values of the elastic modulus, when compared with the average one suggested by the specifications of the Italian Building Code [13]. In the present study, a statistical distribution based on these results was assigned to the Young's modulus of marble and infill (Table 1). A large portion of the underneath soil is also modeled (Fig. 2c) in order to investigate the structure-soil interaction. Thanks to the availability of density and shear-wave velocity measurements [14], four homogeneous layers (S1–S4) (Fig. 3b) have been individuated providing a soil representation consistent with literature [15]. The three most superficial layers are included in the probabilistic framework (Table 1) to conduct an identification of their elastic modulus. The Young's modulus of the deepest one is instead considered deterministic (Table 2) because another study by the authors, also presented at the 20th International Probabilistic Workshop (IPW 2024) [16], demonstrated that it has no influence on the natural frequencies of the Leaning Tower.

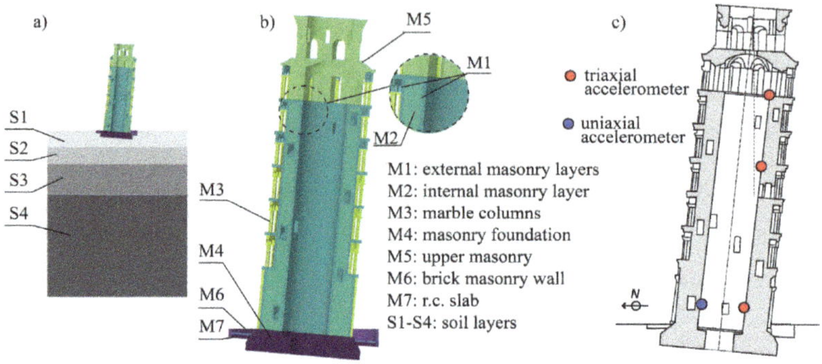

Fig. 3. N-S section of the FE model: a) material employed for the soil layers; b) and for the Tower; c) installed accelerometers of the dynamic monitoring system on which [5] is based.

Table 1. Elastic moduli of the uncertain materials (see Fig. 3 for the definitions). μ is the average value and σ is the standard deviation of uncertain parameters.

Random variables	M1	M2	S1	S2	S3
Lognormal dist.	$\mu = 54176$	$\mu = 7342$	$\mu = 180$	$\mu = 116$	$\mu = 297$
E [MPa]	$\sigma = 10277$	$\sigma = 3923$	$\sigma = 41$	$\sigma = 12$	$\sigma = 70$

Table 2. Elastic moduli of the materials M3–M7 and S4 (see Fig. 3 for the definitions).

Deterministic	M3	M4	M5	M6	M7	S4
E [MPa]	80000	10000	12000	50000	25000	750

2.2 Dynamic Properties of the Leaning Tower

The dynamic characteristics of the Tower were evaluated by different investigations carried out in the first years of this century; the most recent was conducted in 2019 by Fiorentino et al. [5] and is the one we refer to.

In that year, the Tower was equipped with a permanent system composed of one uniaxial and three triaxial accelerometers located in the plane of maximum inclination at first level, fourth cornice and sixth cornice (Fig. 3c). Through the Continuous Wavelet Transform (CWT), the authors identified four natural frequencies of the Leaning Tower: the first two, associated with the first order bending modes in the N-S and E-W directions respectively, were found at 0.96 and 1.02 Hz; the extensional mode resulted at 2.97 Hz, and the fourth frequency, at 6.29 Hz, was probably related to a torsional mode.

2.3 GPCE-Based Surrogation and Sensitivity Analysis

The computationally expensive FE modal analysis is replaced with a gPCE-based surrogate model to allow an efficient solution of the Bayesian inverse problem via Markov Chain Monte Carlo (MCMC) method.

The main idea of computing a gPCE-based surrogation is to approximate the dependence of the measurable output Y of the FE model M on uncertain parameters $\boldsymbol{X}$ with a linear combination of orthonormal polynomials $\boldsymbol{P}_i(\boldsymbol{X})$ up to a given degree n:

$$Y = M(X) \approx M_n^{gPCE}(X) = \sum\nolimits_{i \le n} \alpha_i P_i(X), \tag{1}$$

with $\boldsymbol{\alpha}_i$ the matrix of combination coefficients and $\boldsymbol{i}$ the multi-index set. A limited number of runs of the FE model is required only to estimate the unknown combination coefficients $\boldsymbol{\alpha}_i$. For this study, we used a third-degree basis of Hermite polynomials, and calculated expansion coefficients using a point collocation method based on linear regression on a full tensor grid of 1024 samples. Surrogation was performed using the open source MATLAB/Octave library *sglib*[1]. We checked surrogation accuracy by calculating the mean absolute error (MAE) on a sample of 50 randomly selected vectors of material properties. We set a threshold of 0.01 Hz to deem an approximation acceptable, with a view to use this expansion to perform calibration with experimental data, which accuracy would reasonably be of that magnitude. Tests with lower-degree polynomials yielded unacceptable results, while the selected approximations led to a maximum MAE of 0.002 Hz on higher-order modes, and to a maximum absolute of 0.006 Hz for all modes of all the considered samples. gPCE-based surrogate models allow a fast computation of global sensitivities [11] as well as of the statistics of modal properties. In fact, the Sobol Indices (SIs), i.e. the ratio of the variance attributable to the uncertainties of a given input parameter and the total variance of the model response, can be directly computed from the coefficients $\boldsymbol{\alpha}_i$ of the gPCE. Sobol' Indices vary between 0 and 1, with higher values indicating a higher influence of the input parameters variability on the variability of the model output.

In this study the first ten natural frequencies of the FE model are investigated. Figure 4 represents their respective mode shapes associated to average input parameters.

Sensitivity of natural frequencies to the Young's moduli of the external and internal layers of the walls of the Leaning Tower and of the three most superficial layers of the soil is evaluated. Figure 5 shows that the first six modes are mainly influenced by the first soil layer; the other two soil layers increase their relevance in the vertical mode, while the elastic moduli of masonry are the most relevant parameters starting from the seventh one (that is the second order bending mode).

The identified frequencies of the Tower [5] and their respective FE model frequencies are reported in Table 3; Table 4 shows instead the SIs of the five uncertain parameters for these four frequencies. Second-order SIs, i.e., indices representing the combined influence of parameter couples, were also evaluated but turned out to be negligible.

---

[1] Zander E., 2020. *Sglib*: a Matlab/Octave Toolbox for Stochastic Galerkin Methods https://ezander.github.io/sglib/.

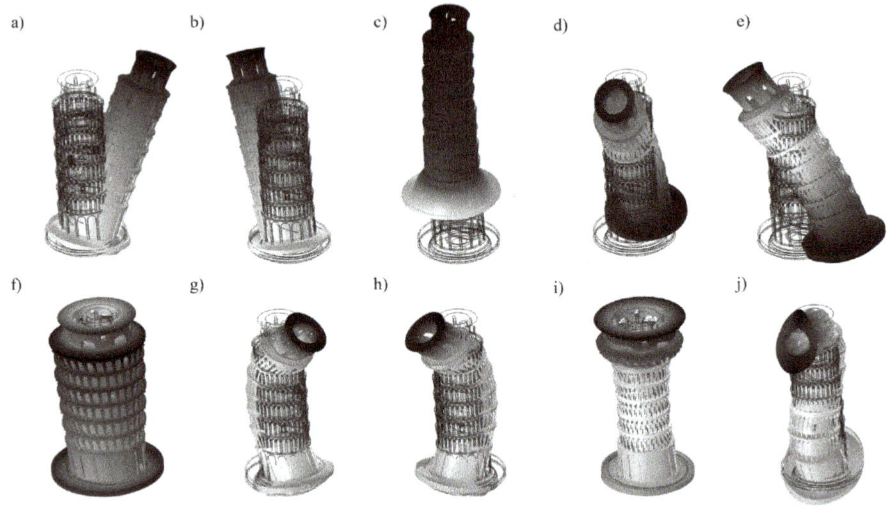

Fig. 4. Mode shapes of the FE model of the Tower: a) 1st mode– first order bending (E-W); b) 2nd mode-first order bending (N-S); c) 3rd mode-extensional; d) 4th mode-foundation movement; e) 5th mode-foundation movement; f) 6th mode-torsional; g) 7th mode- second order bending (N-S); h) 8th mode-second order bending (E-W); i) 9th mode- second order torsional; j) 10th mode- third order bending.

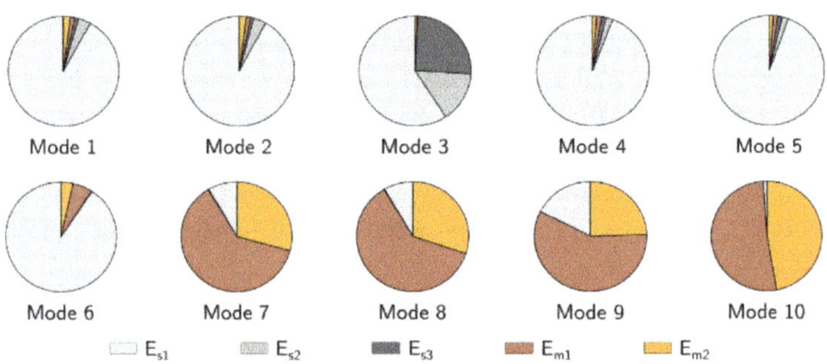

Fig. 5. Sensitivity analysis of the first ten natural frequencies of the FE model of the Tower to the five uncertain parameters.

2.4 Bayesian Approach to the Inverse Problem

The Bayesian approach seeks to estimate the updated density of the uncertain parameters $\boldsymbol{X}$ given a set of observations y_m. Bayes rule takes the form

$$\pi_{X|y_m}(x) = \frac{\pi_{y_m|x}(y_m|x)\pi_X(x)}{\int \pi_{y_m|x}(y_m|x)\pi_X(x)dx}, \tag{2}$$

where $\pi_X(x)$ is the prior probability density of $\boldsymbol{X}$; $\pi_{y_m|x}(y_m|x)$ is the likelihood function; and $\pi_{X|y_m}(x)$ is the density of $\boldsymbol{X}$ conditional on the measurements y_m, i.e., the posterior probability density of $\boldsymbol{X}$. The likelihood of measuring the data y_m given the input

Table 3. Identified natural frequencies of the Tower by Fiorentino et al. [5] F_i and their respective FEM frequencies f_i (μ is the mean valued and V the coefficient of variation).

	1st order bending (N-S)	1st order bending (E-W)	Extensional	Torsional
Identified freq. F_i [Hz]	$F_1 = 0.96$	$F2 = 1.02$	$F_3 = 2.97$	$F_4 = 6.29$
FEM freq. f_i [Hz]	f_2: $\mu = 0.98$ $V = 0.05$	f_1: $\mu = 0.97$ $V = 0.05$	f_3: $\mu = 3.47$ $V = 0.04$	f_6: $\mu = 5.70$ $V = 0.07$

Table 4. Sobol'Indices of the uncertain input parameters for the four natural frequencies identified by [5]. E_{Sx} are the elastic moduli of the soil layers, E_{Mx} of the masonry.

FEM frequencies	SI (E_{S1})	SI(E_{S2})	SI(E_{S3})	SI(E_{M1})	SI(E_{M2})
f_1 (bending E-W)	0.919	0.037	0.011	0.013	0.020
f_2 (bending N-S)	0.920	0.037	0.010	0.013	0.020
f_3 (extensional)	0.592	0.149	0.252	0.002	0.005
f_6 (torsional)	0.905	0.002	0.000	0.058	0.035

parameters x can be computed by the probability of the measurement error to take the value $\varepsilon = y_m - M(x)$. Considering m measurements and mutually independent random errors with pdf π_{ε_i}, the likelihood reads

$$\pi_{y_m|x}(y_m|x) = \prod_{i=1}^{m} \pi_{\varepsilon_i}(y_{m,i} - M(x)). \tag{3}$$

Since no closed form of the posterior can be derived, to solve the inverse problem in Eq. (2) we rely here on the Markov Chain Monte Carlo (MCMC) method. The algorithm of the MCMC generate samples from the posterior distribution, with the help of a random walk in the parameter space, constructing a Markov chain that has the target pdf as its equilibrium distribution [17].

3 Results and Discussion

Bayesian updating has been performed by considering the four experimental natural frequencies identified by Fiorentino et al. [5] (see Sect. 2.2) having assigned a standard deviation of the measurement equal to 2% of each identified frequency.

Prior and posterior distributions of the uncertain parameters are illustrated in Fig. 6 and Table 5: the second and third soil layers show a reduced mean value of the elastic modulus compared with the prior one, while for the other parameters the opposite happens. What is surprising is the increased Young's modulus of the masonry layers; however, it is worth remembering that the sensitivity of the employed natural frequencies to these parameters is very low (see Table 4). Moreover, their coefficient of variation has remained unchanged or even increased. A further updating of E_{M1} and E_{M2} will probably be possible in the next future, thanks to a recent installation of a dynamic monitoring system made of 24 sensors.

Table 5. Prior and posterior distribution of the five uncertain parameters: μ is the mean value and *V* the coefficient of variation. E_{Sx} are the elastic moduli of the soil layers, E_{Mx} of the masonry.

	E_{S1}	E_{S2}	E_{S3}	E_{M1}	E_{M2}
Prior [MPa]	$\mu = 180$ $V = 0.23$	$\mu = 116$ $V = 0.10$	$\mu = 297$ $V = 0.24$	$\mu = 54176$ $V = 0.19$	$\mu = 7342$ $V = 0.53$
Posterior [MPa]	$\mu = 214$ $V = 0.07$	$\mu = 97$ $V = 0.09$	$\mu = 114$ $V = 0.16$	$\mu = 63489$ $V = 0.18$	$\mu = 10404$ $V = 0.56$

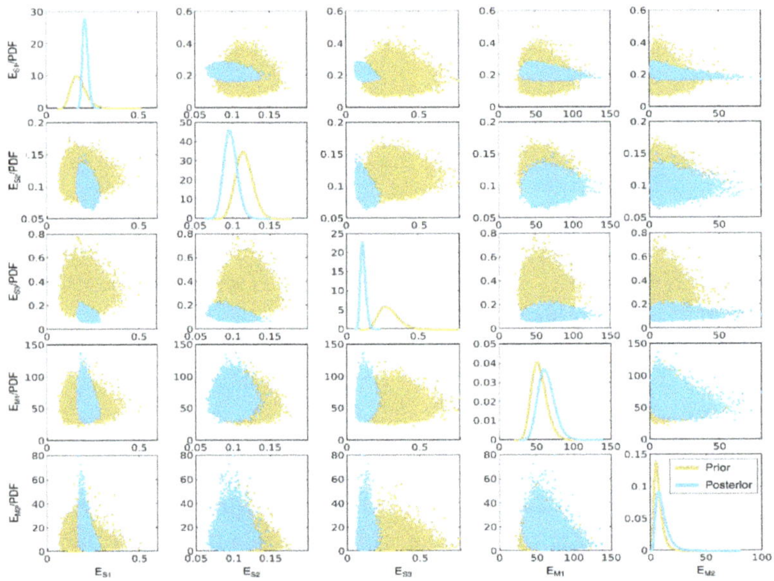

Fig. 6. Prior and posterior distributions of the five uncertain parameters in a pairs scatter plot. E_{Sx} are the elastic moduli of the soil layers, E_{Mx} of the masonry [GPa].

The standard deviation of S1 and S3 is particularly reduced and this result is in line with the sensitivity analysis of Sect. 2.3: the first layer has in fact the most influence on all the analyzed modes and the third layer is quite relevant for the extensional mode; the SIs of S2 are instead lower.

To conclude, Fig. 7 and Table 6 expose the results in terms of frequencies. Each posterior distribution is characterized by a reduced coefficient of variation and its mean values fall approximately within the range of estimated identified frequencies.

We can also notice that the first two natural frequencies of the Leaning Tower, that differ of about 0.06 Hz, become even closer than the initial numerical values which

Table 6. Prior and posterior distributions compared with the experimental values of the four identified natural frequencies Fi: μ is the mean value, V the coefficient of variation and σ_ε the standard deviation of the measurement error of each identified frequency.

	F1 (bending N-S)	F2 (bending E-W)	F3 (extensional)	F4 (torsional)
Prior [Hz]	$\mu = 0.97$ $V = 0.05$	$\mu = 0.98$ $V = 0.05$	$\mu = 3.47$ $V = 0.04$	$\mu = 5.70$ $V = 0.07$
Posterior [Hz]	$\mu = 0.98$ $V = 0.01$	$\mu = 0.98$ $V = 0.91$	$\mu = 3.13$ $V = 0.02$	$\mu = 6.14$ $V = 0.02$
Experimental [Hz]	$\mu = 0.96$ $\sigma_\varepsilon = 0.02$	$\mu = 1.02$ $\sigma_\varepsilon = 0.02$	$\mu = 2.97$ $\sigma_\varepsilon = 0.05$	$\mu = 6.29$ $\sigma_\varepsilon = 0.13$

differed by 0.01 Hz. This outcome proves that the inclination and shape of the Tower, faithfully represented in the model, are not the cause of this difference, and that the FE model needs to be improved to reproduce the correct modal behavior of the Tower.

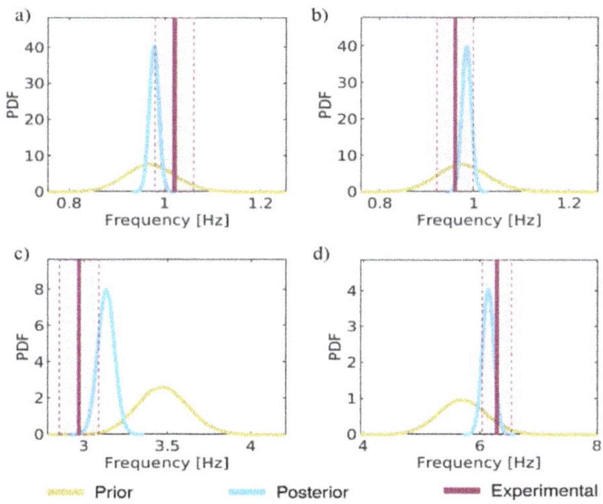

Fig. 7. Prior and posterior distributions with the identified natural frequencies: a) 1st order bending mode (N-S); b) 1st order bending mode (E-W); c) extensional mode; d) torsional mode.

4 Conclusions

In this work, we investigated the dynamic behavior of the Leaning Tower of Pisa taking into account the SSI which has always had a high impact on this monument. The preliminary sensitivity analyses allowed the choice of the set of input parameters which most influence the modal behavior of the Tower leading to the selection of the elastic moduli of three soil layers and of the external and internal layers of masonry walls as parameters worthy of calibration. Initial statistical distributions of these input parameters were attributed on the basis of past in-situ surveys.

The development of a gPCE-based surrogate model, paired with the Bayesian inverse problem solved via MCMC method, has reduced significantly the computation time providing the updating of our five uncertain numerical parameters in order to obtain a FE model suitable for future studies on different scenarios. This outcome was possible thanks to the availability of four natural frequencies of the Tower identified in past dynamic campaigns based on a limited number of sensors. However, this study is only the first step of a broader path of knowledge that will take advantage of the new monitoring system which provides 24 accelerometer measurements, hoping to identify some higher frequencies of the Tower which are, moreover, those on which the mechanical properties of the masonry have the most influence.

Acknowledgments. The authors would like to gratefully acknowledge the Extraordinary Plan for Monitoring and Conservation of Italian Cultural Heritage initiated by the Italian Ministry of Culture (MiC) for its support.

References

1. De Silva, F.: Influence of soil-structure interaction on the site-specific seismic demand to masonry towers. Soil Dyn. Earthq. Eng. **131**, 106023 (2020)
2. Sabia, D., Demarie, G.V., Quattrone, A.: Structural health monitoring of historic masonry towers: the Case of the Ghirlandina Tower, Modena. In: Geotechnical Engineering for the Preservation of Monuments and Historic Sites III, 1st edn., pp. 191–201. CRC Press, London (2022)
3. Landi, F., Marsili, F., Friedman, N., Croce, P.: GPCE-based stochastic inverse methods: a benchmark study from a civil engineer's perspective. Infrastructures **6**, 158 (2021)
4. Bartelletti, R., et al.: Behavior of the leaning tower of Pisa: insights on seismic input and soil-structure interaction. Appl. Mech. Mater. **847**, 454–462 (2016)
5. Fiorentino, G., et al.: Seismic reassessment of the leaning tower of Pisa: dynamic monitoring, site response, and SSI. Earthq. Spectra **35**, 703–736 (2019)
6. Nakamura, Y., Gurler, E.D., Saita, J.: Dynamic characteristics of Leaning Tower of Pisa using microtremor-Preliminary results. In: Proceedings of 25th Japan Conference on Earthquake Engineering, Doboku Gakkai, Tokyo, Japan, vol. 25, pp. 921–924 (1999)
7. Castellaro, S., Mulargia, F.: How far from a building does the ground-motion free-field start? The cases of three famous towers and a modern building. Bull. Seismol. Soc. Am. **100**, 2080–2094 (2010)
8. Atzeni, C., Bicci, A., Dei, D., et al.: Remote survey of the leaning tower of pisa by interferometric sensing. IEEE Geosci. Remote Sensing Lett. **7**, 185–189 (2010)
9. Macchi, G., Ghelfi, S.: Problemi di consolidamento. In: La Torre restituita: gli studi e gli interventi che hanno consentito la stabilizzazione della Torre di Pisa, vol. III, part 1 of Bollettino d'arte, Ministero per i beni e le attività culturali. Libreria dello stato, Roma, IT (2005)
10. Butuc, D., De Falco, A., Resta, C., Squeglia, N., Vignali, L.: A novel FE model for the Tower of Pisa. In: Proceedings of 9th International Conference on Computational Methods in Structural Dynamics and Earthquake Engineering Methods in Structural Dynamics and Earthquake Engineering, Athens, Greece, pp. 2452–2464 (2023)

11. Sudret, B.: Global sensitivity analysis using polynomial chaos expansions. Reliab. Eng. Syst. Saf. **93**, 964–979 (2008)
12. Bartelletti, R., Finzi, L., Sanpaolesi, L.: Investigations on the masonry of the Leaning Tower of Pisa. IABSE reports (1993)
13. Italian Public Works Council: Memorandum for application of Italian building code, N7. Italian Official Journal. Istituto Poligrafico e Zecca dello Stato, Rome, Italy (2019)
14. Lo Presti, D., Jamiolkowski, M., Pepe, M.: Geotechnical characterisation of the subsoil of Pisa Tower. Charact. Eng. Prop. Nat. Soils **2**: 909–946 (2003). Tan et al. (eds)
15. Burland, J.B., Jamiolkowski, M.B.: Leaning tower of Pisa: behaviour after stabilization operations. Int. J. Geoeng. Case Histories **1**, 156–169 (2009)
16. Butuc, D., et al.: Sensitivity analysis on the FE model parameters of the Leaning Tower of Pisa. In: 20th International Probabilistic Workshop (2024)
17. Tierney, L.: Markov chains for exploring posterior distributions. Ann. Stat. **22**, 1701–1762 (1944)

Comparative Analysis of Normality Tests: Integrating Classical and Contemporary Approaches

Cecilia Castro[1], Lucas Henriques[1,2], and Felipe Prata[2](✉)

[1] Centre of Mathematics, Universidade do Minho, 410-057 Braga, Portugal
cecilia@math.uminho.pt

[2] Federal Institute of Alagoas, Maceió, AL 57020-600, Brazil
felipepratalima@gmail.com
https://www.cmat.uminho.pt/user/651

Abstract. This study conducts a systematic evaluation of several normality tests, essential in statistical analyses and validating assumptions in diverse fields, including finance and energy. The primary contribution lies in extending the Jelito-Pitera test to detect deviations in light-tailed distributions, filling a gap in the range of existing normality tests. Additionally, this research conducts a detailed comparison of recent normality tests, specifically focusing on the Jelito-Pitera test and the Kullback-Leibler information-based test. We assess their effectiveness in identifying deviations not only in heavy-tailed but also in light-tailed distributions. Employing Monte Carlo simulations across various sample sizes and significance levels, our work examines the efficacy of well-established tests like Anderson-Darling, Zhang, Shapiro-Wilk, Shapiro-Francia, and Jarque-Bera, alongside these newer methods. The focus is on their performance in different non-normal distributions, aiming to understand their relative strengths and limitations in practical applications. The findings of this research provide a nuanced view of statistical testing methods, emphasizing the adaptation of the Jelito-Pitera test for light tails and offering insights into the comparative effectiveness of both classical and contemporary normality tests under diverse conditions. This study aims to assist researchers and practitioners in selecting the most suitable normality test for their data, thereby improving the precision of data analysis across various research areas.

Keywords: Conditional Dispersion · Data Analytics · Kullback-Leibler Information · Light-tailed Distributions · Simulation Study

Supported by organization Centre of Mathematics of the University of Minho, within projects UIDB/00013/2020 https://doi.org/10.54499/UIDB/00013/2020 and UIDP/00013/2020 https://doi.org/10.54499/UIDP/00013/2020.

J. C. Matos et al. (Eds.): IPW 2024, LNCE 494, pp. 356–365, 2024.
https://doi.org/10.1007/978-3-031-60271-9_33

1 Introduction

The study of normality tests has been a significant focus in statistical analysis for over a century. Beginning with Pearson's chi-squared test in 1900, the development and comparison of these tests have been ongoing. The works [1,2] provide comprehensive overviews of numerous tests, illustrating the normal distribution's central role in statistical theory and applications. This distribution is often the underlying assumption in many statistical procedures, making the validation of normality a fundamental concern for analysts. Normality assumption testing has evolved, with significant contributions from various authors. Pioneering tests like Cramer von Mises and Kolmogorov-Smirnov tests were proposed in the early 20th century, followed by Anderson and Darling, Kuiper, Shapiro and Wilk, and more. The latter half of the century saw the introduction of tests like Vasicek's and Jarque and Bera's tests, focusing on different statistical properties and distribution characteristics [3,4]. Comparative studies continue to be essential in discerning the effectiveness of normality tests under various conditions. Previous research, such as in [5] haves compared several tests through Monte Carlo (MC) simulations, with later studies like [6,7] further explored the power and type I error rates of these tests. These comparisons often reveal contradictions in results, leading to a lack of consensus on the optimal test for specific scenarios. Recent studies, particularly [8], have primarily concentrated on distributions with heavy tails. This focus reflects the complexities encountered in fields like finance, where extreme values are critical, and in biology and social sciences, where data frequently diverges from normal distribution assumptions. Heavy tails are often emphasized because they capture the likelihood of rare, significant events, a key aspect of analysis in these areas. Conversely, light-tailed distributions, despite being less examined, are equally important. In applications such as electricity consumption analysis [9,10], a comprehensive understanding of both heavy and light tails is beneficial. Data in stable and predictable scenarios, such as certain patterns of electricity usage, typically exhibit light-tailed characteristics, indicating a lower likelihood of extreme deviations. Acknowledging the relevance of light tails is critical, particularly in scenarios where data exhibits limited variability. Such distributions, indicative of minimal or tightly controlled extremities, can offer valuable insights into various systems and processes. However, to our knowledge, deviations from normality in light-tailed data have received limited attention in existing literature. Our study addresses this gap, aiming to extend the application of existing normality tests, traditionally focused on heavy tails, to also encompass light-tailed distributions.

In this context, we meticulously examine two contemporary normality tests: the adaptation of the test initially proposed in [11] and refined in [8] for heavy tails, and the Kullback-Leibler Information (KLI) test outlined in [12] for general goodness-of-fit evaluation. A main contribution of our work is the modification of the test from [8] for applicability to light-tailed distributions, significantly broadening its utility in statistical testing.

Moreover, our study offers a critical assessment of whether the KLI test's generalized approach [12], matches the effectiveness of the more specialized heavy-tail-oriented test from [8] when applied to light-tailed distributions. Alongside, we engage in a comparative analysis with several classic normality tests, emphasizing their statistical power and applicability across different distributional scenarios.

To accomplish our objectives, we employ an exhaustive methodological framework involving MC simulations. These simulations enable a comprehensive evaluation of the tests' performance across a spectrum of sample sizes and significance levels, showcasing the adaptability and effectiveness of MC simulations in exploring the efficacy of normality tests under diverse conditions.

This detailed comparative analysis aims to refine the selection criteria for the most suitable normality test based on the specific characteristics of the data under investigation. Consequently, our research contributes significantly to the extant literature on normality tests, offering pragmatic insights for researchers and practitioners to select the most appropriate testing tools for their unique analytical challenges. The structure of the article is as follows: After this introduction, Sect. 2 provides the methodologies and theoretical foundations of KLI and an adaptation of Jaworski and Pitera's test for light-tailed distributions. Section 3, presents a comparative analysis focused solely on the performance of these tests within the context of the Uniform distribution. Section 4 presents the key insights derived from our comparative study, discusses the broader implications of these findings, and outlines potential directions for future research in the field of normality testing, especially concerning light-tailed distributions.

2 Extending and Exploring Normality Tests

Testing for normality is a crucial aspect of statistical analysis, supporting the validity of many statistical models and methodologies. The assumption of normality often serves as a key foundation for inferences and conclusions drawn from data analysis. However, real-world data can exhibit a wide range of distributional characteristics, including heavy-tailed and light-tailed distributions, which necessitate a more nuanced approach to normality testing.

This section explores two significant methodologies in the realm of normality tests developed in recent years: the KLI test and an adaptation of Jaworski and Pitera's test for light-tailed distributions. The KLI test, grounded in the concept of Kullback-Leibler divergence from information theory, offers a robust method for comparing the fit of a theoretical model to the actual data distribution. On the other hand, the Jaworski and Pitera's test, originally focused on heavy-tailed distributions, is adapted in this study to assess light-tailed scenarios. This adaptation addresses a gap in the literature, where light-tailed distributions often receive less attention despite their prevalence in various fields such as quality control and standardized testing.

2.1 Kullback-Leibler Information Test

The KLI test is a prominent example of these advanced methodologies. Let f denote the true density function and $G = \{g(.,\theta) : \theta \in \Omega\}$ be a selected statistical model for the data distribution f, where Ω is a subset of R^p.

The Kullback-Leibler (KL) divergence is a statistical measure used for quantifying the difference between two probability distributions. It assesses how one probability distribution diverges from a second, expected distribution. When f actually belongs to G, the minimal value, $\min_{\theta \in \Theta} D(f, g(.,\theta))$, of the KL divergence is zero, indicating that f and G are identical in terms of their probabilistic predictions. Conversely, a positive KL divergence value implies a discrepancy between f and G, suggesting a misalignment between the actual distribution f and the model G. This divergence forms the basis for constructing a goodness of fit test, which would reject $H_0 : f \in G$ for a large value of $D(f, g)$, where $D(f, g; \theta)$ is given by

$$D(f, g; \theta) = \int_{-\infty}^{\infty} f(x) \log \frac{f(x)}{g(x;\theta)} dx = -H(f) - E_f\{\log(g(X;\theta))\}, \qquad (1)$$

with $H(f)$ representing the entropy of f.

In the context of the KLI test, the estimation of $D(f, g)$ is approached through $DA_{mn} = -HV_{mn} - HA_{mn}$, where HV_{mn} is Vasicek's estimate of $H(f)$ and HA_{mn} is the semi-parametric estimate of $E_f\{\log(g(X;\theta))\}$, given by

$$HA_{mn} = \frac{1}{n} \sum_{i=1}^{n} \log \frac{G(X_{(i+m)}; \hat{\theta}) - G(X_{(i-m)}; \hat{\theta})}{X_{(i+m)} - X_{(i-m)}}, \qquad (2)$$

where $\hat{\theta}$ represents any reasonable equivariant estimate of θ. The necessity for $\hat{\theta}$ to be equivariant ensures that the test is consistent and not contradictory.

The proposed test statistic DA_{mn} is calculated as

$$DA_{mn} = -\frac{1}{n} \sum_{i=1}^{n} \log \left[\frac{n}{2m} \left(G(X_{(i+m)}; \hat{\theta}) - G(X_{(i-m)}; \hat{\theta}) \right) \right], \qquad (3)$$

where DA_{mn} estimates $D(f, g; \theta)$. In this equation, m is a positive integer, $m \leq \frac{n}{2}$, and $X_{(1)} \leq X_{(2)} \leq \cdots \leq X_{(n)}$ are the order statistics, with $X_{(i)} = X_{(1)}$ if $i < 1$ and $X_{(i)} = X_{(n)}$ if $i > n$. If G is the normal distribution function and $\hat{\theta} = (\hat{\mu}, \hat{\sigma})$, with $\hat{\mu} = \bar{X} = \frac{1}{n} \sum_{i=1}^{n} X_i$ and $\hat{\sigma} = s = \sqrt{\frac{1}{n} \sum_{i=1}^{n} (X_i - \bar{X})^2}$, then expression (3) becomes the test statistic DA_{mn} for testing normality. It is crucial to recognize that this test statistic is invariant with respect to location and scale transformations.

A location transformation of a dataset involves adding a constant c to all data points, such that each transformed value Y_i is given by $Y_i = X_i + c$. Under this transformation, the sample mean $\bar{Y}$ becomes $\bar{X} + c$. However, when we compute the transformed test statistic, the addition of c cancels out in the calculation of

$\hat{\mu}$ and the differences $X_{(i+m)} - X_{(i-m)}$, making DA_{mn} invariant to the location transformation.

A scale transformation involves multiplying all data points by a non-zero constant k, so the transformed values are $Y_i = kX_i$. Under this transformation, the sample mean becomes $k\bar{X}$ and the standard deviation becomes ks. In the test statistic, the multiplication by k affects both $\hat{\mu}$ and $\hat{\sigma}$ in such a way that the effect cancels out in the ratio $\frac{G(X_{(i+m)};\hat{\theta})-G(X_{(i-m)};\hat{\theta})}{X_{(i+m)}-X_{(i-m)}}$, thereby preserving the invariance of DA_{mn} under scale transformations.

This property of invariance to location and scale transformations is particularly advantageous in statistical testing, as it allows the test to be applicable to a wide range of datasets without the need for additional adjustments or normalizations. It ensures that the test's conclusions are based solely on the shape of the distribution, rather than being influenced by the specific units or magnitude of the data.

2.2 Adaptation of Jaworski and Pitera's Test for Light Tails

Building upon the exploration of advanced methodologies in normality testing, this subsection focuses on adapting Jaworski and Pitera's test, originally designed for heavy-tailed distributions, to scenarios involving light-tailed distributions. This adaptation is crucial as light-tailed distributions are commonly encountered in various real-world scenarios, yet often receive less attention in statistical literature.

For instance, electricity consumption data in residential areas, characterized by stable and predictable usage patterns, typically exhibit light-tailed distributions. In these contexts, extreme consumption values, either very high or very low, are rare, resulting in fewer outliers compared to a normal distribution. Such patterns are also observed in fields like manufacturing quality control and educational assessments, where measurements and standardized test scores, respectively, often display light-tailed characteristics.

To effectively address these nuances in scenarios involving light tails, we have adapted the tests originally delineated in [11], later refined in [8]. This adaptation fills a gap in the toolkit available for normality testing, particularly in fields where light-tailed distributions are prevalent, and provides a more nuanced approach to assessing data distribution characteristics.

In this adapted framework, the test statistic N, originally designed to measure tail fatness under normality, is reinterpreted to focus on light-tailed scenarios. The test statistic is defined as

$$N = \frac{1}{\rho}\left(\frac{\hat{\sigma}_L^2 - \hat{\sigma}_M^2}{\hat{\sigma}^2} + \frac{\hat{\sigma}_R^2 - \hat{\sigma}_M^2}{\hat{\sigma}^2}\right)\sqrt{n}, \tag{4}$$

where $\hat{\sigma}_L^2$, $\hat{\sigma}_M^2$, and $\hat{\sigma}_R^2$ are the estimated variances of the lower 20%, middle 60%, and upper 20% of the data, respectively, $\hat{\sigma}^2$ is the overall sample variance, n is the sample size, and $\rho \approx 1.8$ is a fixed normalizing constant.

Contrasting with traditional tests like the Anderson-Darling, Zhang, Shapiro-Wilk, Shapiro-Francia, and Jarque-Bera tests, which assess normality through various methodologies, the Jelito-Pitera test offers a distinct approach. It focuses on the direct interpretation of tail heaviness, serving as a valuable alternative for scenarios where tail behavior is of primary concern.

An important mathematical foundation for the Jelito-Pitera test is encapsulated in the following theorem:

Theorem 1. *Let* $X \sim N(\mu, \sigma)$. *Then, as* $n \to \infty$, $N \to N(0, 1)$. *Moreover,* ρ *is independent of* μ, σ, *and* n.

This theorem underlines the robustness of the test, especially as sample sizes increase, indicating its suitability for large datasets.

This test has demonstrated strong performance in analyzing financial return data, often characterized by heavier tails than those predicted by the normal distribution.

In the context of our adaptation of the JP test to detect light tails, we focus on this rapid decrease in probability in the tails. Specifically, for light-tailed distributions, the probability of observing values far from the mean (in either direction) is significantly lower than it would be under a normal distribution. This involves a modification of the rejection region to focus on the left tail of the distribution of the test statistic N defined in (4). Instead of rejecting normality for large values of N, under this adaptation, normality would be rejected for small values of N. In the original test, large values of N indicate heavy tails - a high variance in the outer subsets compared to the middle subset of the data. However, for light-tailed distributions, we are interested in the opposite scenario: the variance in the outer subsets is significantly less than expected under normality. Therefore, our adaptation rejects the hypothesis of normality for small values of N. This shift in focus is mathematically represented by setting a lower threshold for N. When N falls below this threshold, it suggests that the tails of the distribution are thinner than those of a normal distribution, indicative of a light-tailed distribution. This threshold for rejection can be empirically determined through MC simulations, simulating samples from known light-tailed distributions like the Uniform distribution. This approach allows us to establish a threshold based on specific quantiles of the distribution, such as the 5th percentile, providing a clear criterion for identifying distributions with lighter tails than normal. The size of these samples should mimic the size of the data sets we expect to analyze in practice. For each generated sample, we apply the JP test for and calculate the test statistic of interest. We then compile the test statistics from all the generated samples to form a distribution of N under the assumption of a light-tailed distribution. Then we determine the 5th percentile. By following this process, we can empirically establish a threshold that is tailored to detecting light-tailed distributions. This approach ensures that our test is both robust and sensitive to the specific characteristics of light tails, providing a valuable tool for statistical analysis in various real-world scenarios.

The empirical validation of the modified test is presented in Table 1, which summarizes the results of a power study for the JP test for light tails on the Uniform distribution.

Table 1. Power study results for the Jelito-Pitera test variant on the Uniform distribution.

α	n	AD	ZA	SW	SF	JB	JP
5%	50	0.577	0.796	0.749	0.447	0.011	0.903
	100	0.949	0.999	0.995	0.960	0.749	0.998
	250	1.000	1.000	1.000	1.000	1.000	1.000
2,5%	50	0.433	0.654	0.565	0.273	0.000	0.831
	100	0.895	0.996	0.983	0.904	0.058	0.994
	250	1.000	1.000	1.000	1.000	1.000	1.000
1%	50	0.273	0.446	0.342	0.108	0.000	0.708
	100	0.788	0.991	0.940	0.723	0.000	0.981
	250	1.000	1.000	1.000	1.000	0.976	1.000

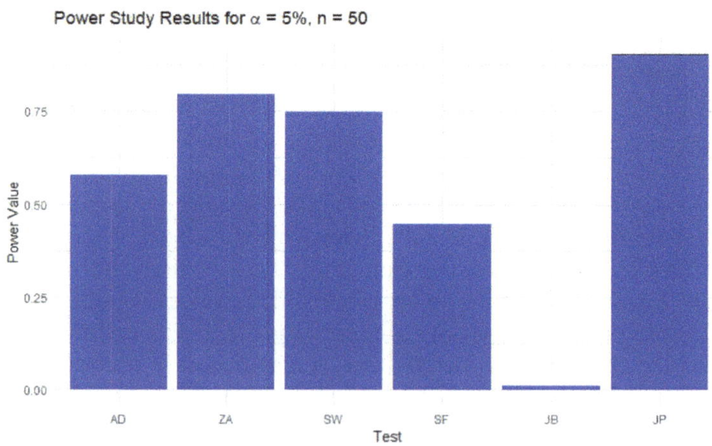

Fig. 1.

3 Comparison of Kullback-Leibler Information Test and Jelito-Pitera Test for Light Tails

In our analysis, we compare the performance of the KLI test (specifically the DA_{mn} statistic) and the JP for light tails in the context of the Uniform distribution. We selected the Uniform distribution as a representative case study because it is paradigmatic of light-tailed distributions, offering a clear and controlled scenario to evaluate these tests' sensitivity to distributions that significantly deviate from normality in a subtle manner.

To quantify the performance of these tests, we conducted MC simulations, generating samples from a Uniform distribution of varied sizes. These simulations enabled us to calculate the power of the tests, that is, the probability of correctly rejecting the null hypothesis of normality when the underlying distribution is, in fact, Uniform.

For this study, the critical values for each normality test were obtained through Monte Carlo simulations with 10^6 replicates. Given the extensive range of data generated through our Monte Carlo simulations, we have chosen to present here the most salient results, particularly focusing on sample sizes and configurations that best illustrate the differential performance of the tests under study. This decision mirrors the approach taken in seminal articles from which our methods were inspired, aiming to provide a clear and concise exploration of the tests' efficacies without overwhelming the reader with exhaustive data.

3.1 Kullback-Leibler Information Test on Uniform Distribution

- For a sample size of n = 20 and a significance level of $\alpha = 0.05$, the power values for the evaluated statistics (SW, AD, DA_{mn}) were as follows: SW = 0.144, AD = 0.151, $DA_{mn} = 0.190$. This indicates that, at this significance level, the DA_{mn} statistic demonstrated the highest power in detecting deviations from normality in the Uniform distribution among these tests for this sample size.

3.2 Jelito-Pitera Test for Light Tails on Uniform Distribution

- Notably, for a sample size of n = 20 at a significance level of $\alpha = 0.05$, the JP test demonstrated a power of 0.203, slightly surpassing the DA_{mn} statistic, which had a power of 0.190.
- The Jelito-Pitera test (JP) for light tails showed excellent performance in rejecting normality for the Uniform distribution across all tested sample sizes (n = 50, n = 100, and n = 250) (Table 1 and Fig. 1).
- For larger sample sizes (n = 250), the JP test completely rejects the normality of the data, indicating its high efficacy in this context.
- The test outperformed most other omnibus tests and the Jarque-Bera test, especially in larger sample sizes.

4 Conclusions and Future Work

Our comparative analysis between the Kullback-Leibler Information test (KLI) and the Jelito-Pitera test (JP) for light tails has revealed nuanced performances, suggesting that both tests are valuable tools in statistical analysis, particularly when dealing with light-tailed distributions such as the Uniform distribution.

The efficacy of both tests in detecting deviations from normality underscores their applicability across diverse analytical contexts. Notably, while the JP test demonstrated slightly higher power in some instances, especially for smaller sample sizes, the difference in effectiveness compared to the DA_{mn} statistic was not substantial. This observation emphasizes the importance of selecting the appropriate normality test based on the specific characteristics of the dataset, including the distribution type and sample size.

Furthermore, this study highlights the need for further research focused on the performance of normality tests within the context of light-tailed distributions. Traditionally, the emphasis in statistical analysis has been on heavy-tailed distributions; however, our findings advocate for a comprehensive examination of how different normality tests perform with distributions exhibiting lighter tails. Such research is vital for developing a deeper understanding of the optimal application of these tests across diverse scenarios.

Expanding simulation studies to include a variety of light-tailed distributions, such as the Log-Normal (with parameters leading to lighter tails), Weibull (with parameters that generate light tails), Gamma (with shape and scale parameters resulting in light tails), and Beta (with parameters emphasizing mass concentration around a central point), could unveil significant insights and potentially lead to the development of new methodologies tailored for these types of data. This broader exploration and comparison of statistical tools in this less commonly studied area would greatly enhance our ability to accurately interpret and analyze data across a wider range of distributional contexts, ultimately benefiting the field of data analysis.

References

1. D'Agostino, R.B.: Goodness-of-Fit-Techniques, vol. 68. CRC Press, London (1986)
2. Thode, H.C.: Testing for Normality, vol. 164. CRC Press, Boca Raton (2002)
3. Vasicek, O.: A test for normality based on sample entropy. J. Roy. Stat. Soc. Ser. B (Methodol.) **38**(1), 54–59 (1976)
4. Jarque, C.M., Bera, A.K.: A test for normality of observations and regression residuals. Int. Stat. Rev. Revue Internationale de Statistique **55**(2), 163–172 (1987)
5. Stephens, M.A.: EDF statistics for goodness of fit and some comparisons. J. Am. Stat. Assoc. **69**(347), 730–737 (1974)
6. Yazici, B., Yolacan, S.: A comparison of various tests of normality. J. Stat. Comput. Simul. **77**(2), 175–183 (2007)
7. Meintanis, S.G.: Goodness-of-fit testing by transforming to normality: comparison between classical and characteristic function-based methods. J. Stat. Comput. Simul. **79**(2), 205–212 (2009)
8. Jelito, D., Pitera, M.: New fat-tail normality test based on conditional second moments with applications to finance. Stat. Pap. **62**(5), 2083–2108 (2021)
9. Engel, R.F., Granger, C.W.J., Rice, J., Weiss, A.: Semiparametric estimates of the relation between weather and electricity sales. J. Am. Stat. Assoc. **81**, 310–320 (1986)
10. Taylor, J.W., Buizza, R.: Neural network load forecasting with weather ensemble predictions. IEEE Trans. Power Syst. **17**, 626–632 (2002)

11. Jaworski, P., Pitera, M.: The 20-60-20 rule. Discrete Continuous Dyn. Syst. Ser. B. **21**(4) (2016)
12. Alizadeh Noughabi, H.: A new estimator of Kullback-Leibler information and its application in goodness of fit tests. J. Stat. Comput. Simul. **89**(10), 1914–1934 (2019)

Exploration of Adaptive Sequential Sampling in the Definition of Surrogate Models for the Rare Event Estimation in Transportation Assets

Carlos Cabanzo[1(✉)], Edward Baron[1], Miroslav Vorechovsky[2], Mitsuyoshi Akiyama[3], Paulo B. Lourenco[1], and Jose C. Matos[1]

[1] Department of Civil Engineering, University of Minho, 4800-058 Guimarães, Portugal
carlos.cabanzo@civil.uminho.pt

[2] Faculty of Civil Engineering, Brno University of Technology, 602 00 Brno, Czech Republic

[3] Department of Civil and Environmental Engineering, Waseda University, Tokyo 169-8555, Japan

Abstract. Adaptive sequential sampling provides a good technique to refine and increase the accuracy of surrogate models, used for reliability analysis, based on the selection of possible future candidates in the input domain (*i.e.*, random variables). In the present research, different methodologies for obtaining the training sample for a surrogate model were explored, considering sample size, distribution of the points, and identification of the failure region. The effects on the reliability of the slope stability under vertical loading based on the safety factors from Bishop's simplified method were obtained. The results reinforce the importance of the characteristics of the training sample used for the application of surrogate models to describe limit states and their accuracy when employed for the computation of the reliability index.

Keywords: Failure surface refinement · Performance of transportation assets · Structure-related parameter uncertainties · Reliability assessment

1 Introduction

Railway systems are significant, and thus the effects due to the failures of their assets are of utmost relevance. Asset management is key to maintaining the desired performance level. One of the many ways used in the industry is to prioritize interventions, to reduce risk related to a given failure mode. Nowadays, the available methodologies for the estimation of failure probability of slopes involve a probabilistic approach considering different sources of uncertainties (*e.g.*, parameter, model) [14]. Surrogate modeling techniques have been

J. C. Matos et al. (Eds.): IPW 2024, LNCE 494, pp. 366–376, 2024.
https://doi.org/10.1007/978-3-031-60271-9_34

employed to address slope stability problems involving various failure modes [4,9,10], emphasizing the importance of the experimental design of the material uncertainties for enhancing the predictive capabilities of the surrogate model. Fragility analysis has also been explored considering uncertainties of mechanical and geometric properties have been applied to the study of the fragility of slopes considering different hazard scenarios [4,13,19]. Likewise, reliability analysis techniques that offer more precise information about the failure probability are usually more demanding regarding computational power, specialized human resources, and quality of information [5,17]. Regardless of the different probabilistic methods used for the reliability analysis of slopes. Monte Carlo-based methods are still widely used due to the reduced computational cost of simulation methods like surrogate models [1,11].

The present case study aims to explore the influence of the sampling used for the development of surrogate models involved in the reliability analysis of soil slopes. The present research is structured as follows. Section 2 contains an overview of the case study. Section 3, presents the different employed training sampling methods. Section 4, discusses the effects on the reliability analysis of the case study. Finally, conclusions drawn during the research are discussed and follow-up studies are presented.

2 Problem Definition

2.1 Case Study and Sensitivity Analysis

The first case study will be based on a soil slope in which a risk-based assessment using surrogate models for the fragility analysis was previously studied [4]. Located in Concordância de Xabregas (Fig. 1), a segment of the Portuguese railway system, near Lisbon. The slope geometry is defined by an inclination between 45° and 60° and an average height of 8 m. The slope material consists of layers of marly limestone and limestone sandstone, commonly found in the region [15].

Methods using analytical solutions based on circular sliding surfaces have been applied with satisfactory reliability and less computational demand [19] than numerical models. The assessment of the slope was based on the Bishop simplified method (BSM [2]) implemented by a MATLAB script to achieve a faster integration with the sampling analysis. A safety factor (FS), the ratio between the ultimate resistance and the total load applied, was used to quantify the stability of the slope. It has been determined that parameter uncertainties related to mechanical properties and pore pressure are the most influential when assessing the stability of a slope [7]. Based on a sensitivity analysis, analysis of covariance (ANCOVA) [20], on the influence of parameter uncertainties on the stability of the asset, it was determined that the parameters with the most influence over the FS are the slope height and the resistant properties of the soil material (Fig. 1). However, since the scope of the present research is the application of the model to two independent variables space, the slope height and the friction angle were selected for the following analyses.

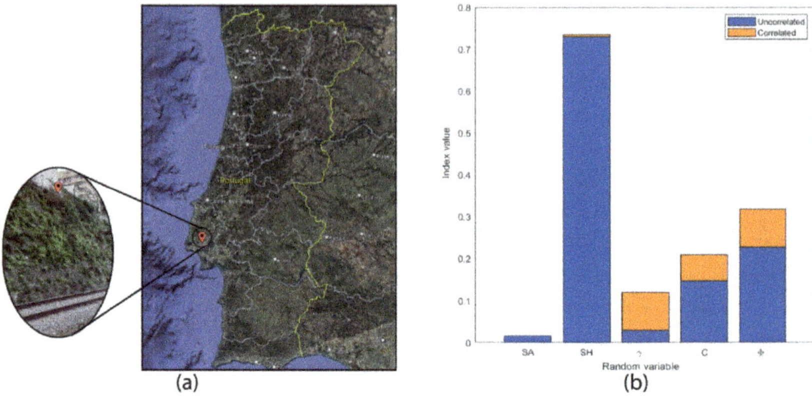

Fig. 1. Case study: (a) Location, (b) ANCOVA analysis for the soil slope analysis; SA-Slope angle, SH-Slope height, γ-Unit weight, c-Cohesion & Φ-Friction angle. Adapted from: [4]

2.2 True Capacity Surface and Validation Set

Traditionally, the G function is assumed to be defined over the input domain and is often smooth which is defined such that its negative values signal a failure event and positive values success. For the given soil stability problem, the capacity function is assumed as a safety factor in which the failure surface is defined by $FS = 1$. The research aims to evaluate the influence of sampling used for the training and refinement of the surrogate model used in the reliability assessment. For the case study, a two-dimensional G function and a 1-dimensional failure surface are defined, and the design point, based on the shortest distance from the origin to the failure surface in the standard Gaussian space of independent variables, will be used for the estimation of failure probability [18]. To this end, a grid of 400 points defined in a range of ± 5 standard deviations, for each of the random variables, was defined, and a surface based on the safety factor was obtained through the BSM (as seen in Fig. 2). Moreover, a validation set of 50 random points around the limit state $FS = 1$ was defined to validate the correct allocation in the failure/safety regions when employing the surrogate model.

3 Surrogate Modelling

The surrogate model will approximate a representation of the true capacity function, which is considerably less expensive to perform when estimating the failure probability of the asset. Surrogate modeling techniques have been proven to be efficient in describing non-linear limit state functions and implemented for the assessment of assets in transportation networks [3,6]. Therefore, a Kriging surrogate model using UQlab [12], was created and validated based on the random variables previously defined. The surrogate model uses a universal trend type, an anisotropic ellipsoidal Matérn 5/2 correlation function, used to define the

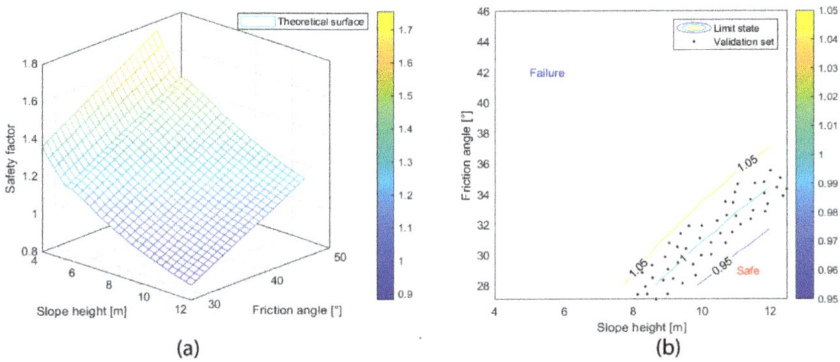

Fig. 2. Failure surface obtained through BSM for a two-dimensional space; (a) Surface plot, (b) Limit state and validation set.

Gaussian process and cross-validation estimation method. For the validation of the surrogate model, the leave-one-out method was used [8].

3.1 Training Sample Based on LHS

The surrogate model is used to assess the stability of the slope in terms of FS. An experimental design using the LHS method is generated and evaluated through the BSM stability analysis which then will be employed for the training of the surrogate model. To compare the effect of the training sample on the training of the surrogate model, two sets consisting of 50 and 200 evaluations of the stability model were used. For comparison, the residual standard error was computed for the validation set, moreover, samples that were misclassified between the safety and the failure regions were identified. Figure 3 shows the capacity surface in terms of safety factors for the range between 5 standard deviations. As can be observed the samples generated by the LHS are concentrated near the mean value (as represented by the red points) thus generating a greater difference for failure points that might provide more valuable information for the reliability analysis. It can be observed that increasing the number of samples without a proper distribution develops in an over-fitting of the surrogate model.

3.2 Training Sample Based on ASSM

Different reliability methods have included different assumptions that brought more efficiency to the solution of reliability problems. Importance sampling (IS) [16] allows the inclusion of error estimation and greatly enhances the reliability analysis. However, the main limitation of the evaluation of the failure probability is the large computational expense associated with the model evaluations required to achieve a reliable estimation. Surrogate models aid in reducing computational costs, there are strategies to refine the models based on already evaluated points in the domain. The adaptive sequential sampling method (ASSM),

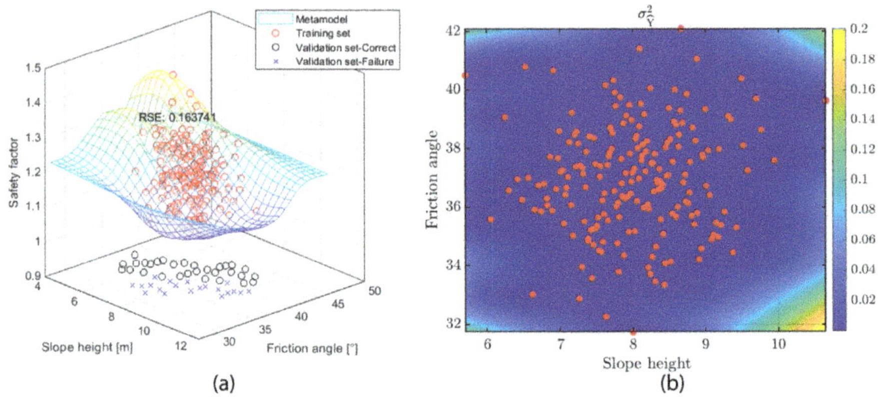

Fig. 3. Kriging metamodel developed based on an experimental design of 200 samples generated with the LHS; (a) Mean values, (b) standard deviation.

allows sequentially extending the experimental design in single steps to maximize the gain of information of the true G function and employing an inexpensive surrogate model, to achieve a quick reliability estimation. Thus, allows not making unnecessary calls of the expensive model in regions where samples are almost sure to be in safety or failure regions.

Exploration Phase. The general process for the application of the sampling is carried out in two phases, in which the first phase of the methodology aims to generate and evaluate samples based on the joint probability of unexplored space p_{out}, to identify the failure region [18]. The number of samples in each layer, n_i, will depend on the upper limit of the joint probability of the layer i and the number of independent random variables in the analysis N_{var}:

$$p_{out,i} = 10^{-i} \tag{1}$$

$$n_i = -N_{\text{var}} \ln \left(\frac{p_{\text{out},i}}{N_{\text{var}}} \right) \tag{2}$$

Samples are generated in advance for a set number of layers, in the present case study 10 layers were generated. To select the optimal distribution of samples, pressure based on Euclidean distance, to all the other samples within the same layer, and then the point with the most pressure, *i.e.*, closest to the other points within the same layer, is removed. Once the point is removed, the pressure is computed again, and this process is then repeated until the desired number of samples in the layer n_i is reached.

The final set of samples in each layer, *i.e.*, candidates, are evaluated by the true model, until failure is achieved $FS < 1$. For this particular case, failure was achieved for the first time for a sample within layer 5 with a probability of unexplored space equal to $p_{out} = 10^{-5}$. After achieving failure, the candidates left in the exploration set, *i.e.*, not evaluated by the true model, will be used in the exploitation phase. The set of samples that were assessed by the true model will be used for the training of a surrogate model, which will be used for comparison with the one obtained after the exploitation phase.

Exploitation Phase. The aim of the present phase is the refinement of the capacity surface. For this scenario, it is assumed as failure every sample in which the resulting safety factor presents a value lower than 1, in which the generation of random new candidates centered along the samples for which failure was found, and that will be added to the list of candidates from the exploratory set. Then, the selection criteria ψ are computed for each one of the candidates, and the selection of the best candidate is performed through the maximum value of the criteria. This process is repeated until the desired level of accuracy is reached.

$$\psi_{\mathrm{c}} = \sqrt{f_{\mathrm{c}} f_{\mathrm{s}}} \left(l_{\mathrm{c,s}}\right)^{N_{\mathrm{var}}} \tag{3}$$

In which the term $\sqrt{f_{\mathrm{c}} f_{\mathrm{s}}}$ relates to the average of the Gaussian probability density of both the candidate c and its nearest known point s, *i.e.*, sample evaluated with the true performance function. While the term $\left(l_{\mathrm{c,s}}\right)^{N_{\mathrm{var}}}$ is an approximation of the volume of the region and $(l_{\mathrm{c,s}})$ is the Euclidean distance between the candidate and its nearest point.

Next in Fig. 4, two plots are presented showing the different variations of the RSE and accuracy for each new best candidate found in the exploitation process. For easy display of information, a maximum of 150 best candidates was used. As can be observed in the figure, after 66 candidates it can be said that the failure surface of interest is well defined, *i.e.*, all points in the validation sample were correctly allocated in failure and safety regions. In addition, by assessing the RSE a similar conclusion can be drawn since the values of RSE do not present changes from this point. For the reliability analysis, the sampling set including 66 candidates will be used as the final training set.

In the exploitation phase, the reliability index for every sequential best candidate was obtained. In Fig. 4, the reliability index and the criteria are presented, as can be seen after the best candidate number 66 the difference between the upper and lower bound to the expected values stabilizes. As expected, it can be observed that there is more variation with the reliability index in the range when the RSE is higher.

Figure 5, presents the surrogate model obtained for the final training sample composed of the initial samples of the exploratory phase and 66 additional samples of the exploitation phase. As can be observed, the sample distribution is more optimal when compared to the results of the LHS, since it allows a closer behavior to the surface obtained using the BSM and can be observed that has a considerable effect on the resulting values of RSE. Moreover, the surrogate model also presents lower values of standard deviation for a wider range of values.

Finally, Fig. 6 shows the comparison between the results from the validation sets for the different training sets. As expected the results from the LHS with 200

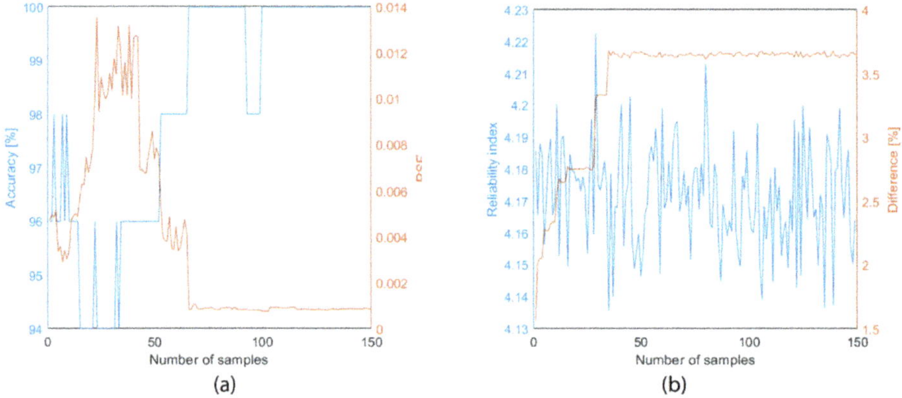

Fig. 4. Influence of the number of samples in the exploitation set on (a) the definition of the failure surface and corresponding RSE; (b) the reliability index

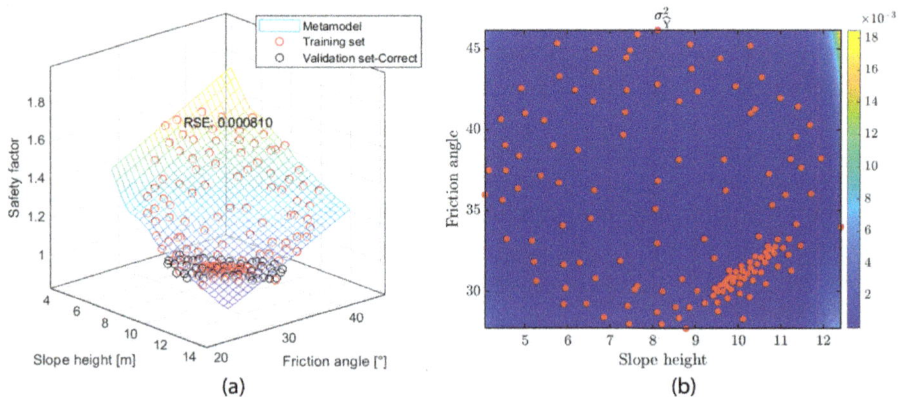

Fig. 5. Kriging metamodel developed based on exploitation set of 66 samples; (a) Mean values, (b) standard deviation.

samples are the furthest apart from the real values, more prominent for values in the failure region. As can be observed there are no considerable differences for the sets including 66 and 150 additional samples from the exploitation phase. However, it can be seen improvements in the region near the limit state by including the samples from the exploitation phase.

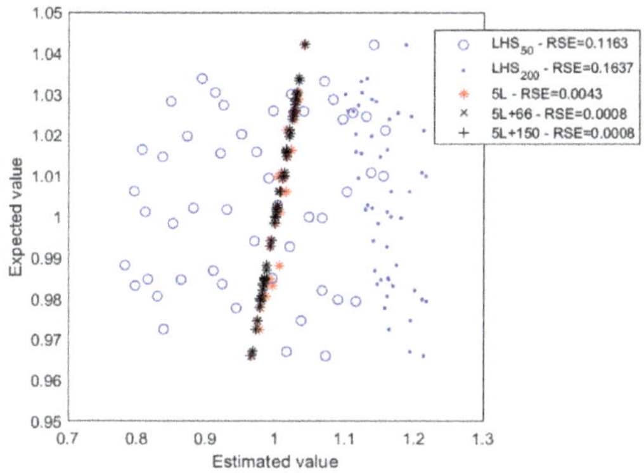

Fig. 6. Comparison between the expected value and the resulting values from the different training sets.

4 Reliability Analysis

To understand the effect on the reliability assessment of the different training sets, a reliability analysis using importance sampling was implemented, in which the method combines the FORM and MC methods for the approximation of the design point, but the exact convergence of the FORM analysis is not required [12,16]. For convergence the following stopping criterion based on β was used:

$$\frac{\widehat{\beta}^{+} - \widehat{\beta}^{-}}{\widehat{\beta}^{0}} \leq 5\% \tag{4}$$

Where the 5% threshold used as the stopping criteria computes the difference between the upper and lower bounds β^{+} and β^{-} over the expected value.

4.1 Comparison of the Design Point

Table 1 presents the different results for the training sets. As can be observed, by distributing the samples in regions near the limit state the reliability index obtained through the surrogate model is closer to the real value. Increasing the number of samples in an experimental design can lead to over-fitting in safe regions, thus not allowing the convergence of the reliability analysis, which was the case for the LHS with 200 samples. Finally, no significant increase in accuracy was identified by increasing the number of samples in the exploitation phase after the stabilization point.

Table 1. Final results for the different training sets

Training set	number of samples	Reliability index	RSE
Bishop simplified method	n/a	4.1846	n/a
Latin Hypercube	50	3.1858	0.116286
Latin Hypercube	200	n/a	0.163741
Exploration set 5 layers	72	4.1827	0.004287
Exploitation set 66 samples	138	4.1940	0.000810
Exploitation set 150 samples	222	4.1543	0.000842

5 Conclusions

Surrogate models are influenced by the training set used for their definition, thus, providing a suitable set that fits the desired goal of the model should be prioritized. In this research simple parameters such as sample size, distribution, and type of sampling were addressed. It was found that increasing the number of samples does not relate to more accuracy since it might lead to over-fitting. Moreover, by identifying and targeting regions closer to the limit state we can increase the reliability of the surrogate model without greatly increasing its computational cost. Finally, it is recommended that a proper identification of the most relevant random variables should be performed to decrease the computational cost of the reliability analysis.

Acknowledgement. This work is financed by national funds through FCT - Foundation for Science and Technology, under grant agreement 2022.09751.BD attributed to the 1st author.

The 3rd author acknowledges financial support from the Czech Science Foundation under project no 22-06684K.

This research was partially conducted under the project "Ferrovia 4.0" (POCI-01-0247-FEDER-046111; Lisboa-01-0247-FEDER-046111), by "Ferrovia" Consortium, and financed by European Regional Development Fund (ERDF), through the Incentive System to Research and Technological development, within the Portugal2020 Competitiveness and Internationalization Operational Program.

This work was partly financed by FCT/MCTES through national funds (PIDDAC) under the R&D Unit Institute for Sustainability and Innovation in Structural Engineering (ISISE), under reference UIDB/04029/2020 (https://doi.org/10.54499/UIDB/04029/2020), and under the Associate Laboratory Advanced Production and Intelligent Systems ARISE under reference LA/P/0112/2020.

References

1. Bardhan, A., Samui, P.: Probabilistic slope stability analysis of heavy-haul freight corridor using a hybrid machine learning paradigm. Transp. Geotech. **37** (2022). https://doi.org/10.1016/j.trgeo.2022.100815
2. Bishop, A.W.: The use of the slip circle in the stability analysis of slopes. Géotechnique **5**, 7–17 (1955). https://doi.org/10.1680/geot.1955.5.1.7. https://www.icevirtuallibrary.com/doi/10.1680/geot.1955.5.1.7
3. Cabanzo, C.M., SantamarÃa, M., Sousa, H.S., Matos, J.C.: In-plane fragility and parametric analyses of masonry arch bridges exposed to flood hazard using surrogate modeling techniques. Appl. Sci. **12**, 1886 (2022). https://doi.org/10.3390/app12041886
4. Cabanzo, C.M., Tinoco, J., Sousa, H.S., Coelho, M., Matos, J.C.: Adaptation of traditional risk-based methodology for slopes to probabilistic-based approach integrating surrogate models. Comput. Geotech. **161**, 105577 (2023). https://doi.org/10.1016/j.compgeo.2023.105577. https://linkinghub.elsevier.com/retrieve/pii/S0266352X23003348
5. Cheng, H., Chen, J., Chen, R., Chen, G., Zhong, Y.: Risk assessment of slope failure considering the variability in soil properties. Comput. Geotech. **103**, 61–72 (2018)
6. Guimarães, H., Matos, J.C., Henriques, A.A.: An innovative adaptive sparse response surface method for structural reliability analysis. Struct. Saf. **73**, 12–28 (2018). https://doi.org/10.1016/j.strusafe.2018.02.001
7. Habibagahi, G., Shahghotian, R.: Reliability analysis of rock slopes using theory of fuzzy sets. In: Numerical Models in Geomechanics - 8th Proceedings of the International Symposium on Numerical Models in Geomechanics, NUMOG 2002, vol. 27, pp. 547–551 (2002). https://doi.org/10.1201/9781439833797-c81
8. Hastie, T., Tibshirani, R., Friedman, J.: The Elements of Statistical Learning. Springer Series in Statistics. Springer, New York (2009). https://doi.org/10.1007/978-0-387-84858-7
9. Ji, J., Zhang, C., Gao, Y., Kodikara, J.: Reliability-based design for geotechnical engineering: an inverse form approach for practice. Comput. Geotech. **111**, 22–29 (2019). https://doi.org/10.1016/j.compgeo.2019.02.027
10. Ji, J., Zhang, C., Gui, Y., Lü, Q., Kodikara, J.: New observations on the application of LS-SVM in slope system reliability analysis. J. Comput. Civ. Eng. **31** (2017). https://doi.org/10.1061/(asce)cp.1943-5487.0000620
11. Jiang, S.H., Huang, J., Griffiths, D.V., Deng, Z.P.: Advances in reliability and risk analyses of slopes in spatially variable soils: a state-of-the-art review. Comput. Geotech. **141** (2022). https://doi.org/10.1016/j.compgeo.2021.104498
12. Marelli, S., Sudret, B.: UQLab: a framework for uncertainty quantification in Matlab, pp. 2554–2563 (2014). https://doi.org/10.1061/9780784413609.257
13. Rossi, N., Bačić, M., Kovačević, M.S., Librić, L.: Development of fragility curves for piping and slope stability of river levees. Water (Switzerland) **13** (2021). https://doi.org/10.3390/w13050738
14. da Silva, S.A.E.S.: Software with probabilistic approach of the slope quality index (SQI) (2015). https://hdl.handle.net/1822/40643
15. de Sousa, M.: Estabilidade de taludes em infra-estruturas rodoviárias engenharia civil (2017)
16. Tabandeh, A., Jia, G., Gardoni, P.: A review and assessment of importance sampling methods for reliability analysis. Struct. Safety **97** (2022). https://doi.org/10.1016/j.strusafe.2022.102216

17. Tang, G., Huang, J., Sheng, D., Sloan, S.W.: Stability analysis of unsaturated soil slopes under random rainfall patterns. Eng. Geol. **245**, 322–332 (2018). https://doi.org/10.1016/j.enggeo.2018.09.013
18. Vořechovský, M.: Reliability analysis of discrete-state performance functions via adaptive sequential sampling with detection of failure surfaces. Comput. Methods Appl. Mech. Eng. **401** (2022). https://doi.org/10.1016/j.cma.2022.115606
19. Wu, X.Z.: Development of fragility functions for slope instability analysis: fragility functions for slope instability analysis. Landslides **12**, 165–175 (2015). https://doi.org/10.1007/s10346-014-0536-3
20. Xu, C., Gertner, G.Z.: Uncertainty and sensitivity analysis for models with correlated parameters. Reliab. Eng. Syst. Saf. **93**, 1563–1573 (2008). https://doi.org/10.1016/j.ress.2007.06.003

Influence of Sample Size on the Bayesian Update of Statistical Hyperparameters of Concrete Compressive Strength

Tânia Feiri[1(✉)], Jan Philip Schulze-Ardey[2], Sebastian Kuhn[1], Udo Wiens[3], and Marcus Ricker[1]

[1] Chair of Structural Concrete, TU Dortmund University, 44227 Dortmund, Germany
tania.feiri@tu-dortmund.de
[2] Institute of Structural Concrete, RWTH Aachen University, 52074 Aachen, Germany
[3] German Committee for Structural Concrete, 10787 Berlin, Germany

Abstract. In structural reliability, the scatter of basic variables, as concrete compressive strength, can influence the failure probabilities and, therefore, the safety levels of new structural components or systems. The current Probabilistic Model Code, postulated by the Joint Committee of Structural Concrete, includes a statistical model for the description of concrete compressive strength by means of four hyperparameters (m, n, s and ν). Normal-Gamma or Lognormal-Gamma distributions are utilised as the conjugate prior distributions for updating the unknown mean and standard deviation of Normal or Lognormal distributions, respectively, describing the concrete compressive strength. This model was based on the preliminary work of Rackwitz conducted during the 1980's, which considered the statistical investigations led by Rüsch during the 1960's. A recent study suggests that some of these hyperparameters could be improved for more accurate safety assessments. However, the updating may depend on the way the concrete compressive strength values are assembled and, particularly, on the sample size configuration. An extensive database of experimental concrete compressive strength values of ready-mixed concretes is utilised to investigate the influence of the sample size on the updating of hyperparameters through Bayesian methods with prior information. The results suggest that an "optimal" sample size depends on the available data and how it is assembled and, therefore, for the updating, certain trade-offs are required. This investigation is relevant to fundament potential adjustments of the existing stochastic model and establish possible guidelines for practitioners to update the stochastic model of concrete strength when new experimental data becomes available.

Keywords: Concrete compressive strength · Stochastic models · Bayesian statistics · Prior information · Reliability analysis

J. C. Matos et al. (Eds.): IPW 2024, LNCE 494, pp. 377–387, 2024.
https://doi.org/10.1007/978-3-031-60271-9_35

1 Introduction

In structural engineering, uncertainties arise from numerous sources [1]. The use of probability-based concepts is the logical basis for dealing with uncertainty and, therefore, it is also the basis for evaluating structural safety. An important cornerstone of probability-based concepts refers to the stochastic description of basic variables that are inherent to the structural behaviour of a structural component or system. Examples are material properties (e.g., concrete compressive strength or yield strength of reinforcing steel), geometric dimensions (e.g., effective depth) or even model uncertainties intrinsic to resistance models (i.e., related to simplified relationships or complex numerical models) and to action effects models (i.e., related to the assessment of load effects and their combinations) [2].

Basic variables are stochastically described through probability distribution functions, which are defined in terms of their parameters (or moments). In turn, these probability distribution functions are used in structural reliability assessments to calculate failure probabilities and corresponding reliability indices β, which are commonly accepted as a measure of structural safety (e.g., [3, 4]). The probability distribution functions adopted for the stochastic description of basic variables have an influence on the resulting safety level of a structural component or system. For example, the scatter of basic variables can influence the failure probability and, therefore, the safety level of a newly designed component. To define the stochastic models for the basic variables, knowledge about the property being investigated (e.g., concrete compressive strength or effective depth) is necessary [4]. Existing stochastic knowledge – commonly designated by *prior* knowledge – can be updated into a posterior belief in case additional information becomes available. This additional information can be given by means of direct observations (e.g., experimental/test strength values) or through indirect information (e.g., information from experts) [5–7]. Bayesian statistics are typically adopted to update prior distribution functions into posterior distribution functions by means of likelihood functions (Fig. 1) [5–8]. These concepts are further described in Sect. 2.

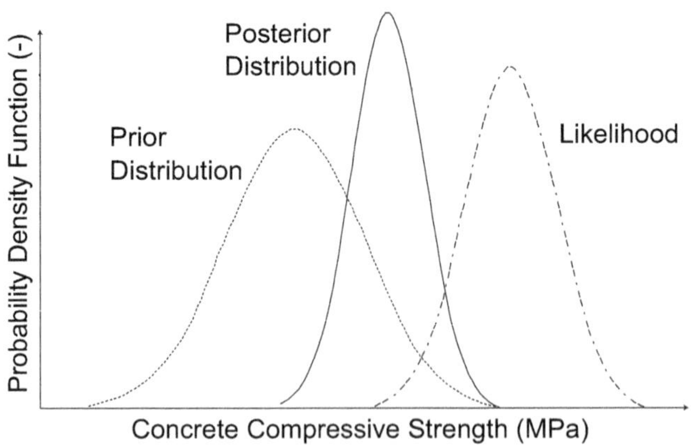

Fig. 1. Application of Bayes' theorem to derive a posterior distribution from a prior distribution based on new strength values (adapted from [9])

In the design of new reinforced concrete structures, the stochastic characterisation of concrete compressive strength plays a central role. The current Probabilistic Model Code (PMC) [10], postulated by the Joint Committee of Structural Concrete (JCSS) (hereafter denoted by JCSS PMC) includes the stochastic model for the description of concrete compressive strength through four hyperparameters (m, n, s and ν) (Table 1). These hyperparameters are the parameters of the Normal-Gamma and Lognormal-Gamma distribution. These distributions characterise the mean and variance of the Normal and Lognormal distributions, respectively, which are used to describe concrete compressive strength (e.g., [5, 8, 10, 11]). The hyperparameters in JCSS PMC were initially derived on the basis of the preliminary work of Rackwitz conducted during the 1980's [5, 8], which considered the results of statistical investigations on concrete compressive strength conducted by Rüsch during the 1960's [12, 13].

A recent study [11] suggested that some of these hyperparameters could be improved for more accurate safety assessments. However, the updating may depend on the way the concrete compressive strength values are assembled and, particularly, on the sample size configuration. In this paper, an extensive database of experimental concrete compressive strength values of ready-mixed concretes is utilised to investigate the influence of the sample size on the updating of concrete hyperparameters through Bayesian statistics.

Table 1. Prior hyperparameters for concrete strength distribution according to JCSS PMC [10]

Concrete grade	Prior parameters(*)			
	m'	n'	s'	ν'
C15	3.40	3.0	0.14	10
C25	3.65	3.0	0.12	10
C35	3.85	3.0	0.09	10
C45	3.98	3.0	0.07	10
C55	-	-	-	-

m' mean value of an equivalent sample of size n'
s' standard deviation of an equivalent sample of size $\nu'+1$
ν' degrees of freedom
(*) hyperparameters derived for cylindric samples (reference dimensions: 300 mm height and 150 mm diameter)

2 Fundamentals of Bayesian Updating Statistics

As explained, for example in [5, 7, 11], in Bayesian updating methods, a posterior distribution $f''(\theta|x)$ results from the combination of existing information expressed by a prior distribution $f'(\theta)$ and additional information attained, e.g., from new experimental data x_i, which is described by a likelihood function $L(A(x)|\theta)$. This relationship is expressed by Eq. (1).

$$f''(\theta|x) \propto f'(\theta) \cdot L(A(x)|\theta). \quad (1)$$

Note that the parameter θ in the likelihood function is a constant value and not a random variable. According to Rackwitz [5, 8], the prior distribution $f'(\mu, h) = f'_{N,\Gamma}(\mu, h|m', s', n', \nu')$ of the expected value μ and the precision (i.e., accuracy) h (with $h = 1/\sigma^2$) is expressed as a Lognormal-Gamma distribution characterised by the four (prior) hyperparameters m', s', n' and ν':

$$f'_{N,\Gamma}(\mu, h|m', s', n', \nu') = \frac{\sqrt{h \cdot n'}}{\sqrt{2 \cdot \pi}} \cdot \exp\left[-\frac{1}{2} \cdot \left(\frac{\mu - m'}{1/\sqrt{h \cdot n'}}\right)^2\right] \cdot \frac{\left(1/2 \cdot \nu' \cdot s'^2 \cdot h\right)^{(\frac{1}{2} \cdot \nu' - 1)}}{2 \cdot \Gamma \cdot (\nu'/2)} \cdot \nu \cdot s'^2 \tag{2}$$

In turn, the posterior distribution $f''(\mu, h) = f''_{N,\Gamma}(\mu, h|m'', s'', n'', \nu'')$ results as:

$$f''_{N,\Gamma}(\mu, h|m'', s'', n'', \nu'') = \frac{f'_{N,\Gamma}(\mu, h|m', s', n', \nu') \cdot L(\mu \cdot h|x)}{\iint f'_{N,\Gamma}(\mu, h|m', s', n', \nu') \cdot L(\mu \cdot h|x) d\mu dh} \tag{3}$$

The denominator in Eq. (3) refers to the marginal distribution and results as a constant value after integration. Since the prior conjugate distribution is assumed as a Lognormal-Gamma distribution, the posterior distribution is also considered as a Lognormal-Gamma distribution, which can be utilised again as a prior distribution for a further update of the hyperparameters when new experimental data is available.

As explained in [11], for the characterisation of concrete compressive strength, Student's-t distributions or log-Student's-t distributions can be adopted. These distributions are the marginal posterior for the normal mean with unknown variance and are utilised to estimate the population hyperparameters of small sample sizes. For sufficiently large sample sizes, this distribution converges towards the "base distribution" (i.e., to the Normal distribution in the case of Student's-t distribution and to Lognormal distribution in the case of log-Student's-t distribution). In the case of Student's-t distribution, it should be highlighted that it has a greater chance for extreme values than a Normal distribution and, therefore, it has heavier (or fatter) tails.

In the context of this investigation, the input variable X refers to concrete compressive strength, which is here assumed to follow a Lognormal distribution. As a result, the marginal distribution is assumed as a log-Student-t distribution [5]:

$$f_X(x|m'', s'', n'', \nu'') = \frac{1}{x - s''} \cdot t_{\nu''} \cdot \left(\frac{\ln(x) - m''}{s''} \cdot \sqrt{\frac{n''}{n'' + 1}}\right) \tag{4}$$

where $t_{\nu''}(\cdot)$ is the probability density function of the Student-t distribution with ν'' degrees of freedom.

The mathematical formulas expressing the relationship between the prior hyperparameters m', s', n', ν' and the posterior hyperparameters m'', s'', n'', ν'' are given by Eq. (5) to Eq. (9) (e.g., [5, 11]):

$$n'' = n' + n \tag{5}$$

$$m'' = \frac{m' \cdot n' + m \cdot n}{n''} \tag{6}$$

$$\nu'' = \left[\nu' + \delta(n')\right] + \left[\nu + \delta(n) - \delta(n'')\right] \tag{7}$$

with:

$$\delta(x) = \begin{cases} 0 \, for \; x = 0 \\ 1 \, for \; x > 0 \end{cases} \tag{8}$$

$$s'' = \frac{1}{\nu''} \cdot \left[\left(\nu' \cdot s'^2 + n' \cdot m'^2\right) + \left(\nu \cdot s^2 + n \cdot m^2\right) - n'' \cdot m''^2\right] \tag{9}$$

Here, the degrees of freedom correspond to $\nu = n - 1$ if both m and s^2 are determined from a sample of size n.

Rackwiz [5] proposed a so-called Maximum Likelihood Estimator (MLE) to estimate the parameters of the prior distribution for the mean and precision (Lognormal-Gamma distribution). MLE can be utilised if the prior information is available in the direct form of data pairs of the sample parameters m_i and s_i from k samples. Due to space limitations, the mathematical formulas inherent to this quantification cannot be here presented; yet, they are thoroughly described, for example, in [5, 11].

3 Experimental Data of Concrete Compressive Strength

3.1 Experimental Strength Values

The experimental strength values utilised in this investigation were measured by the PÜZ Laboratory at the Hochschule Biberach (Germany). The values were retrieved from concrete specimens originated from construction sites and concrete production plants in the region of southern Baden-Württemberg and western Bavaria (Germany) between 2011 and 2022. The tests were conducted on cubic specimens (side length = 150 mm) of different concrete classes (C20/25, C25/30, C30/37, C35/45, C40/50, C45/55 and C50/60). Note that the prior parameters proposed by Rackwitz [5, 8] were derived from cubic samples (side length = 200 mm). Yet, the prior parameters in JCSS PMC [10] were established for cylinders (diameter = 150 mm; height = 300 mm). Since the strength values considered in this investigation were taken from cubic samples (side length = 150 mm), a conversion was necessary [11]. The values were compiled in a database built in *Microsoft Access*. Further details concerning data properties, data handling and statistical treatment of the data are described in [11]. All the calculations were conducted in the statistical software *R* [14].

3.2 Sample Size Configuration

To understand the influence of the sample size on the update of concrete hyperparameters, several configurations were idealised and evaluated (Table 2). Each sample includes strength values measured on specimens originated from the same concrete mixture and the same concrete production plant. For the samples, the mean and the precision were determined. The MLE were then used to derive the parameters of the Lognormal-Gamma distribution, which describes the distribution of the mean (i.e., expected value) and variance of the Lognormal distribution describing the concrete compressive strength.

Table 2. Number of strength values and samples per concrete class

Values per sample		Concrete classes						
		C20/25	C25/30	C30/37	C35/45	C40/50	C45/55	C50/60
5	No. Values	49	180	897	782	112	34	36
	No. Samples	9	31	167	142	20	6	6
10	No. Values	49	140	803	745	91	20	28
	No. Samples	4	12	70	62	7	2	2
15	No. Values	49	92	745	610	56	20	28
	No. Samples	2	5	43	31	3	1	1
20	No. Values	49	140	803	745	91	20	28
	No. Samples	2	3	27	21	1	1	1
25	No. Values	29	30	617	486	25	-	28
	No. Samples	1	1	22	16	1	-	1

4 Discussion

Table 3 presents the prior hyperparameters and corresponding prior information derived from JCSS PMC [10] for all the concrete classes considered in this investigation. Table 4 lists the posterior hyperparameters and corresponding posterior information also derived for all the concrete classes, considering different sample size configurations: 5, 10, 15, 20 and 25 values per sample (Table 2). The results indicate that the sample size configuration influences the derivation of hyperparameters for all the concrete classes. By comparing the results of Table 3 and Table 4, it can be observed that for all the concrete classes, the posterior mean values m'' decrease in comparison to the prior mean values m'. Likewise, the posterior standard deviations values s'' are smaller than the corresponding prior standard deviations values s'. As opposed to this trend, for all the concrete classes, the posterior degrees of freedom values ν' are larger than the prior ν' values. The results of Table 4 indicate that small sample size configurations (5 and 10 values per sample) enable the update (or the derivation) of hyperparameters for all the concrete classes, even when the number of experimental strength values available is limited. For larger sample sizes (i.e., 15, 20 and 25 values per sample), not all the hyperparameters could be derived. Note that for 20 values per sample, the hyperparameters could be derived for four classes: C20/25, C25/30, C30/37 and C35/45. For 25 values per sample, only the classes C30/37 and C35/45 yielded results. This can be explained by the amount of data available – and consequently, samples that could be created – for these classes; the larger the dataset is, the more chances are to generate data samples and, consequently, derive or update hyperparameters.

Particularly interesting to analyse here is the impact of these hyperparameters on the 5% quantile values (Table 4 and Fig. 2). For the **class C20/25**, the 5% quantile value marginally increases between samples with 5 values (= 26.24 N/mm^2) and samples with 10 values (= 26.35 N/mm^2). Then, it suffers a minor decrease to 25.73 N/mm^2 for 15

Table 3. Prior information according to JCSS PMC [10]

Prior information		C20/25	C25/30	C30/37	C35/45	C40/50	C45/55	C50/60
JCSS PMC	m'	3.525	3.650	3.750	3.850	3.915	3.980	-
	N/mm^2	*34.50*	*39.00*	*42.96*	*47.35*	*50.45*	*53.76*	-
	n'	3.000	3.000	3.000	3.000	3.000	3.000	-
	s'	0.130	0.120	0.105	0.090	0.080	0.070	-
	N/mm^2	*6.19*	*6.45*	*6.21*	*5.86*	*5.54*	*5.17*	-
	ν'	10.000	10.000	10.000	10.000	10.000	10.000	-
	5% Quant	**25.34**	**29.36**	**33.57**	**38.37**	**41.88**	**45.71**	**-**

values per sample, after rising again to 26.37 N/mm^2 for 20 values per sample. Despite this minor fluctuation, it can be argued that these values are practically constant for all the sample sizes.

For the **class C25/30**, the 5% quantile value also decreases between samples with 5 values each (= 29.85 N/mm^2) and 10 values each (= 28.87 N/mm^2). Then, the 5% quantile value decreases sharply when samples with 15 values each are considered (= 26.48 N/mm^2). However, these values remain practically constant when 20 values per sample are adopted (= 26.57 N/mm^2).

For the **class C30/37**, the 5% quantile value decreases for all the classes with an increasing sample size. They range between 34.14 N/mm^2 for samples with 5 values each and 31.90 N/mm^2 for samples with 25 values each. However, it can be argued that for samples with 20 and 25 values, the differences between the 5% quantile are minor.

For the **class 35/45**, the 5% quantile value varies marginally between the different sample size configurations, ranging between a minimum of 39.34 N/mm^2 (for 5 values per sample) and a maximum of 39.65 N/mm^2 (for both 15 and 25 values per sample).

For the **class 40/45**, the 5% quantile value increases from 42.75 N/mm^2 (for 5 values per sample) and 43.37 N/mm^2 (for 15 values per sample). Note that no further values could be derived for larger values per sample due to limited data available.

For the class **C45/55**, the 5% quantile value decreases from 46.69 N/mm^2 (for 5 values per sample) to 45.45 N/mm^2 (for 10 values per sample). Like in the previous strength class, no further values could be derived for larger values per sample due to limited experimental data (Table 2).

Finally, for the **class C50/60**, no robust conclusions can be derived since no prior information exists for this class. From Table 2, it can be observed a considerable limitation of the data available.

Table 4. Posterior information according to sample size configuration

Posterior information		C20/25	C25/30	C30/37	C35/45	C40/50	C45/55	C50/60
5 values per sample	m''	3.52	3.635	3.737	3.845	3.912	3.968	3.983
	N/mm^2	*34.186*	*38.314*	*42.319*	*47.041*	*50.248*	*53.053*	*57.158*
	n''	3.409	3.21	3.188	3.161	3.194	3.61	1.036
	s''	0.118	0.111	0.096	0.08	0.073	0.06	0.052
	N/mm^2	*5.286*	*5.595*	*5.353*	*4.970*	*4.807*	*4.033*	*20.969*
	ν''	12.233	12.847	13.159	13.232	12.604	14.896	5.059
	5% Quant	**26.24**	**29.85**	**34.13**	**39.34**	**42.75**	**46.69**	**29.91**
	Dif.(*)	**0.90**	**0.49**	**0.56**	**0.97**	**0.87**	**0.98**	**29.91**
10 values per sample	m''	3.435	3.596	3.723	3.843	3.909	3.921	3.986
	N/mm^2	*31.18*	*36.82*	*41.72*	*46.91*	*50.04*	*50.55*	*54.37*
	n''	48.118	3.718	3.403	3.283	3.345	23	1.211
	s''	0.096	0.114	0.098	0.079	0.069	0.060	0.054
	N/mm^2	*3.09*	*5.27*	*5.27*	*4.79*	*4.44*	*3.21*	*7.66*
	ν''	50.891	16.364	14.177	14.944	15.033	30	13.399
	5% Quant	**26.35**	**28.87**	**33.68**	**39.49**	**43.07**	**45.45**	**42.70**
	Dif.(*)	**1.02**	**−0.49**	**0.11**	**1.12**	**1.19**	**−0.26**	**42.70**
15 values per sample	m''	3.422	3.495	3.709	3.844	3.905	-	-
	N/mm^2	*30.80*	*33.24*	*41.14*	*46.94*	*49.82*	-	-
	n''	33	12.045	3.645	3.259	3.478	-	-
	s''	0.102	0.121	0.1	0.077	0.066	-	-
	N/mm^2	*3.28*	*4.44*	*5.21*	*4.65*	*4.12*	-	-
	ν''	40	20.747	14.983	16.029	19.381	-	-
	5% Quant	**25.73**	**26.48**	**33.19**	**39.65**	**43.37**	**0.00**	**0.00**
	Dif.(*)	**−0.40**	**2.88**	**0.38**	**−1.27**	**−1.49**	**45.71**	**0.00**
20 values per sample	m''	3.432	3.463	3.685	3.836	-	-	-
	N/mm^2	31.083	32.096	40.182	46.533	-	-	-
	n''	97.13	41.125	4.123	3.535	-	-	-
	s''	0.095	0.107	0.105	0.077	-	-	-
	N/mm^2	3.023	3.575	*5.24*	*4.47*	-	-	-
	ν''	53.97	39.753	15.137	18.172	-	-	-
	5% Quant	**26.37**	**26.57**	**32.18**	**39.57**	**-**	**-**	**-**
	Dif.(*)	**1.04**	**-2.79**	**−1.39**	**1.20**	**-**	**-**	**-**

(*continued*)

Table 4. *(continued)*

Posterior information		C20/25	C25/30	C30/37	C35/45	C40/50	C45/55	C50/60
25 values per sample	m''	-	-	3.681	3.838	-	-	-
	N/mm ^{2}	-	-	40.057	46.665	-	-	-
	n''	-	-	4.19	3.427	-	-	-
	s''	-	-	0.108	0.076	-	-	-
	N/mm ^{2}	-	-	*5.36*	*4.50*	-	-	-
	ν''	-	-	14.586	17.788	-	-	-
	5% Quant	**-**	**-**	**31.90**	**39.65**	**-**	**-**	**-**
	Dif	**-**	**-**	**−1.67**	**1.28**	**-**	**-**	**-**

(*)*These values refer to the difference between the prior and posterior 5% Quantile values of the respective concrete strength class*

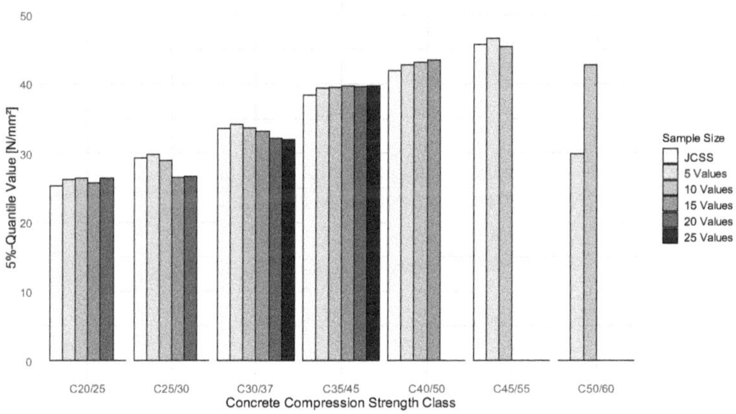

Fig. 2. Variation of 5% quantile values in dependency of the sample size configuration

5 Recommendations

The results discussed in Sect. 4 confirm that the update of the hyperparameters is sensitive to the sample size configuration. However, they do not show an obvious trend that can be recommended to engineering practitioners when new data is available. Therefore, for the updating of hyperparameters different sample size configurations should be investigated in order to find the minimum sample size configuration that leads to constant (or nearly constant) 5% quantile values. Ideally, all the sample size configurations should be analysed. Notwithstanding, at least, samples with 5, 10 and 15 values per sample should be investigated. As a final remark, it is important to highlight the value of experimental strength data; the more data available, the more likely it is to generate different size samples – and ultimately, derive or update the hyperparameters.

6 Conclusions

The results of this investigation led to the following conclusions:

- The results confirm that the update of the hyperparameters is sensitive to the sample size configurations.
- In all the sample size configurations and for all the concrete classes, the posterior values for the mean m'' and the standard deviation s'' decrease in comparison to the respective prior values. Contrary to this trend, the posterior degrees of freedom values ν'' increase in comparison to the prior degrees of freedom values ν'.
- Small sample size configurations (5 and 10 values per sample) enable the update or the derivation of hyperparameters for all the concrete classes even when the number of experimental strength values is limited.
- Hyperparameters could be derived for the classes C30/37 and C35/45 independently of the sample size configuration. The results suggest that large sets of data facilitate the derivation or updating of the hyperparameters for any sample size configuration.
- The recommendation to practitioners is to derive hyperparameters under the consideration of different sample sizes until a nearly constant 5% quantile value is attained. At least 5, 10 and 15 values per sample should be always analysed.

Further investigations should include the impact of these size configurations – and ultimately, the posterior hyperparameters – on the safety level of structural components.

Acknowledgments. The Authors wish to express their gratitude to the German Committee for Structural Concrete (*In German*: Deutscher Ausschuss für Stahlbeton e.V., DAfStb) for funding this investigation.

References

1. Ellingwood, B.: Probability-based codified design: past accomplishments and future challenges. Struct. Saf. **13**(3), 159–176 (1994)
2. Holický, M., Retief, J., Sýkora, M.: Assessment of model uncertainties for structural resistance. Probab. Eng. Mech. **45**, 188–197 (2016)
3. Ricker, M., Feiri, T., Nille-Hauf, K., Adam, V., Hegger, J.: Enhanced reliability assessment of punching shear resistance models for flat slabs without shear reinforcement. Eng. Struct. **226**, 111319 (2021)
4. Melchers, R., Beck, A.: Structural Reliability Analysis and Prediction. John Wiley & Sons, Hoboken (2018)
5. Rackwitz, R.: Predictive distribution of strength under control. Matériaux et Constr. **16**, 259–267 (1983)
6. Caspeele, R.: From quality control to structural reliability: where Bayesian statistics meets risk analysis. Heron **59**(2–3), 79–100 (2014)
7. Caspeele, R., Taerwe, L.: Numerical Bayesian updating of prior distributions for concrete strength properties considering conformity control. Adv. Concrete Constr. **1**(1), 85–102 (2013)
8. Rackwitz, R.: Zur Statistik von Eignungs- und Zulassungsversuchen für Bauteile. Bauingenieur **56**(3) (1981)

9. Neves, L., Wiśniewski, D., Cruz, P.: Bayesian updating, a powerful tool for updating engineering models using results of testing and monitoring. Dolnoslaskie Wydawnictwo Edukacyjne (DWE) (2007)
10. Joint Commitee on Structural Safety. JCSS Probabilistic Model Code (2001). https://www.jcss-lc.org
11. Feiri, T., Kuhn, S., Wiens, U., Ricker, M.: Updating the prior hyperparameters of concrete compressive strength through Bayesian statistics for structural reliability assessment. Structures **58**, 105636 (2023)
12. Rüsch, H.: Zur statistischen Qualitätskontrolle des Betons/on the statistical quality control of concrete/ au sujet du contröle statistique de qualité du béton. Mater. Test. **6**(11), 387–394 (1964)
13. Rüsch, H., et al.: Statistische Analyse der Betonfestigkeit. Beuth (1968)
14. R Core Team: R: A Language and Environment for Statistical Computing, Vienna, Austria (2016). https://www.R-project.org/

Influence of the Time-Dependent Strength Degradation on the Reliability of FRP-Reinforced Concrete Members

Arne Spelter[1(✉)], Jan Philip Schulze-Ardey[2], Tania Feiri[3], Marcus Ricker[3], and Josef Hegger[2]

[1] H+P Ingenieure GmbH (Hegger + Partner), Aachen, Germany
ASpelter@huping.de
[2] Institute of Structural Concrete, RWTH Aachen University, Aachen, Germany
[3] Chair of Structural Concrete, TU Dortmund University, Dortmund, Germany

Abstract. Fibre reinforced plastic (FRP) reinforcements are characterised by a time-dependent material behaviour which results from strength losses. In contrast to basalt and glass reinforcements, carbon reinforcements show only minor strength losses, and the long-term durability is significantly influenced by the type of impregnation. Time-dependent reliability assessments can be used to quantify the influence of the time-dependent strength degradation on the safety level of FRP-reinforced concrete members.

However, in the design of FRP reinforcements, the assumption of the residual strength at the end of service life is normally on the safe side. This paper presents the results of a structural reliability assessment on FRP-reinforced concrete members subjected to bending loads. By using a First-Order-Reliability Method (FORM) and Monte-Carlo Simulations (MC), the time-dependent strength of the FRP reinforcement is investigated for different reference periods (1 to 100 years). Specific time intervals were chosen to significantly reduce the computational effort that is commonly required by the MC. For the action side, the live loads were considered time-independent by means of a Gumbel distribution.

The results of studies in this paper indicate that the time variance has a positive effect on the safety level of FRP-reinforced concrete members when the FRP reinforcement failure becomes governing.

Keywords: Time variance · FRP-reinforced structures · structural reliability · FORM · Monte-Carlo Simulation

1 Introduction

For several decades, the construction industry has been characterised by reinforced concrete structures using steel reinforcement. This construction method combines the relatively high compressive strength of concrete at relatively low costs for formwork and casting with the high strength of reinforcement steel in the tensile zone. Thus, for a long time, it has represented a cost-effective construction method. Today, the building

J. C. Matos et al. (Eds.): IPW 2024, LNCE 494, pp. 388–397, 2024.
https://doi.org/10.1007/978-3-031-60271-9_36

industry is increasingly influenced by the rising construction and energy costs, as well as a shortage of resources and materials. As a result, particular emphasis is placed on saving resources and CO_2 emissions, maintaining, retrofitting, and strengthening existing building structures as well as ensuring high durability of new structures [1, 2].

The high costs for the maintenance, retrofitting and strengthening of existing concrete structures, as well as the demand for CO_2 savings, have facilitated the development of novel reinforcement materials. The application of fibre-reinforced polymer (FRP) reinforcements in concrete structures, represents one of the major innovations in the construction industry in recent decades [3]. FRP reinforcements mainly made of glass or carbon fibres and a polymer matrix are already widely used in practice. They are divided into rebars, strands, sheets, 2D and 3D grids, and preformed textile elements. Bar-shaped reinforcements, as well as sheets and strands have already been used for reinforcement for several years. Today, grids are also used as reinforcement for sandwich constructions [4–6], flexural and shear strengthening of reinforced or prestressed concrete components [7, 8], slabs [9], pedestrian and cycle bridges, and even small road bridges [10–13].

In Germany, the term "carbon reinforced concrete" (CRC) is commonly used for concrete with FRP reinforcement made of carbon or glass fibres, which offers many advantages over conventional steel reinforcement. The concrete cover can be significantly reduced because no corrosion protection for the reinforcement is needed. Thus, only the transfer of bond stresses between concrete and reinforcement needs to be ensured [14]. In addition to saving resources and, consequently, CO_2 emissions, CRC allows for a more flexible design of structural elements due to the variable shaping of soft-impregnated FRP reinforcements. However, when using FRP reinforcement, it should be considered that strength losses can occur during the service life of a component due to external factors as stress, temperature, moisture, and alkalinity. Consequently, the load-carrying capacity of CRC components can decrease over time. These strength losses are caused by damage to the impregnation and filament material. Strength losses should be determined by an appropriate test procedure and considered in standards and in the approval process of non-metallic reinforcements. In design, the strength losses are considered by a reduction factor or directly through the assumption of a reduced long-term strength.

Fibre reinforced plastic (FRP) reinforcements are characterized by a time-dependent material behaviour which results from the strength losses of the reinforcement. In contrast to basalt and glass reinforcements, carbon reinforcements show only minor strength losses, and the long-term durability is significantly influenced by the type of impregnation. Time-dependent reliability assessments can be used to quantify the influence of the time-dependent strength degradation on the reliability level of FRP-reinforced concrete members [15, 16].

In the design of FRP reinforcements for CRC, the residual strength at the end of the intended service is usually estimated on the safe side, i.e., the time-dependent degradation is overestimated. This paper presents the results of a structural reliability assessment of FRP-reinforced concrete members subjected to bending loads. By using a First-Order Reliability Method (FORM) and direct Monte-Carlo Simulations (MC), the time-dependent strength of the FRP reinforcement is investigated for different reference periods (1 to 100 years). Specific time intervals were chosen to significantly reduce the computational effort that is commonly required by the MC. In addition, the influence of uncertainties originated from the determination of the long-term strength was considered. These uncertainties result from extrapolation and time-temperature shift of the long-term durability strength, which is usually determined in long-term tests having a maximum duration of one year. The results indicate that the time variance has a positive effect on the reliability level of CRC members when the FRP reinforcement failure is governing.

2 State of the Art

As described in Sect. 1, FRP reinforcement for concrete structures has several advantages. To investigate this construction method with FRP reinforcement, reliability analysis has been done. The basics are described in this Sect. 2.

Reliability-based evaluations are necessary to prove the (theoretical) value of failure probabilities. To calculate those probabilities and derive partial safety factors, multiple probabilistic methods are available (e.g.; [17]). In the last century (e.g., [18–21]), there were several developments in the field of reliability analysis applied to civil structures. This included the development of different probabilistic methods. The calculation of failure probabilities is often an integration of a high dimensional probability function and, therefore, can be considered a complex mathematical problem [22]. Only simple problems with basic random (or stochastic) variables can be solved with analytic mathematical methods. In general, for multidimensional problems, the accurate solution of the integrals can only be achieved through approximation methods (e.g., FORM) or by means of numerical methods (e.g., MC). For many practicing engineers, these methods can be complex to handle [23]. Plotting the two-dimensional joint probability density function $f_{E,R}(e, r)$ allows for a better understanding of the fundamentals (Fig. 1): The volume of the distribution density is one and divided into a failure and a survival region by the limit state function (LSF) $G = R - E = 0$. The volume of the failure region corresponds to the failure probability P_f and can be calculated by integration of the so-called "convolution integral" according to Eq. (1):

$$P_f = \int_{-\infty}^{+\infty} f_E(e)\left(\int_{-\infty}^{r=e} f_R(r)dr\right)\mathrm{d}e \tag{1}$$

The design point corresponds to the peak of the failure volume (see Fig. 1 (a)). A transformation to standard normal space allows the plotting in a two-dimensional coordinate system with the peak of the density function at the origin surrounded by concentric circles for equal reliability levels (Fig. 1(b)). The design point is located on the perpendicular of the limit state line. The minimum distance between the design point and the origin is the so-called reliability index β, which is considered as a measure of structural safety.

Geometrically, the weighting factors α_E and α_R represent the direction cosine between the coordinate axis and the perpendicular to the LSF. In general, the failure probability is linked to n variables, resulting in an n-dimensional density function, which is divided by an $(n - 1)$ dimensional hypersurface into a failure and survival space. The minimum distance between the limit state surface and the origin of the coordinate system is also defined as the design point.

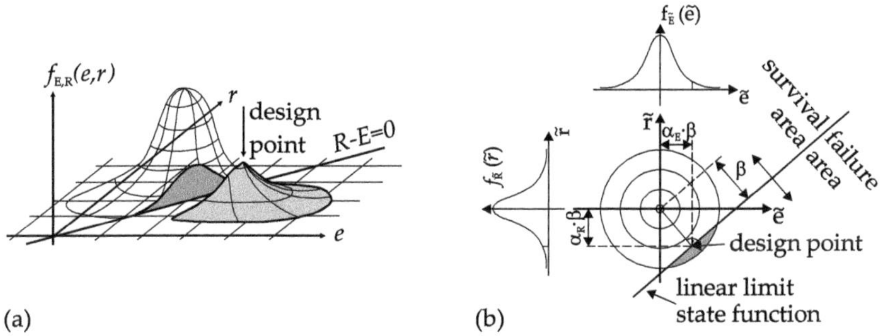

Fig. 1. (a) Two-dimensional density function and (b) transformed density function in standardized Normal space reproduced from [24]

3 Testing Concepts for the Long-Term Durability Investigation

On the basis of the test methods and testing concepts for creep rupture, alkaline resistance, long-term durability of FRP rebars [25–28] as well as experimental investigations [15], two testing concepts to investigate the long-term durability tensile strength of FRP grids and FRP rebars were developed. The objectives are either to derive the long-term durability strength $f_{nm,100a}$ or a reduction factor $\alpha_{nmt,100a}$ of FRP grids (focus on CFRP) in cementitious matrices, after service lives of up to 100 years. Both testing concepts – the "time-to-failure concept" and the "residual capacity concept" are described in detail in [15]. Only the essential boundary conditions for the long-term tests are briefly described here.

The damage mechanisms occurring in concrete structures are reproduced by stress, moisture, and alkalinity in long-term durability tests. Hereby, elevated temperatures are used to accelerate the aging process for these damage mechanisms.

The fibre material and the polymer matrix of the reinforcement strongly influence the choice of stress levels. Thus, higher stress levels are applied to CFRP reinforcements and "stiff" impregnation materials, e.g., epoxy resin or vinyl ester resin, than to FRP reinforcements with glass fibres and "soft" impregnation materials, e.g., polystyrene. Furthermore, the type of reinforcement is important. Due to the lower scatter of strength of FRP rebars compared to FRP grids since, due to the different manufacturing processes, higher stress levels can be applied in the long-term durability tests of FRP rebars.

For the choice of stress levels, pretests with different stress levels (e.g., 75%, 80% and 85% of the reference strength of a GFRP reinforcement and 85%, 90% and 95% of the reference strength of a CFRP reinforcement) are recommended.

For the investigations on CRC (tensile) specimens, a representative cementitious matrix with a pH value of 13 to 14 should be selected. To allow the transferability between different concrete strengths, the lowest strength class should be assessed in long-term durability tests. To exclude an influence on the concrete strength, short-term tests with different concrete strengths under otherwise identical specifications should be carried out.

The two testing concepts differ mainly in the selection of stress levels and the procedures to extrapolate the test data. The testing concept is selected depending on the reinforcement material under investigation. If failure can be achieved due to high load levels as a result of constant loading, the time-to-failure concept is selected. If no failure can be achieved despite very high load levels, as it was the case in [15] for an epoxy resin-impregnated carbon grid, the residual capacity concept can be applied.

4 Reliability Analyses

The long-term durability of FRP reinforcement has a decisive influence on the strength that can be applied to the design of CRC components. The (tensile) strength of some FRP materials is time-dependent, which will be evaluated for an GFRP reinforcement in reliability analyses with FORM and "crude" Monte Carlo method. In previous studies [15] the difference between the results of the FORM und MCM was investigated. As the reliability index differed only slightly, one study with FORM and one with MCM will be presented in this section. First, the statistical parameters of the basic variables that are applied in the reliability analyses are derived. Then, the procedures of the FORM and MC are presented in this section.

For the evaluation of the influence of the long-term durability strength on the reliability of CRC components, reliability analyses were carried out. All reliability analyses were performed with a self-programmed code in the statistic software *R* [29] and performed on one-way slabs. The software code for the reliability analyses FORM and "crude" Monte Carlo method were validated for the time-independent material behaviour. A partial factor of γ_{NM} of 1.3 for the non-metallic reinforcement was applied in all investigations according to [30].

The time-dependent tensile strength was investigated mainly via the reduction factor $\alpha_{\mathrm{nmt,d}}$ in the design of the CRC components and by $\alpha_{\mathrm{nmt,RA}}$ in the reliability analyses (see Fig. 2). The relationship between long-term tensile strength of the FRP reinforcement and time is linear in a double-logarithmic scale. For further details see [15]. Accordingly, the strength decreases rapidly in the first year, so that there is a limited favourable effect on the reliability of CRC components, as the results show. As the tests are carried out on a fully probabilistic basis, the reduction factor $\alpha_{\mathrm{nmt,d}} = 0.6$ is applied for the preliminary investigation to determine related values for the permanent and changeable actions g_{k} and q_{k}. The reduction factor $\alpha_{\mathrm{nmt,RA}}$ is applied in the following reliability analyses. The value $\alpha_{\mathrm{nmt,RA}}$ changes depending on the service life or reference period, respectively (see Fig. 2).

As it can be observed from Fig. 2 (a), the strictly monotonically decreasing function of the reduction factor as a function of the logarithm of time results in an uneven distribution of the reduction factors for a division by years. This means that the reduction factor decreases exponentially. Table 1 summarizes the key parameters of the studies presented below.

The slab dimensions of the studies "G1" and "G2" are 5.00 m × 2.40 m × 0.12 m and 3.50 m × 1.75 m × 0.187 m for study "G3".

The influence of the choice of time intervals was investigated. For this purpose, the development of the reliability index over time is considered. In Fig. 3 the reliability index is plotted over time according to RC2 in Eurocode 0 [31] for the studies "G1" (100-year interval) as well as study "G2" (1-/10-/50-year intervals) for a longitudinal reinforcement ratio $\rho_{\mathrm{l}} = 2.0$ ‰.

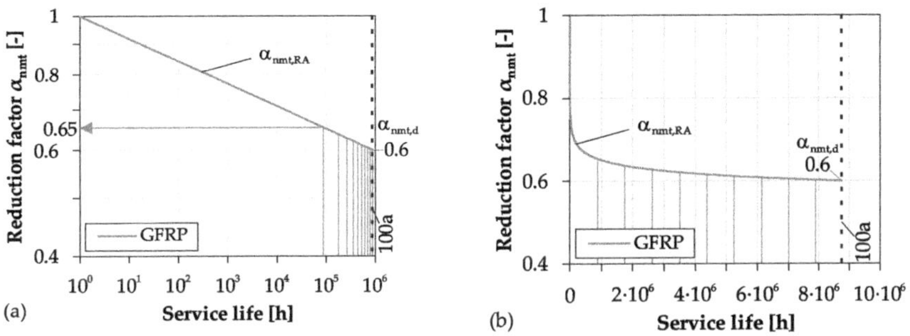

Fig. 2. Illustration of the reduction factor α_{nmt} for 10-year intervals – (a) logarithmic scale; (b) linear scale

Table 1. Overview of the studies with GFRP reinforcement

Study			Textile			Concrete		Static depth
#	FORM	MC	$\alpha_{nmt,d}$	$\alpha_{nmt,RA}$	γ_{nm}	Class	γ_c	d [cm]
G1	X	X	0.6	0.6	1.3	C70/85	1.5	10.5
G2	X	-	0.6	1.0–0.6[a]				10.5
G3	X	X	0.6	1.0–0.6[1]				16.5

[a] one-year intervals

Equations (2) and (3) are used to calculate the reliability index β.

$$P_f(t_L) \approx t_L \cdot P_{f1} \tag{2}$$

$$P = \sum_{i=1}^{n} P_{f1} \text{ with n} = 100 \tag{3}$$

With P_{f1} as the annual failure probability and t_L as the number of time intervals.

The curve of the target reliability index *(t)* according to RC2 is selected in accordance with Eurocode 0 [31] (see Fig. 3). The target reliability index for 1 year is $\beta = 4.7$ and for 50 years $\beta = 3.8$. The corresponding failure probability for β can be determined using Eq. (2). For further details see [15].

For studies "G1" and "G2", the required reliability index according to Eurocode 0 [31] is always reached. The decisive case is the summation of P_{f1} over 100 years (study "G1"), as this study yields the highest failure probability of the smallest reliability index, respectively. By considering the time-dependent development of the tensile strength of the non-metallic reinforcement, a higher reliability is attained, especially in the first years of service life. This is due to the higher existing strength compared to the strength applied in the design or the reduction factor for GFRP reinforcement after 100 years of service live ($\alpha_{nmt,GFRP,d,100a} = 0.6$). Furthermore, the results show that the calculation effort can be significantly reduced. The individual calculations of the respective studies are represented by the symbols in Fig. 3. However, it is not possible to specify uniform intervals since the strength development is subject to a double-logarithmic progression.

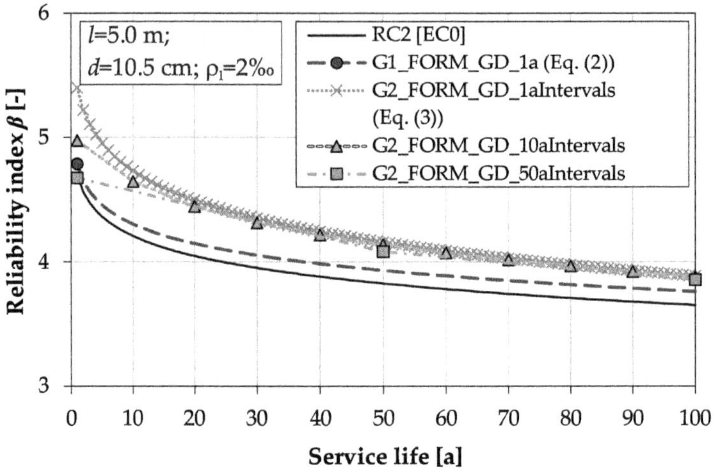

Fig. 3. Reliability index β for studies "G1" and "G2" for $\rho_l = 2.0$ ‰

To reduce the computational effort, in this investigation it was considered 1-year intervals for the first 10 years of the service life of a CRC component, 10-year intervals up to 50 years, and, then, only one 50-year interval. In this case, the variable loads are to be adjusted according to the respective observation period. The results of this proposal are presented in Fig. 4.

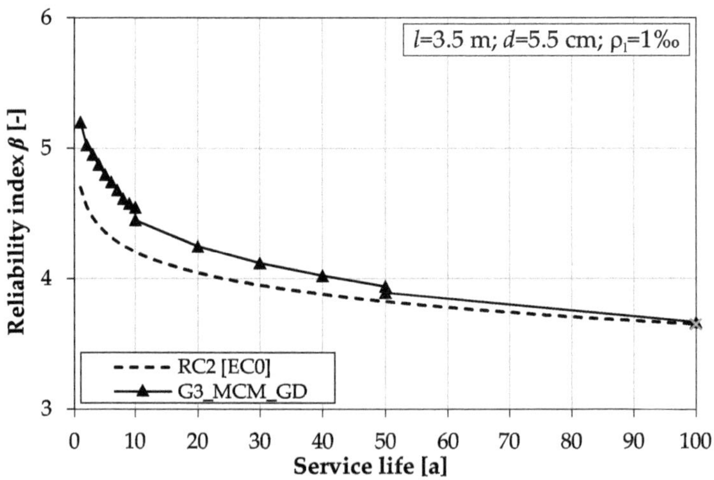

Fig. 4. Staggered investigation of the reliability index with study "G3" for $\rho_l = 1.0$ ‰

The study "G3" (Fig. 4) demonstrates that the computational effort for the FORM or MC can be reduced by an efficient choice of time intervals – here 1-, 10-, 50-year intervals – to be able to properly reproduce the development of the reliability index over time and to take into account the time-dependent material behaviour of the tensile strength of the FRP reinforcement.

5 Conclusion and Outlook

This paper presents the results of a set of reliability analyses, where the influence of the time-dependent strength of FRP reinforcement on reinforced concrete members under bending load was investigated.

In this investigation, the favourable effect of the time-dependent strength development of the non-metallic reinforcement was demonstrated. Reliability analyses were conducted for observation periods of 1, 10, 50 and 100 years. First, the component service life was divided into 1-, 10-, 50-year intervals (study "G2"). Second, the reduction factors associated with the time intervals were determined, and the tensile strength was accordingly reduced.

The consideration of the time-dependent material behaviour the non-metallic reinforcement has a favourable effect on the failure probability. For the design of CRC components, the tensile strength at the end of service live is considered by $\alpha_{\mathrm{nmt,d}}$. At the beginning of the service life, however, the tensile strength is higher than the design strength after 100 years. Due to the double-logarithmic relationship between time dependent strength and time, the strength decreases rapidly in the first years of service life. Thus, the reduction factor due to time dependent degradation of strength is 0.71 for GFRP reinforcement after 1 year (0.6 after 100 years). The favourable effect of considering time dependency, decreases rapidly for increasing service lives.

With an efficient choice of the examined time intervals, the computational effort of the analyses can be significantly reduced. Thus, based on the investigated time intervals, it was suggested that the first 10 years of a CRC component's service life should be examined with 1-year intervals in reliability analyses. Up to 50 years, 10-year intervals could be considered, followed by a single 50-year time interval. This approach has been used for study "G3".

The design for bending of statically determined single span members has already been covered in REMPEL's work [30]. This paper evaluated the influence of the time-dependent degradation of the tensile strength of non-metallic reinforcement. Subsequently, probabilistic investigations for statically indeterminate systems, e.g., two-span beams, are necessary in order to analyse moment redistributions and its influence on their reliability.

References

1. Curbach, M., Hegger, J., Schladitz, F., et al. (Hrsg.): Handbuch Carbonbeton – Einsatz nichtmetallischer Bewehrung. Ernst & Sohn a Wiley brand, Berlin (2023)
2. Müller, E., Scheerer, S., Curbach, M.: Material and space saving carbon concrete elements. Civil Eng. Des. **1**(1), 3–9 (2019). https://doi.org/10.1002/cend.201800001
3. Spelter, A., Wagner, J., Curbach, M., et al.: Carbon reinforced concrete under cyclic tensile loading. ACI Special Publication **345**(1), 1–15 (2021)
4. von der Heid, A.-C.: Zum Tragverhalten von Sandwichelementen aus Textilbeton mit schlaffer und vorgespannter Bewehrung, RWTH Aachen University (2021)

5. Shams, A., Stark, A., Hoogen, F., et al.: Innovative sandwich structures made of high performance concrete and foamed polyurethane. Compos. Struct. **121**, 271–279 (2015). https://doi.org/10.1016/j.compstruct.2014.11.026
6. Stark, A.: Analysis of Prestressed Sandwich Panels with Ultra-High-Performance Concrete Facings. RWTH Aachen University, Aachen (2017)
7. Bergmann, S., May, S., Hegger, J., et al.: Shear strengthening of reinforced concrete T beams using carbon reinforced concrete. ACI Special Publication **345**(1), 169–184 (2021)
8. Herbrand, M., Adam, V., Classen, M., et al.: Strengthening of existing bridge structures for shear and bending with carbon textile-reinforced mortar. Materials **10**(9), 1099 (2017). https://doi.org/10.3390/ma10091099
9. May, S., Michler, H., Schladitz, F., et al.: Lightweight ceiling system made of carbon reinforced concrete. Struct. Concrete **19**(6), 1862–1872 (2018). https://doi.org/10.1002/suco.201700224
10. Bielak, J., Will, N., Hegger, J.: Zwei Praxisbeispiele zur Querkrafttragfähigkeit von Brückenplatten aus Textilbeton. Bautechnik **97**(7), 499–507 (2020). https://doi.org/10.1002/bate.202000037
11. Hegger, J., Goralski, C., Kulas, C.: Schlanke Fußgängerbrücke aus Textilbeton. Beton- und Stahlbetonbau **106**(2), 64–71 (2011). https://doi.org/10.1002/best.201000081
12. Helbig, T., Unterer, K., Kulas, C., et al.: Fuß- und Radwegbrücke aus Carbonbeton in Albstadt-Ebingen. Beton- und Stahlbetonbau **111**(10), 676–685 (2016). https://doi.org/10.1002/best.201600058
13. Kueres, S., Hegger, J.: Reliability analysis of footbridges pre-tensioned with carbon fiber reinforced polymer tendons under flexural loading. Eng. Struct. **203**, 109546 (2020). https://doi.org/10.1016/j.engstruct.2019.109546
14. Rempel, S., Ricker, M., Feiri, T.: Stochastic approach for the material properties of reinforcing textiles for the design of concrete members. Sci. Rep. **11**(1), 21976 (2021). https://doi.org/10.1038/s41598-021-01032-9
15. Spelter, A.: Long-Term Durability of Carbon Reinforced Concrete. Dissertation, RWTH Aachen University, Aachen (2023)
16. Youakim, S.A., Karbhari, V.M.: An approach to determine long-term behavior of concrete members prestressed with FRP tendons. Constr. Build. Mater. **21**(5), 1052–1060 (2007). https://doi.org/10.1016/j.conbuildmat.2006.02.006
17. Ricker, M., Feiri, T., Nille-Hauf, K., et al.: Enhanced reliability assessment of punching shear resistance models for flat slabs without shear reinforcement – first published online. Eng. Struct. **226**, 111319. (2021) https://doi.org/10.1016/j.engstruct.2020.111319
18. Freudenthal, A.M.: Safety and the probability of structural failure. Trans. Am. Soc. Civ. Eng. **121**, 1337–1375 (1954)
19. Faber, M.H., Vrouwenvelder, T.: Probabilistic Model Code – 12th draft (2001)
20. Rackwitz, R.: First Order Reliability Concepts for Design Codes. Joint-Committee CEB/CECM/CIB/FIP/IABSE Structural Safety (1976)
21. Deutsches Institut für Normung (Hrsg.): Grundlagen zur Festlegung von Sicherheitsanforderungen für bauliche Anlagen. Deutsches Institut für Normung. Beuth, Berlin (1981)
22. Melchers, R.E., Beck, A.T.: Structural Reliability Analysis and Prediction. Wiley, Chichester (2018)
23. Feiri, T., Schulze-Ardey, J.P., Ricker, M., et al.: The role of probability-based safety analysis: the perspective of structural product manufacturers. In: Fédération internationale du béton (ed.): Building for the future: Durable, Sustainable, Resilient – Proceedings of the fib Symposium 2023. International Federation for Structural Concrete, Digital Proceedings, Istanbul, Turkey, pp. 1589–1600 (2023)
24. Zilch, K., Zehetmaier, G.: Bemessung im konstruktiven Betonbau. Springer, Heidelberg (2010). https://doi.org/10.1007/978-3-540-70638-0

25. ACI 440.3R-12: Guide Test Methods for Fiber-Reinforced Polymer (FRP) Composites for Reinforcing or Strengthening Concrete and Masonry Structures. Richtlinie, August 2012
26. ASTM D3479/D3479M−12: Standard Test Method for Tension-Tension Fatigue of Polymer Matrix Composite Materials. Norm, August 2012
27. CSA S806-12:2012-03: Design and construction of building structures with fibre-reinforced polymers. Norm, März 2012
28. ISO 10406-1: Fibre-reinforced polymer (FRP) reinforcement of concrete - Test methods. Norm, Ausgabe 2015
29. R Core Team: A Language and Environment for Statistical Computing (2016). https://www.R-project.org/
30. Rempel, S.: Zur Zuverlässigkeit der Bemessung von biegebeanspruchten Betonbauteilen mit textiler Bewehrung. Dissertation, RWTH Aachen University, Aachen (2018)
31. DIN EN 1990:2002: Eurocode: Basis of structural design. Norm

Investigation of the Effect of Artificial Neural Network Performance Parameters and Training Dataset on the Probability Estimate Capacity in Structural Reliability Problems

F. B. Köroğlu[1,2], M. Maguire[1], and E. Aktaş[2(✉)]

[1] Durham School of Architectural Engineering and Construction, University of Nebraska-Lincoln, Lincoln, NE 68182, USA
[2] Faculty of Engineering, Izmir Institute of Technology, Izmir, Türkiye
enginaktas@iyte.edu.tr

Abstract. This study highlights two of the important details of the implementation of artificial neural networks to the structural reliability problems by pointing out the effect of training dataset, and the relationship between the performance parameters (coefficient of determination of train, validation, and test sets) of a network and its probability estimation capacity when it is used as a surrogate model in structural reliability problems. Four numerical examples are covered regarding these key aspects including one that is derived from a real-life reinforced concrete structure. Results have shown that the dataset can affect the probability estimation capacity for complex problems. Furthermore, it is also observed that having a neural network with good performance parameters does not mean that the network always has good probability estimation capacity. However, in order to have a network that can be used for probability estimate purposes, its performance parameters must be at a satisfactory level.

Keywords: Artificial Neural Networks · Structural Reliability · Training Dataset · Performance Parameters

1 Introduction

Implementation of machine learning techniques in structural reliability subject is still an ongoing branch of research to leverage the computational cost of probability estimations. Among the different machine learning techniques, artificial neural networks (ANNs) are promising techniques because of their function approximation capability and flexible architecture.

The application of ANNs to structural reliability problems has been studied by several researchers [1–7]. However, a big portion of these studies are mainly focused on the accuracy or computational efficiency of the proposed methods. The effect of the training dataset or the relationship between the performance parameters of the network and its probability estimation capacity was not examined in detail. In this study, the performance

J. C. Matos et al. (Eds.): IPW 2024, LNCE 494, pp. 398–407, 2024.
https://doi.org/10.1007/978-3-031-60271-9_37

parameters of a neural network refer to the coefficient of determination between the expected output and the output obtained by the neural network for training, validation, and test sets. The term training dataset includes the samples that are used in training, validation, and test sets.

Therefore, this study aims to reveal the relationship between those performance parameters and the probability estimation capacity of a network while also monitoring the effect of training datasets.

2 Coupling of Bayesian Regularized ANN and Crude Monte Carlo Simulation

The classical approach in simulation techniques relies on the evaluation of the limit state function for each sample. However, in structural reliability problems, finite element analysis (FEA) is generally required to perform the analysis of frame-type structures when the limit state function is implicit. Therefore, the computational cost can be prohibitive when finite element analysis is needed to run Crude Monte Carlo Simulation (CMCS).

Bayesian regularized ANNs are coupled with CMCS in this study to perform structural reliability analyses. It is aimed to lower the computational cost of CMCS when finite element analysis is required by using ANNs. The networks were trained by using MATLAB Neural Network Toolbox [8]. Bayesian regularization [9] has been used as the training function to increase the function approximation capacity of ANNs. The architecture of ANNs was determined by trial and error based on the exceedance of the threshold coefficient of determination, $R^2 > 0.98$, of the train and test sets. The maximum number of layers is limited by three whereas the minimum and the maximum number of neurons in each layer are limited by 10 and 50 respectively.

The necessary datasets for training the networks have been created by OpenSeesPy [10] analysis results. The training dataset is divided into the train, validation, and test sets with a percentage of 70, 15, and 15 respectively. After that, the trained network for each problem is used in CMCS to obtain probability estimates.

Several datasets with varying sample sizes have been created to observe the effect of the dataset on the probability estimation capacity. For each dataset, a neural network has been trained and its probability estimate with its performance parameters during the training has been recorded for the scope of this paper. The new datasets are generated by adding new samples to the existing dataset.

The trained ANNs have been called in CMCS runs and each CMCS run with different input samples is repeated 10 times to prevent fluctuation in probability estimates because of randomly generated CMCS samples. This practice became feasible because of the efficiency of trained ANNs which provide 10 CMCS runs within a couple of minutes at maximum.

FEA-CMCS coupling has been used to obtain reference probability estimates. OpenSeesPy is used as a FEA solver, and it is combined with Pystra [11] to perform CMCS.

3 Numerical Examples and Results

Four numerical examples regarding the complexity and dimensionality of the limit state functions are covered to observe the effect of the training dataset and the performance parameters on the probability estimates made by using the coupling of Bayesian regularized ANNs and CMCS.

3.1 Simple Portal Frame

A simple portal frame given in Fig. 1 is subjected to a lateral load, P, acting on the second node. The configuration of the frame is based on the study of [3] and the limit state function is defined as the exceedance of 1 cm of the horizontal displacement in the control node. Statistical properties of the random variables are introduced in Table 1. The relationship between the moment of inertia and area is given in Eq. (1).

$$I_i = \alpha_i A_i^2 \tag{1}$$

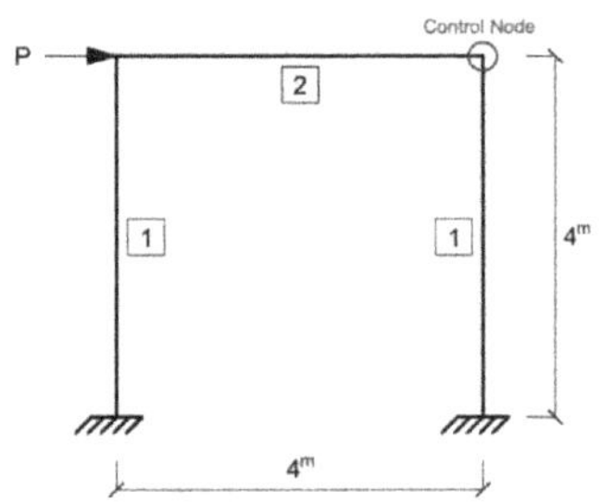

Fig. 1. Schematic view of the simple portal frame.

Table 1. Statistical properties of the random variables for the simple portal frame example.

Random Variable	Mean Value	Coefficient of Variation	Distribution Type	Coefficient α_i
A_1	0.36 m^2	0.10	Lognormal	0.08333
A_2	0.18 m^2	0.10	Lognormal	0.16670
P	20.0 kN	0.25	Gumbel	-

ANNs trained with different sizes of datasets in this example. The negligible amount of fluctuation in probability estimates is depicted in Fig. 2a representing the close approximation to the probability estimate made by FEA-CMCS coupling with 1E6 samples. The performance parameters for the example are presented in Fig. 2b showing that the trained ANNs are excellent in terms of coefficient of determination for the train, validation, and test sets even though fluctuations in probability estimates.

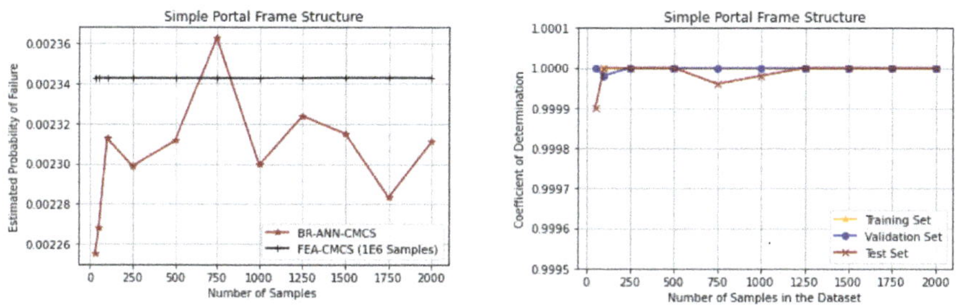

Fig. 2. a) Probability estimates for the simple portal frame. b) Performance parameters of trained ANNs.

3.2 12-Story 3-Bay Frame Structure

12-story 3-bay frame structure covered in the study of [2, 6] is revisited here. The structural configuration is shown in Fig. 3. Statistical properties of the random variables are introduced in Table 2. The relationship between the moment of inertia and area is the same as given in Eq. (1) whereas the limit state function is defined in Eq. (2) where u_A indicates the horizontal displacement of point A.

$$g(A_1, A_2, A_3, A_4, A_5, P) = 0.096 - u_A(A_1, A_2, A_3, A_4, A_5, P) \quad (2)$$

Table 2. Statistical properties of the random variables for the 12-story 3-bay frame structure.

Random Variable	Mean Value	Coefficient of Variation	Distribution Type	Coefficient α_i
A_1	0.25 m^2	0.10	Lognormal	0.08333
A_2	0.16 m^2	0.10	Lognormal	0.08333
A_3	0.13 m^2	0.10	Lognormal	0.08333
A_4	0.20 m^2	0.10	Lognormal	0.26670
A_5	0.15 m^2	0.10	Lognormal	0.20000
P	30.0 kN	0.25	Type I Largest	-

The probability estimate performed by using different ANNs is given in Fig. 4a yielding a close approximation to the reference probability estimate made by FEA-CMCS coupling with 1E5 samples. However, as can be seen in the figure, the probability estimates start to stabilize after 250 samples even though the performance parameters are fluctuating which is shown in Fig. 4b.

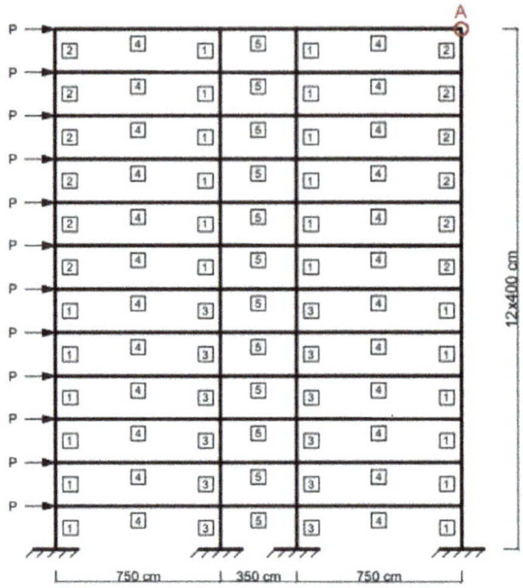

Fig. 3. Structural configuration of the 12-story 3-bay frame structure.

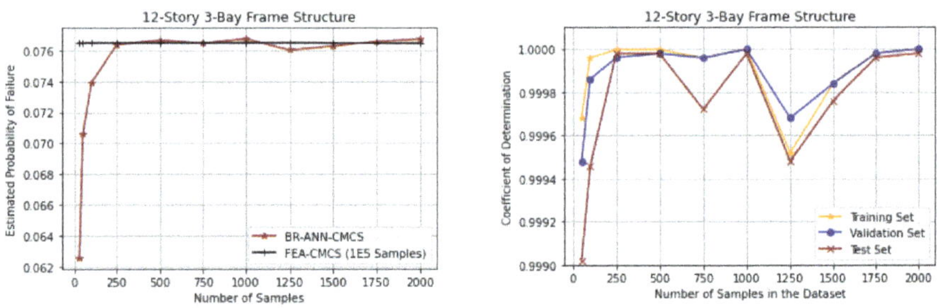

Fig. 4. a) Probability estimates for the 12-story 3-bay frame structure. b) Performance parameters of trained ANNs.

3.3 5-Story 3-Bay Correlated Frame Structure

The correlated frame type structure introduced in several studies [2, 12–14] is revisited here. The structural configuration is shown in Fig. 5. Statistical properties of the random variables are introduced in Table 4. Truncated normal distribution is used to prevent non-physical random variables during the sample generation. The coefficient of correlation between the random variables is provided in Table 3. The limit state function is defined as the horizontal displacement of the control node with a threshold value of 60 mm.

The probability estimate behavior observed in Fig. 6a contradicts common sense in ANN training regarding more data and better function approximation capability. This common sense is generally held, however, in some cases as presented in the figure, there can be a possibility of diverging from the expected probability estimate because of higher sample density in a region in the dataset especially newly added samples affects the statistical properties of the dataset. In this example, it has been observed that the new samples pull down the mean value of loads yielding fewer samples that violate the limit state function. Consequently, leading to the ANNs predict a lower probability of failure

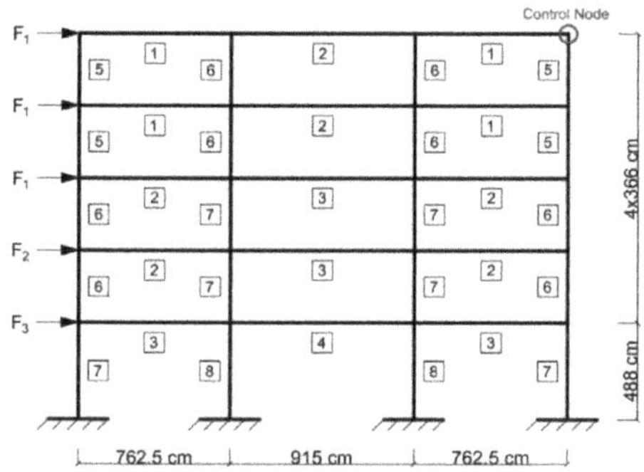

Fig. 5. 5-story 3-bay frame structure configuration.

Table 3. Coefficient of correlation between the random variables for the 5-story 3-bay frame structure.

Random Variables	Coefficient of Correlation
$A_iA_j = I_iI_j = A_iI_j$	0.13
$F_{i,j}$	0.95
$E_{i,j}$	0.90
A_iI_i	0.95

estimate compared to the expected after 6000 samples. This behavior can be observed if ANNs are trained with a low number of samples and are used to small probability estimation. Even though the universal approximation theorem [15] guarantees that ANNs can approximate any function, it also states that they can approximate with arbitrary error and their extrapolation capacity is limited. Therefore, generating a dataset plays a crucial role in ANN training especially the number of samples in the dataset that represent the failure domain. As shown in Fig. 6b, all performance parameters indicate that the ANN is sufficient even though the probability estimate worsens after 6000 samples. Therefore, in ANN-based structural reliability analyses, convergence to a certain probability estimate must be tracked to prevent deceiving results.

3.4 11-Story Real-Life RC Structure

A real-life RC structure having 11-story with varying bays is covered in this example. The structural configuration is presented in Fig. 7 (units are in cm). The shear walls located in grids 4 and 5 having dimensions of 362x25^{cm} are considered as deterministic parameters and modeled with equivalent frame elements. Statistical properties of the random variables are provided in Table 5 where μ represents the Poisson's ratio of concrete. The limit state function is derived based on TSC2018 regarding the maximum

Table 4. Statistical properties of the random variables for the 5-story 3-bay frame structure.

Random Variable	Mean Value	Standard Deviation	Unit	Distribution Type
F_1	133.454	40.04	kN	Lognormal
F_2	88.970	35.59	kN	Lognormal
F_3	71.175	28.47	kN	Lognormal
E_1	2.173752×10^{7}	1.915200×10^{6}	kN/m^2	Truncated Normal [0, +∞)
E_2	2.379736×10^{7}	1.915200×10^{6}	kN/m^2	Truncated Normal [0, +∞)
I_1	0.813443×10^{-2}	1.083440×10^{-3}	m^4	Truncated Normal [0, +∞)
I_2	1.150936×10^{-2}	1.298048×10^{-3}	m^4	Truncated Normal [0, +∞)
I_3	2.137452×10^{-2}	2.506090×10^{-3}	m^4	Truncated Normal [0, +∞)
I_4	2.596095×10^{-2}	3.028778×10^{-3}	m^4	Truncated Normal [0, +∞)
I_5	1.081076×10^{-2}	2.596095×10^{-3}	m^4	Truncated Normal [0, +∞)
I_6	1.415540×10^{-2}	3.461460×10^{-3}	m^4	Truncated Normal [0, +∞)
I_7	2.327853×10^{-2}	5.624873×10^{-3}	m^4	Truncated Normal [0, +∞)
I_8	2.596065×10^{-2}	6.490238×10^{-3}	m^4	Truncated Normal [0, +∞)
A_1	3.125640×10^{-1}	5.581500×10^{-2}	m^2	Truncated Normal [0, +∞)
A_2	3.721000×10^{-1}	7.442000×10^{-2}	m^2	Truncated Normal [0, +∞)
A_3	5.060600×10^{-1}	9.302500×10^{-2}	m^2	Truncated Normal [0, +∞)
A_4	5.581500×10^{-1}	11.163000×10^{-2}	m^2	Truncated Normal [0, +∞)
A_5	2.530280×10^{-1}	9.302500×10^{-2}	m^2	Truncated Normal [0, +∞)
A_6	2.911683×10^{-1}	10.232275×10^{-2}	m^2	Truncated Normal [0, +∞)
A_7	3.730300×10^{-1}	12.093250×10^{-2}	m^2	Truncated Normal [0, +∞)
A_8	4.186000×10^{-1}	19.537500×10^{-2}	m^2	Truncated Normal [0, +∞)

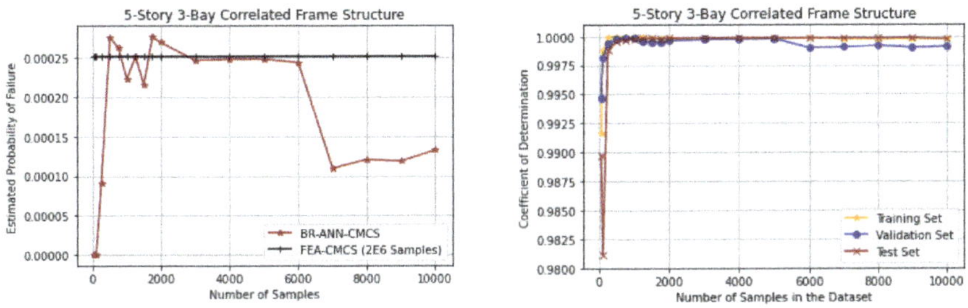

Fig. 6. a) Probability estimates for the 5-story 3-bay correlated frame. b) Performance parameters of trained ANNs.

inter-story drift ratio under the consideration of infill walls and given in Eq. (3).

$$g(X) = 0.008 - \max_{\substack{i = 1\ to\ the \\ maximum\ number\ of\ stories}} \frac{(0.404)\delta_{i,max}^{(X)}}{h_i} \tag{3}$$

where $\delta_{i,max}^{(X)}$ represents the inter-story drift and h_i stands for the corresponding story height.

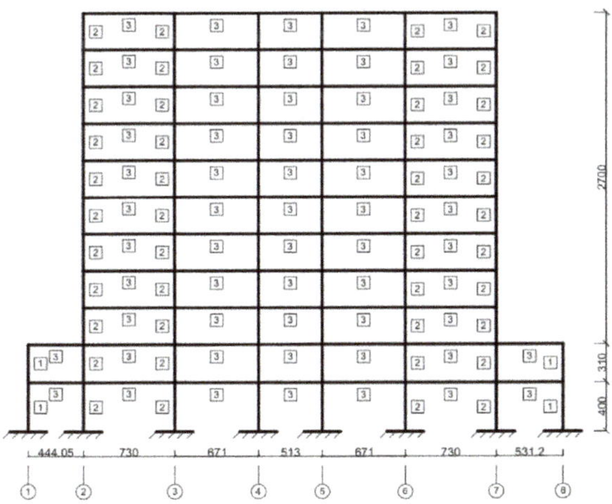

Fig. 7. Structural configuration of the 11-story RC frame structure.

Figure 8a presents the probability estimates for ANNs trained in varying sizes of datasets with high-performance parameters shown in Fig. 8b. The probability estimates are stabilized after 250 samples whereas the performance parameters stabilize after 500 samples. The plots show that the implemented method is suitable for the real-life structural engineering applications.

Table 5. Statistical properties of the random variables for the 11-story RC frame structure.

Random Variable	Units	Mean Value	Coefficient of Variation	Distribution Type
E	kN/m^2	3.18E7	0.10	Normal
μ	-	0.20	0.10	Normal
b1	m	0.70	0.05	Normal
b2	m	1.50	0.05	Normal
b_3	m	0.25	0.05	Normal
h1	m	0.30	0.05	Normal
h2	m	0.25	0.05	Normal
h3	m	0.50	0.05	Normal
P_{11}	kN	283.70	0.20	Lognormal
P_{10}	kN	174.73	0.20	Lognormal
P_9	kN	102.33	0.20	Lognormal
P_8	kN	70.05	0.20	Lognormal
P_7	kN	61.89	0.20	Lognormal
P_6	kN	60.73	0.20	Lognormal
P_5	kN	63.89	0.20	Lognormal
P_4	kN	73.39	0.20	Lognormal
P_3	kN	81.01	0.20	Lognormal
P_2	kN	157.23	0.20	Lognormal
P_1	kN	116.25	0.20	Lognormal

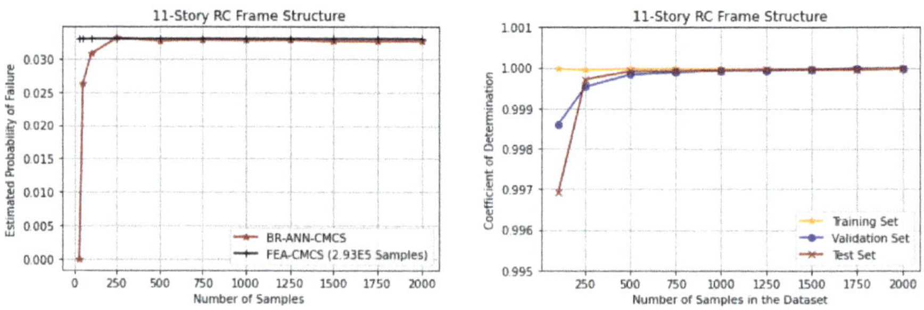

Fig. 8. a) Probability estimates for the 11-story RC frame structure. b) Performance parameters of trained ANNs.

4 Conclusion

The obtained results show that there is a possibility that an increase in the number of samples in the training dataset can negatively affect the probability estimation. Furthermore, high values for performance parameters do not always guarantee the trained ANN can approximate the function that is intended to be replaced by ANN. It is recommended that convergence to a certain probability estimation should be tracked while increasing the training samples of ANN.

References

1. Shao, S., Murotsu, Y.: Structural reliability analysis using a neural network. JSME Int. J. **40**, 242–246 (1997)
2. Beheshti Nezhad, H., Miri, M., Ghasemi, M.R.: New neural network-based response surface method for reliability analysis of structures. Neural Comput. Appl. **31**, 777–791 (2019). https://doi.org/10.1007/s00521-017-3109-2
3. Deng, J., Gu, D., Li, X., Yue, Z.Q.: Structural reliability analysis for implicit performance functions using artificial neural network. Struct. Saf. **27**, 25–48 (2005). https://doi.org/10.1016/j.strusafe.2004.03.004
4. Li, M., Wang, Z.: Deep learning for high-dimensional reliability analysis. Mech. Syst. Signal Process. **139**, 106399 (2020). https://doi.org/10.1016/j.ymssp.2019.106399
5. Lieu, Q.X., Nguyen, K.T., Dang, K.D., et al.: An adaptive surrogate model to structural reliability analysis using deep neural network. Expert Syst. Appl. **189**, 116104 (2022). https://doi.org/10.1016/j.eswa.2021.116104
6. Cheng, J.: Hybrid genetic algorithms for structural reliability analysis. Comput. Struct. **85**, 1524–1533 (2007). https://doi.org/10.1016/j.compstruc.2007.01.018
7. Cheng, J., Xiao, R.C.: Serviceability reliability analysis of cable-stayed bridges. Struct. Eng. Mech. **20**, 609–630 (2005)
8. The MathWorks Inc: MATLAB (2019)
9. Foresee, F.D., Hagan, M.T.: Gauss-Newton approximation to Bayesian learning. In: Proceedings of International Conference on Neural Networks (ICNN 1997), vol. 3, pp 1930–1935 (1997)
10. Zhu, M., McKenna, F., Scott, M.H.: OpenSeesPy: python library for the OpenSees finite element framework. SoftwareX **7**, 6–11 (2018). https://doi.org/10.1016/j.softx.2017.10.009
11. Hackl, J., Caprani, C.: Pystra (2022)
12. Blatman, G., Sudret, B.: An adaptive algorithm to build up sparse polynomial chaos expansions for stochastic finite element analysis. Probab. Eng. Mech. **25**, 183–197 (2010). https://doi.org/10.1016/j.probengmech.2009.10.003
13. Bucher, C.G., Bourgund, U.: A fast and efficient response surface approach for structural reliability problems. Struct. Saf. **7**, 57–66 (1990)
14. Liu, P.-L., Der Kiureghian, A.: Optimization Algorithms for Structural Reliability Analysis. California (1986)
15. Hornik, K.M., Stinchcommbe, M., White, H.: Multilayer feedforward networks are universal approximators. Neural Netw. **2**, 359–366 (1989)

Model Uncertainty Assessment for the Anchorage Length Formulation of *fib* Model Code 2020

Sergio Belluco(✉), Lorenzo Hofer, and Flora Faleschini

Department of Civil, Environmental, and Architectural Engineering, University of Padova, Via Marzolo 9, 35131 Padova, Italy
sergio.belluco@phd.unipd.it

Abstract. In the design of pre-tensioned concrete elements, it is important to avoid the anchorage failure of tendons. This can be done by predicting the anchorage length through any of the empirical or semiempirical models provided by the main design codes. However, the models in these codes are based on experimental evidence gathered many years ago, when the 7-wire strands manufacturing process was significantly different from now. The objective of this contribution is to evaluate the model uncertainty for the anchorage length formulation of the last draft of the *fib* Model Code 2020 and to ensure that all the main influencing factors are properly considered. Since a direct measure of anchorage length is not experimentally feasible, a large data set of flexural tests has been gathered from the scientific literature, and each failure mode has been compared to the corresponding anchorage length prediction, to obtain a censored sample of the model uncertainty from which an appropriate probability distribution could be estimated. Then, the minimum values of the partial safety factors that consider both model and material uncertainties and ensure the achievement of the target reliability in anchorage length predictions have been derived through Monte Carlo Simulations. Finally, these results have been validated through a FORM approach for some case studies and compared to the current *fib* Model Code 2020 proposed partial safety factors, showing that they guarantee the achievement of the expected level of reliability in all the evaluated situations.

Keywords: Bond · Anchorage Length · Prestressed Concrete · Model Uncertainty · *fib* Model Code 2020

1 Introduction

The shear and bending capacity of a pretensioned concrete element is highly influenced by the development of stresses in the bonded tendons and therefore, to ensure a proper and safe design, it is important to avoid anchorage failures and to be able to correctly predict the anchorage length of the tendons themselves. However, since the first introduction of 7-wire strands (which nowadays represent the most common type of prestressing tendon), significant changes occurred in their manufacturing process and

J. C. Matos et al. (Eds.): IPW 2024, LNCE 494, pp. 408–418, 2024.
https://doi.org/10.1007/978-3-031-60271-9_38

current code formulations that have been only slightly modified since their first proposal many years ago may not be accurate anymore. Since the first pioneering studies by Janney [1], strands have increased both in diameter and in strength, and in the 1980s, induction heating furnaces have progressively replaced convection heating furnaces in stress reliever and low-relaxation treatments of the strands [2]. Due to this latter change, the strands are now subjected to lower temperatures than in convection heating, and combustion does not occur: consequently, in induction heating furnaces, many of the surface residues and impurities left from the wire drawing process may no longer be burned or oxidized [2]. These hypotheses find confirmation in the results of many experimental campaigns carried out in the 1980s and 1990s (e.g., [3, 4]), where large scatter in the results or unexpected longer lengths have been reported. Therefore, in subsequent years, the surface condition of the strands gained the attention of the scientific community working on this topic, and different strand surface control procedures were discussed and implemented (as reported, e.g., in [5]).

This contribution aims to evaluate the model uncertainty of the last draft of the *fib* Model Code 2020 formulation for anchorage length (but the same procedure can be applied to any other formulation) and to ensure that all the main influencing variables are properly considered. Then, the minimum values of the partial safety factors that consider both model and material uncertainties and ensure the achievement of the target reliability anchorage lengths have been calibrated through Monte Carlo Simulations (MCS). Finally, the results have been validated through a FORM approach and compared to the current *fib* Model Code 2020 proposed partial safety factors, showing that the latter ensure the achievement of the expected level of reliability in all the evaluated design situations.

2 *fib* Model Code 2020 Anchorage Length Formulation

The current draft of the *fib* Model Code 2020 anchorage length formulation is not substantially different from the previous versions [6, 7], except for some simplifications in the formulation for the basic anchorage length l_b (i.e., the anchorage length of an untensioned tendon). It is briefly recalled that since the *fib* Model Code 1990 [6] the anchorage length l_{pd} is treated as the sum of the transmission length l_{pt} (i.e., the length required to anchor the initial prestress) and the flexural bond length l_{fb} (i.e., the length required to anchor the additional tendon stress due to external loads), and both of these lengths are expressed as functions of l_b, which according to the last draft of the *fib* Model Code 2020, can be computed as:

$$l_b = \gamma_C \cdot \frac{f_{ptd}}{\eta_{p1} \cdot \eta_{p2} \cdot \sqrt{f_{ck}(t)}} \cdot \frac{A_p}{\pi \phi} \tag{1}$$

where γ_C is the partial safety factor for concrete and $f_{ck}(t)$ is the characteristic concrete compressive strength at the time t, A_p and ϕ are the area and the nominal diameter of the prestressing tendon, respectively, f_{ptd} is the tendon design strength. According to the Model Code 2020, the characteristic concrete compressive strength at the time t can be computed as $f_{ck}(t) = \beta_{cc}(t)^{2\backslash 3} \cdot f_{ck}$, with $\beta_{cc}(t)$ being a function to describe the development of strength with time and with f_{ck} being the characteristic concrete compressive strength at the reference time (usually $t = 28$ days). All the η_p coefficients are illustrated in Table 1.

Then, the transmission length l_{pt} can be calculated as:

$$l_{pt} = \alpha_{p1} \cdot \alpha_{p2} \cdot \alpha_{p3} \cdot \frac{\sigma_{pi}}{f_{ptd}} \cdot l_b \tag{2}$$

where σ_{pi} is the initial prestress, and all the α_p coefficients are described in Table 1.

Table 1. Empirical coefficients employed in Eqs. (1) and (2).

Factor	Scope	Value	Case
α_{p1}	Considering the effects of the prestress release method on l_{pt}	1.00	For gradual release
		1.25	For sudden release
α_{p2}	Considering the action effect to be verified, i.e., considering the design situation where l_{pt} has to be adopted	0.50	For verification of transverse stress due to the development of prestress (associated with the 0.05 fractile of the bond stress)
		1.00	For calculation of anchorage length, shear, and moment capacity (associated with the 0.95 fractile of the bond stress)
α_{p3}	Considering the influence of tendon geometry on l_{pt}	0.50	For 7-wire strands
		0.70	For indented and crimped wires
η_{p1}	Considering the type of prestressing tendon	0.36	For 7-wire strands
		0.42	For indented and crimped wires
η_{p2}	Considering the position of the tendon during concreting	1.00	For good casting position
		0.70	For poor casting position

Similarly, the flexural bond length l_{fb} can be calculated as:

$$l_{fb} = \frac{\sigma_{pd} - \sigma_{pe}}{f_{ptd}} \cdot l_b = \frac{\Delta\sigma}{f_{ptd}} \cdot l_b \tag{3}$$

where $\Delta\sigma$ is the additional stress in the tendon due to external loading, computed by subtracting the effective prestress after all losses σ_{pe} from the design tendon stress under the design load σ_{pd}.

Finally, the anchorage length can be estimated as the sum of the Eqs. (2) and (3):

$$l_{pd} = l_{pt} + l_{fb} \tag{4}$$

Observing the formulation proposed in the above Eqs. (2), (3), and (4), the following considerations can be made:

- both Eqs. (2) and (3) are based on the same bond strength formulation, and the increase in bond strength due to the Hoyer effect or the detrimental effect of sudden release on the bond in the transmission length are accounted for with the α_{p1} and the α_{p3} coefficients;
- the attainment of a certain level of reliability in Eq. (2) is performed by adopting partial safety factors and choosing the correct α_{p2} value (which indeed depends on the design situation, and therefore on the probability distribution of the bond stress random variable);
- the attainment of a certain level of reliability in Eq. (3) seems to be performed only by adopting the partial safety factors.

Therefore, it seems reasonable to assume that the coefficient α_{p2} should be related to the basic anchorage length l_b in Eq. (1) rather than to the transmission length l_{pt} in Eq. (2), since α_{p2} is associated with the probability distribution of the bond stress.

3 Model Uncertainty

According to [8], the model uncertainty θ can be expressed in terms of a multiplicative random variable (R.V.) as follows:

$$R_{exp} = \theta \cdot R_{mod} \tag{5}$$

where R_{exp} is the experimental response of a structure and R_{mod} is the corresponding model-predicted response. Therefore, a sample of the model uncertainty can be simply obtained by comparing a set of experimental observations with their corresponding model prediction. However, direct measures of anchorage length are not experimentally feasible, and the most common approach to indirectly assess l_{pd} is to perform multiple bending tests on identical specimens, changing each time the position of the applied flexural load P. In detail, if the flexural test results in a combined flexural/bond failure of the specimen, the distance of P from the end of the member (herein indicated as l_e) is identified as a fully-observed experimental realization of the anchorage length, meaning that the member has reached its flexural capacity exactly at the end of l_{pd}. Conversely, if a flexural (or a bond) failure occurs, l_e represents a left (or right) censored realization of l_{pd}. In these cases, the test is repeated on another specimen, reducing (or increasing) the embedment length l_e, looking for a flexural/bond failure of the member.

Therefore, to assess the model uncertainty of Eq. (4), a dataset of flexural tests of simply supported pre-tensioned beams has been collected from journal articles, scientific reports, and dissertations available in the scientific literature. The dataset, which is briefly summarized in Table 2, was built collecting data from [3, 4, 9–26], including only the elements that showed a flexural, bond, or a combined flexural/bond failure (pure or combined shear failures were excluded as they are not relevant here). Then, to obtain a homogeneous sample of experimental observations, a preliminary filtering operation on the dataset was carried out, excluding from the investigation the specimens that met the following criteria:

- concrete strength class C $\geq$ C100/115 (7 occurrences);
- poor casting position, as defined by the *fib* Model Code 2020 (20 occurrences);

- improved bond conditions due to the epoxy coating of the tendons or the presence of rust on the surface (66 occurrences);
- partially debonded tendons (47 occurrences).

The dataset was therefore reduced from 621 to 481 experimental observations (the number of excluded observations for each campaign is also reported in Table 2).

Table 2. Outline of the collected data set of flexural tests.

Ref.	# of tests [excluded]	Prestress release	ϕ [mm]	f_c [MPa]	σ_{pi} [MPa]	$\Delta\sigma_p$ [MPa]	l_e [mm]
[3]	32 [17]	Sudden	9.5 ÷ 15.2	35 ÷ 46	1357 ÷ 1436	453 ÷ 541	457 ÷ 3353
[4]	21 [0]	Sudden	12.7 ÷ 15.2	35 ÷ 51	1345 ÷ 1427	-	1524 ÷ 4255
[9]	65 [3]	Gradual	9.5 ÷ 17.8	35 ÷ 66	1345 ÷ 1553	409 ÷ 854	813 ÷ 2235
[10]	71 [59]	Sudden	15.2	35 ÷ 102	1395	733	1372 ÷ 7315
[11]	24 [0]	Sudden	15.2	41 ÷ 57	1396	1682	1803
[12]	35 [14]	Sudden	12.7 ÷ 13.3	43 ÷ 57	1234 ÷ 1529	558 ÷ 939	1219 ÷ 1981
[13]	18 [0]	Sudden	12.7	44 ÷ 80	1372 ÷ 1427	553 ÷ 874	2438 ÷ 4572
[14]	43 [0]	Gradual	17.8	64 ÷ 94	1395	522 ÷ 651	991 ÷ 1830
[15]	34 [0]	Sudden	12.7 ÷ 15.2	35 ÷ 55	1400	600	1767 ÷ 2946
[16]	5 [0]	Sudden	15.2	70 ÷ 100	1314	732 ÷ 788	2032 ÷ 3048
[17]	23 [0]	Sudden	12.7 ÷ 15.2	50	1425	-	1200 ÷ 1550
[18]	14 [0]	Sudden	12.7 ÷ 15.2	42 ÷ 75	1403 ÷ 1462	453 ÷ 749	2184 ÷ 4915
[19]	15 [6]	Gradual	12.7	42 ÷ 57	1320 ÷ 1350	574 ÷ 659	1473 ÷ 1854
[20]	34 [10]	Gradual	9.5 ÷ 15.7	31 ÷ 89	871 ÷ 1374	350 ÷ 1000	575 ÷ 1865
[21]	16 [0]	Sudden	13.3 ÷ 15.2	55 ÷ 83	1345	1793 ÷ 2006	1870 ÷ 2221
[22]	16 [0]	Sudden	12.7	40 ÷ 64	1395	634	1473 ÷ 1854
[23]	55 [0]	Sudden	12.7 ÷ 15.2	49 ÷ 103	1397	-	1168 ÷ 2311
[24]	60 [7]	Gradual	12.7 ÷ 15.2	34 ÷ 199	1395 ÷ 1397	404 ÷ 806	365 ÷ 1524
[25]	16 [0]	Sudden	12.7 ÷ 15.2	41	1396	-	1880 ÷ 3607
[26]	24 [24]	Sudden	14.3	58 ÷ 73	1193	-	1600 ÷ 2515

A sample of the model uncertainty was calculated as the ratio of each of the 481 experimental embedment length l_e and the corresponding anchorage length prediction and each realization of θ has been classified as fully-observed, right-censored, or left-censored depending on the observed failure mode of the specimen. To remove any design bias, the anchorage lengths were calculated by setting all partial safety factors equal to 1.0, adopting mean values for all material properties, and setting $\alpha_{p2} = 0.75$, to obtain the mean value of the bond stress in the basic anchorage length. However, a high scatter in the realizations of θ associated with experimental campaigns carried out before 1995 was observed, and it was justified by the changes in the strand production process discussed in the Introduction. Therefore, since it is not possible to know a priori whether the observed high θ is due to some sort of strand surface contamination, the realizations of θ associated with tests performed before 1995 (i.e., the 142 experimental tests collected from [3, 4, 13, 15, 18, 20, 25]) will not be considered in the subsequent part of the investigation. The filtered sample of θ discussed in the next paragraphs (which therefore consists of 339 data) is illustrated in Fig. 1 with black markers, while the 142 realizations excluded for possible surface contamination of the tendons are represented with gray markers. As can be seen, the scatter in the sample of θ obtained from tests performed before 1995 (gray markers in Fig. 1) is significant and may raise concerns about the reliability of anchorages in real structures built before that date. However, the dataset comprises experimental campaigns performed between 1986 [3] and 2020 [22]; tests on members realized before the 1980s are not likely to show this scatter, since they were not influenced by the change in the strand production process described in the Introduction. Note also that the year 1995 should not be intended as an exact value: it was chosen arbitrarily within this study between the year of the last campaign that exhibited such a large scatter (1994, in Ref. [13]) and the year of the first study where this was not observed anymore (2000, in Ref. [10]). Finally, it should be noted that experiments performed in laboratories may not reflect the conditions of real members in existing structures and, indeed, as shown in [15], even one day of weathering of the strands results in a significantly better bond than that of tendons in 'as received' conditions. Thus, the current formulations may be applied to existing structures built in the 1990s, too. However, for the scope of this investigation, the tests performed before 1995 have been excluded.

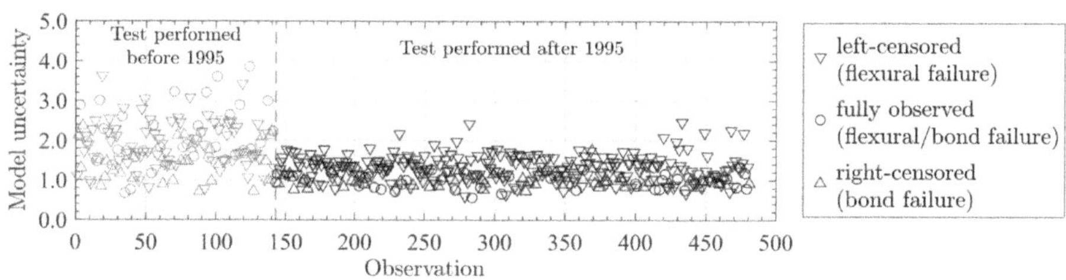

Fig. 1. Sample of θ (filtered sample in black and excluded realizations in gray).

More than 85% of the filtered θ sample consists of censored observations (292 data, identified with black triangular markers in Fig. 1): this happens because it is difficult to experimentally locate the exact point where applying the flexural load P to obtain a combined flexural/bond failure. Thus, for this reason, and since the censoring threshold is different for each observation, it is not straightforward to evaluate possible correlations between the sample of θ and the other variables considered in the models. Therefore, to verify the independence of the model uncertainty, only the uncensored values of the filtered sample of θ (47 observations) are plotted in Fig. 2. As can be noted, no clear trend can be recognized, and therefore, it is assumed that θ is independent from all the other variables in the models.

A lognormal (LN) distribution is commonly accepted to describe the model uncertainty [8]. Therefore, the maximum likelihood estimate (MLE) method was employed to estimate the parameters of the LN distribution of θ considering all the censored data, resulting in $\theta \sim LN(-0.114, 0.246)$, corresponding to a $\mu_\theta = 0.92$, $\sigma_\theta = 0.23$ and a coefficient of variation $V_\theta = \sigma_\theta / \mu_\theta = 0.25$. From the obtained results, a mean value

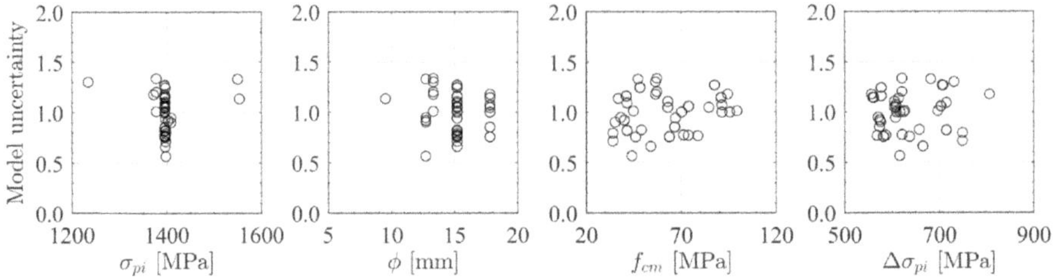

Fig. 2. Verification of the independence of fully observed values in the filtered sample of θ.

below unity and a quite high coefficient of variation can be observed. Nonetheless, these values are close to that obtained in [27] for the model uncertainty of bond strength of ordinary reinforcing bars.

4 Validation of Current Partial Safety Factors

According to [28], Eq. (4) is a model that allows the compute the design value of a resistance R_d through the partial safety factor method, following the Resistance Factor Approach; this means that R_d can be computed as:

$$R_d = R\{\boldsymbol{X_{rep}}; \boldsymbol{a_d}; \boldsymbol{F_{d,R}}\}/\gamma_R \tag{6}$$

where $R\{\cdot\}$ is the response of a resistance model, $\boldsymbol{X_{rep}}$ is a vector of representative values of material properties, $\boldsymbol{a_d}$ is a vector of design values of geometrical properties, and $\boldsymbol{F_{d,R}}$ is a vector of design values of actions dependent on the material properties (note that, in our case, these are the stress in the tendons, since the prestressing force to be anchored to the concrete depends on the strength of the tendons themselves). Finally, γ_R is the partial safety factor associated to a certain reliability index β, which considers the uncertainty in material properties and the uncertainty of the resistance model.

To the authors' best knowledge, there is no clear indication in the main design codes (e.g., *fib* Model Code, Eurocode, JCSS Probabilistic Model Code) that the anchorage length verification should be associated to a different reliability index than the other ULS verifications for the same element. Therefore, in this contribution, the target reliability indexes of $\beta_T = 3.8$ and $\beta_T = 4.3$ are assumed (considering a $t_{ref} = 50$ years and considering consequence classes CC2 and CC3, respectively [28]). Moreover, the scope here is limited to the calibration of the partial safety factor only for the resistance side. Therefore, since only the resistance model R is investigated (and a sensitivity factor $\alpha_R = 0.8$ is adopted [28]), the calibration of the partial safety factor γ_R is performed considering a target reliability index for the resistance $\beta_T^R = \alpha_R \cdot \beta_T$. Consequently, the failure probability P_f^R is defined as the probability of the resistance being lower than its corresponding design value $\left(P_f^R = P[R - R_d \leq 0]\right)$ and the corresponding target value can be computed as $P_f^R = \Phi(-\beta_T^R)$. It is straightforward that P_f^R is lower than the target failure probability of the member $P_f = \Phi(-\beta_T)$, since the latter refers to the probability of a resistance being lower than an action. Therefore, to ensure the achievement of β_T

at the member or structural level, proper partial safety factors should also be applied to the actions being considered (which are out of the scope of the present investigation).

Consistently with the previous definitions, a procedure to calibrate the minimum value of the partial safety factor γ_R based on MCS has been applied and described in the following, starting from the framework described in [29]. First, it should be noted that the model uncertainty θ has been computed considering the mean values of the material properties, while in the RFA approach, representative values are adopted. Therefore, a new random variable Γ is introduced here to describe the global uncertainty due to the randomness of the material properties and to the model uncertainty itself that occurs when applying Eq. (4) [29]. A realization of Γ (denoted with the lowercase γ) can be computed as the ratio of a probabilistic model (considering also θ) and the corresponding deterministic model, that is, in our case, Eq. (4):

$$\gamma = R\{\boldsymbol{X_{prob}}\}/R\{\boldsymbol{X_{rep}}\} \tag{7}$$

where $R\{\cdot\}$ is a realization of the resistance model, $\boldsymbol{X_{prob}}$ indicates a vector of R.V. and $\boldsymbol{X_{rep}}$ the corresponding vector of representative values (e.g., the 5% fractile of the R.V. in $\boldsymbol{X_{prob}}$). Therefore, considering the strength of precast concrete as a lognormal R.V. with $V_{fc} = 0.06$ [30] and θ a lognormal R.V. as described in the previous Section, a sample of Γ can be generated with MCS, and from it, any quantile γ_R corresponding to a certain probability of failure P_f^R can be obtained. These γ_R can be then employed as partial safety factors in Eq. (4) to obtain the target level of reliability β_T^R. The values of γ_R for each consequence class, together with the values of γ_C prescribed in the *fib* Model Code 2020 [30] are reported in Table 3.

Table 3. Partial safety factors for consequence classes CC2 and CC3.

Consequence class	β_T	P_f	α_R	β_T^R	P_f^R	γ_C	γ_R
CC2	3.80	~ 10^{-4}	0.8	3.04	~ 0.0012	1.50	1.35
CC3	4.30	~ 10^{-5}		3.44	~ 0.0003	1.60	1.50

Results in Table 3 are validated with FORM adopting the following limit state function:

$$g = R(\boldsymbol{X_{prob}}) - R_d \tag{8}$$

where $R(\boldsymbol{X_{prob}})$ is the probabilistic model of the anchorage length (considering the model uncertainty θ) and R_d is the design value of the anchorage length computed with Eq. (4), using all the characteristic values of the materials and the partial safety factors reported in Table 3. The FORM has been implemented in MATLAB according to the classical Hasofer-Lind algorithm and adopting a Rackwitz-Fiessler transformation due to the assumption of non-normal uncorrelated R.V. A 7-wire tendon with $\phi = 15.2$ mm, $\sigma_{pi} = 1395$ MPa, and $\Delta\sigma = 500$ MPa, and a gradual prestress release after $t = 3$ days have been assumed as deterministic quantities. Results are illustrated in Fig. 3, where it is shown that the reliability indexes β_{HL} computed with FORM for a certain value of γ varies between concrete classes. This happens because in probabilistic modeling, concrete strength has been assumed to be a lognormally distributed R.V. with a constant V_{fc} (i.e., the standard deviation varies for each concrete class), even though V_{fc} commonly decreases with concrete strength [30]. Conversely, in the formulation reported in Eq. (1), the f_{ck} is always obtained by subtracting 8 MPa from the mean value, that is, assuming a fixed standard deviation σ_{fc} of around 5 MPa. Therefore, for lower concrete strength classes, the adopted f_{ck} in Eq. (1) is smaller than the actual strength value corresponding to the 0.05 fractile, leading to a higher conservativeness of the results; the opposite occurs for higher strength classes, where the adopted f_{ck} is slightly greater than the 0.05 quantile in the probabilistic formulation. For this reason, as can be seen in Fig. 3, in the case of the highest concrete strength classes (C100/115), values of β_{HL} for $\gamma_R = 1.35$ and $\gamma_R = 1.50$ are slightly below the corresponding β_T^R. However, values of γ_C currently proposed by *fib* Model Code 2020 are largely conservative, and computed β_{HL} are always larger than the corresponding β_T^R.

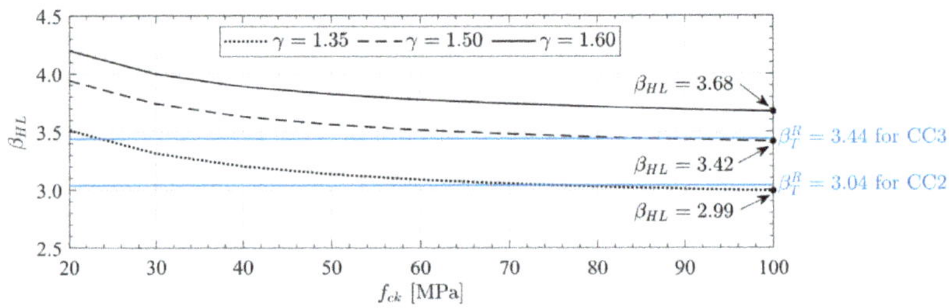

Fig. 3. Reliability indexes obtained adopting the partial safety factors in Table 3.

5 Conclusions

To conclude, in this contribution, the model uncertainty for the anchorage length formulation of *fib* Model Code 2020 has been estimated from a dataset of 339 flexural tests gathered from 12 different experimental campaigns. Then, employing FORM, it has been shown that the partial safety factors γ_C for CC2 and CC3 proposed *fib* Model Code 2020 are larger than the minimum values (obtained with MCS) that guarantee the attainment of the target β_T^R in the anchorage length formulation.

References

1. Janney, J.R.: Nature of bond in pre-tensioned prestressed concrete. ACI J. Proc. **50**(5), 717–736 (1954)
2. Rose, D.R., Russel, B.W.: Investigation of standardized tests to measure the bond performance of prestressing strand. PCI J. **42**(4), 56–80 (1997)
3. Cousins, T.E., Johnston, D.W., Zia, P.: Transfer and development length of epoxy coated and uncoated prestressing strand. PCI J. **35**(4), 92–103 (1990)
4. Russel, B.W., Burns, N.H.: Design Guidelines for Transfer, Development and Debonding of Large Diameter Seven Wire Strands in Pretensioned Concrete Girders. Texas DoT, USA (1993)
5. Osborn, A., Lanier, M., Hawkins, N.: Bond of prestressing strand to concrete. PCI J. **66**(1), 28–48 (2021)
6. CEB-FIP: Model Code 90. Thomas Telford Ltd. London (UK) (1993)
7. *fib*: Model Code for Concrete Structures 2010. Wiley, Berlin (Germany) (2013)
8. JCSS: Probabilistic Model Code. Periodically updated (2001)
9. Alcaraz Emmanuelli, G.D.: Effect of Strand Diameter on Bond, Transfer and Development Length Performance of Prestressing Strands. Ph.D. Dissertation, UF, Florida, USA (2000)
10. Barnes, R.W., Burns, N.H., Kreger, M.E.: Development Length of 0.6-inch Prestressing Strand in Standard I-shaped Pretensioned Concrete Beams. Texas DoT, USA (2000)
11. Bowser, T.M., Murray, C.D., Floyd, R.W.: Bond behavior of 0.6 In. Prestressing strand in BSCA cement concrete. ACI Struct. J. **117**(1), 43–52 (2020)
12. Carrol, J.C., Cousins, T.E., Roberts-Wollmann, C.L.: The use of grade 300 prestressing strand in pretensioned, prestressed concrete beams. PCI J. **62**(1), 49–65 (2017)
13. Cousins, T.E., Stalling, J.M., Simmon, M.B.: Reduced strand spacing in pretensioned, prestressed members. ACI Struct. J. **91**(3), 277–286 (1994)
14. Dang, C.H., Floyd, R.W., Hale, W.M., Martì-Vargas, J.: Measured development lengths of 0.7 in. (17.8 mm) strands for pretensioned beams. ACI Struct. J. **113**(3), 525–535 (2016)
15. Deatherage, J.H., Burdette, E.G., Chew, C.K.: Development length and lateral spacing requirements of prestressing strand for prestressed concrete bridge girders. PCI J. **39**(1), 70–83 (1994)
16. Kahn, L., Dill, J., Reutlinger, C.: Transfer and development length of 15-mm strand in high performance concrete girders. J. Struct. Eng. **128**(7), 913–921 (2002)
17. Kim, E.: Experimental study on development length of prestressing strand in pretensioned prestressed concrete members. J. Korean Soc. Saf. **23**(6), 115–121 (2008)
18. Lane, S.N.: A New Development Length Equation for Pretensioned Strands in Bridge Beams and Piles. U.S. DoT, USA (1998)
19. Larson, K.H., Peterman, R.J., Esmaeily, A.: Bond characteristics of self-consolidating concrete for prestressed bridge girders. PCI J. **52**(4), 44–57 (2007)
20. Mitchell, D., Cook, W.B., Arshad, A., Tham, T.: Influence of high strength concrete on transfer and development length of pretensioning strand. PCI J. **38**(3), 52–66 (1993)
21. Peterman, R.J., Ramirez, J., Olek, J.: Evaluation of Strand Transfer and Development Lengths in Pretensioned Girders with Semi-lightweight Concrete. Indiana DoT, USA (1999)
22. Arezoumandi, M., Looney, K.B., Volz, J.: Development length of prestressing strand in self-consolidating concrete vs. conventional concrete: experimental study. J. Build. Eng. **29**, 101218 (2020)
23. Ramirez, J.A., Russel, B.W.: Transfer, Development, and Splice Length for Strand/Reinforcement in High-Strength Concrete. NCHRP, USA (2008)
24. Ramirez-Garcia, A.T., Floyd, R.W., Hale, W.M., Martì-Vargas, J.R.: Influence of concrete strength on development length of prestressed concrete members. J. Build. Eng. **6**, 173–183 (2016)

25. Shahawy, M.: A critical evaluation of the AASHTO provisions for strand development length of prestressed concrete members. PCI J. **46**(4), 94–117 (2001)
26. Trent, J.: Transfer Length, Development Length, Flexural Strength, and Prestress Loss Evaluation in Pretensioned Self-consolidating Concrete Members. MSc thesis, Virginia Polytechnic Institute, USA (2007)
27. Sykora, M., Holicky, M., Prieto, M., Tanner, P.: Uncertainties in resistance models for sound and corrosion-damaged RC structures according to EN 1992–1-1. Mater. Struct. **48**(10), 3415–3430 (2014)
28. CEN: EN 1990:2023. Eurocode - Basis of structural and geotechnical design (2023)
29. Castaldo, P., Gino, D., Carbone, V.I., Mancini, G.: Framework for definition of design formulations from empirical and semi-empirical resistance models. Struct. Concr. **19**, 980–987 (2018)
30. Sykora, M., Botte, W., Caspeele, R., Diamantis, D., Muttoni, A., Steenbergen, R.D.J.M.: Probabilistic models for resistance variables in *fib* model code 2020 for design and assessment. Acta Polytechnica CTU Proc. **36**, 206–215 (2022)

Modeling Strategies of San Giovanni Baptistery of Pisa Accounting for Uncertainties on Soil and Masonry Mechanical Properties

Giada Bartolini[1(✉)], Anna De Falco[2], and Filippo Landi[2]

[1] Department of Energy, Systems, Territory and Construction Engineering, University of Pisa, Largo Lucio Lazzarino 1, Pisa, Italy
giada.bartolini@phd.unipi.it

[2] Department of Civil and Industrial Engineering, University of Pisa, Largo Lucio Lazzarino 1, Pisa, Italy

Abstract. The comprehension of the structural behavior of historical buildings is pivotal for preserving them through suitable interventions and designing adequate monitoring systems. The complexity lies in articulated geometries, poor knowledge of materials, and often unknown construction sequences, which may have influenced the stress field in a non-linear material such as masonry.

This paper addresses the issues through different modeling strategies accounting for material uncertainties in a probabilistic framework that leverages sensitivity analyses on Finite Element (FE) global models. The prior probability density functions of soil and masonry mechanical parameters are chosen based on expert judgment and available data from experimental campaigns. Response surfaces surrogate numerical models based on general Polynomial Chaos Expansion (gPCE), thus turning burdensome runs into faster analytical evaluations. Modal analyses on the entire FE model of the Baptistery of Pisa are performed to evaluate the sensitivity of masonry and soil mechanical parameters on the variation of the first modal eigenvalues. This aims at understanding the minimum recognizable parameter variation when monitoring natural frequencies, thus guiding the sensors' best positioning.

Keywords: Baptistery of Pisa · Monumental masonry structure · Sensitivity analysis · Uncertainty quantification · gPCE

1 Introduction

Cultural Heritage (CH) buildings are susceptible to natural disasters, like earthquakes, floods, and landslides. In seismic-prone regions, there is a pressing need to enhance the dynamic performance of historic constructions and improve safety. For this purpose, models are useful to make predictions but entail uncertainties related to materials and boundary conditions. In this regard, the soil significantly influences not only the static behavior but also the mode of vibration, and thus, it should be considered in the assessment of seismic safety of buildings. In the literature, models of historical assets

J. C. Matos et al. (Eds.): IPW 2024, LNCE 494, pp. 419–428, 2024.
https://doi.org/10.1007/978-3-031-60271-9_39

generally have a fixed base [1], either when Soil-Structure Interaction (SSI) is ignored, or geotechnical data is not provided. However, investigations into historical masonry structures reveal that the interaction between soil and structure influences the dynamic characteristics of these buildings [2], notably shaping their dynamic properties and seismic response [3–5]. In a relatively small number of cases, seismic assessments take dynamic SSI into account [6]. When it happens, the structural capacity can be computed using complete models [7] or by replacing the soil with appropriate springs to simulate dynamic soil–foundation impedance [8].

The current study focuses on the Baptistery of San Giovanni, part of the monumental complex of the Piazza del Duomo in Pisa, encompassing the Cathedral, the Leaning Tower, and the Monumental Cemetery. The entire complex rests on notably deformable ground, which more than eight centuries ago generated the well-known singularity of the Tower. The Baptistery is a decidedly different construction from the Tower, and it has never experienced structural issues attributable to the ground's characteristics. However, its dynamic behavior could be significantly influenced by the soil, and this aspect is essential for the seismic assessment of the structure. With the help of a high-definition three-dimensional model, we initially examined the variation in the natural frequencies of the structure due to the presence of the soil in comparison to the case of a model with a fixed base. Subsequently, we investigated how epistemic uncertainties related to soil and masonry propagate onto the values of the first eight frequencies of the model. Afterward, we identified the frequencies most influenced by the characteristics of the soil or those of the masonry through a sensitivity analysis on the parameters of the soil layers and masonry within a probabilistic framework. Once the relevant variables were selected and the probability distributions for their values established, the objective was achieved by surrogating the FE model using the gPCE technique, specifically for the first eight frequencies, thus obtaining the Sobol Indices (SIs). These allowed us to discover the extent to which uncertainties in the masonry and soil influence each frequency, and what frequencies are affected by the various materials constituting the building. Finally, response surfaces for natural frequencies are derived as a first step to create a digital twin of the Baptistery.

2 Material and Methods

2.1 Case Study

The Baptistery of San Giovanni is a medieval masonry building located in the Piazza del Duomo in Pisa (Fig. 1a). Its erection took place through a lengthy construction process, starting in 1152 and continuing until 1395. The project was initially designed by Diotisalvi, and later completed by Nicola, and Giovanni Pisano [9]. The cylindrical building has an external diameter of about 34 m and a height of 54 m. On the ground floor, the building creates a circular space that houses the baptismal font and is surrounded by the arcades of the ambulatory, which is topped by masonry cross vaults. On the first level, the women's gallery is covered by barrel vaults featuring a toroidal pattern. The load-bearing structures consist of the outer cylindrical masonry wall and an internal colonnade. From the ground floor the colonnade features 12 supports, including pillars of San Giuliano limestone and columns of granite, extending up to the roof. The masonry structure is layered, with two external leaves of limestone and an internal masonry filling. The building has a double brick masonry dome. The internal structure is shaped like a truncated cone on a dodecahedral base resting on the colonnade. The external hemispherical dome rests on the outer walls and is supported by radial partition walls rising in the space between the two domes (Fig. 1b). The internal dome is connected to the outer wall through radial brick masonry arches. Here, masonry characteristics are not deeply investigated, while a vast experimental campaign was carried out in the past to characterize the underlying soil.

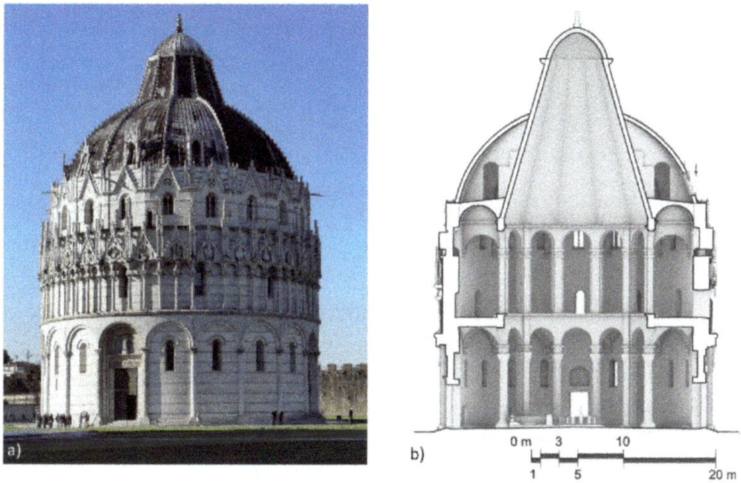

Fig. 1. The Baptistery of San Giovanni: a) photo from north-west; b) the N-S section.

The foundation was investigated through ground penetrating radar and geoelectrical tomography, while the stratigraphy and characteristics of the soil were deduced from different techniques: penetrometer tests with piezocones, continuous core-drillings, and seismic dilatometer tests. The latter, compared with the cross-hole tests carried out near the Tower, showed similar elastic properties of the soil underneath the Tower [10, 11]. In this study, we adopted the same layered substructure of the Tower, as shown in Fig. 2, already used for the FE model of the Tower [12].

2.2 FE Model

An accurate FE model was created based on the terrestrial laser scanning point cloud of the construction. It encompasses all the main structural features, such as the external walls, the internal columns and pillars connected by a circular-shaped arched wall, the two domes, the cross vaults on the first floor, the circular barrel vaults on the upper floor, and the radial arches between the dodecahedral dome and the external walls.

The stratigraphy of the external walls, with a total thickness of about 2.60 m at the base of the monument, was modeled as composed of three-leaf masonry: the 40-cm outer and inner layers of marble masonry, and the inner 180-cm leaf, simulated as an

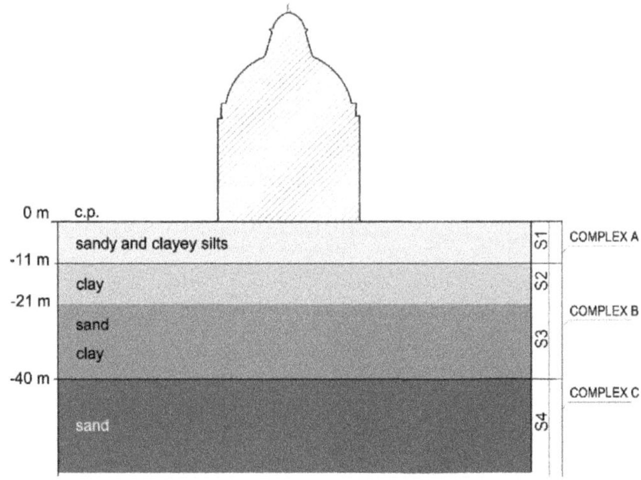

Fig. 2. Subsoil profile with distinction of geological complexes in analogy with the literature about the Leaning Tower [10, 11].

infill of mixed masonry. All decorative elements on the lateral surface were neglected, as well as the first-order half-columns, which instead were taken into account through an equivalent thickness of the wall at the base (Fig. 3a). The presence of the soil was simulated as a cylindrical-shaped solid with a diameter of 170 m and a height of 100 m, corresponding to about five times the horizontal dimension of the Baptistery and twice the vertical one, respectively (Fig. 3b). These dimensions were chosen in such a way that any increase in them does not result in variations in the first two frequencies of the model greater than one-hundredth of a Hertz. The soil stratigraphy was defined as made of four layers as shown in Fig. 2. The computational mesh of the aboveground structure is made of nearly 373000 10-node tetrahedral elements of type serendipity, having minimum and maximum edge lengths of 0.30 m and 0.80 m respectively. A coarse mesh with serendipity tetrahedral elements was chosen to discretize the solid simulating the soil. The displacements of the base node of the cylinder were restrained in three directions, while those of the lateral surface were restrained along the axis orthogonal to the surface itself. The classification of the construction materials is shown in Fig. 3d representing the same vertical section of the model as Fig. 3c. Soil's materials are labeled by S1-S4, whereas those of the structure are identified by M1-M5. The uncertain material characteristics are the Young's moduli of M1, S1, S2, S3, and S4, the deterministic ones are those of M2-M5. In Table 1, the statistical data of the lognormal distributions of the Young's moduli of the soil layers S1-S4 and external layers M1 are reported, while those of deterministic materials are in Table 2. The Poisson's ratio was taken equal to 0.2 for all materials except for M3, for which is equal to 0.36, according to [13].

2.3 gPCE-Based Surrogation and Sensitivity Analysis

A surrogate model of the modal frequencies of the Baptistery, i.e. an analytical formulation able to provide a good approximation of the FE model outputs, is developed

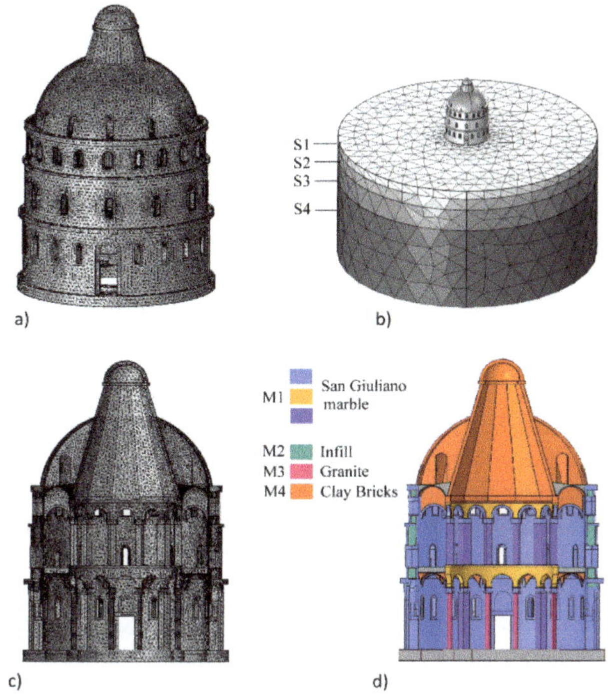

Fig. 3. The FE model of the Baptistery.

Table 1. Mean value (μ) and standard deviation (σ) of the lognormal distribution of the Young's modulus (E).

	M1	S1	S2	S3	S4
E [MPa]	$\mu = 54176$	$\mu = 180$	$\mu = 116$	$\mu = 297$	$\mu = 752$
	$\sigma = 10277$	$\sigma = 41$	$\sigma = 12$	$\sigma = 70$	$\sigma = 110$

Table 2. Elastic moduli of materials M2-M4 (see Fig. 3 for the definitions).

	M2	M3	M4
E [MPa]	7000	80000	1200

to significantly speed up computations, according to the steps indicated in Fig. 4. The surrogate model is obtained through a general Polynomial Chaos Expansion (gPCE) of the FE model response [14]. Without going into mathematical details, we consider M as the numerical model producing the uncertain output Y based on the vector of uncertain parameters $\boldsymbol{X}$. A gPCE form of the model response can be found as a linear combination of orthonormal polynomials $\boldsymbol{P}_i(\boldsymbol{X})$, and it becomes an approximation when we consider only polynomials up to a given degree n

$$Y = M(X) \approx M_n^{gPCE}(X) = \sum_{i \leq n} \alpha_i P_i(X), \tag{1}$$

with $\boldsymbol{\alpha}_i$ the matrix of combination coefficients and $\boldsymbol{i}$ is the multi-index set. The polynomials belong to specific families whose choice depends on the statistical description of the input parameters, i.e. Hermite for Lognormal distribution. Unknown combination coefficients $\boldsymbol{\alpha}_i$ are obtained from a set of known input-output pairs, which knowledge requires running the original model a very limited number of times compared to Monte Carlo methods. After evaluating the difference between the numerical solutions and the surrogate models, a degree n equal to 3 was chosen leading to a set of 1024 samples obtained from a full-tensor grid representation.

To assess sensitivity, we employed Sobol' indices (SIs) [15], which belong to the field of variance-based sensitivity measures. In this way, it is possible to assess the effect of variability in uncertain inputs upon the variability of the output. We consider here the first eight natural frequencies of the model of the Baptistery as the monitored output. The calculation of Sobol' Indices is based on a decomposition of the variance associated with a given output Y of the model, Var(Y), into contributions associated with each input parameter. For an input X, the associated $\mathrm{SI_X}$ is therefore

$$SI_X = \frac{Var_X(Y)}{Var(Y)}, \tag{2}$$

where $\mathrm{Var_X}(Y)$ is the contribution of input X to the variance of output Y. Sobol' Indices vary between 0 and 1, with higher values indicating a higher influence of the input variability on the variability of the output.

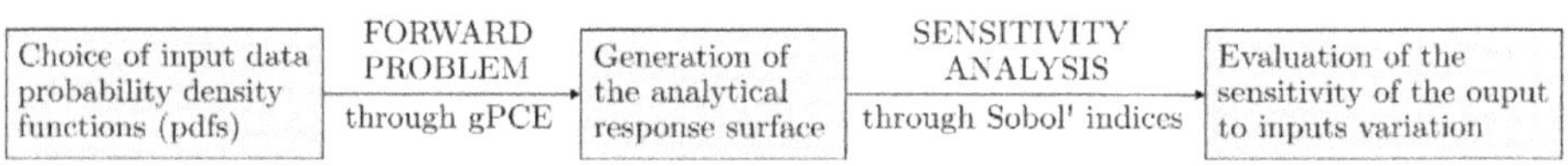

Fig. 4. The adopted probabilistic framework.

3 Results and Discussion

The modal behavior of the numerical model of the Baptistery is shown in Fig. 5, which depicts the mode shapes of the first eight natural frequencies. The first two mode-shapes involve the flexural behavior of the building, the third the radial deformation, the fourth the extensional behavior, and the remaining modes involve the deformation of the cylindrical stem and the dome.

The first significant outcome of the analysis is given by the comparison of the natural frequencies obtained through a model with rigid soil, and the one including the deformable soil layers. The results obtained considering the average values of uncertain input parameters are reported in Table 3. The relevance of the deformability of soil layers is evident, providing frequencies that can be more than 50% lower in the case of deformable soil than in the case of rigid soil.

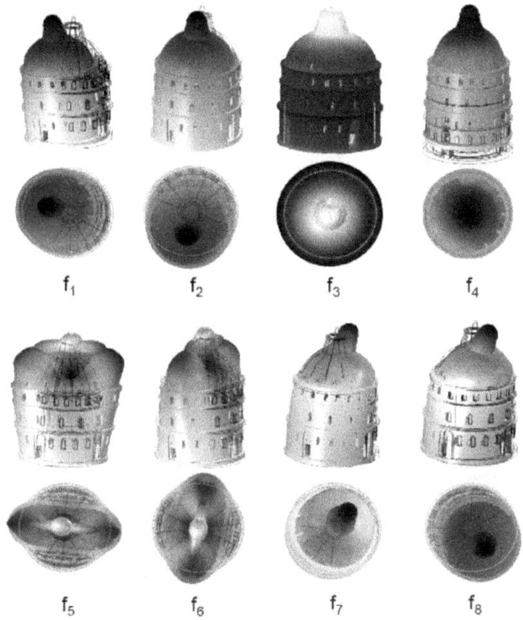

Fig. 5. The mode shapes related to the first eight frequencies of the model of the Baptistery.

Table 3. First eight frequencies obtained by assuming the average values of the input uncertain parameters of the numerical model of the Baptistery in the case of deformable soil (f_{ds}) and in the case of fixed base (f_{fb}).

	f_1	f_2	f_3	f_4	f_5	f_6	f_7	f_8
deformable soil f_{ds} [Hz]	1.63	1.64	3.28	3.72	4.16	4.17	4.21	4.21
fixed base f_{fb} [Hz]	3.78	3.79	4.54	4.56	5.87	5.89	6.47	6.47
f_{fb}/f_{ds}	2.32	2.31	1.39	1.23	1.41	1.41	1.54	1.54

After this preliminary evaluation, the probabilistic framework described in the previous Section has been applied to propagate uncertainties in input parameters and evaluate sensitivities on natural frequencies via the developed FE model. Table 4 summarizes the resulting average natural frequencies and their coefficient of variation (V). The variance associated with the first eight natural frequencies is decomposed in the contributions of each uncertain input according to SIs. The results in terms of partial variances are shown in Fig. 6 and the corresponding SIs are reported in Table 5.

The first soil layer (S1) plays a significant role in the dynamic behavior of the Baptistery, with a SI nearly always greater than 0.5; the second and third soil layers (S2, S3 respectively) have some influence on the first two and fourth frequencies, while are negligible for the others; the soil deepest layer (S4) has an influence close to zero for all the considered frequencies; the external masonry layer (M1) remarkably affects the last four frequencies (f5–f8). This suggests that the deepest soil layer could be excluded from future evaluations, thus reducing the problem dimension.

Table 4. Mean value (μ) and coefficient of variation (V) of the first eight frequencies of the numerical model of the Baptistery.

	f_1	f_2	f_3	f_4	f_5	f_6	f_7	f_8
μ [Hz]	1.62	1.62	3.26	3.68	4.06	4.06	4.26	4.27
V	0.05	0.05	0.07	0.04	0.03	0.03	0.02	0.02

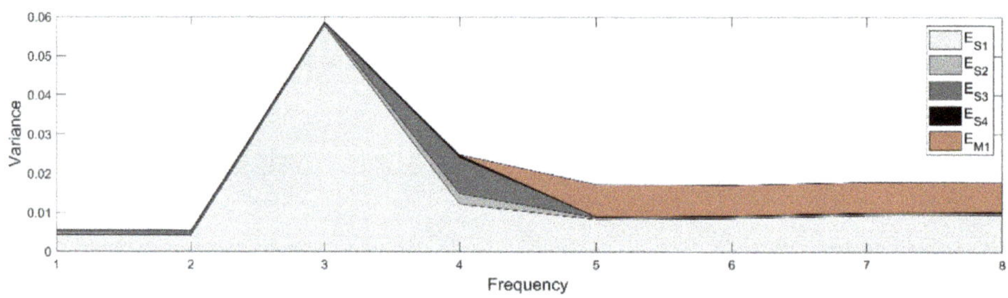

Fig. 6. Partial variance of the first natural frequencies associated with uncertain parameters.

Table 5. Sobol' Indices of the uncertain input parameters for each of the first eight natural frequencies of the model. E_{Si} is the Young's modulus of the i-th soil layer, E_{M1} of the masonry.

Natural frequencies	SI (E_{S1})	SI(E_{S2})	SI(E_{S3})	SI(E_{S4})	SI(E_{M1})
f_1	0.768	0.105	0.116	0.002	0.010
f_2	0.763	0.105	0.119	0.002	0.011
f_3	0.984	0.010	0.002	0.000	0.004
f_4	0.494	0.107	0.372	0.024	0.004
f_5	0.484	0.025	0.013	0.000	0.479
f_6	0.503	0.026	0.013	0.000	0.458
f_7	0.523	0.027	0.013	0.000	0.437
f_8	0.530	0.026	0.013	0.000	0.430

Another significant outcome of the gPCE surrogation is the availability of analytical response surfaces that can provide a direct computation of frequencies without further onerous runs of the FE model. These could be used for model updating procedures in view of the design and installation of monitoring systems. Figure 7 reports the response surfaces of the first (a) and eighth (b) natural frequencies of the FE model.

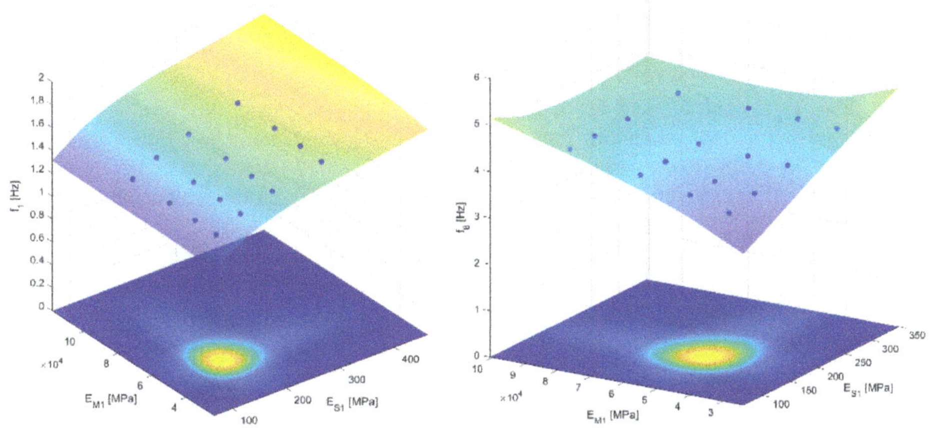

(a) (b)

Fig. 7. Response surface of the first (a) and eighth (b) natural frequencies of the FE model.

4 Conclusions

In this study, the dynamic behavior of the Baptistery of Pisa is investigated in a probabilistic framework accounting for epistemic uncertainties on soil and masonry mechanical parameters. This was achieved by using a highly detailed FE model of the Baptistery that reproduces the layered structure of soil and the different masonry materials in the structure. The obtained results in terms of sensitivity measures and response surfaces will be useful when a monitoring system is installed, to conduct a dynamic identification of parameters for a model that can be effective in assessing the seismic safety of the structure. The frequencies to be employed in the updating process to calibrate the uncertain parameters of the model are identified. Moreover, by knowing which material introduces greater uncertainty in the most representative frequencies of the structure, it will be possible to focus investigations on that material, aiming to reduce uncertainty. Finally, this work could become the first step toward the optimal placement of sensors aimed at detecting structural damage.

References

1. Fathi, A., Sadeghi, A., Emami Azadi, M.R., Hoveidaie, N.: Assessing seismic behavior of a masonry historic building considering soil-foundation-structure interaction (case study of Arge-Tabriz). Int. J. Archit. Herit. **14**(6), 795–810 (2020)
2. Shabani, A., Feyzabadi, M., Kioumarsi, M.: Model updating of a masonry tower based on operational modal analysis: the role of soil-structure interaction. Case Stud. Constr. Mater. **16**, e00957 (2022)
3. Mortezaei, A., Motaghi, A.: Seismic assessment of the world's tallest pure-brick tower including soil-structure interaction. J. Perform. Constr. Facil. **30**, 04016020 (2016)
4. Hökelekli, E., Al-Helwani, A.: Effect of soil properties on the seismic damage assessment of historical masonry minaret–soil interaction systems. Struct. Des. Tall Build. **29** (2020)

5. Erdogan, Y.S.: Discrete and continuous finite element models and their calibration via vibration and material tests for the seismic assessment of masonry structures. Int. J. Archit. Heritage **11**(7), 1026–1045 (2017)
6. de Silva, F., Ceroni, F., Sica, S., Silvestri, F.: Non-linear analysis of the carmine bell tower under seismic actions accounting for soil–foundation–structure interaction. Bull. Earthq. Eng. **16**(7), 2775–2808 (2017). https://doi.org/10.1007/s10518-017-0298-0
7. Casolo, S., Diana, V., Uva, G.: Influence of soil deformability on the seismic response of a masonry tower. Bull. Earth. Eng **15**(5), 1991–2014 (2017)
8. Fiorentino, G., et al.: Leaning tower of pisa: recent studies on dynamic response and soil-structure interaction, Paper No. 11588. In: Proceedings of the 16th European Conference of Earthquake Engineering (16ECEE), Thessaloniki, Greece (2018)
9. Pierotti, P., Benassi, L.: Deotisalvi: l'architetto pisano del secolo d'oro. Pacini Editore, Pisa, Italy (2001)
10. Lo Presti, D., Jamiolkowski, M., Pepe, M.: Geotechnical characterisation of the subsoil of Pisa Tower. In: Characterization and Engineering Properties of Natural Soils, vol. 2, pp. 909–946 (2002)
11. Burland, J.B., Jamiolkowski, M.B.: Leaning tower of pisa: behaviour after stabilization operations. Int. J. Geoeng. Case Hist. **1**, 156–169 (2009)
12. Butuc, D., et al.: Sensitivity analysis on the FE model parameters of the Leaning Tower of Pisa. In: 20th International Probabilistic Workshop (2024)
13. Macchi, G., Ghelfi, S.: Problemi di consolidamento. In: La Torre restituita: gli studi e gli interventi che hanno consentito la stabilizzazione della Torre di Pisa. Vol. III, part 1 of Bollettino d'arte, Ministero per i beni e le attività culturali. Libreria dello stato, Roma, IT (2005)
14. Landi, F., Marsili, F., Friedman, N., Croce, P.: GPCE-based stochastic inverse methods: a benchmark study from a civil engineer's perspective. Infrastructures **6**, 158 (2021)
15. Sudret, B.: Global sensitivity analysis using polynomial chaos expansions. Reliab. Eng. Syst. Saf.. Syst. Saf. **93**, 964–979 (2008)

Probabilistic Quantification of the Structural Capacity in Concrete Discontinuity Regions Using Strut-and-Tie Models

Kito Luyten(✉), Wouter Botte, and Robby Caspeele

Department of Structural Engineering and Building Materials, Laboratory Magnel-Vandepitte for Concrete Research, Ghent University, Ghent, Belgium
kito.luyten@ugent.be

Abstract. In order to achieve a sustainable design or assessment of complex concrete elements, it is important to perform an adequate probabilistic assessment of their structural capacity. In this contribution, the full-probabilistic structural capacity in concrete half-joint girders is investigated by using three different strut-and-tie models, which are an application of the static theorem in limit analysis. Probabilistic Monte Carlo sampling is used to include uncertainties directly into the limit analysis calculations and to investigate the uncertainty propagation in different strut-and-tie models. Moreover, the influence of the strut-and-tie model as well as the reinforcement ratio on the distribution of failure mechanisms is investigated.

Keywords: Structural concrete · discontinuity regions · strut-and-tie models · uncertainties · half-joints

1 Introduction

Being able to accurately determine the strength of complex concrete structures is of primordial importance. In this way, an economic and performance-based design can be achieved that fulfils the safety regulations without being over-conservative, while at the same time adding to its sustainability by a reduced material need.

Mostly, structures are designed by using partial safety factors to account for uncertainties related to loads, material properties and geometric parameters. This design method is an application of a level I reliability method in which the partial safety factors are calibrated by a level II reliability method [1]. This approach works well for linear elements for which the Euler-Bernoulli hypothesis is valid since the used partial safety factors are calibrated based on such linear cases. However, in discontinuity regions (D-regions), the strain distributions are significantly non-linear and strut-and-tie models and stress fields are commonly used to determine the strength of these elements. Both methods are derived from limit analysis for which yet no general probabilistic framework is available to incorporate uncertainties adequately. Hence the partial safety factors derived from linear problems are not necessarily valid in these discontinuity regions and

J. C. Matos et al. (Eds.): IPW 2024, LNCE 494, pp. 429–439, 2024.
https://doi.org/10.1007/978-3-031-60271-9_40

a probabilistic framework needs to be developed to determine equivalent safety factors for discontinuity regions.

One particularly common discontinuity region is the edge of a dapped end beam or Gerber saddle, also known as a half-joint (Fig. 1). This type of joint was often used in bridges across Europe during the 60s and 70s. In a half-joint, the reinforcement detailing is crucial to ensure the force transfer from the linear Bernoulli region (B-region), through the nib, towards the support location [2].

In this contribution, probabilistic calculations using three different strut-and-tie models are executed in order to determine the strength distribution of this half-joint discontinuity region. A deterministic calculation without safety factors is compared to a probabilistic calculation where probability density functions are integrated for the concrete and steel strength. In this way, this research takes a first step towards a necessary probabilistic framework for discontinuity regions.

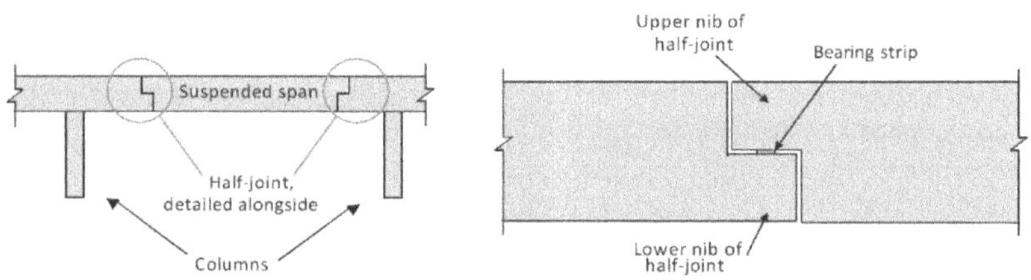

Fig. 1. Half-joint principle for reinforced concrete bridges [2]

2 Strut-and-Tie Method

The calculation according to a strut-and-tie method consists generally of three main steps. First, the B- and D-regions have to be determined. Next, a corresponding strut-and-tie model (STM) needs to be developed. Finally, the individual STM members need to be designed [3].

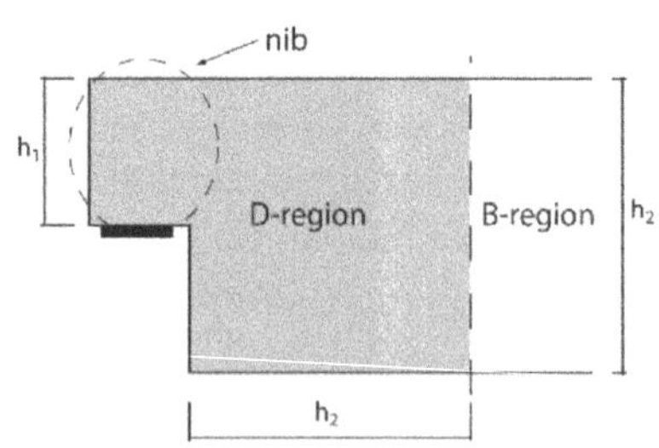

Fig. 2. B- and D-region for half-joint specimen [3]

2.1 B- and D-Region

Defining the B- and D-regions can be done with the principle of Saint-Venant, (Fig. 2). At the interface, the internal forces can be calculated using linear beam theory [3].

2.2 Strut-and-Tie Model

A strut-and-tie model (STM) consists of three elements: compression struts, tension ties and nodes which connect them all into a coherent system.

There is not a single correct STM since this design method is an application of limit analysis, and more specifically of the static method, enabling the determination of a lower bound limit to the structural capacity [4]. So, if a half-joint can withstand a given load according to a certain STM, then it will not collapse under this load, but its real capacity might even be higher.

In this paper, three different STMs are used. Model A is a model considering orthogonal reinforcement, while B is a model considering diagonal reinforcement. Also, the superposition of both models is calculated [3].

2.3 STM Members

Compressive Struts. The acting force in the struts is resisted by the concrete and – if applicable – the amount of compressive reinforcement present. The acting concrete stress, $\sigma_{c,st}$, in the compressive strut can be calculated from Eq. (1), where F_{st} is the acting force in the strut, $A_{s,st}$ the steel area and $A_{c,st}$ the effective concrete area of the strut defined by its width, w, and depth, t. The depth of the strut can be taken equal to the beam's width unless the strut originates at a support, in which case it should be equal to the support's width. The width of a strut is not necessarily constant over its entire length, but it is smallest near the nodes and should hence be verified at that location. For w, the length perpendicular to F_{st} should be used. $\sigma_{s,st}$ can be converted to $\sigma_{c,st}$ by multiplying it with the ratio E_s/E_c [3, 5].

$$F_{st} = \sigma_{c,st}A_{c,st} + \sigma_{s,st}A_{s,st} \tag{1}$$

The acting compressive concrete stress $\sigma_{c,st}$ should remain smaller than the resisting effective concrete strength $f_{c,ult}$, which can be calculated according to Eqs. (2) and (3) [5].

$$f_{c,ult} = k_i \nu' f_{cd} \tag{2}$$

$$\nu' = 1 - \frac{f_{ck}}{250} \tag{3}$$

The value of k_i depends on the stress state. A strut subjected to transversal tension limits k_i to 0.6, while for transversal compression $f_{c,ult}$ equals f_{cd} [5].

Tension Ties. The acting force in the ties F_{tie} is resisted by the available reinforcement and the acting steel stress $\sigma_{s,tie}$ can be determined based on Eq. (4), where $A_{s,tie}$ is the reinforcement steel area [3, 5].

$$F_{tie} = \sigma_{s,tie}A_{s,tie} \tag{4}$$

Nodes. In the nodes of a STM, a multi-directional stress state exists, as at this location different compressive struts and tension ties intersect. The strength of the node depends on the concrete strength and the confinement of this zone, determined by the number of arriving struts and ties. There are four general types of nodes: CCC, CCT, CTT and TTT, where C stand for 'compression' and T for 'tension'. A TTT node does not occur in practice. To the other nodal types, a k-value is attributed, $k_1 = 1.0$, $k_2 = 0.85$ and $k_3 = 0.75$, for CCC, CCT and CTT nodes respectively. If more than three bars come together in one node, this can be converted to a series of closely linked 3 bar nodes according to the modified hydrostatic approach [3, 6]. Typically, the most critical nodes are at the supports and at the locations where the loads are introduced in the concrete element. The acting stress in a node is calculated at each nodal edge separately by using Eq. (5), in which $F_{str/tie,perp}$ is the component of the strut or tie force perpendicular to the nodal edge and $A_{c,node}$ is the nodal area defined by its length and its depth t.

$$\sigma_{c,node} = \frac{F_{str/tie,perp}}{A_{c,node}} \tag{5}$$

3 Calculation Method and Assumptions

3.1 General

Probabilistic calculations are executed for the concrete specimens described in [2, 3], i.e. half-joint girders subjected to a point load in the middle of the beam. In Table 1 the different input parameters are given in accordance to [2, 3] as well as their stochastic distribution based on [7]. The concrete strength and the steel strength are considered probabilistic variables, while the dimensions and reinforcement areas are assumed to be deterministic. It should be noted that the model uncertainty for the resistance is taken on the basis of proposals for B-regions and further research is needed in order to develop a suitable model uncertainty factor for the resistance in case STMs are applied.

Table 1. Stochastic input parameters

Parameter	Distribution	Mean (μ)	COV (δ)	Reference
Acting load, E [kN]	Deterministic	300	–	–
Concrete strength, f_c [MPa]	Lognormal	38 ($f_{ck} = 30$)	0.15	[7]
Steel yield strength, f_y [MPa]	Lognormal	560 ($f_{yk} = 500$)	0.054	[7]
Young's modulus steel, E_s [MPa]	Deterministic	210000	–	–
Resistance model uncertainty, K_R [–]	Lognormal	1.2	0.15	[7]

The length l of the beam between the supports is 3140 mm, its width b is 400 mm and its height h is 700 mm. The concrete cover c amounts to 40 mm and the nib has a height h_n of 325 mm and a length l_n of 260 mm. The support is a steel plate with the following dimensions: 400 mm × 140 mm × 30 mm.

The reinforcement plan for the reference case is as shown in Fig. 3, with 5 ϕ25 bars as bottom reinforcement, 5 ϕ20 bars as top reinforcement, 3 ϕ12 U-bars, 4 ϕ10 double-legged stirrups near the nib and 4 ϕ12 diagonal bars.

In a second case, the concrete strength is reduced to 20 MPa, the U-bars are increased to 3 ϕ16, the diagonal bars to 6 ϕ12 and four-legged stirrups are used.

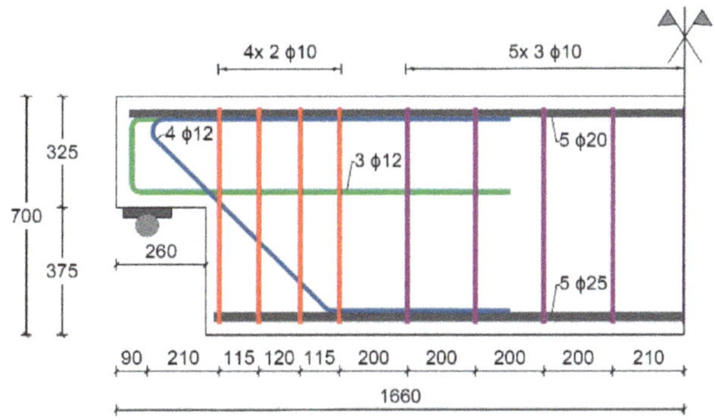

Fig. 3. Reinforcement – Reference case (dimensions in mm)

Models A and B (Fig. 4 and Fig. 5) are chosen as calculation models because they are of a different type, i.e. with orthogonal and diagonal reinforcement, and both are often used in literature. Model AB (Fig. 6) is a superposition of the models A and B [3, 8–10].

3.2 Strut-and-Tie Model A

Model A is an orthogonal model with three main tension ties. A horizontal tie, T_1, at the location of the horizontal part of the U-bar, and two vertical ties, T_2 and T_3, representing the shear reinforcement (Fig. 4).

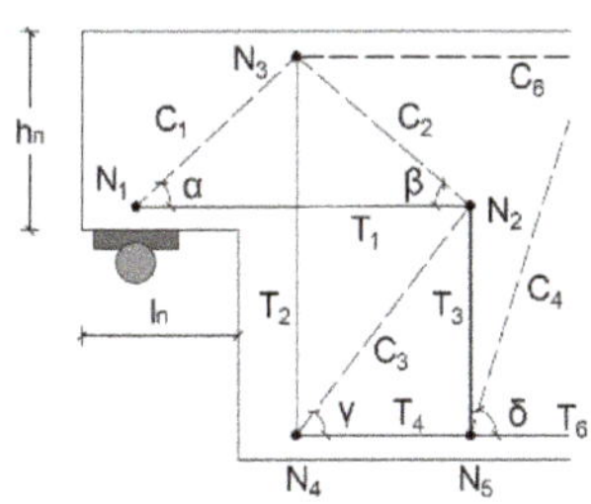

Fig. 4. Model A

First, the locations of the nodes are determined based on the required concrete cover, the desired reinforcement layout and the resulting forces at the interface of the B- and

D-region. T_2 lies between the first two stirrups, T_3 at the location of the 4th stirrup and the strut C_4 is situated between the 4th and the 5th stirrup.

Next, the acting forces in the ties, T_i, and in the struts, C_i, can be calculated based on the static equilibrium in each node. With this, the acting stress in all the members is determined as explained in Sect. 2.3, for T_2 it is assumed that the closest two stirrups contribute to this tension tie [2]. Only the strut width w still depends on the dimensions of the nodes, as it is the projection of the nodal area on the line perpendicular to the line of action of the strut.

Depending on the type of node, a different initial assumption and method is used to define the node's dimensions. Once its dimensions are known, the acting stress on each nodal area can be calculated with Eq. (5).

The nodal area at the support is defined by Eq. (6), where the support's length b_0 is diminished with twice the concrete cover d_1, measured from the rebar's centre line, to account for the inclination of strut C_1. The nodal areas created by a tension tie have a length u based on Eq. (7), with n the number of rebar layers, s the distance between the axes of these layers and φ the rebar diameter [5]. If two out of the three nodal sides are known with these equations, the third one can be calculated with the Pythagorean theorem (e.g. N_1 & N_4). Otherwise, a hydrostatic stress state is assumed. This means that the stress is equal in the entire node and that the intersecting strut and ties are all perpendicular to the nodal areas. The unknown dimensions are determined from the hydrostatic assumption via Eq. (8) [11]. Nodes with more than three intersecting forces can be split into nodes with three sides and according to [6] these nodes will be hydrostatic as well (e.g. N_2 & N_3).

$$a_1 = b_0 - 2\,d_1 \tag{6}$$

$$u = 2\,d_1 + (n - 1)s + \varphi \tag{7}$$

$$\frac{F_i}{a_i} = \frac{F_j}{a_j} \tag{8}$$

3.3 Strut-and-Tie Model B and AB

Model B shown in Fig. 5 is a diagonal model due to its diagonal tension tie and model AB (Fig. 6) is the superposition of both models. Similarly as in model A, the locations of the nodes are first determined. Then, the acting forces and stresses in the ties T_i and struts C_i are calculated as well as the nodal dimensions.

3.4 Resistance Calculation

To determine the resistance of each model, first the acting stress is calculated in every element as described in Sect. 2.3. With the acting stresses, $\sigma_{Ed,i}$, known, the safety of each element is calculated with Eq. (9), where $\sigma_{Rd,i}$ is the resistance in that element, determined by Eq. (2) for concrete and by the yield strength, f_{yd}, for steel.

$$S_i = \frac{\sigma_{Rd,i}}{\sigma_{Ed,i}} \tag{9}$$

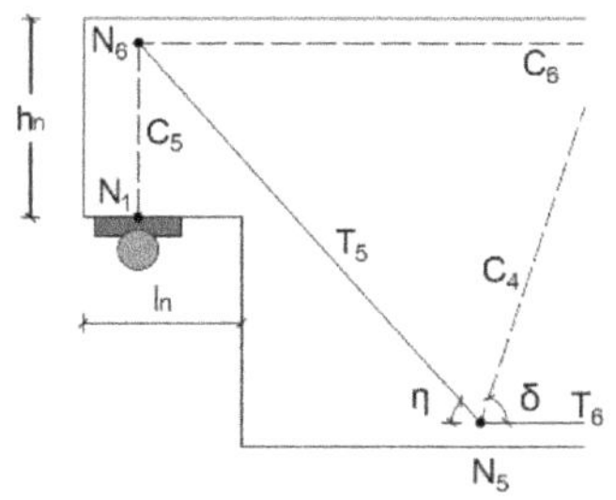

Fig. 5. Model B

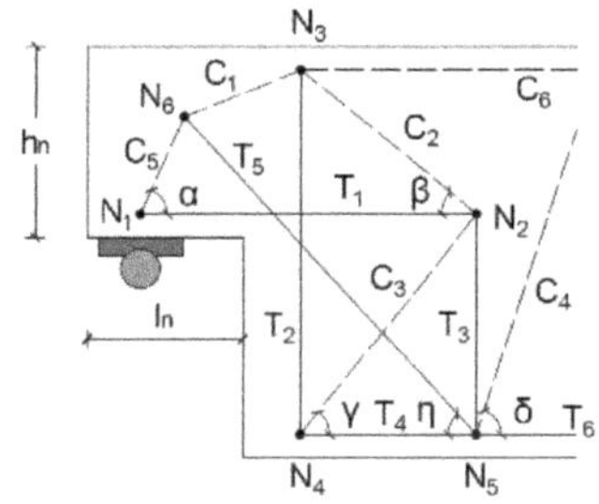

Fig. 6. Model AB

The lowest safety over all the elements yields the global safety since the STMs are isostatic and are hence a series system. The total resistance, R_{tot}, of the system is calculated by Eq. (10) in which E is the acting external load, which can still be multiplied with $\min(S_i)$ before the structure fails.

$$R_{tot} = E\, min(S_i) \tag{10}$$

3.5 Monte Carlo Simulations

For the probabilistic calculations, 100 000 Monte Carlo simulations are executed. In this part, no safety factors are applied. For the mean concrete strength and mean steel yield strength, the following formulas are used based on their characteristic strength:

$$f_{cm} = f_{ck} + 8 \tag{11}$$

$$f_{ym} = f_{yk} + 60 \tag{12}$$

Monte Carlo simulations are executed, considering the three stochastic input parameters as given in Table 1. With these values, the STMs explained in Sects. 3.2 and 3.3 are calculated and the corresponding resistances are determined. This gives a distribution of the resistance of models A, B and AB.

The model uncertainties K_R and K_E, respectively for the resistance and the load model, are generally integrated through the following limit state equation:

$$g(X) = K_R R - K_E E \tag{13}$$

However, since only the resistance is investigated in this contribution, it suffices to multiply the calculated resistance R_{tot} with K_R to integrate the resistance model uncertainty in the probabilistic results.

4 Results

The resistance of a half-joint with a reinforcement scheme as shown in Fig. 3 is calculated with models A, B and AB in a deterministic manner based on the characteristic values without safety factors, as well as in a probabilistic manner.

As can be seen in Fig. 7 STM B has a larger resistance than model A due to the optimal usage of the diagonal rebars, but it also has a larger standard deviation. Model AB has an even larger resistance by using all the reinforcement present, but exhibits also the largest variability. In Fig. 8 the resistance of the case study with an increased reinforcement amount and decreased concrete strength is shown and an overview of the failure mechanisms is given in Table 2.

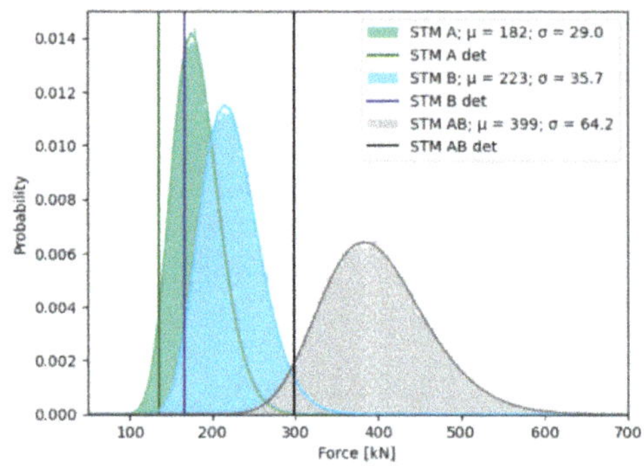

Fig. 7. Resistance - Reference case

Table 2. Failure mechanisms

		STM A	STM B	STM AB
Reference case	Deterministic	T_2	T_5	T_2
	Probabilistic	T_2	T_5	T_2, T_5 or N_1
Increased reinforcement amount	Deterministic	N_1	T_5	N_1
	Probabilistic	T_1, T_2 or N_1	T_5 or C_4	T_5 or N_1

In the reference case, only 2% of the samples of model AB fail due to concrete crushing in node N_1; all the other samples of the models fail due to yielding of the reinforcement. Hence, implementing a variation in the concrete or steel quality by performing a probabilistic calculation has no significant effect on the predominant failure mechanism of the models for this case. Moreover, Model A appears to be always the weakest and AB the strongest for the same set of input variables. When the strength reserves of the tension ties, the nodes and the concrete struts lie closer to each other on the contrary, different failure mechanisms occur based on the material quality. This is the case for the second calculation example.

In the second case, models A and AB fail at the same moment according to the deterministic calculations due to concrete failure in node N_1, while STM B fails due to steel yielding. The calculated resistances are given in Table 3. The deterministic calculations are performed with the characteristic values of the stochastic variables.

Table 3. Resistance STMs

		Resistance [kN]		
		STM A	STM B	STM AB
Reference case	Deterministic	135	166	298
	Probabilistic 5% quantile	139	171	304
	Probabilistic mean value (stand. dev.)	182 (29.0)	223 (35.7)	399 (64.2)
	Experimental value [2]	402		
Increased reinforcement amount	Deterministic	234	248	234
	Probabilistic 5% quantile	257	255	270
	Probabilistic mean value (stand. dev.)	344 (56.7)	334 (53.3)	393 (83.8)

Despite model B having the highest resistance in the second case study when being calculated deterministically, this is no longer the case when the total resistance distributions (Fig. 8) are taken into account. It is therefore important not to rely purely on deterministic calculations based on characteristic values to determine the strength of a half-joint, but also to consider the probability density functions of the stochastic variables to correctly account for uncertainties.

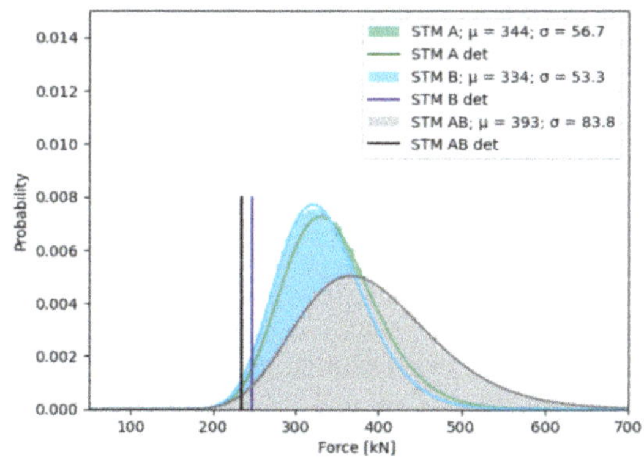

Fig. 8. Resistance - Increased reinforcement amount and decreased concrete strength

Considering the outcomes of the Monte Carlo simulations, the probability that a certain model yields the highest resistance can be calculated. Model AB has a 75.08% chance of being the strongest, model B a 16.88% chance and model A a 8.04% chance.

The variation in the material strength determines the failure mechanism. When the samples have a significantly low concrete strength, models A and AB fail due to concrete crushing in node N_1 due to high concrete stresses at this location, while STM B fails due to yielding of tension tie T_5. For these characteristics STM B is the strongest.

However, if the concrete strength is high, yielding of the rebars determines the resistance of both STM A and B. Only model AB still fails due to concrete crushing because of its low steel stresses. For these samples, STM A provides a higher resistance than STM B, but model AB is then even stronger than model A.

This highlights the importance of a probabilistic calculation as in this second case study the variations of the steel and concrete strength have a significant impact on the resistance of the STMs due to the shift in failure mechanism.

5 Conclusions

STMs are applications of the static method and always provide a lower bound on the effective strength of the half-joint girders. In this contribution, two strut-and-tie models A and B were calculated individually and the superposition AB was calculated as well. This is however not a simple addition of the resistances of STM A and STM B, for example, the concrete at the location of node N_1 is in both models compressed, but the concrete strength in model AB is not twice as large. Hence a more complex superposition should be searched for. It is essential to be able to make a founded judgment on how much each model effectively contributes to the actual resistance of the structure.

Furthermore, the probabilistic calculations of the second case study show that model B, which proved to be the strongest according to the deterministic calculations, is not the strongest in all the calculated Monte Carlo samples. In most of the cases, it is the weakest. Model AB on the contrary showed the highest resistance for all the samples in the first case study and was the strongest as well in over 75% of the samples in the second case.

The probabilistic calculation of the STMs proved to be essential to assess the probability of the different failure mechanisms, which has a primordial effect on the structure's resistance.

References

1. Vereecken, E., Botte, W., Droogne, D., Caspeele, R.: Reliability-based calibration of partial factors for the design of temporary scaffold structures. Struct. Infrastruct. Eng. **16**(4), 642–658 (2020). https://doi.org/10.1080/15732479.2019.1670214
2. Desnerck, P., Lees, J.M., Morley, C.T.: Impact of the reinforcement layout on the load capacity of reinforced concrete half-joints. Eng. Struct. **127**, 227–239 (2016). https://doi.org/10.1016/j.engstruct.2016.08.061
3. Desnerck, P., Lees, J.M., Morley, C.T.: Strut-and-tie models for deteriorated reinforced concrete half-joints. Eng. Struct. **161**, 41–54 (2018). https://doi.org/10.1016/j.engstruct.2018.01.013
4. Passos Sério Lourenço, M.F., et al.: fib Bulletin 100. Design and assessment with strut-and-tie models and stress fields: from simple calculations to detailed numerical analysis. In: The International Federation for Structural Concrete (fib) (2021). https://doi.org/10.35789/fib.BULL.0100
5. European Committee for Standardisation (2004) EN 1992-1-1 Eurocode 2: Design of concrete structures—Part 1-1: General rules and rules for buildings. European Committee for Standardisation

6. Schlaich, M., Anagnostou, G.: Stress fields for nodes of strut-and-tie models **116**(1), 24217 (1990)
7. Caspeele, R., Sykora, M., Taerwe, L.: Influence of quality control of concrete on structural reliability: assessment using a Bayesian approach. Mater. Struct. **47**(1–2), 105–116 (2014). https://doi.org/10.1617/s11527-013-0048-y
8. Schlaich, J., Schäfer, K.: Design and detailing of structural concrete using strut-and-tie models **69**(6), 113–125 (1991)
9. Granata, M.F., Messina, D., Colajanni, P., La Mendola, L., Lo Giudice, E.: Performance of a historical cantilever reinforced concrete bridge with half-joint degradation. Structures **37**, 561–575 (2022). https://doi.org/10.1016/j.istruc.2022.01.039
10. Campione, G., Granata, M.F., Papia, M., Zizzo, M.: Degraded gerber saddles in RC bridges. J. Perform. Constr. Facil. **37**(2), 04022083 (2023). https://doi.org/10.1061/JPCFEV.CFENG-4016
11. Zilch, K., Zehetmaier, G.: Bemessung im konstruktiven Betonbau: Nach DIN 1045-1 (Fassung 2008) und EN 1992-1-1 (Eurocode 2). Springer, Heidelberg (2010). https://doi.org/10.1007/978-3-540-70638-0

Reassessment of Existing Concrete Bridges with Full Probabilistic Methods – Case Studies and Discussion of Relevant Input Parameters

Christian Kainz[1(✉)], Thomas Braml[1], Manfred Keuser[2], Alexander Soukup[2], and Stefan Küttenbaum[3]

[1] University of the Bundeswehr Munich, 85577 Neubiberg, Germany
christian.kainz@unibw.de
[2] BUNG Ingenieure AG, 81379 Munich, Germany
[3] BAM – Federal Institute for Materials Research and Testing, Div. 8.2, Berlin, Germany

Abstract. This article presents two case studies for a full-probabilistic assessment of existing concrete bridges combined with the use of the finite element method for the evaluation of the internal forces. Since the structures already exist, additional information about material parameters and loading conditions can be collected. With a sensitivity analysis, the contribution of each parameter to the total uncertainty is examined and the relevance of their actual value and variation is shown. The first case study is a single-span reinforced concrete frame bridge on the German highway A92 constructed in the year 1985. With a span of around 5 m, but a width of 27 m and a skewness of 70 gon, the bridge is modelled as a three-dimensional finite element model. The second case study is a two-span reinforced concrete frame bridge from 1986 with span lengths of about 7,50 m each. The superstructure is a 45 cm thick slab and the base plate has a thickness of 80 cm. The assessment is performed based on the semi-probabilistic safety concept according to the codes first and secondly with a full-probabilistic calculation. In this article the selection of parameters for the relevant input data are presented and discussed. The results of the sensitivity analysis are used for the setup of measurements to acquire important data for the modelling in the critical limit states.

Keywords: bridge assessment · existing structures · reliability-based format · adjusted partial safety factors

1 Introduction

Bridge infrastructure is suffering from significantly increasing traffic loads and a continuous aging of existing structures. In many cases, the permissible loads must be restricted, or a replacement of structures is planned and carried out. Despite having considerable safety reserves according to design regulations, these structures often cannot utilize these reserves in the reassessment process according to current codes. This is mainly due to a lack of information about the actual condition of the structures. There are currently

J. C. Matos et al. (Eds.): IPW 2024, LNCE 494, pp. 440–451, 2024.
https://doi.org/10.1007/978-3-031-60271-9_41

almost no standardized rules for measurement procedures and for a direct use of the measurement results in static calculations. As a result, many structures face unnecessary use restrictions, leading to unreasonable closures and resource consumptions through premature demolition and replacement works.

2 Input Parameters for the Calculation

2.1 Limit States and Input Parameters for the Full-Probabilistic Assessment of Reinforced Concrete Bridges

The limit state function for bending failure of a rectangular cross section can be found in FISCHER [1] for beam elements. Adapted for the use with finite shell elements it is given in Eq. 1.

$$g = U_{R,M} \cdot \left\{ \left(a_{s1} \cdot f_y + U_{E,N} \cdot \sum n \right) \cdot \left[d - \frac{a_{s1} \cdot f_y + U_{E,N} \cdot \sum n}{2 \cdot \kappa \cdot \alpha \cdot f_c} \right] + U_{E,N} \cdot \sum n \cdot \left(\frac{d_1 - d}{2} \right) \right\}$$
$$- U_{E,M} \cdot \sum m - U_{E,N} \cdot \sum n \cdot \frac{d - d_1}{2} \quad (1)$$

The parameters for this limit state are described in Table 1.

Table 1. Description of the parameters for the bending limit state function

Symbol	Dimension	Name
$U_{R,M}$	[–]	Model uncertainty – resistances, moment
$U_{E,M}$	[–]	Model uncertainty – actions, moment
$U_{E,N}$	[–]	Model uncertainty – actions, normal force
a_{s1}	[m^2/m]	Area of the longitudinal reinforcement
f_y	[MN/m^2]	Yield strength of the reinforcement
n	[MN/m]	Normal force
d	[m]	$d = h - c_{nom} - d_{Bü} - 0{,}5\ d_{St}$
d_1	[m]	Distance of longitudinal rebars from edge
h	[m]	Hight of the elements
c_{nom}	[m]	Concrete cover
$d_{Bü}$	[m]	Diameter of rebars of shear reinforcement
d_{St}	[m]	Diameter of longitudinal reinforcement
κ	[–]	$\kappa = 0{,}5 \cdot \alpha_R / k_a$
k_a	[–]	Coefficient for the location of the resulting force in the compression zone
α_R	[–]	Coefficient of the compression zone
α	[–]	Robinson-Factor, $\alpha_{cc} = 0{,}85$
f_c	[MN/m^2]	Concrete compression strength
m	[MNm/m]	Bending moment, see Eq. (14) and (15)

2.2 Use of Results from Finite Element Calculations and Consideration of Skewness

Usually, the Eurocodes provide characteristic values for the use within the semi-probabilistic safety format. Therefore, the results like internal forces are based on characteristic values. Whereas in a full probabilistic calculation the mean values and variation are of interest. For a practical use of the results based on characteristic values, there is the need for a simply applicable transformation with factors. The factor ζ is introduced to transform a quantile value X_k to the mean value μ, see Eq. (2). A detailed derivation for this conversation process is shown by STEFFENS in [2].

$$\mu_x = \zeta \cdot X_k \tag{2}$$

For a normal distribution the distribution function F(x) is given by Eq. (3).

$$F(x) = \frac{1}{\sigma_x \cdot \sqrt{2\pi}} \cdot \int_{-\infty}^{x} \exp\left[-\frac{1}{2} \cdot \left(\frac{x - \mu_x}{\sigma_x} \right)^2 \right] dx \tag{3}$$

There is no analytical solution available for the integral in Eq. (3). Therefore, a transformation into the standard normal distribution is beneficial. The density function ϕ^{NN} (y) and the distribution function Φ^{NN} (y) have a mean of zero and a standard deviation of one. The quantile values of the standard normal distribution are available in tables or can be calculated with numerical integration. The quantile values X_q are obtained with the inverse function, see Eq. (4) to (7).

$$X_q = F^{-1}(q) \tag{4}$$

$$Y_q = \Phi^{-1}(q) \tag{5}$$

$$X_q = \mu_x + \sigma_x \cdot \Phi^{-1}(q) \tag{6}$$

$$X_q = \mu_x \cdot \left(1 + v_x \cdot \Phi^{-1}(q)\right) \tag{7}$$

For the case of given characteristic values, the mean value is calculated according to Eq. (8).

$$\mu_x = \frac{X_k}{\left(1 + v_x \cdot \Phi^{-1}(q)\right)} \tag{8}$$

A transformation factor ζ is deduced and given in Eq. (9) for a normal distributed value. For the log.-normal distribution and the Gumbel distribution, the transformation factors are calculated similarly, see Eq. (10) and (11).

$$\zeta_{NV} = \frac{1}{(1 + v_x \cdot \Phi^{-1}(q)} \tag{9}$$

$$\zeta_{LNV} = \frac{1}{\exp\left[-0,5 \cdot v^2 + v \cdot \Phi^{-1}(q)\right]} \tag{10}$$

$$\zeta_{GUMBEL} = \frac{1}{\left\{1 - \nu \cdot \frac{\sqrt{6}}{\pi} \cdot \left(\gamma + \left[\ln(-\ln q)\right]\right)\right\}} \tag{11}$$

The conversion of internal forces for skew structures is another essential task. The skewness influences the load distribution throughout the entire structure, which makes a differentiated calculation necessary for elements with an inclined coordination axis. Commercial FE-software includes this conversation in the design process internally, but the actual calculation can often not be seen or exported. Without this conversion, incorrect assumptions about structural integrity are made during a full probabilistic calculation based on the given forces from the FE-software in combination with the available limit state functions, which are typically derived for beam elements. Inclined elements are subjected to forces not only in the longitudinal and cross directions but also in inclined directions. Figure 1 shows the resulting forces and a definition of the parameters.

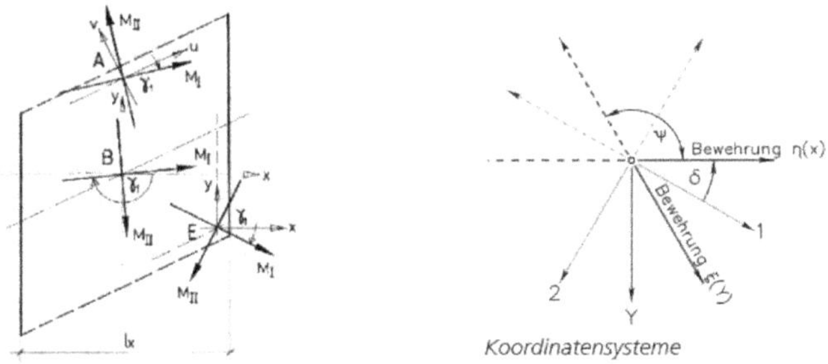

Fig. 1. Main moments in skew plates [3] and definition of the coordinate system of the reinforcement (German: "Bewehrung") steel [4]

The corresponding angle δ to the moment m_I is given by Eq. (12).

$$\tan\delta = \frac{2 \cdot m_{xy}}{\left(m_x - m_y\right) + \sqrt{\left(m_x - m_y\right)^2 + 4 \cdot m_{xy}^2}} \tag{12}$$

The extreme values of the bending moments (main bending moments $m_{1,2}$) can be calculated by Eq. (13). The results are defined with $m_1 \geq m_2$.

$$m_{1,2} = \frac{1}{2} \cdot \left(m_x + m_y\right) \pm \frac{1}{2}\sqrt{\left(m_x - m_y\right)^2 + 4m_{xy}^2} \tag{13}$$

Due to practical reasons, the direction of the reinforcement is usually not identical to the mechanical direction of the main bending moments $m_{1,2}$. While the rebars are placed perpendicular to each other in bridges with big skewness (<60°), the rebars are placed parallel to the edges of the abutments and the gradient in bridges with little skewness (>60°) [4]. The calculation of the design moments in direction of the reinforcement m_η and m_ξ is documented in [3] and shown in Eq. (14) and (15).

$$m_\eta = \frac{1}{\sin^2\psi}\Big[m_1 \sin^2(\delta + \psi) + m_2 \cos^2(\delta + \psi)$$

$$\pm\ |m_1 sin\delta \sin(\delta+\psi) + m_2 \cos\delta \cos(\delta+\psi)|] \tag{14}$$

$$m_\zeta = \frac{1}{\sin^2\psi}\Big[m_1 \sin^2\delta + m_2 \cos^2\delta \\ \pm |m_1 sin\delta \sin(\delta+\psi) + m_2 \cos\delta \cos(\delta+\psi)|] \tag{15}$$

The normal forces in direction of the reinforcement n_η and n_ξ are calculated in the same way with the main normal forces $n_{1,2}$.

3 Case Study 1 – Bridge BW81-2

3.1 Description of the Structure

The bridge 81/2 is located on the A 92 highway and was constructed in 1985. The 40 cm thick superstructure of the reinforced concrete frame with rigidly connections to the abutment walls crosses the river at an angle of 63°. The span length is 5.16 m. The total width of the structure is 27.00 m, with each substructure having a road width of 10.00 m. The 30 cm thick pavement layer in this section consists of a 25 cm thick concrete pavement, a 4.5 cm thick cast asphalt protective layer and a 0.5 cm thick waterproofing layer, applied directly on the superstructure.

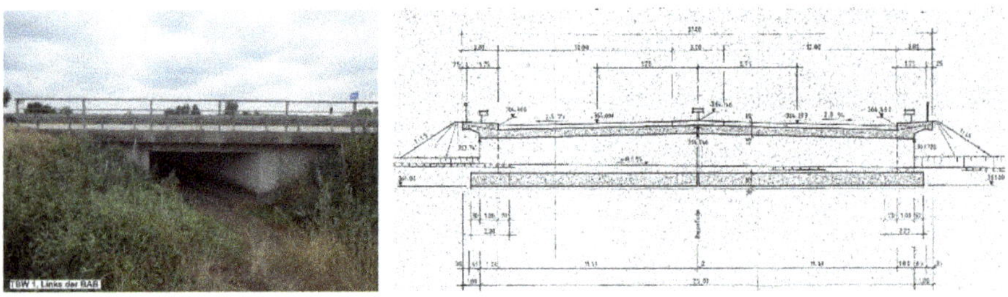

Fig. 2. Photo and cross section of the bridge BW81-2

The reinforced concrete slab increases to 0.45 m thickness towards the center. The cantilever arms have a thickness of 0.30 m and taper to 0.23 m at the edges, see Fig. 2. The load of the abutments (thickness 0.60 m) is transferred through a 0.75 m thick base plate into the soil. The original design of the structure was done in 1986 for a traffic load of Bridge Class 60 according to DIN 1072:1967-11 [5] and Military Load Class 100/50. For all components, concrete of type B25, equivalent to the current concrete strength class C20/25 according to the reassessment guideline [6], and reinforcing steel of type BSt 420/500 RK were used.

3.2 Semi-probabilistic Assessment

The bridge is modeled as a three-dimensional finite element surface model, see Fig. 3. The reassessment was carried out according to DIN technical reports 101 and 102 [7, 8] for the traffic load model LM1 of DIN-FB 101. The weight of the concrete is assumed with $\gamma_c = 25\ kN/m^3$. The characteristic load of the road surface is given as 6,25 kN/m^2 and of the rim beams with 5,88 kN/m^2. For the temperature loading, a constant temperature change of $T_N = -27K$ and a linear temperature gradient of $T_M = -8K$ is given in the current codes.

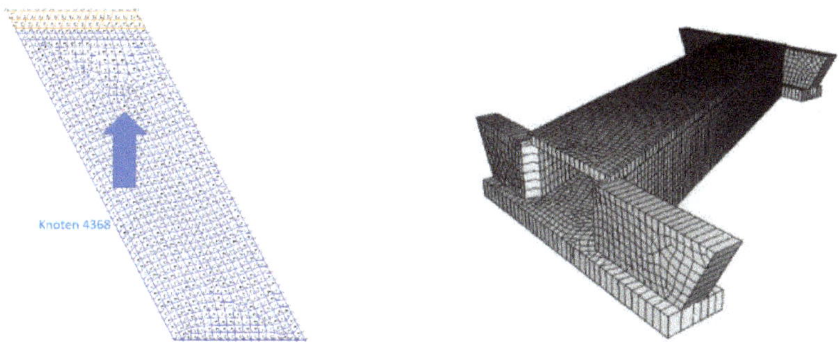

Fig. 3. Finite element model for the superstructure of the frame bridge BW81-2, created by *BUNG Ingenieure AG.*

Deficiencies in the lower reinforcement were identified in both longitudinal and transverse directions during the calculation. The existing reinforcement in the longitudinal direction in the field area of Ø14–17.5 (8.8 cm^2/m) is insufficient for fulfilling the requirements, as is the reinforcement of Ø12–19.5 (5.8 cm^2/m) in the frame corner. The existing reinforcement level in the transverse direction (both upper and lower layers) is very low with Ø10–20 (3.9 cm^2/m) and the load-bearing capacity thus partly significantly exceeded. Additionally, the shear reinforcement of Ø16–50 (8.0 cm^2/m) in the critical section has deficiencies. Three representative spots were identified during the semi-probabilistic reassessment:

1. FE node 4368, bending with normal force at the bottom in midspan.
2. FE node 2516, bending with normal force at the top at the frame corner.
3. FE node 4666, shear failure in around 1,00 m distance from the abutment.

In this contribution, the first representative spot is further investigated. The internal forces from the finite element calculation for a bending failure at FE node 4368 are shown in Table 2.

Table 2. Characteristic values of the internal forces from the FEM model for a bending failure at FE node 4368

Component	Combination factor ψ	m_x [kNm]	m_y [kNm]	m_{xy} [kNm]	n_x [kN]	n_y [kN]	n_{xy} [kN]
Dead weight	1,00	10	4	−5	26	18	10
Rim beams	1,00	0	−1	−1	1	4	1
Pavement	1,00	5	3	−2	−6	−4	−2
Creep/ Shrink	1,00	5	4	0	300	496	201
Temperature T_N-	0,80	18	15	−2	420	690	298
Temperature T_M-	0,80	33	36	2	19	8	−5
Traffic load TS	1,00	44	25	−9	−40	−16	−4
Traffic load UDL	1,00	6	4	−2	−8	−2	−2

3.3 Full-Probabilistic Assessment

For the full probabilistic assessment, the First Order Reliability Method (FORM) and the commercial software *STRUREL* are used. For the use of the internal forces from the finite element calculation, the results must be transferred for a skew layout of the reinforcement bars, which is not necessarily identical to the directions of the internal forces. While this process is implemented into the commercial finite element software, for the probabilistic assessment the relevant input parameters must be selected carefully. Table 3 gives the input parameters for the internal forces and Table 4 contains the parameters of the structure, e.g., the geometry.

The internal forces due to the dead load of the construction as well as creep and shrinkage are modelled with a normal distribution. For the variable loads from temperature and road traffic, a Gumbel distribution is selected and the model uncertainties are assumed to be log.-normally distributed (LNV). The area of the longitudinal reinforcement a_{sl} and the yield strength of the reinforcement f_y as well as the height h of the elements are modelled with a normal distribution (N), while the other geometric parameters are represented by (deterministic) constants due to their small influence on the load bearing capacity.

A sensitivity analysis informs about the contribution of each parameter to the reliability of the structure. Figure 4 shows the sensitivity factors α for the limit state function for bending, see Eq. (1). The influence of the normal forces from creeping and shrinking n_{CS} and from constant temperature change n_{TN} are predominant. The model uncertainties $U_{R,M}$, $U_{E,N}$ and $U_{E,M}$ also have a large portions, while the forces due to traffic loads (m_{TS}, m_{UDL}) or the material properties (f_y, a_s) have only small sensitivity factors. These results are reasonable due to the large magnitudes of the internal forces, see Table 2.

Based on the results of the sensitivity analysis and the big impact of creepage, shrinking and temperature with the load assumptions from the codes, a parameter study is conducted in order to assess, what the influence of a reduction of these parameters on the reliability is. An important aspect when dealing with internal constraint forces is,

Table 3. Mean values and coefficient of variation for the internal forces from the FEM model for a bending failure at FE node 4368

Component	Distribution type	Quantile q	C.O.V	Factor ζ	m_η [kNm/m]	$n\eta$ [kN/m]
Dead weight G1	N	95%	0.06 [1]	0.91	9	23
Rim beams G2	N	95%	0.10 [9]	0.86	0	3
Pavement G3	N	50%	0.25 [10]	1.00	4	−2
Creep/ Shrink CS	N	50%	0.18 [11, 12]	1.00	4	567
Temperature T_N-	GUM	98%	0.15 [2]	0.72	14	666
Temperature T_M-	GUM	98%	0.15 [2]	0.72	22	4
Traffic load TS	GUM	98%	0.15 [13]	0.72	37	−6
Traffic load UDL	GUM	98%	0.15 [13]	0.72	5	−1

that a increasing cracking of the reinforced concrete structure can lead to an reduction of the constraint forces, which is a major difference to external loads, e.g. self-weight or traffic loads. The Eurocode 2 [11] and the DIN technical report 102 [8] allow to consider a stiffness reduction for temperature effects by assuming the stiffness of the cracked concrete with 60% of the stiffness of uncracked concrete. With a more detailed verificaion, a reduction to 40% of the stiffness is allowed. In practical calculations with finite element software, this stiffness reduction is often approximated by multiplying the partial safety factor for temperature actions γ_T with a factor of 0.60 and then using a partial safety factor of $\gamma_T = 1.35 \cdot 0.60 = 0.81$ during the calculation. The normal force from creeping and shrinking n_{CS} is reduced likewise as the normal force from temperature for the parameter study in Fig. 5, which shows the influence of a reduction of the normal forces due to creep/shrinking n_{CS} and constant temperature change n_{TN-} on the reliability of the structure.

When considering the large values of the normal forces from creeping and shrinking n_{CS} and from the constant portion of the temperature change n_{TN}, only a very small reliability is obtained. It is important to note that a reliability analysis is very dependent on the input parameters, so that a small calculated reliability index can indicate a low actual reliability of the structure but also a calculation model, that is not consistent with the actual bridge construction. With a reduction n_{CS} and N_{TN} by 40 percent, which equals a remaining stiffness of 60%, the reliability index for this failure mode is $\beta = 2.35$. A reduction of 60% to a remaining stiffness of 40% leads to a reliability index

Table 4. Mean values and coefficient of variation for the other parameters for a bending failure at FE node 4368

Component	Distribution type	C.O.V [–]	Mean value	Standard deviation
$U_{R,M}$ [–]	LNV	0.070 [1]	1.025	0.072
$U_{E,M}$ [–]	LNV	0.100 [1, 14]	1.000	0.100
$U_{E,N}$ [–]	LNV	0.050 [1, 14]	1.000	0.050
a_{s1} [m^2/m]	N	0.050	$8.8 \cdot 10^{-4}$	$4.4 \cdot 10^{-5}$
f_y [MN/m^2]	N	0.05 [15, 16]	457	22.9
h [m]	N	0.011	0.450	0.005 [16]
c_{nom} [m]	const	–	0.040	–
$d_{Bü}$ [m]	const	–	0.010	–
d_{St} [m]	const	–	0.014	–
k_a [–]	const	–	0.400	–
α_R [–]	const	–	0.800	–
α_{cc} [–]	const	–	0.850	–
f_c [MN/m^2]	const	–	28	–

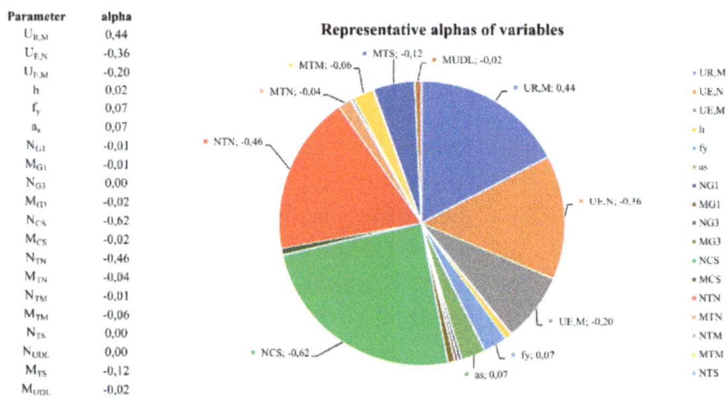

Fig. 4. Results of the sensitivity analysis for bending failure at node 4368

for this failure mode of $\beta = 3.87$. The reference period of the traffic load model LM1 according to DIN technical report 101 is 50 years.

3.4 Measurement Concept

Based on the results from the semi-probabilistic and full-probabilistic calculations, a measurement campaign on the existing structure is planned and conducted. It is assumed that the 30 cm thick pavement layer of the superstructure could lead to a reduced temperature impact on this frame bridge compared to the code regulations. Because of the

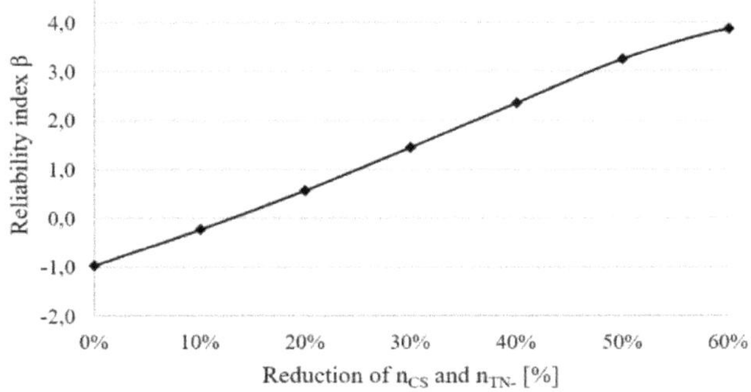

Fig. 5. Parameter study for the constraining forces

influence of the constraining forces on the reliability, a monitoring of the temperature is installed. As part of the project, the idea was to conduct temperature measurements on the superstructure and in the base plate to determine actual component-specific temperature differences over an annual cycle directly on the component, compared to the normatively assumed temperature differences for design. For this purpose, a narrow borehole (approx. d = 12 mm) is required at one location on the superstructure and one location in the base plate to insert the temperature sensor (approx. 7 mm diameter) to the desired depths of the component. The boreholes are sealed with a high-strength, treated concrete substitute after placing the sensors to avoid falsifying the measured values. Cables from the sensors lead to a transmission box attached to the bridge substructure.

Additionally, the detection of the reinforcement (geometry and position) with radar (bottom side in the span) and with ultrasound (upper side in frame corners) is planned. The measurements are currently in preparation and are intended to start early 2024. The results from the measurements should then be used in the reassessment process.

4 Case Study 2 – Bridge BW79-1

The bridge BW79/1 was constructed in 1986 and carries the federal highway A92 over a lake. Each of the two substructures is a two-span reinforced concrete bridge as a closed frame with rigid connections to the abutment walls. The superstructure crosses the lake with a skewness angle of 80 degrees. Each span has a length of 7,43 m resulting in a total length of 14,86 m. The overall width of the structure is 26,30 m, with a traffic lane width of 10,00 m in each direction. The original design of the structure was in 1986 for a traffic load of Bridge Class 60 according to DIN 1072:1967-11 [5] and Military Load Class 100/50.

The bridge is modeled as a three-dimensional finite element surface model similar to case study 1. Deficiencies in the lower and upper reinforcement in the transverse direction were identified during the semi-probabilistic calculation. The existing transverse reinforcement (both upper and lower layers) is very small with Ø10–20 (3.9 cm^2/m) and is sometimes significantly exceeded. Shear reinforcement is not present in the structure, although existing spacing bars can be considered as shear reinforcement in the calculations. Nevertheless, the proof is not fulfilled. Figure 6 shows the finite element model of the bridge.

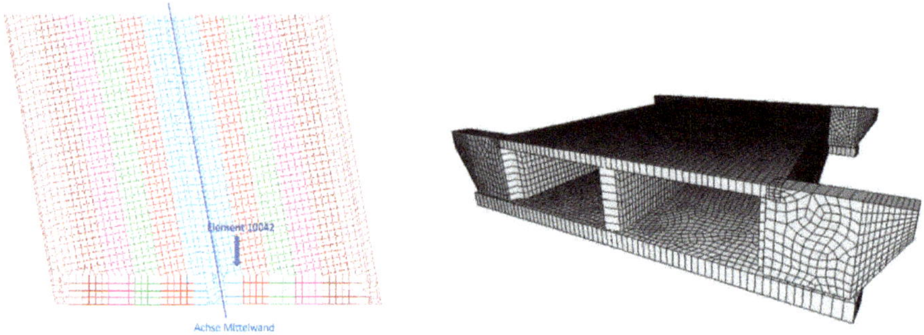

Fig. 6. Finite element model for the superstructure of the two-span bridge BW79-1

Like observed in case study one, a big portion of the internal forces is due to creeping and shrinking as well as the constant temperature change. Because of the similarities in the assessment process of this case study, a measurement concept comparable to case study one has been developed. The identified structural weaknesses might also be relevant for frame bridges with comparable dimensions and materials.

5 Summary and Conclusion

Full probabilistic methods deliver information about the reliability of the examined structure, but they require specialized knowledge. The input parameters must be selected carefully. For complex three-dimensional finite element models, the commercially available software workflow is currently still very limited and a manual exchange of results is necessary, which causes considerable efforts. Transformation factors are a practical tool to use characteristic results from finite element models as input parameters in full-probabilistic calculations.

Using the example of two representative reinforced concrete frame bridges, deficits during the reassessment are presented. The effects of creeping and shrinking as well as temperature can be identified as critical aspects for the existing small to medium span concrete frame bridges. An object-specific load model from on-site measurements can be used for a more realistic assessment of existing infrastructures. However, the inspections must be planned individually – based on the relevant parameters.

References

1. Fischer, A.M.: Bestimmung modifizierter Teilsicherheitsbeiwerte zur semiprobabilistischen Bemessung von Stahlbetonkonstruktionen im Bestand, Dissertation. Technische Universität Kaiserslautern, Kaiserslautern (2010)
2. Steffens, N.: Sicherheitsäquivalente Bewertung von Brücken durch Bauwerksmonitoring. Shaker Verlag, Düren (2019). ISBN: 9783844070514
3. InfoGraph GmbH, Aachen
4. Holst R., Holst, K.H.: Brücken aus Stahlbeton und Spannbeton – Entwurf, Konstruktion und Berechnung, 6. Auflage. Ernst & Sohn, Berlin (2014). ISBN 978-3-433-02953-4
5. DIN 1072:1967-11: Straßen- und Wegbrücken. Lastannahmen, Berlin

6. Bundesministerium für Verkehr, Bau und Stadtentwicklung: Richtlinie zur Nachrechnung von Straßenbrücken im Bestand (Nachrechnungsrichtlinie) (2011)
7. DIN-Fachbericht 101:2009-03. Einwirkungen auf Brücken, Berlin
8. DIN Fachbericht 102:2009-03. Betonbrücken, Berlin
9. Vejdirectoratet: Reliability-Based Classification of the Load Carrying Capacity of Existing Bridges – Report 291. Road Directorate, Ministry of Transport, Denmark (2004)
10. Nowak, A.S., Szerszen, M.M.: Bridge load and resistance models. Eng. Struct. **20**, 985–990 (2003)
11. DIN EN 1992-2:2010-12: Eurocode 2: Bemessung und Konstruktion von Stahlbeton- und Spannbetontragwerken - Teil 2: Betonbrücken - Bemessungs- und Konstruktionsregeln; Deutsche Fassung EN 1992-2:2005 + AC:2008. Beuth, Berlin
12. Küttenbaum, S.: Zur Validierung von zerstörungsfreien Messverfahren für die probabilistische Beurteilung von Bestandsbauwerken mit gemessenen Daten, Dissertation. Universität der Bundeswehr München, München (2021)
13. Braml, T., Wurzer, O.: Probabilistische Berechnungsverfahren als zusätzlicher Baustein der ganzheitlichen Bewertung von Brücken im Bestand. Beton- und Stahlbetonbau 107. Heft **10**, 654–668 (2012). https://doi.org/10.1002/best.201200037.Ernst&Sohn,Berlin
14. Braml, T.: Zur Beurteilung der Zuverlässigkeit von Massivbrücken auf der Grundlage der Ergebnisse von Überprüfungen am Bauwerk. Dissertation. Universität der Bundeswehr München. VDI – Verlag, Düsseldorf (2010)
15. Deutscher Ausschuss für Stahlbeton: Heft 307 – Zur Bemessung der Schubbewehrung von Stahlbetonbalken mit möglichst gleichmäßiger Zuverlässigkeit. Beuth, Berlin (1979)
16. Voigt, J.: Beitrag zur Bestimmung der Tragfähigkeit bestehender Stahlbetonkonstruktionen auf Grundlage der Systemzuverlässigkeit. Dissertation. Universität Siegen (2014)

Reliability Assessment of the Operating Mechanism of a Movable Bridge

Franka E. M. Swaalf[1](✉), Coen H. H. van Battum[1], Wim M. G. Courage[1], and Raphaël D. J. M. Steenbergen[1,2]

[1] Reliable Structures, TNO, Delft, The Netherlands
lisa.swaalf@tno.nl
[2] Ghent University, Ghent, Belgium

Abstract. Large movable bridges in the Netherlands are often equipped with an electromechanical operating mechanism (for opening, closing, and locking) consisting of a panama wheel system. Many of these operating mechanisms are reaching the end of their design lifetime. Currently, there is no standard to assess the structural safety of existing operating mechanisms. The standard for new-built operating mechanisms is used, but it is unknown to which level of safety it leads. The goal of this research is to recalibrate the partial factor for the wind load effect. The partial factor for wind is calibrated to meet target probabilities set for both consequence class 2 and 3 and a reference period of 50 years according to EN 1990. In the static ultimate limit state, the tooth root stress of the panama wheel is assumed to be critical. The analysis focuses on the load combination holding the bridge deck in a fixed position by the operating mechanism. A dedicated wind load model was developed taking into account the number and duration of openings over the design life of the bridge. The ISO 6336-3 standard was used to model the strength of the teeth. Three typical movable bridges are used as case studies. Results show that the partial factor for wind load of the aforementioned scope could be set to 1.5 for maximum opening durations with possible reductions for shorter opening durations. Future research will focus on modelling of dynamic load combinations and calibrating partial factors for the failure mechanism fatigue.

Keywords: Movable Bridges · Partial Safety Factors · Code Calibration · Wind Load

1 Introduction

In the Netherlands, an extensive network of waterways and movable bridges exists. Bascule bridges, drawbridges, vertical-lift bridges, and swing bridges are common types of movable bridges. Central to the functionality of these movable bridges is the operating mechanism: a mechanism designed for opening, closing, and locking. An instrumental part for opening and closing of movable bridges is the operating mechanism which is either electromechanical or electrohydraulic for large movable bridges. Well-known electromechanical operating mechanisms in the Netherlands are the straight or curved rack and pinion and the panama wheel.

J. C. Matos et al. (Eds.): IPW 2024, LNCE 494, pp. 452–458, 2024.
https://doi.org/10.1007/978-3-031-60271-9_42

Many movable bridges in the Netherlands are reaching the end of their design lifetime. Based on a reassessment of their structural safety, a decision is made whether the bridge or its operating mechanism should be renovated or replaced. However, for existing operating mechanisms in movable bridges, there is no specific standard to assess their structural safety. Instead, the standard for new-built structures is used, but it is unknown to which level of safety it leads. This can potentially lead to uneconomical decisions regarding renovation or replacement of the operating mechanism.

As a first step towards a standard for existing operating mechanisms, the standard for new operating mechanisms (NEN 6786) is revisited. The goal of this research is to recalibrate the partial factors of electromechanical operating mechanisms of movable bridges of the *ultimate limit state* for the failure mechanism *yielding* of the tooth root of the teeth of the panama wheel, and loading combination *holding (the bridge deck) in a fixed position*. The loads and strength of the panama wheel are modelled according to the wind load model developed in this research and ISO 6336-3 [1], respectively.

This paper focuses on bascule bridges with an electromechanical operating mechanism containing a panama wheel. In particular, the partial factor is derived based on the reliability of the bending stress in the tooth root of the panama wheel connected to the drivetrain by open gear transmission with its driving pinion. The panama wheel is assumed to be critical because replacement could lead to a complete redesign of an electromechanical operating mechanism. The partial factor for wind is calibrated for consequence classes (CC) 2 and 3 according to NEN-EN 1990 [2]. To derive the partial factor for wind, three bascule bridges in the Netherlands are used as case studies.

2 Method

2.1 Calibration of Partial Factors

The calibration process of partial factors for the operating mechanisms of movable bridges is shown schematically in Fig. 1 and consists of the following general steps. 1) Establish the area of application of the partial factors resulting in a subpopulation of operating mechanisms. 2) Formulate a limit state function given the area of application in step 1. 3) Select a case study from the defined subpopulation and derive relevant parameters to the limit state function. 4) Use an initial set of partial factors to calculate the value of the unity check. If the unity check is not equal to 1, adjust geometric or material properties of the operating mechanism in such a way that it is equal to 1 within reasonable bounds. 5) Determine the value of the reliability index (β) based on the limit state function, parameters of the stochastic variables in the limit state function, and the chosen partial factors. 6) Compare the obtained reliability index to the target reliability index (for instance based on NEN-EN 1990 [2]). If the obtained reliability index is not sufficiently close to the target reliability index, iteration of steps 3 to 5 is required. This process results in a set of partial factors for an individual case study.

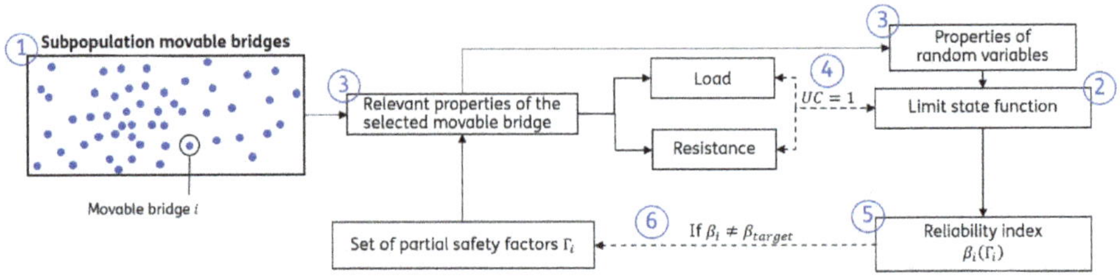

Fig. 1. Schematic overview of calibration of partial factors for movable bridges and their operating mechanism.

2.2 Comprehensive Limit State Function

Operating mechanisms of movable bridges can be designed using ISO 6336-3 [1]. The limit state function in Eq. (1) is based on ISO 6336-3 [1]. Random variables in the limit state function are shown in bold font.

$$\boldsymbol{Z} = \boldsymbol{m_R} \cdot \boldsymbol{\sigma}_{Flim} \cdot \prod_{k=1}^{n} Y_k - \boldsymbol{m_S} \frac{2\boldsymbol{T}(\boldsymbol{W}, \boldsymbol{G})}{z \cdot b \cdot m_n^2} \cdot \prod_{i=1}^{m} Y_i \cdot \prod_{j=1}^{p} K_j \tag{1}$$

in which $\boldsymbol{m_R}$ is the model uncertainty based on uncertainty in input of resistance parameters; $\boldsymbol{\sigma_{Flim}}$ the nominal stress number for bending; Y_K the set of Y-factors from ISO 6336-3 [1]; $\boldsymbol{m_S}$ the model uncertainty based on the uncertainty of the load parameters; $\boldsymbol{T}(\boldsymbol{W},\boldsymbol{G})$ the torque on the gear as function of wind load ($\boldsymbol{W}$) and self-weight ($\boldsymbol{G}$); z, b, and m_n are the number of teeth, the face width, and normal module of the panama wheel, respectively; $\boldsymbol{Y_i}$ the set of Y-factors from ISO 6336-3 [1] and K_i the set of K-factors from ISO 6336-1 [3].

2.3 Generalized Limit State Function

To investigate the influence of the wind load ($\boldsymbol{W}$) and self-weight ($\boldsymbol{G}$) on the reliability of the teeth of the panama wheel for the failure mechanism *yielding* and the load combination *holding in fixed position*, a FORM calculation is performed based on the limit state function shown in Eq. (1). The resulting influence factors (α-factors) for the wind load $\boldsymbol{W}$ and self-weight $\boldsymbol{G}$ are respectively −0.8 and 0.01 for all three cases studied. The low influence factor for self-weight can be explained by the fact that these bascule bridges are balanced around their main hinge point. Thus, a general limit state function only based on wind load can be constructed.

In Eq. (1), torque can be reduced to torque due to wind load only. K- and Y-factors on the loading-side can be accounted for in one factor s_{prod}. The Y-factors on the resistance-side can be written as r_{prod}. The model uncertainties $\boldsymbol{m_R}$ and $\boldsymbol{m_S}$ capture the geometrical uncertainties. The result of these simplifications is shown in Eq. (2).

$$\boldsymbol{Z} = \boldsymbol{m_R}\boldsymbol{\sigma}_{Flim} r_{prod} - \boldsymbol{m_S}\boldsymbol{T}(\boldsymbol{W}) s_{prod} \tag{2}$$

Consequently, Eq. (2) can be reduced to a resistance R, independent of the dimensions of the bridge. Wind load can be reduced to wind speed $\boldsymbol{v}$, the size factor c_s, dynamic factor c_d, and model uncertainty $\boldsymbol{\Theta}$ as shown in Eq. (3). This model uncertainty takes into account the variation and bias in the parameters of the A. Davenport wind loading chain (with exception of the wind speed $\boldsymbol{v}$) [4].

$$\boldsymbol{Z} = \boldsymbol{m_R R} r_{prod} - \boldsymbol{m_s} v^2 c_s c_d \boldsymbol{\theta} s_{prod} \tag{3}$$

in which r_{prod} and s_{prod} are constant values, and the remaining stochastic variables are defined in Table 1.

The resistance is derived from $R_d = S_d$ with S the torque due to wind load on the panama wheel $\boldsymbol{T}_{\boldsymbol{wd,wind,panamawheel}}$ is calculated based on Eq. (4).

$$\boldsymbol{T}_{wd,wind,panamawheel} = iS_{Fmin}\boldsymbol{T}_{wk,wind,motoraxle}\gamma_w K_{Fi}\psi \tag{4}$$

in which i is the transmission ratio between the motor axle and panama wheel, S_{Fmin} is the factor on strength equal to 1.65 based on NEN 6786 [2], γ_w is the partial factor for wind, K_{Fi} is the factor based on the consequence class of the movable bridge, and ψ is the duration factor. The input from Table 1 is used for the calculation of the limit state.

Table 1. Parameters for the general limit state function shown in Eq. 3.

Parameter	Distribution	Mean	Standard deviation	COV
$\boldsymbol{m_R}$	LN	1.00	0.013	0.013
R (for CC3, $K_{Fi} = 1.1$, $\gamma_w = 1.8$)	LN	322.60	22.58	0.07 [5]
$\boldsymbol{m_S}$	LN	1.00	0.033	0.033
$\boldsymbol{v}$	Distribution based on Sect. 2.4	–	–	–
$\boldsymbol{\Theta}$	NOR	0.80	0.22	0.28

2.4 Wind Load Model

In order to perform a comprehensive probabilistic calculation, a distribution function for wind speed has to be formulated. This is in contrast to the design reports of the case-study bridges, where a semi-probabilistic calculation is performed in which the characteristic 10-min average wind speed at 10 m height is used to calculate the wind load (v_{m10}).

To determine the distribution of the maximum wind speeds, 20 years of measured 10-min averaged wind speed from the Cabauw wind station are used [7]. Cabauw is located in Wind Area III of the Netherlands, which is where the three case-study bridges are also located. On the 10-min average measured wind speeds, a 3-parameter-Weibull distribution is fitted as can be seen in Fig. 2.

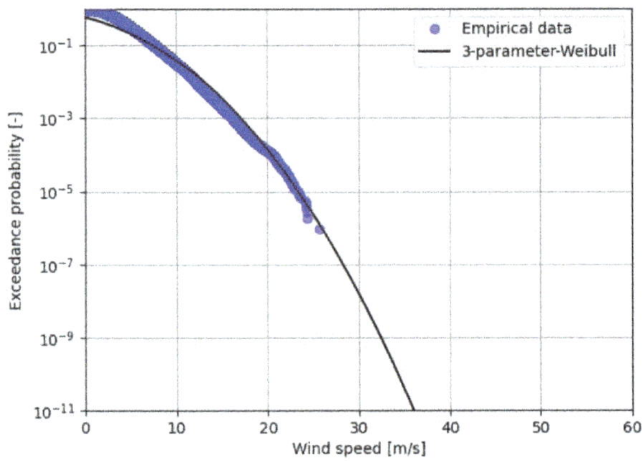

Fig. 2. Fitted 3-parameter Weibull distribution (momentary distribution) to 20 years of 10-min average wind speeds at Cabauw station.

For the load combination *holding in fixed position* (by the operating mechanism), an opening regime must be defined. An opening regime consists of a number of openings per year, an opening duration, a steering wind, and a non-availability. For this case 1 opening per year (N) and 480 min (T) is taken in accordance with NEN 6786 [2]. It is important to note that this opening duration is for calamities and different from the regular bridge openings that are significantly shorter in duration. The steering value of the wind is the speed above which the bridge ceases to open. The steering value and the non-availability of the bridge are directly related to each other, and are equal to respectively 15.9 m/s and 12 h for Dutch movable bridges based on NEN 6786 [2].

First, the maximum wind speeds occurring during one opening (thus one year) are derived. Consequently, this distribution is transformed to the distribution for the 50 year maximum wind speed for that opening regime, according to Eq. (5). Both distributions are shown in Fig. 3.

$$F_{50} = (F_1)^{50} \tag{5}$$

in which F_{50} and F_1 are the 50- and 1- year maximum cumulative distribution functions, respectively for the wind speed on the opened bridge deck given the opening regime. It has been assumed that the annual maximum wind speeds are independent. In the limit state function, this distribution of maximum wind speeds combined with the opening regime for 50 years is used to calculate the reliability for a reference period of 50 years (v in Table 1).

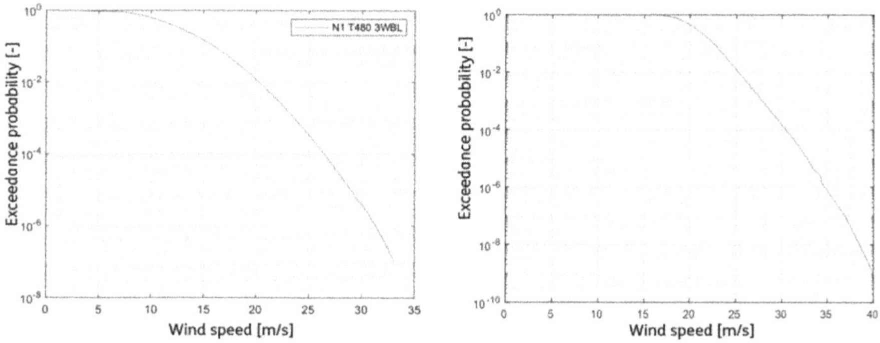

Fig. 3. Distribution of maximum wind speeds for opening regime *holding in fixed position* with wind measurements from Cabauw station for 1 year (left) and for 50 years (right).

3 Results

Based on the generalized limit state function and a design that meets the requirement that the unity check is equal to 1 (see Sect. 2.1 step 4), the reliability index was calculated for different values of the partial factor for the wind load and a reference period of 50 years. The results are shown in Table 2. Due to the use of the factor K_{Fi} (from EN 1990 [2]), the reliability indices of consequence class 3 are larger compared to consequence class 2.

Table 2. Resulting reliability indices for consequence classes 2 and 3 for the partial factor of the wind load.

Consequence Class (CC)	Partial factor wind	β_{Tref50}
CC3	1.2	3.7
CC3	1.5	4.4
CC3	1.8	5.0
CC2	1.2	3.4
CC2	1.5	4.1
CC2	1.8	4.7

4 Conclusion and Discussion

The partial factor of the wind load has been recalibrated based on yielding of the tooth root stress of the panama wheel for the load combination *holding in fixed position* and three cases of bascule bridges in the Netherlands with electromechanical operating mechanisms. The wind load was calculated based on 20 years of wind measurements in The Netherlands in combination with an openings regime of 1 opening per year for a duration of 480 min. This duration is not applicable to regular openings, but resembles a calamity. The strength of the panama wheel has been modelled according to ISO 6336-3 [1].

For CC3, a reference period of 50 years, and a partial safety factor for the wind load of 1.8, the reliability index is 5.0. The reliability index is thus higher than the required reliability level for a new movable bridge designed according to CC3, for which a value of 4.3 applies according to EN 1990 [2]. The preliminary results of this paper indicate that the partial factor for wind for the aforementioned situation could be equal to 1.5. For the loading combination *holding in fixed position*, the contribution of the partial factor for self-weight to the reliability index for the tooth root stress is not significant.

The generalized limit state function resulted in the same reliability indices as the comprehensive limit state function. In the generalized limit state function, the dimensions and material parameters of the specific bridges are not included. Only random variables representing the wind load and the strength remain. This leads to the conclusion that the preliminary partial factors can be generalized for all balanced movable bridges with electromechanical drivetrains for the failure mechanism *yielding*, and loading combination *holding in a fixed position* based on the bending stress in the tooth root of the panama wheel.

Future research will focus on an extension of the wind load model by including other wind areas and other opening regimes. Furthermore, a dynamic model will be developed to calibrate partial factors for dynamic load combinations, corresponding critical parts and the failure mechanism *fatigue*. Based on the knowledge gained in calibration of the partial factors for new-built operating mechanisms, assessment criteria for a standard applicable to existing operating mechanisms will be devised.

References

1. ISO 6336-3:2019. ISO, Genève (2019)
2. EN 1990+A1+A1/C2, Delft (2019)
3. ISO 6336-1:2019. ISO, Genève (2019)
4. Davenport, A.G. : Wind loading of structures. National Research Council of Canada, Division of Building Research (1960)
5. JCSS: JCSS Probabilistic Model Code - Part 3: Material Properties (2001). Author F (2010) Contribution title. In: 9th International Proceedings on Proceedings, pp. 1–2. Publisher, Location
6. NEN 6786-1:2017+C1:2021. NEN, Delft (2021)
7. KNMI: Meteo profiles - validated and gapfilled tower profiles of wind, dew point, temperature and visibility at 10 minute interval at Cabauw. KNMI. https://dataplatform.knmi.nl/dataset/cesar-tower-meteo-lc1-t10-v1-0. Accessed 2021

Reliability-Based Calibration of EN 1991-2 Traffic Load Models for Inner-City Areas

Benjamin Cerar[1(✉)], Liesette la Gasse[1], Ane de Boer[2], and Raphaël D. J. M. Steenbergen[1,3]

[1] Netherlands Organization for Applied Scientific Research (TNO), Delft, The Netherlands
benjamin.cerar@tno.nl

[2] Engineering Office - City of Amsterdam, Amsterdam, The Netherlands

[3] Faculty of Engineering and Architecture, Ghent University (UGhent), Ghent, Belgium

Abstract. In the inner-city area, the maximum vehicle weight of heavy transport vehicles is often enforced to lower limit values compared to the main road network. In the case of the inner city of Amsterdam, yearly permits are required for vehicles heavier than 7.5 ton and special limited time permits for vehicles heavier than 30 ton. Currently, load models for assessment and design are derived with traffic load data for the main road network. To prevent unnecessary costs and material use in redesign or renovation of existing bridges in these areas, it is necessary to consider realistic traffic loads. Data regarding the properties of observable transport vehicles in inner-city areas is consequently limited, thus making modeling of the traffic load and load effect in these areas difficult. This can be problematic when a safety assessment of transport infrastructure needs to be performed. Here, we use a Weigh-In-Motion dataset from a measurement station in the outskirts of the city of Amsterdam, along with data from cameras set up in the inner city to create a database of vehicles that is representative for the enforced zone area. We then computationally simulate the traffic load effect and use a Bayesian inference approach to fit distribution functions to the simulated load effect. Consequently, we perform a reliability analysis to calibrate the characteristic values of EN 1991-2 Load Models LM1 and LM2 so that they reflect the reliability requirements prescribed by NEN 8700, which is the normative document in the Netherlands for the assessment of existing structures.

Keywords: Probabilistic heavy traffic load modeling · structural reliability · code calibration · inner city · WIM-ROAD

1 Introduction

The city of Amsterdam faces a significant infrastructural challenge. Many of the inner city's historic bridges and quay walls were built two or three centuries ago on wooden pile foundations. They were not designed for current day heavy traffic. The inner city has about 200 bridges and 150 km quay walls loaded by motorized traffic. In 2010, it was estimated that a substantial number of these structures are in a poor condition, due to a combination of (too much) heavy traffic and overdue maintenance. Exactly how many

J. C. Matos et al. (Eds.): IPW 2024, LNCE 494, pp. 459–471, 2024.
https://doi.org/10.1007/978-3-031-60271-9_43

of these structures are in a poor state of repair is unknown, which is why all these assets are currently investigated, inspected and tested. Research has been done with regards to the condition of the wooden piles and bridge decks, which results today in a small number of structure renovations. Also the assessment procedures of bridge decks and foundations are improved.

Structures, like bridges and viaducts, are considered safe, when they comply with minimum reliability requirements. In the European Union these requirements are set out in the Eurocode NEN-EN1990 [1] and its National Annex. They are expressed by a minimum reliability index (β) which is related to a maximum failure probability in a given reference period. The minimum reliability requirements in the Eurocode are defined for three consequence classes that account for the consequences of failure. Based on these consequences of failure for most road bridges, Consequence Class 2 (CC2) of the Eurocode are applicable in the inner city of Amsterdam. The CC2 case with a reliability index of 3.8 for a reference period of 50 years is considered. These reliability requirements have been set for newly designed structures. In the Dutch design codes, the NEN 8700 [2] series, additional requirements for existing bridges are defined e.g. to assess whether an existing bridge should be disapproved or not, the so-called disapproval level. For CC2, this relates to a reliability level of $\beta = 2.5$ for a reference period of 15 years. The city of Amsterdam uses a requirement with a reliability level in between the one for disapproval and renovation of structures. However, only the reliability level for disapproval $\beta = 2.5$ is considered, which was found to be the governing case.

To prevent the need of performing a complete reliability analysis for every individual structure, the semi-probabilistic assessment has been introduced in normative documents via partial factors. The partial factor for traffic loads in the Eurocode (EN1991-2 [3]) for CC2 is $\gamma_s = 1.35$. The partial factor for traffic loads for disapproval of existing bridges in the Netherlands for CC2 is $\gamma_s = 1.10$. Combined with the characteristic value of the traffic loads, a load model is defined in the normative codes so that it can be used to design a new bridge or assess whether an existing bridge complies with the required reliability level.

The Engineering Office of the city of Amsterdam is developing an (semi-probabilistic) assessment framework for Amsterdam's bridges and quay walls to validate the structural safety of infrastructure. This framework comprises assessment of (material) strengths and loads of existing structures.

Road bridges in Europe are typically designed and assessed using traffic load models that are based on the Eurocode (NEN-EN 1991-2 [3]). These traffic load models were originally calibrated using traffic load measurements on the highway A6 in France at Auxerre (Bruls et al. [4, 5]). Auxerre traffic is characterized as very heavy and representative of long-distance traffic in Europe. The urban road network consists of inner-city and urban access roads which are not part of the national highway network. The traffic load composition of highway roads is therefore not representative for these urban roads. Using these traffic load models calibrated for the highway network, the structures in the urban road network are often unjustly disapproved. It is therefore important to take into account the actual traffic on these roads.

In the Dutch design code for actions, NEN 8701 [6], a modified traffic load model for existing bridges in the urban road network is provided, which was based on the research presented in [7]. Since direct measurements in the urban or inner city road network were not available at that time, this model was based on a dataset of traffic load measurements of the highway A16 near Moerdijk. The scope of application of the NEN 8701 [6] traffic load model is on structures characterized by a maximum weight limit according to regulations, respectively 50 and 60 tons for lorries and cranes. The databases for the highway location were filtered by a maximum gross vehicle weight of 65 tons to consider possible overloading of vehicles.

Because the traffic load model in NEN 8701 [6] is still based on traffic data from the highway road network, the traffic composition of this dataset is expected to be conservative for the urban road network, including the inner city areas of Amsterdam. Secondly, the city of Amsterdam has decreased the regular yearly permit limit for heavy vehicles in the inner city from 50 ton to 7.5 ton. Vehicles with a weight between 7.5 and 30 ton can ask for a yearly permit. Vehicles with a weight above the 30 ton need a special (route-specific) permit that is valid for a limited time. Therefore the traffic load model given in NEN 8701 [6] is no longer representative for the inner city of Amsterdam.

The assessment standards do allow for the actual use of an asset to be considered, given that it is based on traffic data measured with sufficient accuracy and reliability. The project "Measuring Traffic Loads" has performed traffic load measurements at one location and intends to map out a traffic load spectrum of the actual traffic throughout Amsterdam. Within the project, the focus is on determining a normative traffic load model based on the actual use of bridges in the inner city of Amsterdam. In this paper, the methodology and results are presented of the calibration of a location-specific traffic load model for bridges in the inner city of Amsterdam where vehicles up to 30 tons are allowed by yearly permits. Traffic load measurements within the urban road network in the city of Amsterdam were used to obtain a representative vehicle database for the inner city area. This calibration of the design load values has been obtained using a full-probabilistic and reliability-based approach instead of a calibration of the characteristic value with specific return period (in NEN-EN 1991-2 a return period of 1000 years is used for traffic loads).

2 Methodology

2.1 Heavy Traffic Measurement Methods

Regarding the publications of COST 323 [8], NMi [9], OIML R134 [10] and finally the ISWIM Guide for Users of Weigh-In-Motion [11], a WIM-ROAD system in the inner-city is not suitable, because of the accelerating and braking of vehicles and, in particular, heavy vehicles. Additionally, the installation of a WIM-ROAD system needs a straight approaching lane of at least 100 m, with no irregularities on the road surface, which is not possible in the inner city of Amsterdam.

The idea of the Measuring Traffic Loads project team of Amsterdam is based on getting a lot of vehicle load measurements of all vehicle types at a location which is fit for a WIM-ROAD system that are strongly related to the vehicles representative for the inner city of Amsterdam. The WIM-ROAD measurement data are divided in 13 different categories according to the EUR13 category system [12]. The WIM-ROAD system is located at the Westpoortweg, which is located in an industrial area of Amsterdam. This location is also part of the provincial road that connects Amsterdam with IJmuiden located on the shore of the North Sea, which is a busy provincial road in the Netherlands with 500,000 vehicle passages above 3.5 ton in each direction (counted in 2016). The passage frequencies (Table 1) also show that there are enough low and midsize freight vehicles, which are typical for the vehicle composition in the inner city. All together, these aspects make it a good location for the WIM-ROAD system.

The system is installed in both traffic lanes for both driving directions and consists of 8 Lineas quartz pressure sensors and one induction loop detector per lane, installed in the road surface (2 over the lane width). The measurement data from the 8 pressure sensors is combined to obtain accurate axle load measurements. The induction loop detectors are used to derive the vehicle passages for the measurements and also for an accurate link between the measured axle loads and the front and backside camera shots, including the Automatic Number (or License) Plate Recognition (ANPR). Data was recorded for one year (2021). The results of this measurement campaign are discussed in more detail in [13–15].

2.2 Location-Specific Data

In this study, the traffic load models are calibrated for a specific part of the inner city area of Amsterdam, the Wallen, where yearly permits are required for vehicles heavier than 7.5 ton and special limited time and route-specific permits are required for vehicles heavier than 30 ton. However, the axle load data used for the reliability calculations in this study is based on Weigh-in-Motion (WIM) station data, collected at the outskirts of the city where the regular weight limits of 50 and 100 tons respectively apply. Additionally, as mentioned before, the vehicle composition at the Westpoortweg is expected to be different from that on the inner city roads. Therefore the WIM-data should be resampled to be representative for the inner city area.

The automatic control of the legal limit in the inner city is done by Automatic Number (or License) Plate Recognition (ANPR) cameras oriented at the front side of the vehicle which are located all around the inner city of Amsterdam. Coupling the license plate information from the cameras with the data from the Dutch Vehicle Authority (RDW), the number of axles and the maximum allowable weight by law of the passing vehicles is known. However, vehicles with separate lorries have a different license plate for the lorry at the backside of the vehicle. With only a camera oriented at the front license plate, the relevant information of the lorries is unknown. Therefore, a camera setup oriented at the backside of the vehicle is added at 2 main entrance locations for this specific inner city area, so all passing vehicles including lorries are known including their number of axles and weight limit for a period of 5 weeks. Based on this information the EUR13 categories of the observed vehicles are determined and the frequencies of occurrence of the categories at the inner city roads can be derived. Using these frequencies derived from the ANPR data in the inner city area, we resample WIM data based on the frequency of observing particular heavy vehicle categories in the inner city area in order to use vehicle data and a traffic composition for the reliability analysis that are representative for the inner city area.

The frequencies of occurrence in the inner city from measurements are presented in the third and fourth column of Table 1. The frequencies used to resample the database for the inner city situation were chosen such that conservative values are used for the most governing vehicle types for the relevant load effects. These chosen frequencies are shown in the fifth column of the table. The sixth and seventh columns show the data for the Westpoortweg measurement location. The vehicle types that are not relevant for the inner city were disregarded before computing the frequencies, therefore these frequencies are not equal to the actual observed traffic at the Westpoortweg. They can however be used to show the relative difference to the inner city traffic composition. One can specifically see that the vehicle categories most relevant for the inner city EUR13-1 and 13-2 2 axle vehicles only comprise a small part of the Westpoortweg vehicles. The percentages proposed in EN 1991-2 Table 4.7 for local traffic [3] are shown in the last column. It can be observed that the percentages observed in the inner city of Amsterdam do not comply with these proposed values, especially the three and five axle vehicles. The three axle vehicles form a relatively larger part of the observed inner city traffic, than the five axle vehicles, compared to the EN1991-2 proposed values. All five axle vehicle passages show a low allowable weight by law. The camera shots show a light vehicle followed by a light lorry (allowable by law GVW < 23 ton).

As described in Sect. 2.1, the images of heavy vehicles were collected over a period of one year. Based on that information, we count the total number of vehicles that drive in the inner city area and the relative frequency of each of the vehicle categories, highlighted in Sect. 2.1. Using the vehicle category percentages, we then perform resampling of the WIM population dataset in order to obtain a sample of the vehicle axle data that are representative for the inner city area. The principle of stratified sampling is used in order to obtain samples where the proportion of each vehicle category to the total vehicles in the stratified sample is equal to the sampling percentage. For the stratified sampling, we utilize the Python programming language and the related Python pandas library for data and statistical analysis.

Since a different legal weight limit is enforced at the measurement location of the WIM-data we have to additionally cap our resampled database to a certain maximum that is representative for the maximum vehicle weight limit of 30 tons (with yearly permit) in the inner city area. Since we have insufficient information regarding the effectiveness of enforcing traffic regulations in the city of Amsterdam, we consider three different limit values: 30, 35 and 39 tons. Vehicles in the database that exceed these weight limits are replaced with vehicles with a weight below the limit. This results in three separate databases, each consisting of vehicles complying with one of the considered weight limits.

2.3 Traffic Load Effect Simulation

The simulation of the traffic load effect based on the resampled WIM data is performed using a computational procedure developed by the Risk & Reliability research group of TNO. An influence line of a simply-supported beam with varying span length was used to model the bridge. Both the bending moment at mid-span and the shear force near the supports was considered. It was found that the moment at mid-span for a bridge of 5 m length was governing and therefore only these results are presented.

During this simulation process for computing the traffic load effect it was assumed that only a single vehicle is present on the structure at the same time and that there is

Table 1. Percentages of vehicles types according to the EUR categorization in the inner city of Amsterdam.

Vehicle type		Inner city [%]			Westpoortweg		EN1991-2 Table 4.7 local traffic
Number of axles	EUR13 category	Camera Location 1	Camera Location 2	Chosen for calibration	Number	[%]	[%]
2 axles	1	–	0.1	38.2	2887	<1	80
	2	44.2	43.2		40713	4	
	12	31.1	20.5	30	271467	25	
3 axles	1	4.4	6.4	5	8185	1	5
	3	14.2	22.2	22	41658	4	
	7	–	0.1	0.1	34270	3	
	12	–	0.05	–	9398	1	
4 axles	1	3.6	4.7	2	12223	1	5
	4	2.0	2.4	2.4	46251	4	
	5	0.2	0.05	–	39140	4	
	8	–	0.05	–	241169	22	
5 axles	5	0.1	0.2	0.2	2114	<1	10
	9	0.1	–	0.1	339108	31	
Total	–	100	100	100	1091223	100	100

only a single traffic lane. This situation is representative for the inner city bridges with a short span and limited width, allowing for only one heavy vehicle passing at the same time. Therefore vehicle distances were not considered.

For obtaining extreme values of the traffic load effect we assume that 250 days of heavy traffic are normative for one year. Additionally, it is assumed that there are either 25,000 or 5,000 observable load events during one year (N_{obs}) as maximum and minimum values for the inner city area, based on the measured number of passings. The daily maximum load effect is computed by randomly sorting simulated vehicle load effect values into 250 segments of size $N_{\text{obs}}/250$ (either 100 or 20 vehicles per day) and selecting the maximum from the values of the load effect in each segment.

2.4 Reliability-Based Calibration of Traffic Load Models

Introduction. To derive a traffic load model for the inner city of Amsterdam a reliability-based calibration of the loads and load effects is performed. As a basis, Load Model 1 (LM1) and Load Model 2 (LM2), introduced in the EN 1991-2 [1] were used. LM1 requires the calibration of a tandem axle load (Q_k) and a uniformly distributed load (q_k). LM2 requires only the calibration of a single axle load.

Procedure for Calibrating LM1. The calibration of the LM1 traffic load model requires two parameters; the tandem axle load Q_k and uniformly distributed load q_k. Therefore choice was made to first derive a value for Q_k, based on a simplified procedure for the tandem axle configurations of the WIM-ROAD measured data and secondly deriving the related q_k based on a full probabilistic procedure making sure the reliability target for CC2 disapproval is met. The following steps are taken to perform a probabilistic calibration of Q_k and q_k.

Computing the Value of the Tandem Axle Load. The characteristic value of Q_k is computed such that the reliability requirements for the CC2 disapproval scenario are met. First the tandem axles were identified in the resampled datasets where the inter-axle distance is restricted between 50 and 140 cm, similar to the assumption in the background document to [6]. Second, a distribution function was fitted to the empirical distribution of the tandem axle load using Bayesian inference, considering the statistical uncertainties explicitly. Lastly, a Monte Carlo simulation procedure was used to account for dynamic amplification effects, model uncertainty, and extrapolating to a design value according to CC2 disapproval. The characteristic value is then determined by reducing the design value with a partial factor for the traffic load that corresponds to the CC2 disapproval scenario, $\gamma_s = 1.10$.

Computing the Value of the Uniformly Distributed Load. The characteristic value of q_k is calibrated such that the target reliability is met. The probability of failure is derived with the First-Order Reliability method (FORM) using the limit state function $Z = R - E$ with R the resistance of the bridge and E the load effects due to traffic. In our approach, we consider the q_k as an unknown parameter that is determined by taking a set of trial values $q_{k,i}$, and computing the corresponding reliability index β_i. The value of q_k should then be set to such $q_{k,I}$ when β_i is greater or equal to the target reliability index. This procedure is highlighted below.

The structural resistance $R_{d,i}$ is calculated for each trial value of $q_{k,i}$ by setting $R_{d,i} = E_{d,i}$, where $E_{d,i}$ is the design load effect computed as follows:

$$R_{d,i} = E_{d,i} = \gamma_T \cdot E(Q_k \text{ “} + \text{”} q_{k,i}), \tag{1}$$

Where γ_T is the partial safety factor for the traffic load (=1.10 for CC2 disapproval), and $E(Q_k$ “+” $q_{k,i})$ is the bending moment at the mid span of a simply supported structure, computed by the principle of the superposition of loads. Q_k is the selected characteristic value of the tandem axle load and $q_{k,i}$ is the (still unknown) characteristic value of the uniformly distributed load. The mean value of the resistance $\mu_{R,i}$ is then computed using Eq. (2), as follows:

$$\mu_{R,i} = R_{d,i} \cdot \exp(\alpha_R \cdot \beta_t \cdot V_R), \tag{2}$$

where α_R is the sensitivity factor for the distribution of the resistance random variable (=0.8), β_t is the target reliability index for CC2 (=3.8) and V_R is the coefficient of variation for the distribution of the resistance random variable. $V_R = 0.20$ is assumed to account for the uncertainties in the bridge resistances considering the masonry structure of many of the bridges and their wooden pile foundation. However it was found that the resulting traffic load model was not very sensitive to this assumed value. α_R has been set to 0.8 and β_t to 3.8 since the bridges will be checked on the resistance side using Eurocodes 2–7, for which $0.8 \cdot 3.8$ was chosen in the calibration of design capacity.

The limit state function Z used for computing minimum value of the uniformly distributed load q_k given in the LM1 model with all considered uncertainties is defined as given in Eq. (3). In Table 2, the distributions of parameters, shown in Eq. (3), and their parameters are given.

$$Z = R - m_S \cdot \theta_{\text{dyn}} \cdot \theta_{\text{trend}} \cdot E_{\text{traffic,ref}}. \tag{3}$$

$E_{\text{traffic,ref}}$ is the empirical distribution of the daily maximum of the traffic load effect, scaled to the reference period t_{ref} (=30 years, assuming 250 working days in a year) assuming independence of the daily maxima, calculated using Eq. (4).

$$E_{\text{traffic,ref}} = E_{\text{traffic,day,max}}{}^{tref/250}. \tag{4}$$

We select the value of the characteristic uniformly distributed load q_{k} is the $q_{\text{k},i}$ so that the required reliability for the assessment case is satisfied, i.e.:

$$q_{\text{k}} = q_{\text{k},i}(\beta_i = \beta_{\text{t}}). \tag{5}$$

Procedure for Calibrating LM2. The following steps are taken for calibrating the characteristic value of the single axle load Q_{k}, which is the parameter of Load Model 2.

Computing the Value of the Axle Load. The procedure for calculating the value of Q_{k} is similar to the calibration of the value of q_{k} in LM1 and utilizes FORM to calculate the reliability index for a set of trial values of the axle load $Q_{\text{k},i}$. The design value of the resistance is calculated as follows:

$$R_{\text{d},i} = E_{\text{d},i} = \gamma_S \cdot E(Q_{\text{k},i}), \tag{6}$$

Where $E(Q_{\text{k},i})$ is the bending moment at the mid span of a simply supported structure with a span of L = 5 m. The following limit state function Z is used for computing the value for the single axle load Q_{k} as given in the LM2 model:

$$Z = R - \theta_{\text{dyn}} \cdot \theta_{\text{trend}} \cdot Q_{\text{ref}}, \tag{7}$$

The parameters of the variables, presented in Eq. (7) are given in Table 3.

The distribution function of the axle loads Q_{ref} for the considered reference period, shown in Eq. (7), is calculated based on the fitted distribution to the measured axle loads from the resampled dataset Q as follows:

$$Q_{\text{ref}} = Q \cdot \exp(t_{\text{ref}} \cdot N_{\text{obs}} \cdot N_{\text{axle,avg}}), \tag{8}$$

where t_{ref} is the reference period (=30 years), N_{obs} is the number of vehicles in one year (=25,000), and $N_{\text{axle,avg}}$ is the average number of axles per vehicle in the resampled dataset. This approach for scaling to the reference period, as shown in Eq. (8), was chosen by assuming independence of passages of individual vehicles.

Table 2. Parameters of the variables used in the limit state function for the calibration of the characteristic value of the uniformly distributed load q_k given in the LM1 model.

Variable	Description	Distribution	Mean μ	Std. dev. σ	Remark
R	Structural resistance [in kNm]	Lognormal	Calibrated using Eq. (2)	0.20 μ	
m_S	Model uncertainty related to the traffic load effect	Normal	1.0	0.10	In accordance with JCSS probabilistic model code [17]
θ_{dyn}	Uncertainty related to dynamic effects	Normal	1.1	0.05	In accordance with the literature study in [16] and the background for the Eurocode EN1991-2 for global effects
θ_{trend}	Uncertainty related to trend for a reference period 30 years	Normal	1.0	0.07	Agreed for local traffic

Table 3. Parameters of the variables used in the limit state function for the calibration of the characteristic value of the single axle load Q_k given in the LM2 model.

Variable	Description	Distribution	Mean μ	Std. dev. σ	Remark
R	Structural resistance [in kN]	Lognormal	Calibrated	0.20 μ	
θ_{dyn}	Uncertainty related to dynamic effects caused by single axles	Deterministic	1.4	—	In accordance with [16] for local effects
θ_{trend}	Uncertainty related to trend for a reference period 30 years	Normal	1.0	0.07	Agreed for local traffic

3 Results

3.1 Overview of Case Studies

The reliability analyses were performed based on different reliability requirements, given different consequence classes. In this paper, we present the results of reliability analysis and consequently calibrated parameter of LM1 and LM2 for the CC2 disapproval scenario (target reliability index $\beta_t = 2.5$, reference period 30 years, partial factor $\gamma_T = 1.10$).

For the calibration of LM1, the calculation is performed based on the results of the simulation of the bending moment at mid span for various structural spans, ranging from 5 m to 20 m, and number of observations per year $N_{obs} = 5{,}000$ and $N_{obs} = 25{,}000$. However, we observed that the highest values of the model parameters are obtained for the shortest span ($L = 5$ m) and $N_{obs} = 25{,}000$. Consequently, only the results based on this situation are presented in this paper. LM2 is calibrated based on the assessment situation that leads to highest required value of the characteristic axle load. As for LM1, this was observed for $N_{obs} = 25{,}000$.

3.2 LM1 Model

In Table 4 the minimum values of tandem axle load Q_k, featured in Load Model 1 are presented for the CC2 disapproval assessment scenario. The results given in each row are presented, considering a maximum GVW of 30 tons, 35 tons, and 39 tons, respectively. The values of the tandem axle load Q_k were derived using the resampled datasets with corresponding maximum GVW. It was found that the vehicles types most dominant in the tail distribution function of the tandem axle loads are the three axle EUR13-3 vehicles, which are standard three axle rigid trucks with one axle in the front and a tandem axle at the rear. For larger weight limits like 39 tons also four axle EUR13-4 rigid trucks become present in the tail of the tandem axle loads.

Given that the model is calibrated for an inner city area where special limit time and route-specific permits are required for vehicles heavier than 30 ton, we consider that the presence of such vehicles has a very low probability of occurrence. However, since we have insufficient information regarding the effectiveness of enforcing traffic regulations in the city of Amsterdam, we select a conservative value of the characteristic tandem axle load that takes into account possible overloading of vehicles. Hence, the value $Q_k = 360$ kN is selected for the tandem axle load (180 kN per axle).

With the selected characteristic value of the tandem axle load, we perform the calibration of the structural resistance, as described in Sect. 2.4 by varying the value of the uniform-distributed load (UDL) q_k. For each value of q_k, a FORM analysis is performed. In Fig. 1, the computed reliability index for each value of q_k is presented when the maximum gross vehicle weight is capped at 30, 35, and 39 tons. The minimum required value for q_k is reached when the corresponding reliability index is at least equal to the target reliability value ($\beta = \beta_t$). A conservative value corresponding to the 390 kN maximum vehicle weight is selected – $q_k = 5\text{kN/m}^2$. This value for the uniformly distributed load is select in combination with the previously derived tandem axle load of 360 kN to result in a traffic load model that complies with the required reliability target. Although a value

Table 4. Minimum design and characteristic values of the LM1 tandem axle load given the maximum vehicle weight (GVW) limit, corresponding to the reliability requirements for the CC2 disapproval scenario.

GVW limit [ton]	Design value Q_d [kN]	Characteristic value [kN]
30	338	307
35	399	363
39	430	391

of 4 kN/m^2 would also be suitable, it was increased to keep it in line with the maximum pedestrian traffic load model in the inner city. The derived value for q_k has no direct physical relation with the vehicle weight.

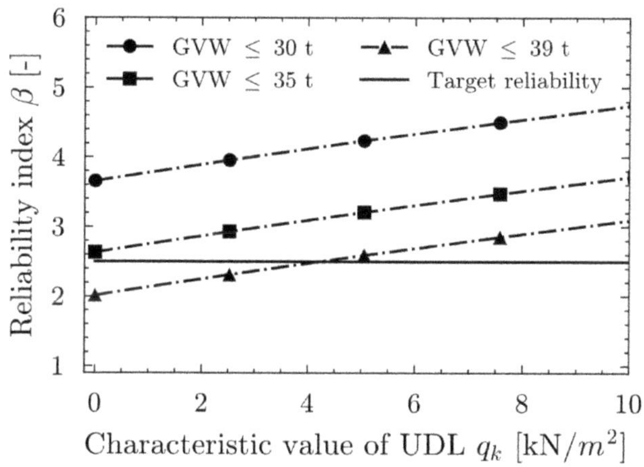

Fig. 1. Reliability indices computed for selected value of the q_k parameter of the LM1 model. Results are shown for the CC2 disapproval scenario, span length $L = 5$ m and $N_{obs} = 25{,}000$. The characteristic tandem load value $Q_k = 360$ kN is used.

3.3 LM2 Model

In Fig. 2, the computed reliability index is presented for different characteristic values of the axle load and different GVW limits. As in the case of the LM1 model, the parameter of LM2 model is selected with the notion that special limited time and route-specific permits are required for vehicles above 30 ton for entering the city area under consideration. The presence of such vehicles is therefore unlikely, however, since we have insufficient data regarding effectiveness of enforcement of this policy, the value for the LM2 model parameter is chosen conservatively, based on GVW limit of 35 ton. In this case 220 kN is chosen as the characteristic value of the axle load. It can be observed that the characteristic value for the single axle weight converges to a certain value just above

220 kN for this measurement data. This makes sense since for larger GVW limits vehicles tend to get more axles instead of heavier axles weights above a certain value.

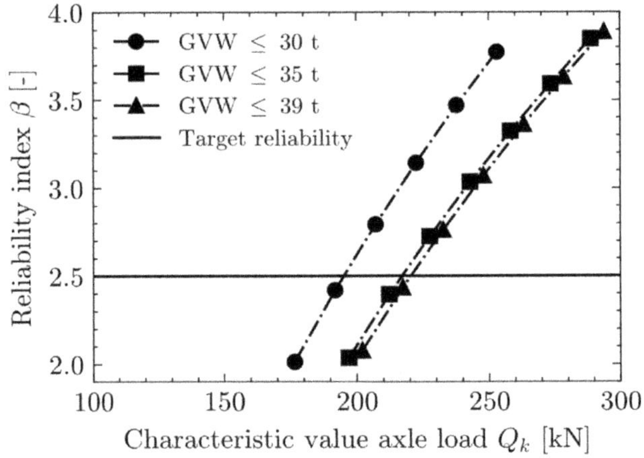

Fig. 2. Reliability indices computed for selected value of the Q_{k} parameter of the LM2 model. Results are shown for the CC2 disapproval scenario, span length $L = 5$ m and Nobs $= 25000$.

4 Conclusions

The traffic load models for the design and assessment of road bridges that are given in design codes are typically based on traffic data on the highway road network. When such models are used for assessing road bridges in regional and inner city roads, this typically leads to a conservative assessments and unjust disapproval of these bridges. Even more conservatism arises if such traffic load models are used for bridges on municipal roads in inner city areas, where access to heavy traffic is nowadays strictly limited and regulated.

In this paper, we present results of a calibration of location specific traffic load models (Load Model 1 and Load Model 2 following the EN 1991-2 approach) for the inner city of Amsterdam using Weigh-In-Motion (WIM-ROAD) traffic data from a provincial road on the outskirts of the city as well as a vehicle detection system in the inner city as it was shown that the vehicle composition in the inner city very much deviates from the WIM-ROAD location and also from the assumed values in EN1991-2 Table 4.7 for local traffic [1]. The calibration is performed in a reliability-based manner, meaning that the selected parameters of the traffic load models satisfy the prescribed reliability demands. The CC2 disapproval scenario, as defined in NEN 8700 [2], is considered.

The chosen parameters of the traffic load models show a substantial decrease in comparison to the traffic load models currently used in the Netherlands for inner city areas [6]. This means that when the proposed traffic load models will be used in practice, it will lead to an optimization of the maintenance of road bridges in this part of the inner city of Amsterdam, as well as ensuring the required structural safety is fulfilled for the duration of the assessment reference period (30 years).

Acknowledgements. The work presented in this paper was carried out as part of a project funded by the municipality of Amsterdam. The authors wish to express their gratitude to the Program Bridges and Quay Walls of the Engineering Office of the municipality of Amsterdam for their support of the pilot project "Measuring Traffic Loads". Additionally, we thank other project team members Maarten Soudijn, Sebastiaan van Rossum and Frederick van Waarde for their contribution.

References

1. NEN: Eurocode – Grondslagen van het constructief ontwerp, NEN-EN 1990:2002, Nederlands Normalisatie Instituut, Delft (2002)
2. NEN: Beoordeling van de constructieve veiligheid van een bestaand bouwwerk bij verbouw en afkeuren – Grondslagen, NEN 8700:2011+A1:2020. Nederlands NormalisatieInstituut, Delft (2020)
3. NEN: Eurocode 1: Belasting op constructies – Deel 2: Verkeersbelasting op bruggen, NEN EN 1991-2:2003, Nederlands Normalisatie Instituut, Delft (2003)
4. Bruls, A., et al.: ENV 1991 – Part 3: The main models of traffic loads on road bridges. Background studies. IABSE Colloquium on Basis of Design and Actions on Structures; Background and Application of Eurocode 1, Delft (1996)
5. Bruls, A., et al.: ENV1991 Part 3: Traffic loads on bridges: Calibration of road load models for road bridges. IABSE Colloquium on Basis of Design and Actions on Structures; Background and Application of Eurocode 1, Delft (1996)
6. NEN: Beoordeling van de constructieve veiligheid van een bestaand bouwwerk bij verbouwen en afkeuren – Belastingen, NEN 8701:2011+A1:2020, Nederlands Normalisatie Instituut, Delft, Netherlands (2020)
7. la Gasse, L.C., Steenbergen, R.D.J.M., Allaix, D.L.: Traffic load model for road bridges in the urban road network. Heron **64**(3), 297–318 (2019)
8. Jacob, B., O'Brien, E.J., Jehaes, S. (eds.): Weigh-in-Motion of Road Vehicles: Final Report of the COST 323 Action. LCPC, Paris, France (2002)
9. NMi: International WIM standard - Specifications and test procedures for Weigh-in-Motion Systems. Nederlands Meet Instituut, Dordrecht, Netherlands (2016)
10. OIML: Automatic instruments for weighing road vehicles in motion and measuring axle loads. International Organization of Legal Metrology, Paris, France (2009)
11. van Loo, H., Žnidarič, A.: Guide for Users of Weigh-In-Motion, Version 1.2., International Society for Weigh-In-Motion (ISWIM) (2021)
12. MetroCount: MTE User Manual – Classification Schemes v 4.03 (2013)
13. Soudijn, M., van Rossum, S., de Boer, A.: Measuring heavy traffic with WIM-ROAD and WIM-BRIDGE systems in an urban environment. [Paper presentation], IABSE Congress Ghent 2021: Structural Engineering for Future Societal Needs, Ghent, Belgium, Congress Proceedings, pp. 331–340, 22–24 September 2021. ISBN: 978-3-85748-176-5
14. Soudijn, M., van Rossum, S., de Boer, A.: Traffic load measurements with WIM-ROAD and WIM-BRIDGE systems in an urban environment. [Dutch, Paper presentation], CROW In-fradagen 2022, Boskalis, Rotterdam, The Netherlands, 10 November 2022
15. Soudijn, M., van Rossum, S., de Boer, A.: Measuring heavy traffic using alternative systems in an urban environment. [Paper presentation], IALCCE2023 Politecnico di Milano, Milan, Italy, Symposium Proceedings, pp. 2873–2880, 2–6 July 2023. ISBN 978-1-003-32302-0
16. Steenbergen, R.D.J.M., Morales Napoles, O., Vrouwenvelder, A.C.W.M.: Algemene veiligheidsbeschouwing en modellering van wegverkeerbelasting voor brugconstructies (update van TNO Rapport 98-CON-R1813). TNO rapport TNO-060-DTM-2011-03695-1814 (2012)
17. JCSS: Probabilistic Model Code, European Convention for Constructional Steelwork (2001)

Reliability-Based Evaluations for Non-metallic Reinforced Concrete Structural Members

Jan Philip Schulze-Ardey[1](✉), Tânia Feiri[2], Marcus Ricker[2], Martin Claßen[1], and Josef Hegger[1]

[1] Institute of Structural Concrete, RWTH Aachen University, Aachen, Germany
jschulze@imb.rwth-aachen.de

[2] Chair of Structural Concrete, TU Dortmund University, Dortmund, Germany

Abstract. Modern technological and research developments of textile-reinforced concrete (TRC) have led to extensive applications in building and civil engineering structures. Examples of TRC can be found in retrofitting of existing buildings, facade slabs or bridges. Despite its potential, the widespread use of TRC still remains limited. This is partly explained by the lack of a consistent design framework since conventional design methods utilised for other materials (e.g., steel-reinforced concrete) cannot be directly applicable to TRC structural members. The use of high-order reliability methods to determine the exact failure probability (and safety levels) and derive partial safety factors for TRC is of great importance in structural engineering. Reliability-based assessments of TRC structural members can be efficient approaches to comply with essential design requirements as acceptable low failure probabilities, as well as economic and sustainable structures. To this, stochastic models for the basic variables on the resistance side (e.g., model uncertainty, effective depth, and properties of textile reinforcement) must be derived. In this context, large databases of experimental investigations are available for numerous basic variables (e.g., concrete properties). This paper shows the results of a thorough reliability analysis of a selected TRC structural member under bending load. In this investigation, different reliability-based methods were adopted with the support of an open-source software available in "R".

Keywords: Textile-reinforced Concrete · Reliability-based methods · Reliability Analysis · Partial Factor

1 Motivation

Nowadays, the building sector is increasingly influenced by fast-rising construction costs as well as high energy costs. Additionally, there is a huge pressure to reduce resources and material consumption as well as to minimise greenhouse gas emissions. Thus, the building and construction industry not only aim at strengthening existing building structures as well as ensuring high durability for new buildings but also need to consider the impact of decision and construction choices on the environment. In this regard, multiple research initiatives have been focusing on the more efficient use of resources and materials, and on the development of innovative structural materials or solutions

J. C. Matos et al. (Eds.): IPW 2024, LNCE 494, pp. 472–481, 2024.
https://doi.org/10.1007/978-3-031-60271-9_44

(e.g., composite structures out of non-metallic-reinforced-concrete slabs in combination with timber beams). This includes the use of non-metallic reinforcements for structural concrete and the development of new concrete shapes for structural members with non-metallic reinforcement grids. In this context, non-metallic reinforcements refer to textile-reinforced concrete (TRC) and carbon-reinforced concrete (CRC) made out of different non-metallic materials and reinforcement shapes (e.g., rebars, grids, among others). In the last few years, progresses in the field of non-metallic reinforcements have led to the adoption of extensive applications in building and civil engineering structures [1]. Despite these modern advances, a wide spread adoption of TRC remains limited. This is partly due to the lack of a consistent design framework since conventional design methods utilised for other materials (e.g., steel-reinforced concrete) cannot be directly applicable to TRC structural members.

To ensure a reliable and safe application of non-metallic reinforcement in concrete structures, new design approaches and models need to be proposed, verified and validated. High-order reliability methods – based on the reliability-based concepts – offer additional capabilities to unlock the potential of TRC. They enable the estimation of exact failure probabilities (and corresponding safety levels) and support the derivation of partial safety factors. Yet, for the structural analyses of TRC, stochastic models for the basic variables on the resistance side (e.g., model uncertainty, effective depth, and properties of non-metallic reinforcement) must be established (e.g., through existing models proposed in literature such as in JCSS Probabilistic Model Code or through experimental data).

This paper shows the potential of using high-order reliability methods for the design of non-metallic reinforced concrete members subjected to bending loads by using a First-Order Reliability Method (FORM) and a direct Monte-Carlo Simulation. The calculations conducted in the context of this investigation were carried out in the open-source statistic software *R* [2].

2 State of the Art

2.1 Basics of Non-metallic Reinforced Concrete

It is widely acknowledged that modern buildings are subject to a wide range of requirements. Additionally, demands for more economic and ecologic solutions as well as enhanced concrete surface quality (e.g., [3]) are more common thanks to the improved technical knowledge that sector has received in the last few years. Furthermore, aspects concerning durability should be equally addressed. Non-metallic reinforced concrete optimally combines the high tensile load-bearing capacity of non-metallic technical fibers with the high compressive strength of concrete. These excellent properties enable the realisation of non-metallic reinforced concrete structures in multiple applications. Examples of TRC can be found in new and existing structures [4–7] and include facade slabs and sandwich constructions [8–10], slabs [11], bridges [5, 12–14] among others (e.g., garages, facades).

The most common non-metallic reinforcements consist of alkali-resistant (AR) glass or carbon fibers. AR-glass and carbon fibers are multifilament yarns with several hundred filaments per yarn. One of the aspects to highlight on these fibers is the dependence of strength on the defects, which leads to lower strengths for longer lengths. This scale effect is due to the higher probability of imperfections in long samples [15]. Additionally, the strength of a yarn is lower than the strength of a fiber bundle due to the failure mechanisms of the fiber bundle, different test lengths (100 mm filaments, 500 mm yarns) and different test speeds. The tensile strength of AR-glass is in the order of 1,000 to 2,000 MPa and the modulus of elasticity is 74,000 to 80,000 MPa. The stiffness of AR-glass is, therefore, around 40% of the stiffness of steel.

Textile concrete normally consists of a special concrete mix, typically characterised by a standard maximum grain size of less than 8 mm and impregnated high-strength textiles as reinforcement elements. It should be noted here that special concretes – also known as fine concrete – have been specially developed for textile concrete, which are often classified as high-strength concretes. As these concretes are often outside the standards provisions, standard concretes that comply with EN 206 [16] can be also utilised. For the sake of reliability analyses, a structure can be analysed at the micro and meso level, whereby the meso level is considered in this work. Important parameters of the meso level are the fracture stress, the modulus of elasticity and the stress-strain curve of the textile concrete. Material laws are developed through experimental investigations and reproduced in the form of stress-strain curves.

Additionally, there are differences in the material laws of a pure fiber strand and a textile concrete expansion body. While the stress-strain curves of the pure fiber strand can be described as linear-elastic, the material law of the expansion body is divided into the following three states for TRC: State I: Uncracked concrete, State IIa: Crack formation phase, and State IIb: Completed crack formation phase. State IIb of the expansion body runs parallel to the pure fiber strand. For the component design adopted in this investigation, the stress-strain curve of the pure fiber strand is utilised to describe the material behavior of the textile reinforcement. Related literature includes effective values for the textile fracture stress and the modulus of elasticity based on various influences taking into account the number of fiber strands of the textile reinforcement mesh and the fiber strand length as they occur in a component [17, 18]. The effective values result from the textile mean values of the standardized pure fiber strand test under consideration of a roving by multiplication with the factor α_{eff} (e.g., [18]). This conversion with effective tensions is carried out on the carbon textile and AR-glass textile. Figure 1 shows the reduction by the effective factor in the form of the stress-strain curve.

For bending design with and without axial force, Rempel (e.g., [18]) distinguished between component and composite design. Component design is suitable for general cases, whereas composite design can be carried out using interaction diagrams developed specifically for a certain structure and, therefore, is not universally valid. The interaction diagrams are individually determined based on experimental investigations. Therefore, in this work, the component design for bending design will be pursued.

2.2 Basics of Structural Safety

The probabilistic analyses conducted in this investigation are grounded on the fundamental principles of structural reliability. Here, a failure probability is assessed by evaluating whether the limit state functions (LSF) (i.e., ultimate and/or serviceability limit states) are exceeded [20]. Generally, for the simple case, the limit state indicates the margin of

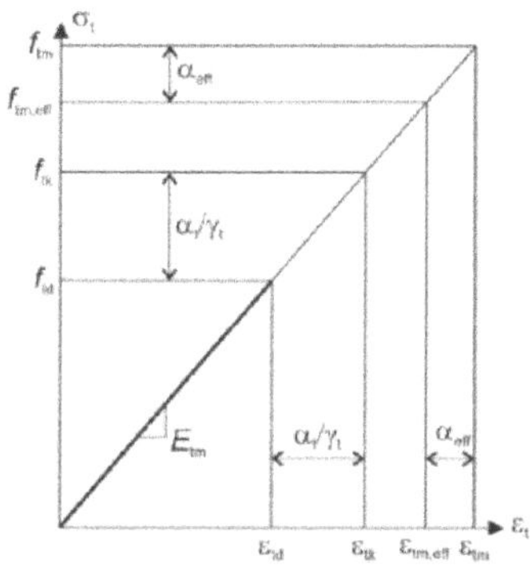

Fig. 1. Stress-strain curve of the non-metallic reinforcement according to [19]

structural safety between the resistance side (R) and the action effect (E) side. A limit state function $g(\cdot)$ and a failure probability P_{f} can be defined as (Eq. 1 and Eq. 2):

$$g(\cdot) = R(X) - E(X) \tag{1}$$

$$P_f = P[g(\cdot) < 0] \tag{2}$$

Both $R(\cdot)$ and $E(\cdot)$ are functions of basic (random) variables $\boldsymbol{X}$ (e.g., material strength, geometric dimensions and model uncertainties). The notation $g(\cdot) < 0$ denotes the failure region. Likewise, $g(\cdot) = 0$ and $g(\cdot) > 0$ indicate the failure surface and safe region, respectively. The mean and standard deviation of the limit state function $g(\cdot)$ can be determined from the elementary definition of the mean and the variance. The mean of $g(\cdot)$ is given as (Eq. 3):

$$\mu_g = \mu_R - \mu_E \tag{3}$$

where μ_{R} and μ_{E} are the means of R and E, respectively.

The significant role played by uncertainties in resistance models (i.e., related to simplified relationships of complex numerical models) and action effects (i.e., related to the assessment of load effects and their combinations) (e.g. [21]) is also relevant for structural reliability analyses. Model uncertainties affect material and action models applied in engineering practice such as those based on the partial factor method in DIN EN 1990 [22] (Eq. 4).

$$\gamma_E \cdot \theta_E \cdot E(x_1, x_2, \ldots, x_n) \leq \theta_R \cdot R(x_1, x_2, \ldots, x_n)/\gamma_R \tag{4}$$

where θ_R and θ_E are the model uncertainties and γ_{R} and γ_{E} refer to the partial factor for R and E, respectively.

Reliability-based assessments are necessary to prove the (theoretical) value of failure probabilities. To calculate those probabilities and derive partial safety factors, multiple probabilistic methods are available to the communities of practice [23] as a result of the remarkable scientific and computational developments emerging over the last few years

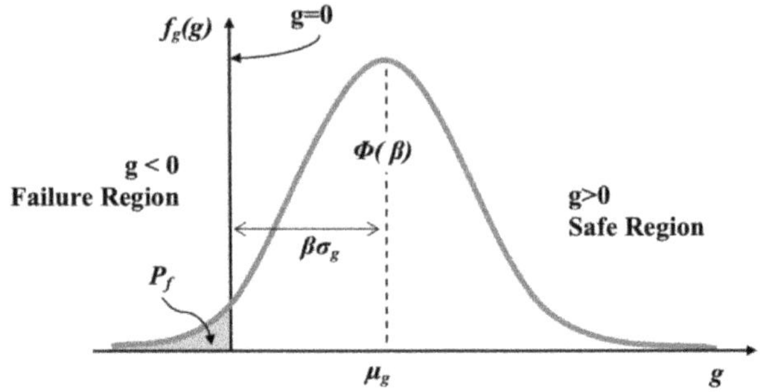

Fig. 2. Probability density for the limit state function $g(\cdot)$ according to [20]

[24–27]. The calculation of failure probabilities is often an integration of a high dimensional probability function and, therefore, can be considered a complex mathematical problem [28]. Only simple problems with basic (random) variables can be solved with analytical methods. In general, for multidimensional problems, the accurate solution of the integrals can only be achieved through approximation methods (e.g., First Order Reliability Methods - FORM) or by means of numerical methods (e.g., direct Monte Carlo Simulations). For many practitioners, these methods can be somewhat complex to handle [29]. Plotting the two-dimensional joint probability density function $f_{\mathrm{E,R}}(e, r)$ allows for a better understanding of the fundamentals (Fig. 2). The volume of the distribution density is one and divided into a failure and a survival region by the limit state function (LSF): $g(\cdot) = R - E = 0$. The volume of the failure region corresponds to the failure probability P_{f} and can be calculated by integration of the so-called "convolution integral" (Eq. (5)):

$$P_f = \int_{-\infty}^{+\infty} f_E(e) \left(\int_{-\infty}^{r=e} f_R(r) dr \right) \mathrm{d}e \tag{5}$$

The so-called "design point" corresponds to the peak of the failure volume. A transformation to standard normal space allows for the plotting in a two-dimensional coordinate system with the peak of the density function at the origin surrounded by concentric circles for equal reliability levels. The "design point" is located on the perpendicular of the limit state line. The minimum distance between the "design point" and the origin is designated as reliability index β, which is considered as a measure of structural safety. Geometrically, the weighting factors α_{E} and α_{R} represent the direction cosine between the coordinate axis and the perpendicular to the LSF. In general, the failure probability is linked to n variables, resulting in a n-dimensional density function, which is divided by a $(n - 1)$ dimensional hypersurface into a failure and survival space. The minimum distance between the limit state surface and the origin of the coordinate system is also defined as the "design point".

2.3 Database for Structural Analysis

The existing JCSS Probabilistic Model Code [30] provides a compilation of statistical parameters and distribution types of relevant basic variables for different materials on the resistance side, namely for steel-reinforced concrete. However, this document lacks stochastic information for basic variables of new structural solutions such as non-metallic reinforced concrete. Current findings are available for the basic variables of the bending design of non-metallic reinforced concrete components; for the bending design with and without axial force, reference is made to the work of Rempel [17, 18]. More recent stochastic models concerning concrete properties can be found in [31]. For the reliability analysis of other design models (e.g., shear with and without shear reinforcement [32]) large experimental databases are available.

3 Reliability Analyses

In Germany, the textile-reinforced concrete construction method has not been introduced by the building authorities yet, despite being already adopted in practical buildings, as mentioned in Sect. 2. Nonetheless, other countries have made standards available for FRP rebars (e.g., [33, 34]). Thus, in the context of this investigation, the following boundary conditions were specified for the reliability analysis of the bending design:

- Use of component design for bending with and without normal force as well as for normal force alone;
- In contrast to the bending design of reinforced concrete components, rules are supplemented, adapted or modified for non-metallic reinforcement (e.g., use of the breaking stress of the reinforcement instead of the yield stress.);
- In addition to DIN EN 1992-1-1 [35], other areas of application can be developed (e.g. more filigree components, new component shapes).

In the context of these boundary conditions, the concepts of reliability-based theory enable detailed parameter studies for the analysis of relevant components. As mentioned above, for the bending load-bearing capacity of non-metallic reinforced concrete components, component design is utilised to characterise the basic variables involved in the selected application examples, although other design concepts specified in the literature (e.g., composite design) could be considered. Component bending design is best suited for a general consideration during reliability analyses. The component-by-component bending design of carbon concrete components is usually carried out iteratively through the variation of the strain plane and is based on models utilised in the design of reinforced concrete components. In practice, the bending verification is carried out after an iterative determination of the strain plane and the finding of equilibrium by solving the following inequality (Eq. 6):

$$M_{Ed} \leq M_{Rd} \tag{6}$$

For non-metallic reinforced concrete components, the bending resistance M_{Rd} mainly depends on different stochastic variables. For the resistance, the following variables were considered: for the concrete (compression strength f_{cm}, maximum strain $\varepsilon_{\mathrm{cm}}$) and for the carbon fiber reinforce polymers (CFRP) reinforcement (tensile strength f_{nm}). Additionally, the effective depth d and the resistance model uncertainty for bending θ_R were also considered. For the effects, the permanent g and live loads q as well as the model uncertainty of loads θ_E were considered as stochastic variables. Due to the lack of a minimum thickness requirement to ensure the durability of carbon concrete components, thin constructions with low depths are also possible, whereby the absolute tolerances of the reinforcement layers must be limited to a certain level. For the other basic variables, the information from the JCSS Probabilistic Model Code [30] was utilised with the exception of the effective depth. For the characterization of this variable, information on the characteristic values of carbon concrete components were determined for the application examples analogous to the explanations in [17]. These considerations are summarised in Table 1.

Table 1. Input basic variables of the reliability analyses

Variable	Unit	Distr. Type*	X_k	μ_X	V_X
g	kN/m	N	g_k	$1.05g_k$	0.05
q	kN/m	GUM	q_k	$0.75g_k$	0.24
θ_E	–	LN	1.0	0.10	0.10
f_{cm}	MPa	LN	90	98	0.06
ε_{cm}	‰	LN	2.6	2.6	0.15
f_{nm}	MPa	N	2,432	2,800	0.08
d	mm	N	3 … 114	4 … 120	**
θ_R	–	LN	1.0	0.10	0.10

*N = Normal distribution, LN = Log-Normal distribution, GUM = Gumbel distribution.
**depending on absolute value of height with characteristic value of d_{eff} in accordance with [17].

Generally, the partial safety factor takes a prominent position in the common safety format (e.g., Eurocodes). The code calibration of these partial safety factors for non-metallic reinforced concrete is done in accordance with literature (e.g., [28]). In this section, reliability analyses are conducted for the derivation of partial safety factors γ_{nm} of non-metallic reinforcement under bending loads. The reliability analyses are evaluated for concrete members with non-metallic reinforcement grids. All the reliability analyses were conducted with a FORM and direct Monte Carlo simulations in a self-programmed code in the programming language *R* [2]. The results show a marginal difference between results of methods (FORM or Monte Carlo) (less than 1%). For the described reliability analyses, the results are shown in Fig. 3.

As it can be observed from Fig. 3, with low reinforcement ratio $\rho_L \leq 3‰$ and high reinforcement ratio $\rho_L \geq 6‰$, the curves of the reliability index are almost parallel to the horizontal axis. In the range of reinforcement ratio $3‰ < \rho_L < 6‰$, the curves show a different behavior. This can be explained by the different conditions considered in the probabilistic analyses where either a concrete failure or a reinforcement failure occurred.

The partial safety factor $\gamma_{nm} \geq 1.3$ can be calculated independently of the existing reinforcement ratio ρ_L when the reliability index β is greater than the target value. The curve of the target reliability index for 50 years ($\beta = 3.8$) for a reliability class 2 (RC2) was selected in accordance to DIN EN 1990 [22].

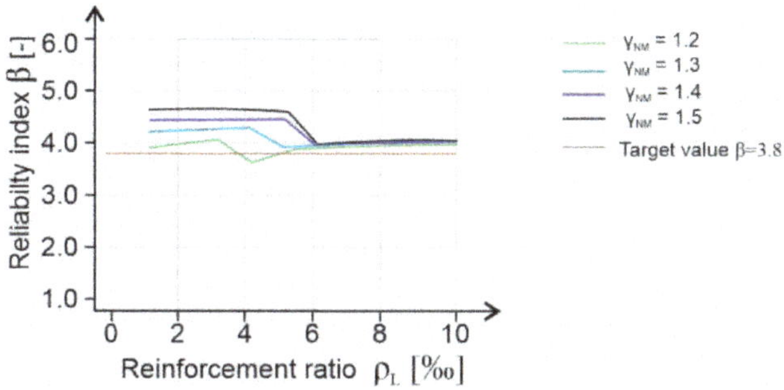

Fig. 3. Example of parameter study for different reinforcement rations to derive γ_{nm}

4 Conclusion and Outlook

This paper presents the results of a set of reliability analyses, where the influence of partial safety factor for non-metallic reinforcement on reinforced concrete members under bending load was investigated. The results indicate that the partial safety factor can be derived to $\gamma_{\mathrm{nm}} = 1.3$ to fulfil the design provisions for bending loads. For this investigation, a compilation of statistical parameters and distribution types of relevant basic variables on the resistance side were considered from JCSS Probabilistic Model Code. Yet, for the stochastic models of non-metallic reinforcement, data retrieved from other relevant literature can be adopted.

This investigation highlighted the potential of using high-level reliability methods for the design of innovative structural solutions, when a robust design framework is still lacking.

In the future, other design models (e.g., shear design) should be addressed. Additionally, further stochastic models for the basic variables of other non-metallic reinforcement types should be derived.

References

1. Beckmann, B., Bielak, J., Bosbach, S., et al.: Collaborative research on carbon reinforced concrete structures in the CRC/TRR 280 project. Civ. Eng. Des. **3**, 99–109 (2021). https://doi.org/10.1002/cend.202100017
2. R Core Team: A Language and Environment for Statistical Computing (2016). https://www.R-project.org/
3. von der Heid, A.-C.: Sandwichelemente für Eastsite VIII - Mehrwert durch Textilbeton. Tudalit Magazin **15**, 12 (2016)
4. Rempel, S., Will, N., Hegger, J., et al.: Grundlagen des Bewehrens. In: Curbach, M., Hegger, J., Schladitz, F., et al. (eds.) Handbuch Carbonbeton: Einsatz nichtmetallischer Bewehrung, pp. 99–134. Ernst & Sohn a Wiley brand, Berlin (2023)
5. Helbig, T., Unterer, K., Kulas, C., et al.: Fuß- und Radwegbrücke aus Carbonbeton in Albstadt-Ebingen. BuSt **111**, 676–685 (2016). https://doi.org/10.1002/best.201600058
6. Bergmann, S., May, S., Hegger, J., et al.: Shear strengthening of reinforced concrete t beams using carbon reinforced concrete. ACI Spec. Publ. **345**, 169–184 (2021)

7. Herbrand, M., Adam, V., Classen, M., et al.: Strengthening of existing bridge structures for shear and bending with carbon textile-reinforced mortar. Materials **10**, 1099 (2017). https://doi.org/10.3390/ma10091099
8. Heid A-C von der: Zum Tragverhalten von Sandwichelementen aus Textilbeton mit schlaffer und vorgespannter Bewehrung, RWTH Aachen University (2021)
9. Shams, A., Stark, A., Hoogen, F., et al.: Innovative sandwich structures made of high performance concrete and foamed polyurethane. Compos. Struct. **121**, 271–279 (2015). https://doi.org/10.1016/j.compstruct.2014.11.026
10. Stark, A.: Analysis of Prestressed Sandwich Panels with Ultra-High-Performance Concrete Facings, RWTH Aachen University (2017)
11. May, S., Michler, H., Schladitz, F., et al.: Lightweight ceiling system made of carbon reinforced concrete. Struct. Concr. **19**, 1862–1872 (2018). https://doi.org/10.1002/suco.201700224
12. Bielak, J., Will, N., Hegger, J.: Zwei Praxisbeispiele zur Querkrafttragfähigkeit von Brückenplatten aus Textilbeton. Bautechnik **97**, 499–507 (2020). https://doi.org/10.1002/bate.202000037
13. Hegger, J., Goralski, C., Kulas, C.: Schlanke Fußgängerbrücke aus Textilbeton. BuSt **106**, 64–71 (2011). https://doi.org/10.1002/best.201000081
14. Kueres, S., Hegger, J.: Reliability analysis of footbridges pre-tensioned with carbon fiber reinforced polymer tendons under flexural loading. Eng. Struct. **203** (2020). https://doi.org/10.1016/j.engstruct.2019.109546
15. Jesse, F., Curbach, M.: Verstärken mit Textilbeton. In: Bergmeister, K., Fingerloos, F., Wörner, J.-D. (eds.) Beton-Kalender 2010: Brücken - Betonbau im Wasser, 99. Jahrgang, pp. 457–565. Ernst & Sohn, Berlin (2010)
16. Deutsches Institut für Normung e.V.: Beton – Festlegung, Eigenschaften, Herstellung und Konformität: Deutsche Fassung EN 206:2013+A2:2021 91.100.30(DIN EN 206:2021-06) (2021)
17. Rempel, S.: Zur Zuverlässigkeit der Bemessung von biegebeanspruchten Betonbauteilen mit textiler Bewehrung. Dissertation, RWTH Aachen University (2018)
18. Rempel, S., Ricker, M., Hegger, J.: Safety concept for textile-reinforced concrete structures with bending load. Appl. Sci. **10**, 7328 (2020). https://doi.org/10.3390/app10207328
19. Deutscher Ausschuss für Stahlbeton: DAfStb-Richtlinie Betonbauteile mit nichtmetallischer Bewehrung, Gelbdruck-Entwurf: Entwurf November 2022 (2022)
20. Choi, S.-K., Canfield, R.A., Grandhi, R.V.: Reliability-Based Structural Design. Springer, London (2007). https://doi.org/10.1007/978-1-84628-445-8
21. Holicky, M., Diamantidis, D., Sykora, M.: Reliability levels related to different reference periods and consequence classes. Beton- und Stahlbetonbau **113**, 22–26 (2018). https://doi.org/10.1002/best.201800039
22. Deutsches Institut für Normung e.V.: Eurocode 0: Grundlagen der Tragwerksplanung: Deutsche Fassung EN 1990:2002 + A1:2005 + A1:2005/AC:2010 91.010.30(DIN EN 1990:2010-12) (2010)
23. Ricker, M., Feiri, T., Nille-Hauf, K., et al.: Enhanced reliability assessment of punching shear resistance models for flat slabs without shear reinforcement: first published online. Eng. Struct. **226**, 111319 (2021). https://doi.org/10.1016/j.engstruct.2020.111319
24. Freudenthal, A.M.: Safety and the probability of structural failure. Trans. Am. Soc. Civ. Eng. **121**, 1337–1375 (1954)
25. Faber, M.H., Vrouwenvelder, T.: Probabilistic Model Code: 12th draft (2001)
26. Rackwitz, R.: First Order Reliability Concepts for Design Codes. CEB Bulletin d'Information (1976)
27. Deutsches Institut für Normung (ed): Grundlagen zur Festlegung von Sicherheitsanforderungen für bauliche Anlagen, 1. Aufl. Beuth, Berlin (1981)

28. Melchers, R.E., Beck, A.T.: Structural Reliability Analysis and Prediction, 3rd edn. Wiley, Chichester (2018)
29. Feiri, T., Schulze-Ardey, J.P., Ricker, M., et al.: The role of probability-based safety analysis: the perspective of structural product manufacturers. In: Fédération internationale du béton (ed.) Building for the Future: Durable, Sustainable, Resilient: Proceedings of the fib Symposium 2023, vol. 349, pp. 1589–1600 (2023)
30. Joint Committee on Structural Safety: Probabilistic Model Code 2010 - Part 3: Material Properties and Resistance Models (2002)
31. Feiri, T., Kuhn, S., Wiens, U., et al.: Updating the prior parameters of concrete compressive strength through Bayesian statistics for structural reliability assessment. Structures **58**, 105636 (2023). https://doi.org/10.1016/j.istruc.2023.105636
32. Bielak, J.: Shear in slabs with non-metallic reinforcement. Dissertation, RWTH Aachen University (2021)
33. Canadian Standards Association: Design and construction of building structures with fibre-reinforced polymers (CSA S806-12:2012–03) (2012)
34. American Concrete Institute: Guide for the design and construction of structural concrete reinforced with fiber-reinforced polymer (FRP) bars 440.1R-15(ACI-440.1R-15) (2015)
35. Deutsches Institut für Normung e.V.: Eurocode 2: Bemessung und Konstruktion von Stahlbeton- und Spannbetontragwerken – Teil 1-1: Allgemeine Bemessungsregeln und Regeln für den Hochbau: Deutsche Fassung EN 1992-1-1:2004 + AC:2010 91.010.30; 91.080.40(DIN EN 1992-1-1:2011-01) (2011)

Sensitivity Analysis on the FE Model Parameters of the Leaning Tower of Pisa

Dionis Butuc[1], Anna De Falco[1], Filippo Landi[1], Carlo Resta[2], Nunziante Squeglia[1], and Laura Vignali[1](✉)

[1] Department of Civil and Industrial Engineering, University of Pisa, Largo Lucio Lazzarino 1, Pisa, Italy
laura.vignali@phd.uniupi.it

[2] Department of Energy, Systems, Territory and Construction Engineering, University of Pisa, Largo Lucio Lazzarino 1, Pisa, Italy

Abstract. The behavior of cultural heritage structures is often investigated or predicted using numerical models. The several parameters defining each model are always affected by uncertainties, and their effects on the outputs should be investigated. In this context, a sensitivity analysis on model parameters provides insight into the dependence of outputs on changes in specific inputs, enabling the selection of the variables that are most suitable for calibration. In the case of modal sensitivity analysis, this approach can also supply useful information for the optimal placement of monitoring sensors. For these purposes, the present paper evaluates the sensitivity of the modal eigenvalues of the model of the Tower of Pisa to the mechanical parameters of the masonry and different layers constituting the underlying soil. This preliminary study allows the detection of modal frequencies that are most influenced by the stiffness of the soil and those purely relying on the masonry, providing guidance for the future model identification based on structural health monitoring data.

Keywords: sensitivity analysis · gPCE · modal analysis · Leaning Tower of Pisa · Cultural Heritage

1 Introduction

The safeguarding of historical buildings requires understanding their structural behavior, often evaluated through reliable numerical models capable of assessing safety in the presence of different and future scenarios. In this context, inaccuracies and assumptions can mislead simulation results, thus underlining the need to improve the reliability of models by reducing uncertainties as much as possible. The first step in this direction is precisely to identify the sources of uncertainty and understand its effects on the predictive capacity of the models.

Model uncertainties can derive from both an inaccurate knowledge of the shape of the structure and the values of material parameters, and the boundary conditions. Regarding this latter source, it's worth emphasizing that considering fixed boundary

J. C. Matos et al. (Eds.): IPW 2024, LNCE 494, pp. 482–491, 2024.
https://doi.org/10.1007/978-3-031-60271-9_45

conditions is widely used for analysis and design purposes when geotechnical data is not provided [1]. Nevertheless, studies on historical masonry structures demonstrate that soil-structure interaction influences the dynamic characteristics of structures [2], significantly affecting their dynamic properties and seismic behavior [3–6].

The Tower of Pisa rests on extremely deformable soil, which caused the current situation and significantly influences its static and dynamic behavior [7]. This structure, which has always raised concerns among scholars regarding its safety, has been studied continuously for several decades through various experimental campaigns and simulations [8], providing important indications on the possible values of the parameters to be adopted for materials and soil (Fig. 1).

Fig. 1. A view of the Leaning Tower of Pisa next to the Cathedral.

This work represents the first step of a broader study aimed at reducing uncertainties in the parameters of the Tower's model through dynamic identification. In the context of conservation and non-destructive investigations, the dynamic characteristics of the Tower are crucial for calibrating the model through ambient vibrations. We therefore analyzed the first ten natural frequencies of a high-definition model specifically created for the assessment of the monument's safety [9], and on these we carried out the propagation of uncertainties deriving from the mechanical characteristics of the structural materials and the underlying soil.

The choice of uncertain variables focused on the elastic modulus of the material constituting the outer layers of the masonry of the Tower and those of four layers of soil identified through various experimental campaigns [10, 11]. A-priori probability density functions were assigned based on in-situ tests conducted by the committees responsible for the safety of the monument over time. The influence of each stochastic variable in determining the first ten natural frequencies of the Tower was then obtained through Sobol' Indices (SIs) computed from a Generalized Polynomial Chaos Expansion (gPCE)-based surrogate model. Two main advantages arise from surrogation: Sobol' Indices can be calculated directly from the expansion coefficients [12], and the surrogate model can be used instead of the original one in any situation where multiple, rapid evaluations are needed, such as in the case of real-time model updating based on monitoring data.

The information obtained as a result of this work can assist in modal updating based on acceleration records to minimize model uncertainties. It can also guide the optimal placement of dynamic monitoring sensors (e.g., accelerometers) based on the expected experimental mode shapes that we anticipate observing.

2 Materials and Methods

2.1 Experimental Tests on Soil and Masonry

The Leaning Tower and the soil of Piazza del Duomo both have a long history of monitoring and experimental investigations, which became extensive during the second half of the twentieth century due to concerns for the increasing inclination of the Tower. Among the several tests conducted in the past, attention is focused on those related to the mechanical characterization of masonry, specifically, aiming to define the Young's modulus of the different materials. Thanks to endoscopy and core drillings, it was ascertained that the masonry wall of the Tower is composed of three layers: two external layers of squared San Giuliano marble blocks and a mixed masonry infill. The Young's modulus of the latter was obtained by dilatometer tests performed in core sampling boreholes [13]; with average values of the superstructure infill measuring 7342 MPa (st. dev.: 3923 MPa). On the external marble walls 32 flat jack tests were conducted on both the external and internal faces of the Tower at four different tiers [13] (Fig. 2). The average Young's modulus of these layers is 54176 MPa, and the standard deviation is 10277 MPa.

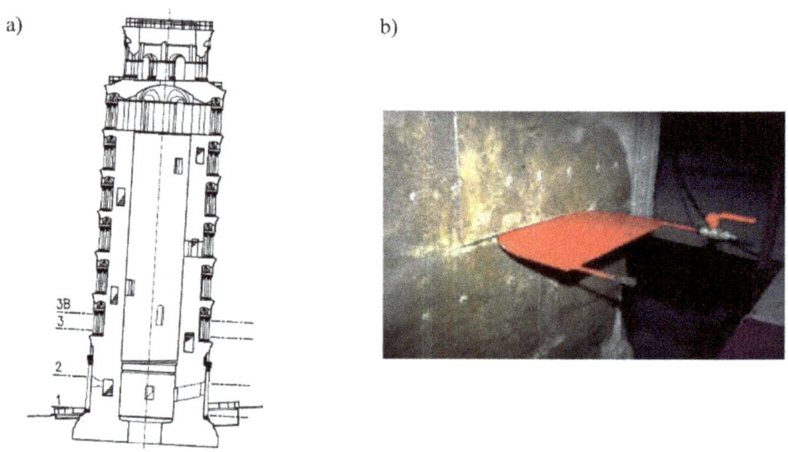

Fig. 2. Flat jack tests: a) N-S section representing the four tiers where 32 tests were performed (from [13]); b) photo of a flat jack test (from [8]).

We focused part of the investigation on the influence of this parameter within the model of the monument.

An estimate of the soil layers' Young's can be obtained from literature in-situ tests providing information about shear wave velocity (V_s) and density (ρ) profiles (Eq. 1 with ν equal to 0.5).

$$E = 2 \cdot (1 + \nu) \cdot \rho \cdot V_s^2 \tag{1}$$

In this context, several experimental campaigns were conducted by Committees responsible for the preservation of the Tower. Density was estimated from borehole surveys performed around the monument, and shear wave velocity was collected from three cross-hole tests conducted in 2000 (see Fig. 3).

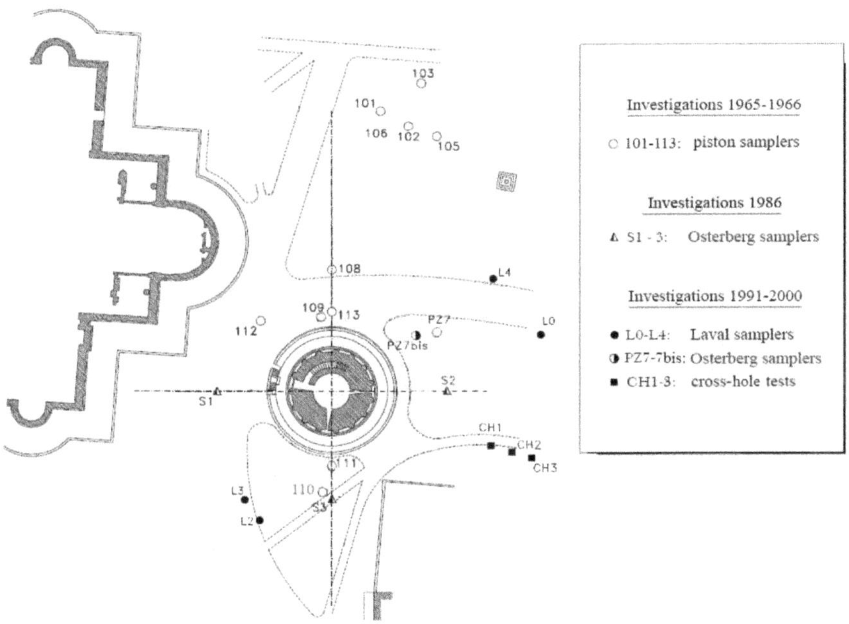

Fig. 3. Location of some of the soil investigations carried out during the twentieth century. Density and S-wave velocity profiles adopted in this study are derived from borehole surveys and cross-hole tests, respectively. Adapted from [10].

The analysis of these data resulted in a four-layer soil profile (S1–S4) characterized by homogeneous values of the elastic modulus. This subdivision, which is consistent with the literature subdivision in three complexes [10, 11], provides a refinement of the soil characterization as the added layer coincides with a discontinuity in the complex B (see Fig. 4). Table 1 summarizes the estimated average and standard deviation values of the soil layers' Young's moduli.

2.2 Finite Element Model and Description of Material Properties

An accurate finite element (FE) model of the Leaning Tower from a refined point cloud was realized by the authors to gain insight into open questions regarding the Tower's

Table 1. Mean (μ) and standard deviation (σ) of the elastic moduli of the soil layers.

	E_{S1}	E_{S2}	E_{S3}	E_{S4}
μ [MPa]	180	116	297	752
σ [MPa]	41	12	70	110

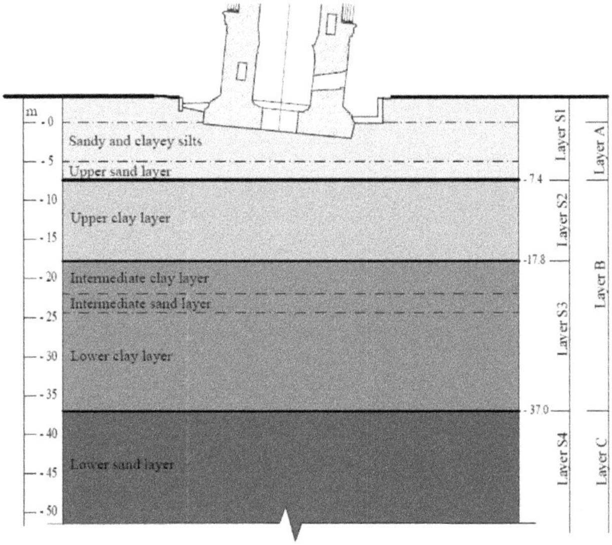

Fig. 4. Subsoil profile with distinction between geological complexes from literature (Layers A–C) and the four ones (Layers S1–S4) identified as layers with homogeneous characteristics of shear wave velocity and unit weight.

structural health [9]. The model reproduces the current inclination of the Tower and its curvilinear shape, the presence of the internal stairways and the layered structure of the wall; a large portion of the soil is also represented (Fig. 5). In this study, the model was slightly simplified to reduce computational time and used to perform a sensitivity analysis of the modal behavior of the monument to the variability in the elastic moduli of masonry and soil layers.

To do so, we shifted to a probabilistic framing of the problem and assigned statistical distributions based on experimental results (see Sect. 2.1) to each of the investigated mechanical properties, i.e. the elastic moduli of the soil layers and of the external wall layers. Deterministic values were instead given to the remaining parameters (cfr. [9]). Tables 2 and 3 collect the values of all the employed elastic moduli.

2.3 Sensitivity Analysis and GPCE-Based Surrogation

To assess sensitivity, we employed Sobol' Indices (SIs) [14], which belong to the field of variance-based sensitivity measures. In this way, it is possible to assess the effect of

Table 2. Elastic moduli of the materials M2–M7 (see Fig. 5 for the definitions).

Deterministic	M2	M3	M4	M5	M6	M7
E [MPa]	7000	80000	10000	12000	50000	25000

Table 3. Elastic moduli of the uncertain materials (see Fig. 5 for the definitions). μ is the average value and σ is the standard deviation of uncertain parameters.

Random variables	M1	S1	S2	S3	S4
Lognormal dist.	$\mu = 54176$	$\mu = 180$	$\mu = 116$	$\mu = 297$	$\mu = 752$
E [MPa]	$\sigma = 10277$	$\sigma = 41$	$\sigma = 12$	$\sigma = 70$	$\sigma = 110$

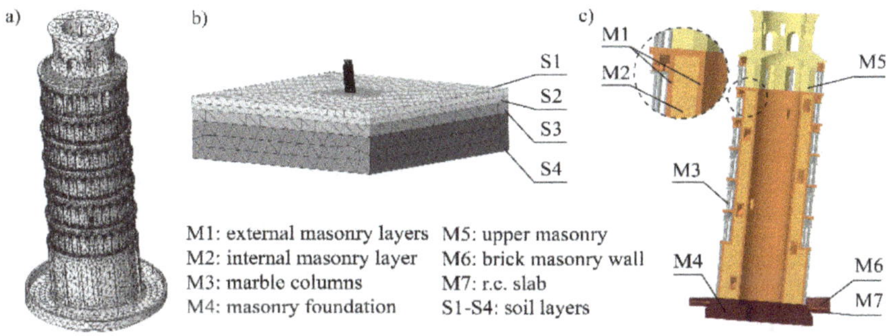

Fig. 5. Numerical model of the Tower: a) FE mesh; b) general overview of the FE model and materials employed for the soil; c) N-S section: materials used for the Tower.

variability in uncertain inputs upon the variability of the output. We considered the first ten natural frequencies of the model of the Tower as the monitored output. Calculation of Sobol' Indices is based on a decomposition of the variance associated to a given output Y of the model, Var(Y), into contributions associated to each input parameter. For an input $\boldsymbol{X}$, the associated $\mathrm{SI_X}$ is therefore:

$$SI_X = \frac{Var_X(Y)}{Var(Y)}, \tag{2}$$

where $\mathrm{Var}_{\mathbf{X}}(Y)$ is the contribution of input $\boldsymbol{X}$ to the variance of output Y. First-order Sobol' Indices, describing the influence of a single input variable, vary between 0 and 1, with higher values indicating a higher influence on variability of the output. We developed a surrogate model of the modal frequencies of the Tower, i.e. an analytical formulation which, for the considered input range, provides a good approximation of the FE model outputs. To derive the surrogate model, we relied on a technique known as general Polynomial Chaos Expansion (gPCE) [15]. Let M be the numerical model, producing an uncertain output Y based on a vector of uncertain parameters $\boldsymbol{X}$. Its gPCE form is a linear combination of orthonormal polynomials $\boldsymbol{P}_i(\boldsymbol{X})$, and it becomes an approximation when we consider only polynomials up to a given degree *n:*

$$Y = M(X) \approx M_n^{gPCE}(X) = \sum_{i \le n} \alpha_i P_i(X), \quad (3)$$

with $\boldsymbol{\alpha_i}$ the matrix of combination coefficients and $\boldsymbol{i}$ is the multi-index set. The polynomials belong to specific families which choice depends on the statistical description of the input parameters. Unknown combination coefficients $\boldsymbol{\alpha_i}$ are obtained from a set of known input-output pairs, which knowledge requires running the original model a very limited number of times compared to Monte Carlo methods. For this study we employed a third-degree basis of Hermite polynomials, after transformation of input distributions into standard normals. Combination coefficients were obtained by collocation on a full-tensor gaussian grid of 1024 samples. Accuracy was assessed using a threshold of 0.01 Hz (coherent with real-world accuracy on eigenfrequency measurements) on the mean absolute error (MAE) of all 10 frequencies on 50 ulterior randomly selected samples. This also led to the choice of expansion order. Surrogation and calculation of SIs was performed using the open source Octave/MATLAB library *sglib*[1].

3 Results and Discussion

The first result that can be derived from the described probabilistic framework concerns the way in which uncertainty relating to the input parameters affects output prediction. In this regard, Table 4 summarizes the resulting average natural frequencies and their coefficient of variation (V).

Table 4. Mean value (μ) and coefficient of variation (V) of the first ten frequencies of the numerical model of the Leaning Tower.

	f_1	f_2	f_3	f_4	f_5	f_6	f_7	f_8	f_9	f_{10}
μ [Hz]	0.97	0.98	3.44	4.61	4.62	5.71	10.24	10.27	13.32	18.09
V	0.05	0.05	0.04	0.06	0.06	0.07	0.04	0.04	0.04	0.03

To better understand the results of Table 4, it is important to summarize the modal behavior of the numerical model of Tower: the first two natural frequencies are referred to first order bending mode shapes, along the E-W direction and the N-S direction (along which the Tower has the maximum inclination) respectively; the third one is an extensional mode shape; the fourth and fifth natural frequencies refer to modes of vibration governed by the movement of the foundation; the sixth is a torsional mode shape; the seventh and eighth are second order bending mode shapes; the ninth is a second order torsional mode shape and the last is a third order mode shape.

Looking at Table 4, we can first notice that the curvilinear shape and the inclination of the Tower along the N-S direction seem to have no influence on the analyzed frequencies.

[1] Zander E., 2020. *Sglib*: a Matlab/Octave Toolbox for Stochastic Galerkin Methods https://ezander.github.io/sglib/.

The frequencies related to homologous bending mode shapes (f_1–f_2, f_7–f_8) differ indeed by 1–2 hundredths of a Hertz, which in is line with the typical variability of FE model frequencies due to the asymmetry in the mesh and the numerical approximations. The other outcome is that the higher frequencies are characterized by lower values of *V;* this is in agreement with the results of the sensitivity analysis, that are displayed in Table 5 in terms of first-order Sobol' Indices.

Table 5. Sobol' Indices of the uncertain input parameters for each of the first ten natural frequencies of the model. E_{Sx} are the elastic moduli of the soil layers, E_{M1} of the masonry.

Natural frequencies	SI (E_{S1})	SI (E_{S2})	SI (E_{S3})	SI (E_{S4})	SI (E_{M1})
f_1	0.937	0.038	0.012	0.000	0.013
f_2	0.939	0.038	0.010	0.000	0.013
f_3	0.577	0.147	0.259	0.015	0.002
f_4	0.953	0.019	0.013	0.001	0.014
f_5	0.953	0.019	0.013	0.001	0.014
f_6	0.940	0.002	0.000	0.000	0.058
f_7	0.115	0.003	0.000	0.001	0.881
f_8	0.115	0.003	0.000	0.001	0.881
f_9	0.230	0.000	0.000	0.000	0.770
f_{10}	0.021	0.001	0.000	0.000	0.978

For the first six modal eigenvalues (some of their relative mode shapes are shown in Fig. 6), the elastic modulus of the topsoil layer (E_{S1}) is the most influential mechanical parameter with a SI always higher than 0.90, except for the third mode (the extensional one). In this case, the impact of E_{S1} is reduced in favor of the elastic moduli of the third (E_{S3}) and second layer (E_{S2}), which influence is no more negligible. The Young's modulus of the external masonry layers (E_{M1}) becomes of primary importance in the sensitivity of the remaining natural frequencies, as evidenced by its SI which is always higher than 0.77. Seventh, eighth, and ninth mode shapes are shown in Fig. 7. It should be noted that, as expected, the fourth soil layer (about 40 m deep) has no influence on the natural frequencies of the Leaning Tower. Therefore, for future analyses, the model of the soil could be simplified by imposing a deterministic value to the elastic modulus E_{S4}. Total-order SIs were also calculated, showing in all cases very small (≤ 0.001) differences with first-order ones, thus suggesting a very limited influence of parameter interaction.

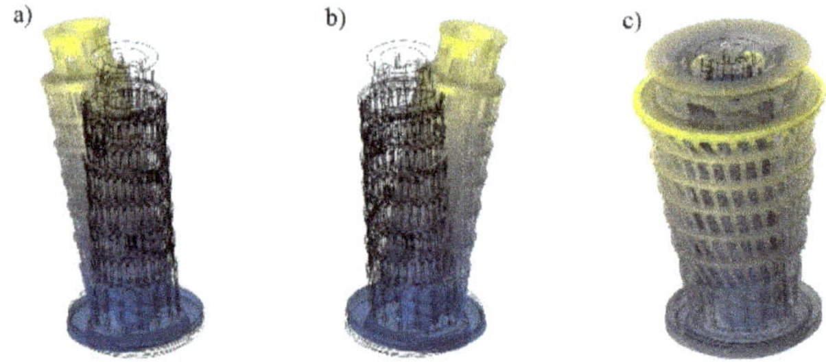

Fig. 6. Mode shapes of the numerical model of the Leaning Tower: a) 1st mode – bending (E-W); b) 2nd mode – bending (N-S); c) 6th mode – torsional.

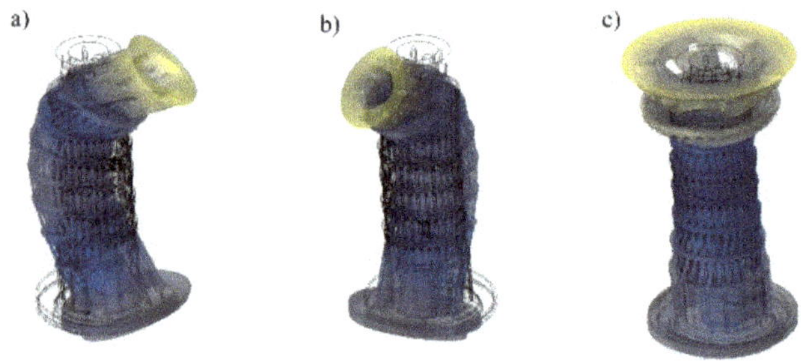

Fig. 7. Mode shapes of the numerical model of the Leaning Tower: a) 7st mode – second order bending (N-S); b) 8th mode – second order bending (E-W); c) 9th mode – second order torsional.

4 Conclusions

In this study, we investigated the dynamic behavior of the Leaning Tower of Pisa, taking into account epistemic uncertainties on soil and masonry mechanical parameters. This was achieved using a highly detailed FE model of the Tower that reproduces the layered structure of the walls and the soil. Data from several past experimental campaigns were collected and analyzed to characterize masonry and soil materials. Two material types were identified for the walls of the Tower, and four for the underlying soil. Uncertainties in the elastic moduli of five of the investigated materials were propagated through the FE model when evaluating the first ten natural frequencies. To facilitate this process, we developed a gPCE-based surrogate model, significantly reducing computational time and allowing for the direct evaluation of sensitivity to input parameters.

The analysis shows that the deepest layer of the soil has no influence on the modal behavior, suggesting a potential reduction of uncertain parameters in future applications. Furthermore, we observe that the first six natural frequencies of the model are primarily influenced by the stiffness of the soil, while for the higher frequencies, the masonry of the outer layers of the Tower becomes the most significant. This finding offers valuable insights for the design of structural health monitoring system (SHM) and provides guidance for future model identification based on SHM data, aiming to reduce uncertainties.

This study underscores the importance of sensitivity analyses as the initial crucial step towards establishing a robust and reliable structural model for the knowledge and preservation of cultural heritage.

References

1. Fathi, A., Sadeghi, A., Emami Azadi, M.R., Hoveidaie, N.: Assessing seismic behavior of a masonry historic building considering soil-foundation-structure interaction (case study of Arge-Tabriz). Int. J. Archit. Herit. (2020)
2. Shabani, A., Feyzabadi, M., Kioumarsi, M.: Model updating of a masonry tower based on operational modal analysis: the role of soil-structure interaction. Case Stud. Constr. Mater. **16**, e00957 (2022)
3. de Silva, F., Ceroni, F., Sica, S., Silvestri, F.: Non-linear analysis of the Carmine bell tower under seismic actions accounting for soil–foundation–structure interaction. Bull. Earthq. Eng. **16**(7), 2775–2808 (2017). https://doi.org/10.1007/s10518-017-0298-0
4. Mortezaei, A., Motaghi, A.: Seismic assessment of the world's tallest pure-brick tower including soil-structure interaction. J. Perform. Constr. Facil. **30**, 04016020 (2016)
5. Hökelekli, E., Al-Helwani, A.: Effect of soil properties on the seismic damage assessment of historical masonry minaret–soil interaction systems. Struct. Des. Tall Build. **29,** e1694 (2020)
6. Erdogan, Y.S.: Discrete and continuous finite element models and their calibration via vibration and material tests for the seismic assessment of masonry structures. Int. J. Archit. Herit., 1–20 (2017)
7. Fiorentino, G., et al.: Seismic reassessment of the Leaning Tower of Pisa: dynamic monitoring, site response, and SSI. Earthq. Spectra **35**, 703–736 (2019)
8. Macchi, G., Ghelfi, S.: Problemi di consolidamento. In: La Torre restituita: gli studi e gli interventi che hanno consentito la stabilizzazione della Torre di Pisa. Volume III, part 1 of Bollettino d'arte, Libreria dello stato edition, Roma, Italy (2005)
9. Butuc, D., De Falco, A., Resta, C., Squeglia, N., Vignali, L.: A novel FE model for the tower of Pisa. In: Proceedings of 9th International Conference on Computational Methods in Structural Dynamics and Earthquake Engineering Methods in Structural Dynamics and Earthquake Engineering, Athens, Greece, pp. 2452–2464 (2023)
10. Lo Presti, D., Jamiolkowski, M., Pepe, M.: Geotechnical characterisation of the subsoil of Pisa Tower. Charact. Eng. Prop. Nat. Soils (2003). Tan et al (eds)
11. Burland, J.B., Jamiolkowski, M.B.: Leaning Tower of Pisa: behaviour after stabilization operations. Int. J. Geoeng. Case Hist. **1**, 156–169 (2009)
12. Sudret, B.: Global sensitivity analysis using polynomial chaos expansions. Reliab. Eng. Syst. Saf. **93**, 964–979 (2008)
13. Bartelletti, R., Finzi, L., Sanpaolesi, L.: Investigations on the masonry of the Leaning Tower of Pisa. IABSE reports (1993)
14. Saltelli, A., Ratto, M., Andres, T., Campolongo, F., Cariboni, J., Gatelli, D., Saisana, M., Tarantola, S.: Global Sensitivity Analysis: The Primer. Wiley, Chichester (2008)
15. Xiu, D.: Numerical Methods for Stochastic Computations: A Spectral Method Approach. Princeton University Press, Princeton (2010)

Substructure Analyses of the Baptistery of Pisa as a Method to Address Uncertainties on Materials and Construction Phases of Monumental Structures

Giada Bartolini[1](✉), Anna De Falco[2], and Filippo Landi[2]

[1] Department of Energy, Systems, Territory and Construction Engineering, University of Pisa, Largo Lucio Lazzarino 1, Pisa, Italy
giada.bartolini@phd.unipi.it

[2] Department of Civil and Industrial Engineering, University of Pisa, Largo Lucio Lazzarino 1, Pisa, Italy

Abstract. The study of monumental structures is a challenging task due to various factors. This includes the difficulty of gathering information for their mechanical characterization and the high variability of non-standard materials and employed construction techniques. Besides, the knowledge of the construction sequence is often partial, despite its significance in identifying potential past intermediate stages when the structure withstood precarious equilibrium conditions potentially leading to crack and deformation patterns.

In this context, material uncertainties play a significant role in determining the performance of masonry historical structures. This paper applies probabilistic methodologies to evaluate the static behavior of a relevant case, notably a substructure of the Baptistery of Pisa that includes a dodecahedral dome resting on an arched drum supported by 12 pillars. Mazars' material damage law is used to simulate the non-linear behavior of masonry, and uncertainty affecting masonry mechanical properties is considered by appointing suitable density functions according to prior engineering expertise. The uncertainties propagation is based upon general Polynomial Chaos Expansion (gPCE) response surfaces, which remarkably reduces the computational burden in the forward problem-solving of the non-linear FE model. The sensitivity of the response in terms of crack propagation is assessed through Sobol' indices, which identify the most relevant input parameters and, thus worthy of careful consideration. This work significantly contributes to the comprehension of the construction sequence of the Baptistery, thus paving the way for further evaluations entailing staged construction analyses of the entire monumental structure.

Keywords: Monumental masonry structure · Mazars' material · Sensitivity analysis · Uncertainty quantification · gPCE

J. C. Matos et al. (Eds.): IPW 2024, LNCE 494, pp. 492–501, 2024.
https://doi.org/10.1007/978-3-031-60271-9_46

1 Introduction

The structural analysis and diagnosis of historical constructions is a challenging activity, especially when structures are placed in seismic-prone regions that make them potentially at risk. The challenge primarily stems from reproducing complex geometries, often associated with computationally demanding models, uncertain material properties, limited knowledge of history-related phenomena and the construction process, and the general difficulty in gathering information about the past history of the structure [1]. However, identifying the construction sequence, especially intermediate stages in which the structure might have been called to withstand precarious equilibrium conditions, is crucial to understanding the origins of current deformative and crack patterns. Similarly, given the high dispersion of mechanical property values in historical masonry [2], it would be advisable to include these epistemic uncertainties in structural models.

Scientific literature provides various modeling strategies [3]. However, the difficulty in characterizing materials limits the available modeling options. FE models often represent the optimal solution for their favorable balance between model accuracy, computational efficiency, and the requirement for relatively limited inputs. Nonetheless, FE models inherently possess a deterministic structure.

In recent years, probability theories with the inclusion of material uncertainties in numerical models have increasingly gained ground. Classical approaches entail assigning probability density functions to uncertain input parameters and employing sampling techniques, such as Monte Carlo methods, to draw sets of samples corresponding to deterministic calls of the FE model [4]. Anyway, achieving precise predictions often requires numerous model evaluations, which makes the problem often unmanageable, especially in the case of complex monumental structures with nonlinear material laws for masonry. Thereby, it is convenient to rely on the generation of metamodels, such as gPCE-based response surfaces, which analytically approximate the response of the original numerical model so speeding up the direct problem solving [5]. In addition, such proxy models can be also employed for the inverse problem, with the updating of the input parameters' prior distributions based on real data [6].

In the literature, only very few cases specifically address the modeling of historical construction along with the effects of history-related phenomena and material uncertainties [7, 8].

This paper employs a probabilistic setting incorporating material uncertainty to assess the plausibility of a relevant substructure of the Baptistery of Pisa, representative of a potential intermediate past configuration. Furthermore, interpreting the construction sequence paves the way for further staged construction analyses of the entire structure, thus enhancing the reading of the global current conditions of the monument.

This work is organized as follows: Sect. 2 illustrates the case study, the FE model of the substructure, and the adopted probabilistic framework to assess and calibrate masonry properties; Sect. 3 summarizes the main results and lessons learned from this work; Sect. 4 provides the conclusions and directions for future works.

2 Materials and Methods

2.1 Case Study

The medieval Baptistery of San Giovanni, the largest baptistery in Italy, is part of the monumental complex of Piazza del Duomo in Pisa. It is a cylindrical stone masonry building of considerable size, 54 m high, with a diameter of 34 m. The external walls, 2.63 m thick at the base, are made of San Giuliano marble (Fig. 1a). The internal structure is composed at the base of 8 granite columns and 4 stone masonry pilasters arranged in a circle, defining the ambulatory (Fig. 1b). The women's gallery, accessed by a staircase within the thickness of the external wall, is located halfway up, supported by masonry cross vaults, and covered by toroidal-shaped barrel vaults. The roofing system includes two brick masonry domes: the internal one features a pyramidal trunk shape with a dodecahedral base of approximately 20 m wide, resting on the 12 internal stone masonry pillars of the women's gallery, while the external hemispherical dome rests on the external walls. 12 radial masonry arches in the upper part of the building connect the drum of the prismatic dome to the external walls, thus helping to counteract its horizontal thrust.

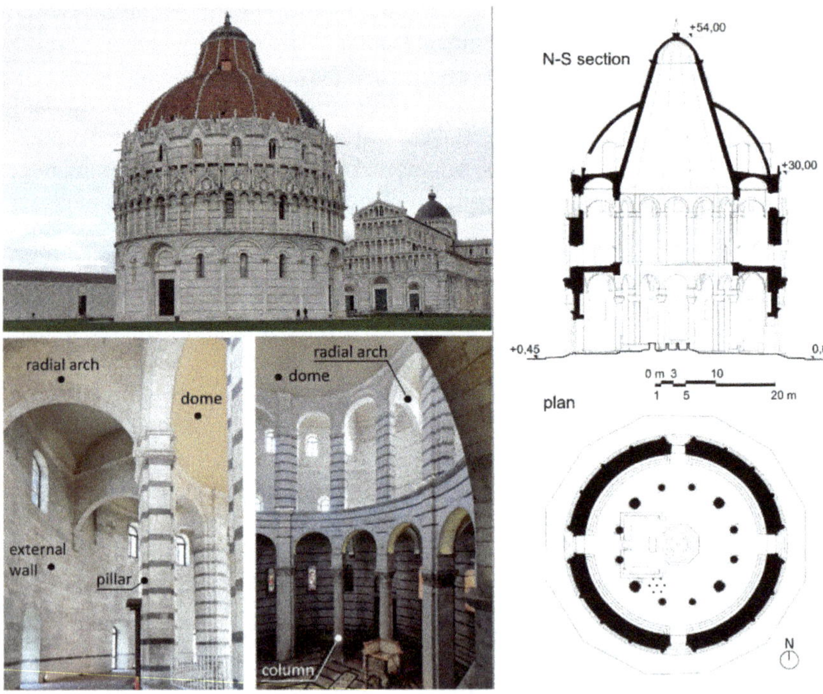

Fig. 1. Baptistery of Pisa: a) View of Cathedral Square; b) N-S vertical section.

The construction started in 1152 and it was completed in 1363. During this period, the monument underwent a lengthy construction process, characterized by a largely unknown sequence of phases, followed by a century of substantial abandonment. It also seems that the external dome was built later than the internal one. Later, in the XVI and XIX centuries, significant restorations took place, primarily involving the substitution of external decorative elements [9, 10].

The study of this structure has recently revealed crack and deformation patterns, currently under investigation to understand their causes, which are to be sought both in the inherent characteristics of the structure and the construction sequence. The potential effects of the building process on the condition are to be sought more in the years of its construction than in subsequent interventions. It is thought that the building sequence of the two domes, the upper arched drum, and the radial arches connecting the drum with the external circular walls might have influenced the development of the current crack pattern at the internal edges of the dodecahedral dome.

In this study, we aim to investigate the equilibrium condition of possible configurations during construction, regarding the internal dome and the radial arches.

2.2 FE Model of the Substructure

To better understand the construction sequence of the upper part of the Baptistery, a model has been created assuming that the dodecahedral dome was built before the radial arches. Under this hypothesis, the configuration of the Baptistery could have consisted only of the building up to the first level floor and the internal dodecahedral dome supported by the 12 pillars of the women's gallery.

After some preliminary evaluations about the displacement at the top of the pillars, aiming to reduce the computation burden, the final FE model of the substructure was made of the 12 s-order pillars (those of the women's gallery), the arched drum, and the dodecahedral dome only (Fig. 2). The boundary conditions of the FE model of the substructure were set fixing the node at the base of the second-order pillars. The model was subject to the self-weight of the structural elements involved. The computational mesh of the model was created by means of Comsol Multiphysics and is composed of approximately 190000 10 nodes solid tetrahedral elements of type serendipity, having minimum and maximum edge lengths of about 0.4 m and 0.8 m respectively (Fig. 2a). Material assignments relied on the work of [11] that identified the various materials used in the Cathedral Square monuments, although no experimental campaign is available. In particular, the eight pillars are made of San Giuliano marble, and the dodecahedral dome of brick masonry (Fig. 2b).

The choice of masonry mechanical properties was made based on those obtained from the experimental in-situ campaigns carried out for the Tower's masonry [12] considering that all materials were extracted from the same quarries at that time [11]. Table 1 summarizes the deterministic values adopted in this study (Fig. 2a).

The material constituting the arched drum of the dome is not easily identifiable due to the presence of a circular barrel vault with the filling partially covering it (Fig. 1b). Therefore, the drum's material was considered uncertain, and it was assumed as the value to be calibrated in this study.

In this paper, the non-linear, isotropic Mazars' constitutive law was used to simulate masonry material [13].

Although originally developed for concrete, this law has already been successfully used to simulate damage development in masonry structures [14]. The state variable *d*

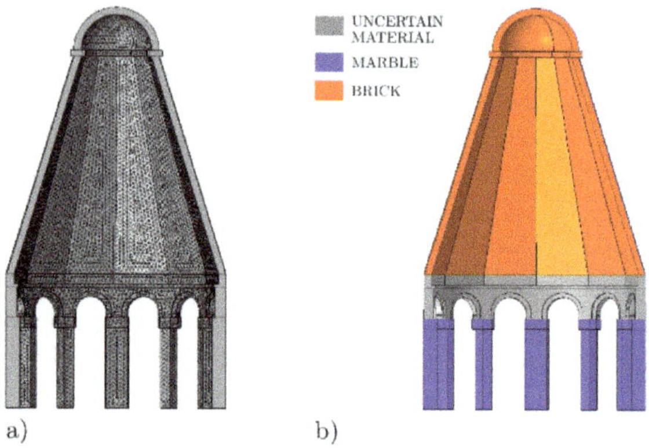

Fig. 2. FE model of the substructure: a) computational mesh; b) assigned materials.

Table 1. Materials assigned to the FE model.

Material	f_m [MPa]	f_t [MPa]	E_m [MPa]	ν	ρ [kg/m^3]
Marble	6	0.5	50000	0.36	2600
Brick	2.6	0.05	1200	0.20	1800

tracks damage evolution by updating the original stiffness tensor Λ_0 to a damaged one Λ_d.

$$\Lambda_d = (1 - d) \cdot \Lambda_0 \tag{1}$$

Damage, ranging from 0 (undamaged material) to 1 (completely damaged material), depends on the maximum positive strains reached at a point and is irreversible, meaning that no stiffness recovery is allowed. It can be expressed as a combination of damage due to tension and compression (indirectly, through Poisson's ratio)

$$d = \alpha_t d_t + \alpha_c d_c \tag{2}$$

where $\alpha_t, \alpha_c \in [0, 1]$ are the combination coefficients

$$\alpha_t = \sum_{i=1}^{3} \frac{\langle \varepsilon_i^{(t)} \rangle \langle \varepsilon_i \rangle}{\tilde{\varepsilon}^2}, \alpha_c = \sum_{i=1}^{3} \frac{\langle \varepsilon_i^{(c)} \rangle \langle \varepsilon_i \rangle}{\tilde{\varepsilon}^2} \tag{3}$$

with εi the principal strain in each point, along direction i; $\varepsilon_i^{(t)}$ and $\varepsilon_i^{(c)}$ the eigenvalues of strain tensors obtained by considering only their tensile and compressive components respectively. Macaulay brackets correspond to the ramp function. $\tilde{\varepsilon}$ is the equivalent strain relative to the local principal strain εi

$$\tilde{\varepsilon} = \sqrt{\sum_i \langle \varepsilon_i \rangle^2} \tag{4}$$

Damage components d_t and d_c are calculated as follows

$$d_t(k) = 1 - \frac{k_0(1 - A_t)}{k} - A_t e^{-B_t(k-k_0)} \tag{5}$$

$$d_c(k) = 1 - \frac{k_0(1 - A_c)}{k} - A_c e^{-B_c(k-k_0)} \tag{6}$$

Scalar parameters A_t, B_t, A_c, B_c can be derived from experimental uniaxial tests on the material (or equivalent simulations). Figure 3 displays the ideal behavior of a specimen subjected to uniaxial tension-compression testing.

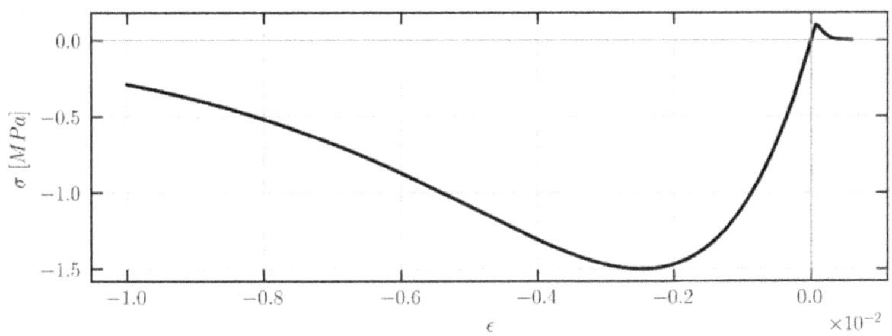

Fig. 3. The stress-strain Mazars' uniaxial constitutive law.

2.3 FE Model of the Substructure

To assess the plausibility of the described configuration, we set up a probabilistic framework for the calibration of the uncertain material, that is the material of the drum of the internal dome. In particular, assuming that the crack pattern observed at the internal edges of the dodecahedral dome originated from the investigated structural configuration, the aim is to identify the parameters of the material leading to a damage pattern consistent with the observations.

The steps of the procedure are summarized as follows: (i) choice of prior probability density functions (pdfs) for the input random variables (RVs); (ii) surrogation of the model response, in terms of crack length; (iii) global sensitivity analysis based on Sobol' indices (SIs); (iv) calibration of the FE model uncertain parameters based on measurements of crack length. In the FE model, cracks are identified by the highly damaged regions having a minimum Young's modulus reduction of 80%.

Probabilistic Description of the Arched Drum's Input Parameters. The first step of the probabilistic setting is the assignment of the prior pdfs to the mechanical properties of Mazars' material, namely f_{m}, f_{t} and E_{m}, based on the engineering judgment and available information. Specifically, the so-called squared stone blocks masonry of [15] is considered, assigning uniform probability distributions for f_{m} and E_{m}, while f_{t} was taken equal to a fraction of the compressive strength. Table 2 reports statistical data of the distribution assumed for the uncertain parameters of the drum's material.

Table 2. Parameters and distributions for SAs and corresponding polynomials for the gPCE.

Parameter	Distribution	Interval	Polynomial
E_m [MPa]	uniform	[2500 5000]	Legendre
f_m [MPa]	uniform	[6.2 10]	Legendre

Intervals were taken from those provided in the [15] with an increased upper limit of 1.2 to consider the good quality of mortar.

f_t was taken equal to 5% of f_m, as already adopted in previous studies [2, 12].

gPCE-Based Surrogate Model Generation and SIs Calculation for Sensitivity Analysis (SA). The uncertainty propagation leveraged on the generation of a gPCE-based response surface [16]. The choice of fourth-degree approximation required a total of 25 deterministic calls of the numerical model, considering a full tensor grid. The estimated PCE coefficients were usefully employed to perform a global sensitivity analysis allowing the computation of SIs [5]. The method, belonging to general SA methods [17], allows the decomposition of the variation of the model output into different contributions caused by different input parameters [18].

Bayesian Inference on the Input Parameters. Bayesian methods were used to solve the inverse problem, namely calibrating the uncertain input parameters with real data from on-site observations, as present for civil engineering problems [19]. In this case, the posterior distribution of input RVs did not have a closed form, and the Markov Chain Monte Carlo method, was employed [20]. The reference value of the observed crack length that was considered for the updating is around 3.90 m.

3 Results and Discussion

Figure 4a shows an example of the resulting damage map obtained for the investigated sample, which highlights a pattern consistent with the observed one at the dome's edges. The first outcome of the solution of the forward problem is provided by the SIs of the random variables f_m and E_m (and indirectly f_t). In particular, SI resulted equal to 0.97 for f_m, and 0.03 for E_m, denoting the key role of f_m and the low relevance of E_m in the response. This outcome is also confirmed by the response surface shown in Fig. 4b, which demonstrates that the greater variation in the length of cracks is mainly due to f_m (and f_t), while the E_m leads to a smaller change in the crack's length.

The solution of the inverse problem is illustrated in Fig. 5 in terms of prior and posterior distributions of f_m and E_m. Overall, we notice a significant reduction of uncertainty associated with the input parameters with respect to priors. The posterior of E_m is characterized by a peak around values of 2600 MPa, while that of f_m is shifted towards the upper limit of the investigated interval, close to 10 MPa, and consequently, to a tensile strength of roughly 0.5 MPa.

Critical discussion about the plausibility of the assumed configuration can be initiated starting from this result. The tensile strength of the drum in this configuration is governed by bed-joint sliding. In fact, the mechanism of interlocking and friction between the stones allows the masonry to transmit significant tension stresses, so developing crosswise tensile strength [21]. Sliding resistance f_{v} can be evaluated through the well-known Mohr-Coulomb failure criterion

$$f_v = f_{v0} + \mu\sigma_n. \tag{7}$$

Here, f_{v0} is the shear strength of masonry in the absence of normal stresses; μ is the friction coefficient, usually taken equal to 0.4 for masonry [22]; σ_{n} is the compression stress. The upper limit of f_{v0} is equal to 0.34 MPa for the squared stone block masonry category, taking into account good quality mortar [15]. This leads to a sliding resistance of 0.41 MPa. Considering the height and length of the stone blocks and their arrangement, we can conclude that this value would lead to an equivalent crosswise tensile strength not so different from the previous value obtained by the updating. This outcome shows that this configuration would be consistent with the presence of the highest mechanical properties for the drum, according to [15]. Therefore, the hypothesis that the dodecahedral dome was built after the construction of the radial arches remains the most likely one.

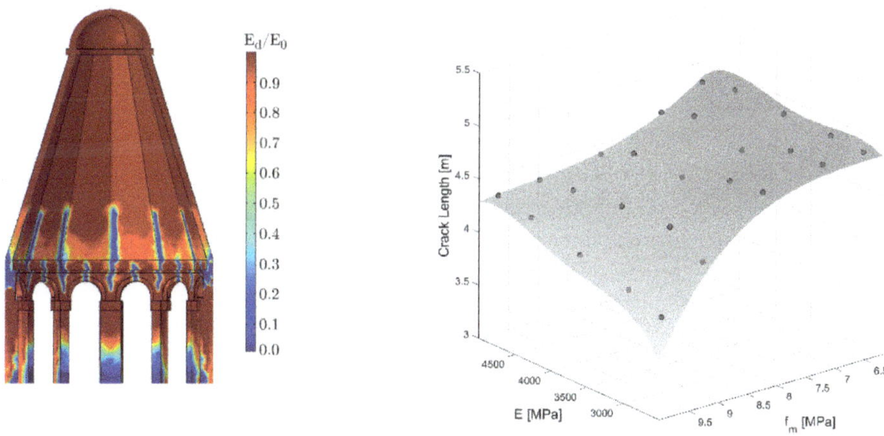

(a) (b)

Fig. 4. Damage at the dome's edges (a) and response surface for the crack length (b).

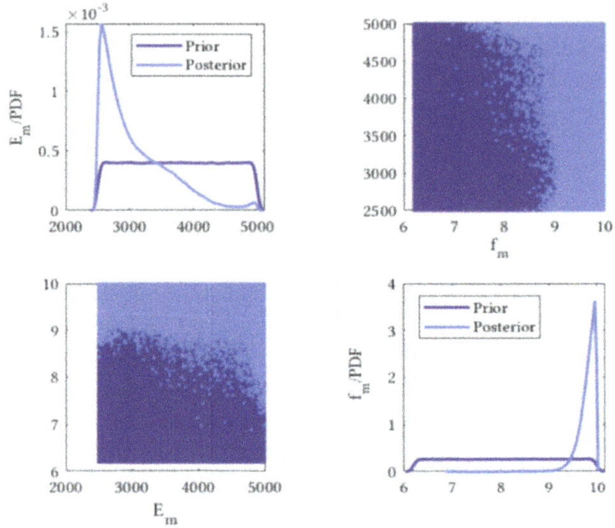

Fig. 5. Results of Bayesian updating, prior and posterior distributions of input parameters in a pairs scatter plot.

4 Conclusions

In this study, the problem of identifying construction sequences of historical monuments is addressed within a probabilistic framework for a relevant case study, the Baptistery of Pisa. Analyzing the historical documentation about the Baptistery and the detected crack pattern on the inner corners of the dome, the knowledge of the configuration assumed by the structure over the centuries is still incomplete.

The non-linear analysis of a substructure of the whole FE model under the self-weight has led to the identification of the material properties of the arched drum, highlighting that the equilibrium on the hypothesized intermediate configuration can be achieved only assuming a significantly high strength value (around 0.5 MPa), hardly attainable for the existing masonry category.

The proposed probabilistic framework resulted in a useful tool to address uncertainties on materials and construction phases of monumental structures. For the case study, the hypothesis that the dodecahedral dome was built after the construction of the radial arches remains the most likely one. Future studies will focus attention on this construction sequence, addressing the problem of FE model calibration.

References

1. Sánchez, A.R., Meli, R., Chávez, M.M.: Structural monitoring of the Mexico City Cathedral (1990–2014). Int. J. Archit. Herit. **10**(2–3), 254–268 (2016)
2. Sykora, M., Cejka, T., Holicky, M., Witzany, J.: Probabilistic model for compressive strength of historic masonry. In: Steenbergen, van Gelder, Miraglia, Vrouwenvelder (eds.) Safety, Reliability and Risk Analysis, pp. 2645–2652. CRC Press, Boca Raton (2013)
3. Roca, P., Cervera, M., Gariup, G., Pela', L.: Structural analysis of masonry historical constructions. Classical and advanced approaches. Arch. Comput. Methods Eng. **17**(3), 299–325 (2010)
4. Marek, P., Brozzetti, J., Gustar, M., Elishakoff, I., Reviewer: Probabilistic assessment of structures using Monte Carlo simulations. Appl. Mech. Rev. **55**(2), B31–B32 (2002)
5. Sudret, B.: Global sensitivity analysis using polynomial chaos expansions. Reliab. Eng. Syst. Saf. **93**(7), 964–979 (2008)
6. Rus, G., Chiachío, J., Chiachío, M.: Logical inference for inverse problems. Inverse Probl. Sci. Eng. **24**(3), 448–464 (2016)
7. Drougkas, A., et al.: Numerical modeling of a church nave wall subjected to differential settlements: soil-structure interaction, time-dependence and sensitivity analysis. Int. J. Archit. Herit. **14**(8), 1221–1238 (2020)
8. Saloustros, S., Pelà, L., Roca, P.: Nonlinear numerical modeling of complex masonry heritage structures considering history-related phenomena in staged construction analysis and material uncertainty in seismic assessment. J. Perform. Constr. Facil. **34**(5), 04020096 (2020)
9. Caleca, A.: La dotta mano: il Battistero di Pisa. Edizioni Bolis Bergamo, Italy (1991)
10. Pierotti, P., Benassi, L.: Deotisalvi: l'architetto pisano del secolo d'oro. Pacini, Pisa, Italy (2001)
11. Franzini, M., Lezzerini, M.: The stones of medieval buildings in Pisa and Lucca (Western Tuscany, Italy). 4 - "Agnano Breccias" from Mt. Pisano. Eur. J. Mineral. **14**(2), 447–451 (2002)
12. Bartelletti, R., Finzi, L., Sanpaolesi, L.: Investigations on the masonry of the leaning tower of Pisa. IABSE reports (1993)
13. Mazars, J.: Application de La Mecanique de l'endommagement au comportement non lineaire et a rupture du beton de structure. Ph.D. thesis, Université Pierre et Marie Curie – CNRS (1984)
14. De Falco, A., Mori, M., Sevieri, G.: Mazars' damage model for masonry structures: a case study of a church in Italy. In: Proceedings of the COMSOL Conference in Lausanne, COMSOL Multiphysics (2018)
15. Italian Public Works Council: Memorandum for application of Italian building code, N7, Italian Official Journal. Istituto Poligrafico e Zecca dello Stato, Rome, Italy (2019)
16. Xiu, D.: Numerical Methods for Stochastic Computations: A Spectral Method Approach. Princeton University Press, Princeton (2010)
17. Saltelli, A., Ratto, M., Tarantola, S., Campolongo, F.: Sensitivity analysis practices: strategies for model-based inference. Reliab. Eng. Syst. Saf. **91**(10), 1109–1125 (2006)
18. Saltelli, A., Sobol', I.M.: About the use of rank transformation in sensitivity analysis of model output. Reliab. Eng. Syst. Saf. **50**(3), 225–239 (1995)
19. Landi, F., Marsili, F., Friedman, N., Croce, P.: GPCE-based stochastic inverse methods: a benchmark study from a civil engineer's perspective. Infrastructures **6**(11), 158 (2021)
20. Tierney, L.: Markov chains for exploring posterior distributions. Ann. Stat. **22**(4), 1701–1762 (1994)
21. Chen, S., Bagi, K.: Crosswise tensile resistance of masonry patterns due to contact friction. Proc. R. Soc. A Math. Phys. Eng. Sci. **476**(2240), 20200439 (2020)
22. Borri, A., Cangi, G., De Maria, A.: Caratterizzazione meccanica delle murature (anche alla luce del recente sisma in Emilia) e interpretazione delle prove sperimentali a taglio. In: Proceedings of the XV ANIDIS Conference, Padova, Italy, 30 June–4 July 2013 (2013)

Tail Length for Estimating Traffic Load Effects

Frederik P. Bakker(✉), Roman Lenner, and Nico de Koker

Department of Civil Engineering, University of Stellenbosch, Stellenbosch, South Africa
18302866@sun.ac.za

Abstract. Block maximum extreme value theory is commonly used for estimating lifetime maximum road traffic load effects. These effects originate from various loading situations, implying that block maxima may be drawn from different populations with distinct distributions. This poses a challenge when fitting extreme value distributions, as the events should be independent and identically distributed. To address this issue and isolate the critical load effect component, it has become standard practice to employ a censored fit to the upper tail of observed block maxima. The length of this tail significantly influences extrapolated load effects. However, determining the appropriate tail length remains uncertain. In this study, we investigate fitting to different tail lengths by simulating mixtures of extreme value distributions. Our findings show that an excessively short tail increases the statistical uncertainty of the fit, while an overly long tail risks introducing bias by incorporating events from non-critical load effect components. Consequently, the optimal tail length depends on both the sample size and the relative distributions of the different load effect components. This serves to discourage the use of certain accepted tail lengths without an assessment of the specific situation.

Keywords: Traffic Load Effects · Extreme Value Theory · Tail Length

1 Background

To ensure a road traffic bridge is acceptably safe, it is necessary to establish a suitable lifetime maximum traffic load. While bridges are designed to last several decades, the data collected from Weigh In Motion sensors usually covers shorter periods, often just a few months or years. Consequently, estimating the maximum load a bridge may experience throughout its lifespan involves extrapolation beyond the available data range. In such cases, the focus is typically on characterising the most severe load effects. A common method involves fitting a censored extreme value distribution to the highest block maxima, as outlined by O'Brien et al. [1].

1.1 Extreme Value Theory

The distribution of block maxima $F(x)$ approach one of three limiting forms as the number of events per block increases. These asymptotic forms can be represented by the

J. C. Matos et al. (Eds.): IPW 2024, LNCE 494, pp. 502–511, 2024.
https://doi.org/10.1007/978-3-031-60271-9_47

Generalised Extreme Value (GEV) distribution [2];

$$F(x) = \begin{cases} \exp\left[-\exp\left(-\frac{x-u}{b}\right)\right] & k = 0 \\ \exp\left[-\left(1 - k\frac{x-u}{b}\right)^{\frac{1}{k}}\right] & k \neq 0. \end{cases} \tag{1}$$

The GEV distribution has a location, scale and shape parameter denoted by u, b and k, respectively. The k parameter corresponds to the asymptotic form with $k > 0$ indicating the Reverse Weibull distribution, $k = 0$ the Gumbel distribution, and $k < 0$ the Fréchet distribution.

The asymptotic form is dependent on the tail behavior of the parent distribution. The block maxima from a parent with an exponential tail, such as the normal distribution, converge to a Gumbel distribution. Maxima from a parent that decays with a polynomial tail will converge on a Fréchet distribution. A bounded parent distribution yields Reverse Weibull distributed extremes.

1.2 Traffic Load Effects

Road Traffic load effect data appear to follow a normal distribution [1, 3], which has led to its use by some authors [4, 5]. However, many of these observations may not be relevant when characterising the lifetime maximum. Therefore, instead of using all the observations, one can consider the maxima from a suitable block and employ extreme value theory.

If the distribution of the block maxima $F(x)$ is defined, the value of a severe load effect x corresponding to a return period of r blocks $x_{|r}$, such that the non-exceedance probability is given by $1 - 1/r$, can be determined as

$$x_{|r} = F^{-1}\left(1 - \frac{1}{r}\right). \tag{2}$$

An estimate $\widehat{x_{|r}}$ can therefore be found by selecting an appropriate extreme value distribution F and fitting to a sample of n maximums as described by Castillo [2].

Traffic block maxima have been shown to correspond to either the Reverse Weibull or the Gumbel distributions [6–8]. The choice of the Gumbel distribution is substantiated by its representation of maximums drawn from a normal population, aligning well with observations of traffic load effects [9]. The upper limit on x implied by the Reverse Weibull distribution has been justified by the physical upper bound on the load effects that trucks can exert on a bridge [1].

The application of the Fréchet distribution, however, lacks any justification and results in unrealistic extrapolations [1]. Despite this, observations of traffic load effect maxima often appear to be Fréchet [3, 10, 11].

This apparent Fréchet form may arise from a mixture of load effect components with different distributions [7, 12]. This is illustrated in Fig. 1 where artificially generated samples, made up of the maxima of two Gumbel distributions appear to be Fréchet.

Similarly, the block maxima of traffic load effects may exhibit a Fréchet form because they can be made up of a mixture of different types of events - for example, different types of vehicles (low-loader vs mobile cranes), various axle configurations (5-axle vs 7-axle vehicles), or vehicle convoys (single vehicle vs two vehicles) [7, 10]. These different events result in different load effect components that are unlikely to be identically distributed [10]. In this case, fitting $F(x)$ to the observed block maxima violates the assumption that events are identically distributed. The fit could therefore be dominated by observations that are of little interest and extrapolating could introduce significant error into the analysis.

This problem is not unique to traffic load effects and has been an issue in various cases where maxima could be generated from a variety of different mechanisms. When faced with a mixture of maximum wind speeds generated from N storm types $x = \max\{x_1, x_2, \ldots, x_N\}$, Gomes and Vickery [13] considered the maximum of each component separately, the distribution of x is then given by

$$G(x) = \prod_{i=1}^{N} F_i(x). \tag{3}$$

Each of the N components are thus accounted for using its own distribution function, which is estimated by identifying and fitting to the block maximums x_i of the relevant component.

Caprani et al. [7] applied this approach to traffic load effects by identifying components based on the number of vehicles involved in a loading event. However, events with the same number of vehicles could still consist of multiple components [6], which may not always be easily identified [11, 14]. In such cases, the critical load effects component must be found using a different approach.

1.3 Fitting to the Tail

When extrapolating, the load effects in the tail are of greatest interest because these observations are more likely to belong to the critical load effect component. This has led various authors [3, 15–17] to employ a censored fit to the upper tail, as suggested by Castillo [2]. This can be used as an alternative to, or in conjunction with, Eq. (3) when further components cannot be identified.

To use a censored fit, the tail length t must be selected, which O'Connor [18] and Van Der Spuy and Lenner [19] showed to have a significant impact on the fitted distribution.

O'Brien et al. [10] stated that the tail length can be chosen by engineering judgement when the cumulative distribution is seen to change at a particular probability level. This may not always be practical in cases where load effects from multiple stations, spans, and lanes need to be considered [3, 17] and is not ideal due to the subjectivity of relying on the judgement of the researcher.

Alternatively, O'Brien et al. [10] suggest using $t = 2\sqrt{n}$ or $t = 0.3n$, based on recommendations by Castillo [2] and Enright [16] respectively. The reason behind Castillo [2]'s choice of $t = 2\sqrt{n}$ is unclear [20], but in a paper discussing a hypothesis test to find the asymptotic form, where the tail is also defined as the upper $2\sqrt{n}$, Castillo et al. [21] explicitly states that there is no specific reason for the choice of $2\sqrt{n}$. It should also be noted that Castillo [2] was advocating for a fit to the tail in the case of incomplete convergence to the asymptotic extreme value distribution and not for finding the critical component, as required to determine critical traffic load effects.

Enright [16] questioned the choice of $2\sqrt{n}$ because of the small quantity of data used, especially as n increases. He performed long-run simulations and compared the simulation results with extrapolations based on fitting to the upper $2\sqrt{n}$ or $0.3n$ from subsets of the simulated data. He found little difference in the accuracy of the extrapolated results and concluded that fitting to the upper $0.3n$ was more appropriate. While this may have been the case for the situation that was analysed, it is unclear how appropriate fitting to a fairly large percentage of the upper tail would be as the quantity of load effects from other components in the tail increases.

2 Performance of Different Tail Lengths

2.1 Demonstration

To explore some of the issues involved with tail fitting in mixed populations, two mixtures were generated. In the first mixture, two Gumbel distributions, $F(x_1|u_1 = 100, b_1 = 6)$ and $F(x_2|u_2 = 100, b_2 = 2)$, were used. The mixture was formed by sampling $n = 1000$ values from each distribution and considering only the maximum $x = \max\{x_1, x_2\}$ of each realisation. This was repeated with u_2 increased to 110, to create the second mixture. While for both mixtures the x_1 component will become critical as x increases due to its higher variability, the different u_2 values mean that the quantity of the non-critical component in the tail varies between the mixtures.

The Gumbel distribution was then fitted to the full sample, the upper $2\sqrt{n}$, and the upper $0.3n$ of x using maximum likelihood estimation. The two mixtures and the various fits are shown in Fig. 1a and Fig. 1b, where both mixtures appear to exhibit a Fréchet form.

These Figures show why fitting to the tail is reasonable; the fits to the full sample do not capture the critical component and are inordinately affected by the non-critical x_2 component. The impact of this component also increases as u_2 increases because more observations come from the non-critical component.

For the first mixture, with $u_2 = 100$, Fig. 1a shows that fitting to the upper $0.3n$ resulted in the best representation of the critical component because enough data was used and most of this data was drawn from the critical component. Fitting to the upper $2\sqrt{n}$, on the other hand, used a smaller proportion of the data resulting in a poorer representation of the critical component. This was not the case in the second mixture, with $u_2 = 110$, shown in Fig. 1b, where using the upper $2\sqrt{n}$ resulted in a better representation of the critical component than the upper $0.3n$, because the upper $0.3n$ contained more data from the non-critical component.

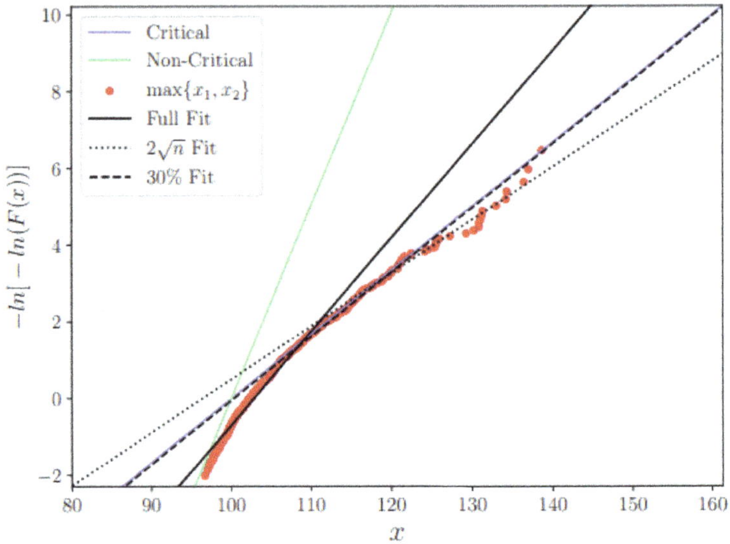

Fig. 1a. Fitting to the tail of a mixture with $u_2 = 100$.

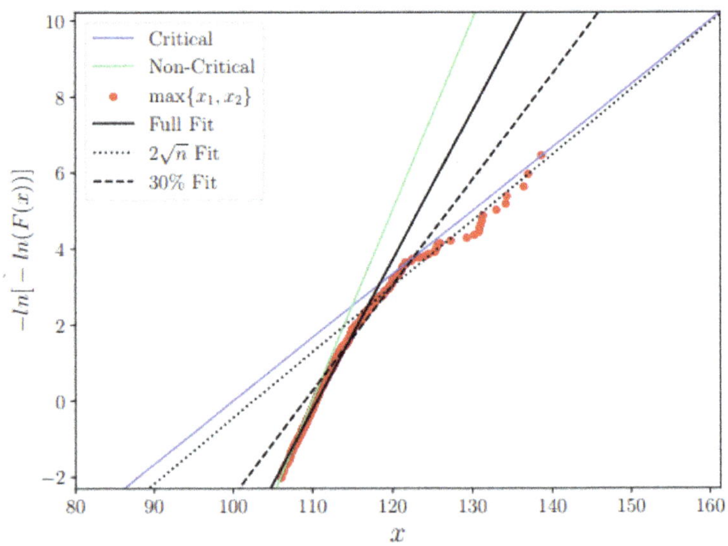

Fig. 1b. Fitting to the tail of a mixture with $u_2 = 110$.

2.2 Determination of Root Mean Square Error

To further study the performance of certain tail lengths for a wide range of scenarios, we developed a method to determine the error versus t implied for a given mixture of two Gumbel Distributions.

Two samples of size n were drawn from Gumbel distributions $F(x_1|u_1, b_1)$ and $F(x_2|u_2, b_2)$ where $b_1 > b_2$ so that x_1 is the critical component for extrapolation. Then for each possible tail length $t = \{2, 3, \ldots, n\}$ the value of an extrapolated value $x_{|r}$ was estimated by fitting a Gumbel distribution to the upper t values of the $\max\{x_1, x_2\}$. This gives a set of extrapolated values for each tail length $\{\hat{x}_{|r,t=2}, \hat{x}_{|r,t=3}, \ldots, \hat{x}_{|r,t=n}\}$. This was repeated S times and the Root Mean Square Error (RMSE) for each t was calculated.

This yields a relationship between the tail length and the error in estimating an extrapolated value using a mixture of two different distributions. It was found that $S = 10\,000$ was sufficient to obtain stable RMSE values.

To make these simulations relevant to typical cases encountered when analysing actual traffic load effects, $n = \{75, 250, 750, 2500\}$ were considered. These n values relate to a record of between 3 months and 10 years if daily maximums were observed and a year consists of 250 economic days [1]. This would mean that a 100-year return period is given by $r = 250\,\text{days/year} \times 100\,\text{years} = 25\,000\,\text{days}$. Other high return periods were tested, and it was found that, while the RMSE inevitably increased for higher return periods, the general trend in RMSE was consistent. Therefore, a return period of $r = 25\,000$ was considered in all cases.

2.3 Tail Length Errors

Varying Distribution Parameters. The results of this procedure when the location parameter of the non-critical component u_2 was varied are shown in Fig. 2, which also indicates what constitutes a tail length of $2\sqrt{n}$ and $0.3n$.

In all cases when t was small, a similar RMSE was obtained. The RMSEs then diverge from one another as t increases, and the likelihood that $x_2 = \max\{x_1, x_2\}$ differs.

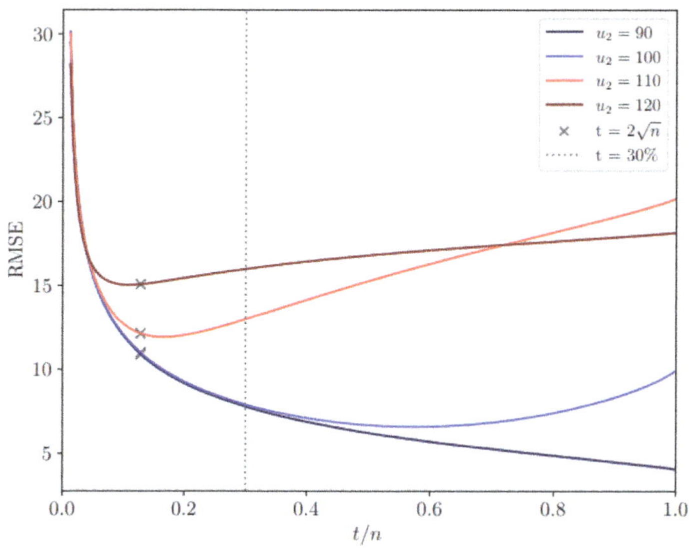

Fig. 2. Tail Fit RMSE, $u_1 = 100$, $b_1 = 6$, $b_2 = 2$, $n = 250$, $r = 25\,000$(100-years).

When $u_2 = 90$, for example, the likelihood that $x_2 = \max\{x_1, x_2\}$ was low for any t, and so a large tail was acceptable because more data could be used without risking fitting to the non-critical component. Conversely, when $u_2 = 120$, the RMSE decreased with t until a reasonable quantity of data was used to obtain a decent fit, then it increased as the likelihood of fitting to data from the non-critical component increased.

It is interesting to compare the RMSE for $u_2 = 110$ and $u_2 = 120$. In this case, while both exhibited a similar trend with increasing t, the $\text{RMSE}|u_2 = 110 < \text{RMSE}|u_2 = 120$ for smaller tail lengths but $\text{RMSE}|u_2 = 110 > \text{RMSE}|u_2 = 120$ for larger tail lengths. This is because the underestimation of the shape parameter was offset by the overestimation of the location parameter as more data from the non-critical component was used.

When b_2 was varied, then the closer b_2 was to b_1, the less beneficial a small tail length was. This is because when t was large, the likelihood that non-critical values were used was offset by b_2 being nearer to b_1, improving estimates of b_1. Therefore, $t = 2\sqrt{n}$ performed better than $t = 0.3n$ when b_2 was more different from b_1.

Varying Sample Size. The performance of different tail lengths was also assessed for different sample sizes; the results can be seen in Fig. 3 and Fig. 4 where $u_2 = 110$ and $u_2 = 90$ respectively.

When $u_2 = 110$, the likelihood that $x_2 = \max\{x_1, x_2\}$ is fairly high. In this case, using a smaller t for larger sample sizes results in a lower RMSE because there is still enough data in the tail to obtain fairly good estimates without risking fitting to data from the non-critical component. However, as the sample size decreases, using a larger proportion of the data is better because fitting to too few values does not produce good estimates. Thus, because $t = 2\sqrt{n}$ uses a decreasing proportion of the sample with increasing n, it performed well, resulting in a RMSE near the minimum in all cases.

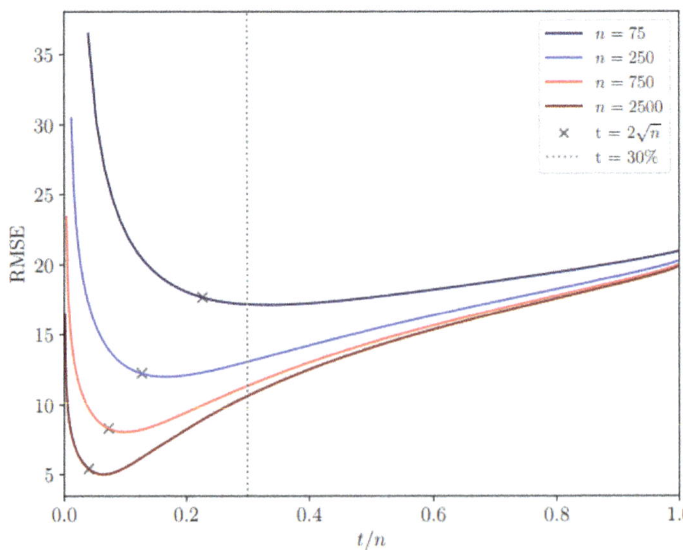

Fig. 3. Tail Fit RMSE, $u_1 = 100$, $b_1 = 6$, $u_2 = 110$, $b_2 = 2$, $r = 25\,000$(100-years).

When different sample sizes were examined with $u_2 = 90$, as shown in Fig. 4, a larger tail size consistently led to a lower RMSE. This is due to the reduced likelihood that $x_2 = \max\{x_1, x_2\}$. Consequently, using $t = 0.3n$ is better than $t = 2\sqrt{n}$, although the disparity in their RMSE reduces for larger sample sizes.

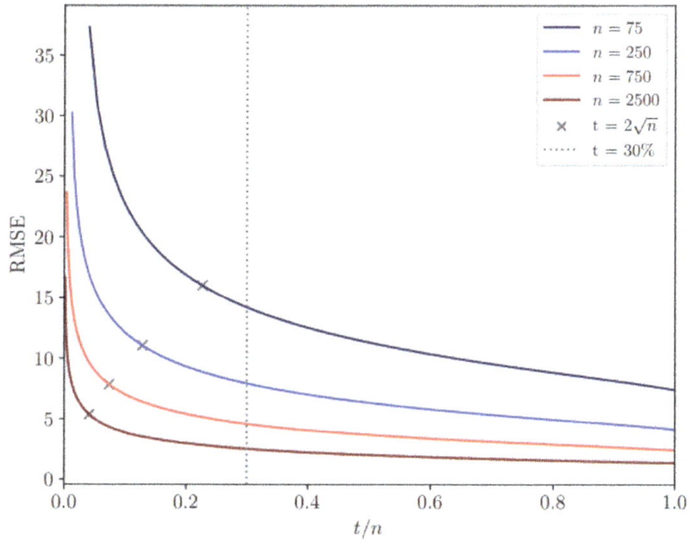

Fig. 4. Tail Fit RMSE, $u_1 = 100$, $b_1 = 6$, $u_2 = 90$, $b_2 = 2$, $r = 25\,000$(100-years).

3 Discussion

The simulations were only performed for scenarios involving a mixture of two Gumbel distributions, which cannot capture the complexity of real-world situations where maximum traffic load effects could be composed of many components that may not be Gumbel distributed. However, the findings from these simulations provide some insights into tail length selection.

Fitting to the upper $2\sqrt{n}$ appears to be sensible when there is a strong indication that load effects in the upper tail originate from sources with significantly different distributions. Using only the upper $2\sqrt{n}$ ensures that the quantity of data used is minimised while still being sufficiently large that fits are not dominated by statistical uncertainty.

If there is less risk that the upper tail of the block maxima originates from a mixture of load effect components, the upper $2\sqrt{n}$ is not a good choice for tail length because of the large proportion of relevant data that is not used. Instead, using the upper $0.3n$ could be considered. This utilizes a larger proportion of the data but still censors the fit to reduce the impact of non-critical load effect components.

Enright [16] found little difference in using the upper $2\sqrt{n}$ versus the upper $0.3n$. This may have been because the particular mixture meant that the optimal tail length was between $2\sqrt{n}$ and $0.3n$ or that the number of simulations he performed were insufficient to differentiate between them.

The simulations we performed demonstrate that selecting an appropriate tail length depends on whether the upper tail of the block maxima contains data from a mixture of load effect components. Determining whether this is the case is difficult in practice because one might only observe $\max\{x_1, x_2, \ldots, x_N\}$. Therefore, the composition of the upper tail is unknown.

It is likely that the engineering judgement approach suggested by O'Brien et al. [10] would consistently yield better results than any particular tail length function, which can only consider the sample size. If, however, the components can be identified, then applying the composite approach by Caprani et al. [7] would probably give the best estimates.

4 Conclusion

A set of block maximum traffic load effects is generally not identically distributed. Despite this, the simplicity of fitting an extreme value distribution to the maxima makes it an attractive option for predicting severe traffic load effects.

It was found that, if using the traditional approach of fitting only to the upper tail, care should be exercised when selecting a tail length. Enough data should be used to obtain fairly reliable estimates without risking fitting to data from non-critical load effect components that would bias the fit. A tail length of $2\sqrt{n}$ appears to be a reasonable choice if the minimum quantity of data is to be used but is not optimal when the upper tail does not contain many values from non-critical load effect components.

Determining whether this is the case is difficult because it depends on the characteristics of the different load effect components, which are unknown and differ depending on the situation. Therefore, no tail length function can consistently perform well in all situations and a reasonable choice should be made on a case-by-case basis, introducing a degree of subjectivity to the analysis.

References

1. O'Brien, E., Nowak, A., Caprani, C.: Bridge Traffic Loading: From Research to Practice. CRC Press, Boca Raton (2021)
2. Castillo, E.: Extreme Value Theory in Engineering. Academic Press, Boston (1988)
3. Van Der Spuy, P.F.: Derivation of a traffic load model for the structural design of highway bridges in South Africa, Ph.D. thesis, Stellenbosch University, Stellenbosch (2020)
4. Anitori, G., Casas, J.R., Ghosn, M.: Wim-based live-load model for advanced analysis of simply supported short-and medium-span highway bridges. J. Bridge Eng. **22**, 04017062 (2017)
5. Nowak, A.S.: Live load model for highway bridges. Struct. Saf. **13**, 53–66 (1993)
6. Caprani, C.C., O'Brien, E.J.: Statistical computation for extreme bridge traffic load effects (2006)
7. Caprani, C.C., O'Brien, E.J., McLachlan, G.J.: Characteristic traffic load effects from a mixture of loading events on short to medium span bridges. Struct. Saf. **30**, 394–404 (2008)

8. O'Connor, A., O'Brien, E.J.: Traffic load modelling and factors influencing the accuracy of predicted extremes. Can. J. Civ. Eng. **32**, 270–278 (2005)
9. Zhou, X.Y.: Statistical analysis of traffic loads and traffic load effects on bridges, Ph.D. thesis, Universite Paris-Est (2013)
10. O'Brien, E.J., et al.: A review of probabilistic methods of assessment of load effects in bridges. Struct. Saf. **53**, 44–56 (2015)
11. Stap, T.: Critical load effects on bridges using heavy vehicle weigh-in-motion data (2022)
12. Palutikof, J.P., Brabson, B., Lister, D.H., Adcock, S.: A review of methods to calculate extreme wind speeds. Meteorol. Appl. **6**, 119–132 (1999)
13. Gomes, L., Vickery, B.: Extreme wind speeds in mixed wind climates. J. Wind Eng. Ind. Aerodyn. **2**, 331–344 (1978)
14. Dai, B., Xia, Y., Li, Q.: An extreme value prediction method based on clustering algorithm. Reliab. Eng. Syst. Saf. **222**, 108442 (2022)
15. O'Brien, E.J., Enright, B., Getachew, A.: Importance of the tail in truck weight modeling for bridge assessment. J. Bridge Eng. **15**, 210–213 (2010)
16. Enright, B.: Simulation of traffic loading on highway bridges (2010)
17. Leahy, C., O'Brien, E.J., Enright, B., Hajializadeh, D.: Review of HL-93 bridge traffic load model using an extensive WIM database. J. Bridge Eng. **20**, 04014115 (2015)
18. O'Connor, A.J.: Probabilistic traffic load modelling for highway bridges, Ph.D. thesis, Trinity College Dublin (2001)
19. Van Der Spuy, P.F., Lenner, R.: Establishing the effect of the tail length on extrapolation when fitted to WIM data. In: Proceedings of the International Conference on Sustainable Materials, Systems and Structures (SMSS2019) Challenges in Design and Management of Structures, Rovinj, Croatia, pp. 1–6 (2019)
20. O'Brien, E.J., O'Connor, A., Arrigan, J.: Procedures for calibrating Eurocode traffic load model 1 for national conditions. In: Proceedings of the 6th International Conference on Bridge Maintenance, Safety and Management, pp. 8–12. Springer, Heidelberg (2012)
21. Castillo, E., Galambos, J., Sarabia, J.M.: The selection of the domain of attraction of an extreme value distribution from a set of data. In: Hüsler, J., Reiss, R.D. (eds.) Extreme Value Theory. LNS, vol. 51, pp. 181–190. Springer, New York. (1989).https://doi.org/10.1007/978-1-4612-3634-4_16

Uncertainty Quantification of Soil-Structure Interaction in Tunnel Linings by Polynomial Chaos Expansion

Lukáš Novák[1](✉), Angelos Gakis[2], Michael Křížek[1], Drahomír Novák[1], and Panagiotis Spyridis[3]

[1] Brno University of Technology, Brno, Czech Republic
novak.l@fce.vutbr.com
[2] Dr. Sauer & Partners, Salzburg, Austria
[3] University of Rostock, Rostock, Germany

Abstract. The paper is focused on uncertainty quantification of soil-structure interaction in tunnel linings using a surrogate model in form of Polynomial Chaos Expansion (PCE). Tunnel design is a complex and complicated task since it is strongly associated with a great number of load and material uncertainties. Moreover, modelling the soil-structure interaction multiplies the complexity and nonlinearity of a tunnel engineering problems. In order to handle such uncertainties, finite element method with random input variables has proven to be a very accurate tool. The probabilistic analysis is typically performed by Monte Carlo simulation (MC), simulating uncertainties according to their complete probability distributions and statistical correlations. The computational burden of MC represents the main obstacle to its use in complex numerical models and it is therefore not practical for industrial applications. The solution can be an efficient approximation of the original mathematical model by computationally efficient analytical function – a surrogate model. In this study, the surrogate model in form of PCE is utilized, allowing for analytical post-processing (statistical and sensitivity analysis). Uncertainty quantification is focused on estimation of spatial variability of internal forces caused by the soil-structure interaction.

Keywords: Uncertainty Quantification · Polynomial Chaos Expansion · Spatial Variability · Tunnel Linings · Concrete Structures

1 Introduction

With the rapid advancement of a computer technology and the growing demand for cost-effective and complex design solutions in the construction industry, numerical modeling has emerged as the preferred approach in various structural and heavy civil engineering domains, including tunnel engineering. Finite element (FE) methods have become essential tools, proving their efficacy in analyzing complex underground structures by advanced constitutive materials and detailed geometrical simulations. Additionally, numerical modeling facilitates the analysis of intricate geometries, encompassing non-circular cross-sections, unique interfaces, and junctions.

J. C. Matos et al. (Eds.): IPW 2024, LNCE 494, pp. 512–519, 2024.
https://doi.org/10.1007/978-3-031-60271-9_48

In tunnel design and assessment, uncertainties arising from knowledge gaps, randomness in soil/rock performance, structure properties and their interactions, or mathematical indeterminism necessitate the consideration of safety formats and risk-based decisions. The following four analysis methodologies in literature and design standards quantify these uncertainties in terms of failure probabilities or structural reliability measures:

- Level I: Uncertain parameters are modelled by a single nominal value, e.g. characteristic values with partial safety factor design.
- Level II: Uncertain parameters are implicitly modelled with a probability distribution described by two characteristics, e.g. mean and variance or characteristic value, i.e.. Semi-probabilistic methods.
- Level III: Uncertain quantities are modelled by their distribution functions, and correlations, leading to a calculated failure probability, i.e. full probabilistic Monte Carlo simulations.
- Risk-based methods: The consequences of failure are also accounted as a design criterion.

Despite the prevalence of Level I methods amongst practitioners, more advanced approaches are gaining favor due to their efficiency and rationalization in engineering verifications. Nevertheless, risk-based and probabilistic methods face challenges in their implementation due to their reliance on multidisciplinary expertise and computational power. "Level II" methods are advantageous since they offer a balance between an accuracy and an efficiency as was shown in the previous work of the authors [1].

This study continues in exploration of the possibilities of advanced analysis methods in tunnelling, particularly it is focused on the Level III method. The full probabilistic approach of costly mathematical models is unfortunately not feasible in practical applications and thus we investigate possibilities of surrogate modeling for uncertainty quantification instead of crude Monte Carlo simulation. The paper initially introduces theoretical background of polynomial chaos expansion and formal model descriptions. Comparative calculations are then demonstrated using a plane-strain FE model from a realistic tunnel project in soft soil and obtained results based on surrogate modelling are compared to a MC with original FE model as a reference solution.

2 Polynomial Chaos Expansion

Considering a model response $Y = f(\mathbf{X})$ to be a random variable referenced as a quantity of interest (QoI) with finite variance σ^2, the polynomial chaos expansion of QoI is in the following form [2]:

$$Y = f(\mathbf{X}) = \sum_{\alpha \in N^M} \beta_\alpha \psi_\alpha(\boldsymbol{\xi}), \tag{1}$$

where M is the number of input random variables, β_α are unknown deterministic coefficients and ψ_α are multivariate basis functions orthonormal with respect to the joint probability density function (PDF) of $\boldsymbol{\xi}$. The basis functions must be selected in dependence to probability distributions of components of the input random vector $\mathbf{X}$ which must be transformed to the associated standardized variables ξ [3].

Further, it is necessary to truncate PCE to a finite number of terms P. Truncated set of basis functions $\mathcal{A}^{M,p}$ is typically dependent on given maximal polynomial order p and M as follows:

$$\mathcal{A}^{M,p} = \{\alpha \in N^M : |\alpha| = \sum_{i=1}^{M} \alpha_i \leq p\} \tag{2}$$

Deterministic coefficients β_α can be obtained by intrusive and non-intrusive approaches. Non-intrusive methods utilize the original mathematical model (e.g. FE) as a black-box, which allows for their easy applications in combination with commercial software and thus is typically preferred for industrial applications. The most popular type of the non-intrusive approach is based on a simple linear regression.

The original mathematical model must be evaluated to obtain a vector of results corresponding to generated sample points. Once the basis functions are created and experimental design (ED) is prepared, PCE coefficients can be estimated by ordinary least square (OLS) regression method. Unfortunately, a truncated PCE solved by OLS is not highly computationally efficient and cannot be employed for practical examples with large number of input random variables due to the *curse of dimensionality*. The solution is an additional reduction of the truncated basis set by any model-selection algorithm such as Least Angle Regression (LAR) [4]. LAR automatically detects the most important basis functions for given ED and create the so called sparse PCE. For further reduction of computational cost, it is beneficial to employ advanced sampling schemes for a sequential enrichment of ED [5]. In this paper, UQPy implementation [6] of an algorithm for construction of a non-intrusive sparse PCE based on LAR is employed for numerical examples. Powerful post-processing is main advantage of PCE over to other surrogate models. First of all, thanks to the explicit form of PCE it is possible to obtain a leave-one-out error Q^2 analytically without additional computational demands which can be further used for an adaptive construction of PCE approximation. Besides the analytical derivation of Q^2, it is possible to derive also the first four statistical moments directly from deterministic coefficients [7]. Estimated statistical moments can be further utilized for a global sensitivity analysis [8].

3 Numerical Model

3.1 Model Set-Up

A 2D plane-strain FE model (see Fig. 1) has been developed and analyzed using Abaqus FE code, featuring the excavation of a sprayed concrete lined tunnel with a cross-sectional area of 50 m^2 in a soil medium at a depth of 20 m to the tunnel axis. The tunnel is about 7 m high and 9.5 m wide and consists of 4 radii.

The analyses included three steps:

- a geostatic step where the initial soil pressures were estimated from the density of the soil material and the specified coefficient of lateral earth pressure,

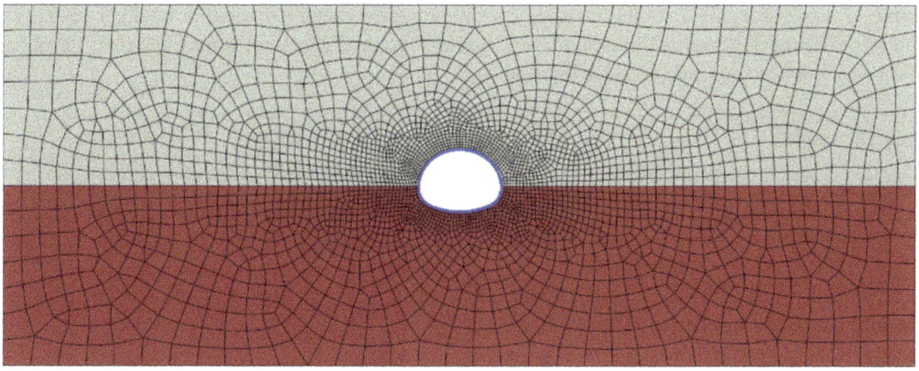

Fig. 1. Mesh of the finite element model.

- a relaxation step simulating the stress relief and 3D arching effects during excavation. This has been realized through the stiffness reduction method, where the soil within the excavation was assigned lower parameters (Young's modulus, Poisson's ratio) allowing the soil to deform to a new equilibrium before the lining installation (refer to details of this modelling approach and methodology in [9, 10],
- the undrained excavation step, where the elements inside the excavation were removed in a single step and the tunnel support was activated.

The FE mesh included 3740 four-node plane-strain elements for the soil (type CPE4R) and 86 beam elements (type B21) for the lining. The soil elements were assigned a non-linear Morh-Coulomb plasticity. The sprayed concrete lining elements were modelled with a linear-elastic material model. To account for the stiffer soil response during unloading and reloading stress paths, the soil elements below the tunnel axis were assigned a three times higher stiffness.

3.2 Random Input Parameters

The geological model and the range of parameters used have been selected based on an extensive survey of available geotechnical information in [11, 12] and they are anticipated to be representative of a typical shallow tunneling project in urban environment. In summary we assumed the following uncertain parameters: Young's modulus of concrete E_c [13] coefficient of lateral earth pressure K_0, undrained shear strength S_u, ratio between the Young's modulus of the soil and S_u and relaxation factor κ, while all other parameters are kept as deterministic. An overview of the uncertain parameters used is given in Table 1. The concrete liner was modelled with a constant thickness of 300 mm and a Poisson's ratio of $\nu = 0.2$.

Table 1. Input stochastic model parameters, indicating the mean values, the standard deviations and (in parenthesis) the coefficients of variation for Normal and Lognormal variables or upper and lower bounds for Uniform distributions.

Input Variable	Mean [units]	Std (CoV) [bounds]	Distribution
Young's modulus E_c	13 [GPa]	1 (0.08)	Lognormal
E_s/S_u ratio	1000	150 (0.15)	Normal
Undrained shear strength s_u	250 [kPa]	50 (0.2)	Normal
Lateral stress coefficient K_o	1 [–]	[0.4, 1.6]	Uniform
Relaxation factor κ	0.5 [–]	[0.1, 0.9]	Uniform

3.3 Methodology

The FE model of the concrete lining consists of 86 finite elements and three QoIs in each element (internal forces N, V and bending moments M) are approximated by separate individual PCEs, i.e. totally 86×3 surrogate models were constructed and further utilized for UQ of the whole structure. Each PCE is constructed by the adaptive sparse algorithm implemented in UQPy Python package and it is based on ED containing 50 samples generated by Crude Monte Carlo. The maximum polynomial order is automatically selected as $p \in [3, 10]$ and a sparse basis set is identified by LAR. Polynomial basis consists of Legendre and Hermite polynomials associated to Uniform and Normal random variables respectively, input Young's modulus is thus first transformed to a Gaussian space by an iso-probabilistic transformation. Note that a computational cost of construction of a single PCE is less than a second and thus it is possible to construct surrogate models for all finite elements in a few minutes and further use them for UQ without any additional computational cost. In this study, the main task is a statistical analysis – an estimation of local mean values $\mathbb{E}[\mathrm{Y}|x]$and standard deviations $\sigma[\mathrm{Y}|x]$ of internal forces (normal and shear) and bending moments. Design and assessment of structures is typically based on extreme quantiles estimated from mean values, variances, and prescribed reliability indices dependent on specific types of structures [14]. Therefore, for the sake of generality, we estimate also quantiles corresponding to $\pm 3\sigma$ in the example.

4 Numerical Results and Discussion

The numerical results are depicted in Fig. 2. The statistical moments and quantiles of internal forces obtained from PCE (right column) are compared to a reference solution obtained by MC with 1000 simulations (middle column) and also an estimation from ED generated by MC and utilized for a construction of PCE containing 50 samples (left column). As can be seen from identical shapes of internal forces shown in Fig. 2, PCE leads to consistent results. Detailed insight to numerical results can be seen in Fig. 3. It is clear that although a direct estimation from 50 samples leads to a significant under-estimation of mean values and variance (clearly visible in Fig. 3 for normal forces), PCE based on the identical samples, and thus identical computational cost, significantly improves the accuracy of an estimation. This is caused by the fact, that PCE considers the whole probability distributions of input variables through basis functions instead of limited point-wise information from samples. Errors can be clearly seen in Fig. 3 showing relative absolute errors between estimations and the reference solution. The most significant improvement can be seen in both local mean values and standard deviations of normal forces, while the least significant improvement is for shear forces. However, extreme values of relative errors are associated to locations with shear forces close to zero, and thus such errors are not visible in Fig. 2.

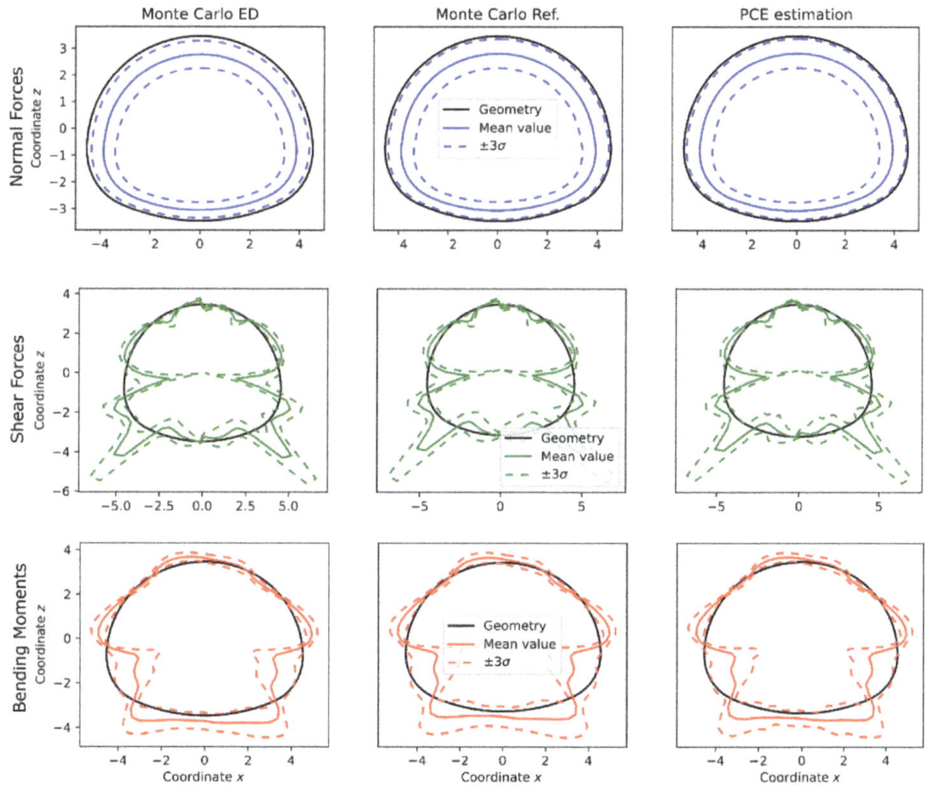

Fig. 2. Geometry of the concrete liner (black) together with estimations of local mean values (solid) and extreme quantiles (dashed) of normal forces (blue), shear forces (green) and bending moments (red) by PCE (right), Monte Carlo with 1000 samples (middle) and 50 samples (left).

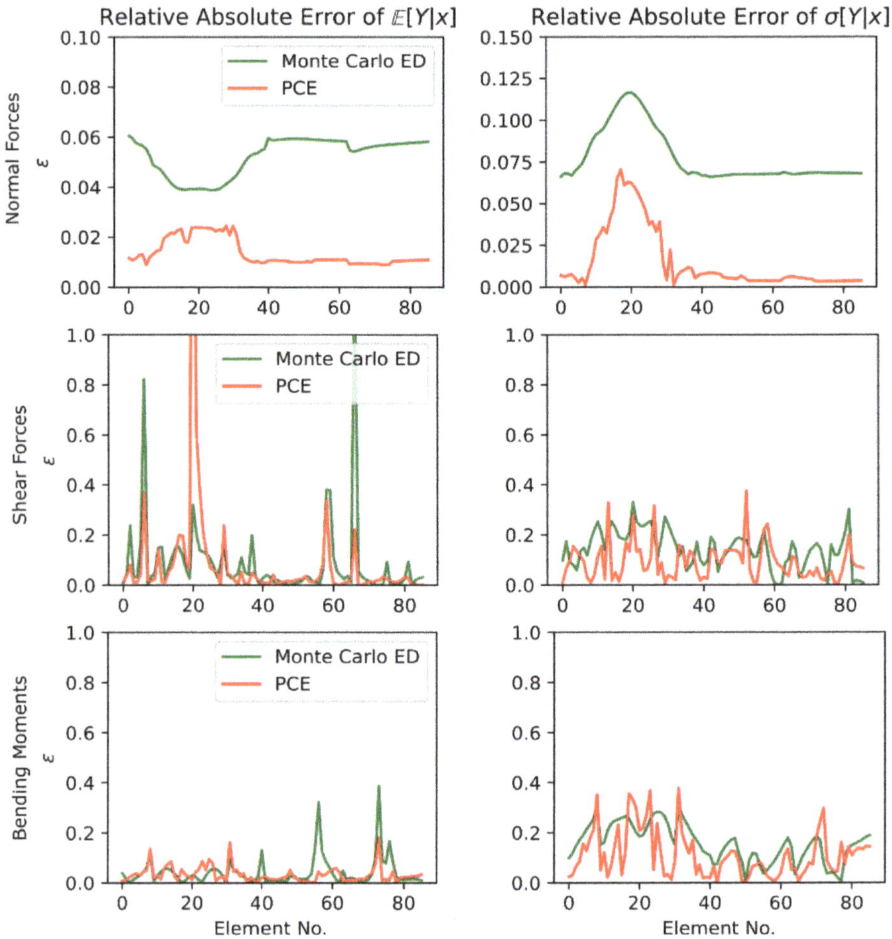

Fig. 3. Relative absolute error of local mean values (right) and standard deviations (left) of normal forces (top), shear forces (middle) and bending moments (bottom) for each element number (both methods are based on 50 samples in ED)

5 Conclusions

The paper explores possibilities of PCE for UQ as a LEVEL III method for design and assessment of tunnel linings. The main task was an estimation of uncertain internal forces caused by a soil-structure interaction affected by input random parameters. It was shown that PCE is able to greatly reduce the number of numerical simulations in a comparison to a standard crude MC approach while achieving satisfactory accuracy of estimated statistical moments. Moreover, thanks to a negligible computational cost of adaptive sparse solvers for construction of PCE, it was possible to create an independent PCE for each finite element of the tunnel liner. The obtained set of PCEs was further utilized for an analytical post-processing – an estimation of local mean values and standard deviations. Naturally, besides accurate UQ of QoIs, PCE can be also utilized further as a standard surrogate model for various additional simulations (e.g., a reliability analysis), and thus it offers an additional value in comparison to standard Monte Carlo analysis though it might lead to comparable estimation of statistical moments.

Acknowledgments. Theoretical background of the study was supported by Czech Science Foundation under project No. 24-10892S, the practical implementation of the numerical algorithm was part of the project DELTA supported by Technological Agency of Czech Republic under project No. TM04000012. The international collaboration of authors was supported by bilateral Czech-Germany mobility project No. 8J24DE002 supported by Ministry of Education, Youth and Sports of Czech Republic and DAAD.

References

1. Spyridis, P., Novák, L., Novák, D., Bergmeister, K.: Advanced semi-probabilistic methods for the design and assessment of concrete tunnel linings. In: Proceedings of Expanding Underground - Knowledge and Passion to Make a Positive Impact on the World- Proceedings of the ITA-AITES World Tunnel Congress, WTC 2023 (2023). ISBN: 978-100334803-0
2. Ghanem, R.G., Spanos, P.D.: Stochastic finite Elements: A Spectral Approach. Springer, Berlin (1991). https://doi.org/10.1007/978-1-4612-3094-6
3. Xiu, D., Karniadakis, G.E.: The Wiener-Askey polynomial chaos for stochastic differential equations. SIAM J. Sci. Comput. **24**(2), 619–644 (2002)
4. Efron, B., Hastie, T., Johnstone, I., Tibshirani, R.: Least angle regression. Ann. Statist. **32**(2), 407–451 (2004)
5. Novák, L., Vořechovský, M., Sadílek, V., Shields, M.D.: Variance-based adaptive sequential sampling for polynomial chaos expansion. Comput. Methods Appl. Mech. Eng. **386**, 114105 (2021). https://doi.org/10.1016/j.cma.2021.114105
6. Tsapetis, D., et al.: UQpy v4.1: Uncertainty quantification with Python, SoftwareX 24 (2023)
7. Novák, L.: On distribution-based global sensitivity analysis by polynomial chaos expansion. Comput. Struct. **267**, 106808 (2022)
8. Sudret, B.: Global sensitivity analysis using polynomial chaos expansions. Reliab. Eng. Syst. Saf. **93**(7), 964–979 (2008)
9. Potts, D.M., Zdravkovic, L.: Finite Element Analysis in Geotechnical Engineering: Theory & Application, vol. 2. Thomas Telford (2001)
10. Jones, B.: Soft Ground Tunnel Design. CRC Press, Boca Raton (2021)
11. Bond, A., Harris, A.: Decoding Eurocode 7. Taylor & Francis (2008)
12. Phoon, K.K.: Reliability-Based Design in Geotechnical Engineering Computations and Applications. Taylor and Francis, Routledge (2008)
13. John, M., Mattle, B.: Factors of shotcrete lining design. Tunn. Tunn. Int. **35**(10), 42–44 (2003)
14. CEN: prEN 1992-1-1 Eurocode 2: Design of concrete structures - Part 1-1: General rules - Rules for buildings, bridges and civil engineering structures (2021)

Value of Information Applied to the Post-fire Assessment of Concrete Slabs

Ruben Van Coile[1] (✉), Mengying Peng[1], Florian Put[1], and Balša Jovanović[1,2]

[1] Department of Structural Engineering and Building Materials, Ghent University, Ghent, Belgium
ruben.vancoile@ugent.be
[2] Department of Civil Engineering, KU Leuven, Leuven, Belgium

Abstract. Post-fire assessment of concrete structures currently relies heavily on expert judgement. Thus, whether or not tests are performed as part of the post-fire assessment and which test method is chosen depends mainly on the subjective preferences of the assessing engineer. In this contribution a rational decision-making framework based on the Value of Information concept is demonstrated using a simplified example application of a fire exposed concrete slab in a residential building. The example application highlights that gathering data on the post-fire status of the concrete slab is likely to be cost-effective also when the assessor is very confident about the status of the structural element post-fire. This demonstration thus suggests that the current reliance on expert judgement is not reasonable.

Keywords: Concrete · Fire · Post-fire assessment · Value of Information

1 Introduction

Fires are rare events, but when they do occur, they can significantly challenge the structure. As concrete is a non-combustible material with low thermal diffusivity, concrete structures are commonly considered to have a good fire performance. Concrete is nevertheless susceptible to permanent damage from fire [1]. Conducting a comprehensive evaluation of concrete structures after a fire is crucial to assess their condition and make informed decisions on the most appropriate course of action. Such post-fire evaluations are commonly done through a combination of tests and calculations, guided to a very large degree by expert judgement [1]. Whether or not tests are performed and which tests are preferred thus commonly depends on the personal preferences of the assessing engineer. In this contribution, a rational decision-making framework is presented based on the Value-of-Information (VoI) concept, as introduced in e.g., [2]. The VoI application is demonstrated through a simplified example application. To the best of the authors' knowledge, the work presented in this paper constitutes the first ever application of the VoI methodology within the field of fire safety engineering.

J. C. Matos et al. (Eds.): IPW 2024, LNCE 494, pp. 520–528, 2024.
https://doi.org/10.1007/978-3-031-60271-9_49

2 Decision Framework

2.1 Problem Statement

The permanent damage induced by fire to concrete members realizes itself as cracks, deformations, a loss of capacity, and a loss of ductility [3]. The concrete structure may nevertheless be repairable, and it is even possible that the residual capacity is sufficient also without post-fire repair. This residual capacity can however not be readily observed from a post-fire inspection and must be inferred through calculation, testing, and expert judgement. For an engineer responsible for the post-fire assessment of the load bearing capacity, the question thus becomes whether or not to obtain additional test data, and which test method to apply. Currently, no structured decision framework is available which tackles this question.

2.2 Decision Making on Data Collection Using the Value of Information

Up-front decision making on data collection can be tackled through the VoI approach. Evaluation of the VoI entails a pre-posterior analysis of the benefits from performing tests (i.e., an assessment before the tests are executed). In other words, it is an evaluation of how the experiment is expected to reduce uncertainty [4]. VoI is a central concept in other engineering and safety disciplines, such as the monitoring of structures [2], where VoI is used to make a prior assessment of the benefit of collecting additional information. The concept of VoI is so far absent in Fire Safety Science and Engineering, with exceptions in relation to forest fire risk management, e.g. [5].

In essence, the VoI can be expressed by Eq. (1), with x denoting the test method considered (i.e., recognizing that there may be different options), $E[.]$ indicating the expected value operator, C denoting the cost, θ_0 the prior assessment of the uncertain parameters and $\theta_{1,x}$ the posterior assessment of the uncertain parameters after application of the test method x.

$$VoI_x = E[C|\theta_0] - E\left[C|\theta_{1,x}\right] \quad (1)$$

Expressing Eq. (1) in words, the VoI is the difference between the expected cost prior to obtaining additional test data and the expected cost after obtaining additional test data. This assessment is done before the test is conducted. As such the evaluation "after obtaining additional test data" is a pre-posterior assessment whereby all possible outcomes of the test and their associated likelihood are taken into account. The difference in expected cost results from the test data (i.e., the improved information on the parameters θ) influencing decision making. Where additional information does not influence the decision making, the VoI will be zero. The magnitude of the VoI clarifies the allowable cost for obtaining the experimental data. Thus, a net VoI can be defined by Eq. (2), where C_x is the cost of performing the test x. Amongst all possible options, the test with the highest net VoI_x is to be preferred (assuming the cost C_x is feasible to pay upfront and ignoring questions on tolerability, see the risk acceptance framework in [6]). When there are no options with a positive net VoI_x, obtaining additional test data is not cost-effective.

$$net\ VoI_x = VoI_x - C_x \quad (2)$$

3 Example Application – Fire Exposed Concrete Slab in a Residential Building

3.1 Case Description

A compartment fire occurs in a single apartment in a multi-story residential building, whereby each apartment is its own fire compartment and the ceiling slab of one apartment is the floor slab of the next. The fire burns up to burnout, consuming all the fuel inside the compartment. The structure survives the fire, and the assessing engineer needs to determine whether structural strengthening of the concrete ceiling slab is necessary, or whether rehabilitation for aesthetics and durability is sufficient. When the temperature of the reinforcement in the concrete slab exceeds 600 °C a permanent reduction in yield stress is incurred (model presented in [7] based on [8, 9]. Thus, in the following the maximum reinforcement temperature exceeding 600 °C is adopted as simplified criterion for the need for structural repair.

The fuel load consumed by the fire and the opening factor (ventilation conditions) during the fire are uncertain, and thus so are the temperatures attained within the concrete ceiling slab. The reinforcement axis distance to the exposed surface is known (25 mm). The engineer considers two possible testing techniques: (i) a fast rebound hammer test which can be applied by the engineer at limited cost, and (ii) a discoloration test which requires an external company to drill a core from the fire-exposed slab. Further specifics on these test techniques are listed in Table 1.

Table 1. Properties of the considered test techniques (simplified consideration).

Characteristic	Rebound hammer	Discoloration test
Ease-of-use and cost	Can be applied by the assessing engineer at a limited cost	Requires an external company to drill a concrete core from the exposed slab
Quantity (indirectly) assessed by the test	Maximum temperature attained at a depth of 15 mm	Maximum depth attained by the 300 °C isotherm
Measurement uncertainties	Assumed value[a]	Assumed value*

[a] The same measurement uncertainty was assumed for both test methods: a bi-variate normal distribution with a mean value of zero and an uncertain standard deviation modelled by a normally distributed variable (for the depth: mean value of 2 mm and standard deviation of 0.5 mm; for the temperature: mean value of 25 °C and standard deviation of 5 °C).

The cost of structural repair is denoted as C_R and is known upfront. If the structure is in need of repair but is not repaired because this need was not recognized as part of the assessment, then damages are considered to materialize upon reuse and a "failure cost" C_F will be incurred (i.e., total cost upon later discovery). Also, this cost is considered to be known upfront. Note that the consideration of deterministic costs C_R and C_F does not reduce the generality of the description further. For uncertain costs it suffices to interpret C_R and C_F further as the expected values of these costs. This statement is an application of the standard Von Neumann-Morgenstern utility theorem and is thus subject to the same limitations.

The above cost model is visualized in Fig. 1. The decision to opt for structural repairs or to leave the structure as-is depends on the probability P_s that the structure is "safe" and not in need of repair. The threshold for P_s is defined by the break-even point whereby both decision branches within Fig. 1 result in the same expected cost, i.e., where Eq. (3) holds. In the following it is assumed that $C_F \approx 10\ C_R$. The decision threshold is thus defined by a P_s of 90%. In other words, if the probability that the reinforcement temperature remained below 600 °C exceeds 90% then the most cost-effective decision is to leave-as-is. Otherwise, the slab should be structurally repaired.

$$C_R = (1 - P_s)C_F \rightarrow P_s = 1 - \frac{C_R}{C_F} \tag{3}$$

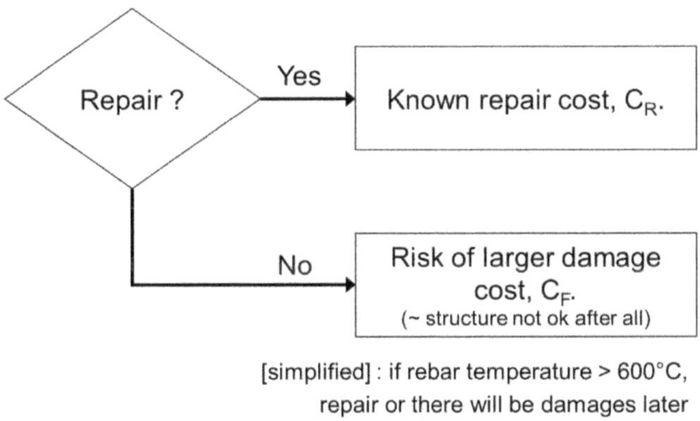

Fig. 1. Visualization of the decision problem: repair the concrete slab and incur a repair cost C_R or decide to leave-as-is but run the risk of incurring the larger failure cost C_F if the structure was in need of repairs after all (i.e., if the reinforcement temperature did exceed 600 °C).

3.2 Prior Decision Without Measurements

A prior assessment is made considering prior uncertainty for the opening factor and fuel load. The Eurocode parametric fire curve [10] is used to model the compartment fire. The parameters for the fire model are described in Table 2. The opening factor O is defined by Eq. (4), as specified in the JCSS Probabilistic Model Code [11], with O_{nom} here equal to 0.05. For simplicity, and in agreement with current practice, no model uncertainty is considered for the fire model. The heat transfer into the concrete slab is calculated considering the Eurocode material models [12], adopting the lower bound for the thermal conductivity, using a 1D heat transfer model. The reinforcement temperature is taken as the concrete temperature at the reinforcement axis position, in agreement with common practice [13].

The resulting prior distribution for the reinforcement maximum temperature is listed in Fig. 2, based on 10^4 simulations. There is a prior probability of only 4% that the reinforcement temperature exceeded 600 °C (i.e., 96% probability that the slab does not need structural repairs). Considering the threshold probability of 90% specified in 3.1, the prior decision is to leave-as-is and accept the 4% probability of incurring the failure cost C_F. The expected cost based on the prior data thus equals $0.04{\cdot}C_F = 0.4{\cdot}C_R$.

Table 2. Parameters describing the fire and heat transfer models, their distribution and (prior) mean and coefficient of variation (COV).

Parameter	Distribution	(mean value; COV)
Fire load density, q_f [MJ/m^2]	Gumbel	(780; 0.3)
Correction factor for the ventilation conditions η [–]	Truncated lognormal (range 0–1)	(0.2; 0.2)
Height of the compartment [m]	Deterministic	(3; –)
Floor area of the compartment [m^2]	Deterministic	(100; –)
Thermal inertia [J/(m^{2}s$^{0.5}$K)]	Deterministic	(1450; –)

$$O = (1 - \eta)O_{nom} \tag{4}$$

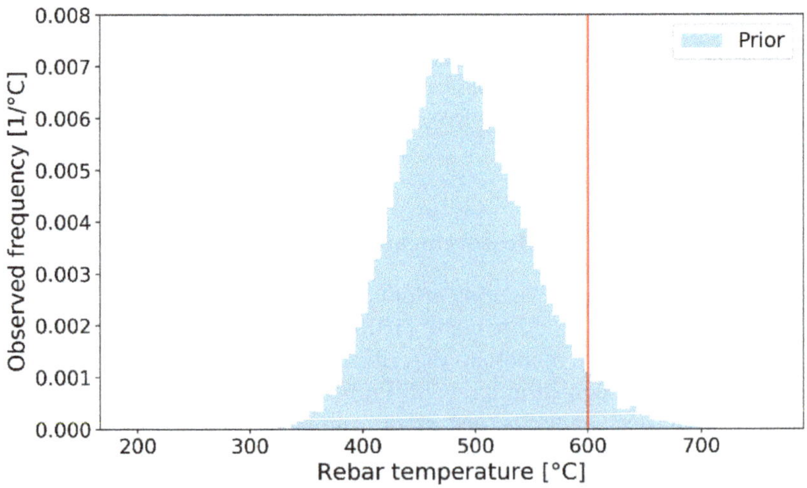

Fig. 2. Prior observed probability density for the maximum reinforcement temperature.

3.3 Posterior Decision Taking into Account Test Data

The VoI evaluation relies on a pre-posterior evaluation of the expected cost when additional data is obtained using test method x, see Eq. (1). This expected cost for test method x combines the expected cost given a specific test result R_j with the prior probability $P[R_j]$ of this test result being obtained, i.e. Equation (5). In other words, for each possible outcome R_j of the test method, a repair decision is made in accordance with Fig. 1 and the decision threshold specified in Sect. 3.1 (i.e., repair if the probability P_s of the reinforcement maximum temperature being below 600 °C is lower than 90%). This repair decision corresponds with an expected cost, as elaborated in Sect. 3.2 for the prior decision. To assess the expected cost for the test method itself, the expected costs for the different possible test outcomes R_j need to be combined. This is done considering the (prior) probability $P[R_j]$ of obtaining this specific test result. It is called a prior probability because it is assessed using the prior information. Within this paper, the range of possible measurement points was discretized into zones of equal probability (in concreto: 100 zones each with probability 0.01). Thus $P[R_j]$ is the same for all R_j and the evaluation of Eq. (5) reduces to taking the average of the expected cost for all considered possible measurements for a given test method. For completeness, please note that it could be considered more appropriate to write Eq. (5) using an integral. Here, a summation has been used for simplicity and because it corresponds with how the calculation was effectively done within this paper.

$$E\left[C|\theta_{1,x}\right] = \sum_j E\left[C|\theta_{1,x}, R_j\right] \cdot P\left[R_j\right] \tag{5}$$

The evaluation of the expected cost given the test outcome R_j takes into account that the test outcome provides information to update the assessment of the uncertain parameters q_f and η. This constitutes a standard application of Bayes' theorem, conveniently represented here by Eq. (6) where specific notation required to distinguish between the probability density functions and discrete probability values has been omitted for simplicity. For any value of R_j, posterior distributions of q_f and η are obtained. This is achieved using the Markov Chain Monte Carlo method employing the Metropolis algorithm for updating. The process takes into account measuring uncertainties of the assessment techniques. A more detailed description of the adopted Bayesian updating procedure is presented in [14]. Based on these distributions, as mentioned in Sect. 3.2, the distribution of the reinforcement temperature can be calculated.

$$P\left[q_f, \eta|R_j\right] = \frac{P\left[R_j|q_f, \eta\right] \cdot P\left[q_f, \eta\right]}{P\left[R_j\right]} = \frac{P\left[R_j|q_f, \eta\right] \cdot P\left[q_f, \eta\right]}{\sum_{q_f, \eta} P\left[R_j|q_f, \eta\right] \cdot P\left[q_f, \eta\right]} \tag{6}$$

Results are presented in Fig. 3 for specific assumed measurement results for the depth of the 300 °C isotherm (i.e., the discoloration test). When the discoloration test indicates a depth of 40 mm, the distribution for the steel maximum temperatures shifts to the left and the probability of the steel temperature exceeding 600 °C becomes negligibly small. A discoloration test indicating a depth of 67 mm shifts the distribution to the right, but the measurement result also reduces uncertainty (i.e., results in a narrower distribution) which together still gives a reduction in the assessed probability of exceeding 600 °C relative to the prior (0.02 instead of 0.04). Higher measurement values such as 70 mm further shift the distribution to the right and result in an increase of the assessed exceedance probability. A tabulated view of select measurements, corresponding probabilities of exceeding the 600 °C limit and the associated repair decision are listed in Table 3 for the discoloration test. A similar table was constructed for the rebound hammer test.

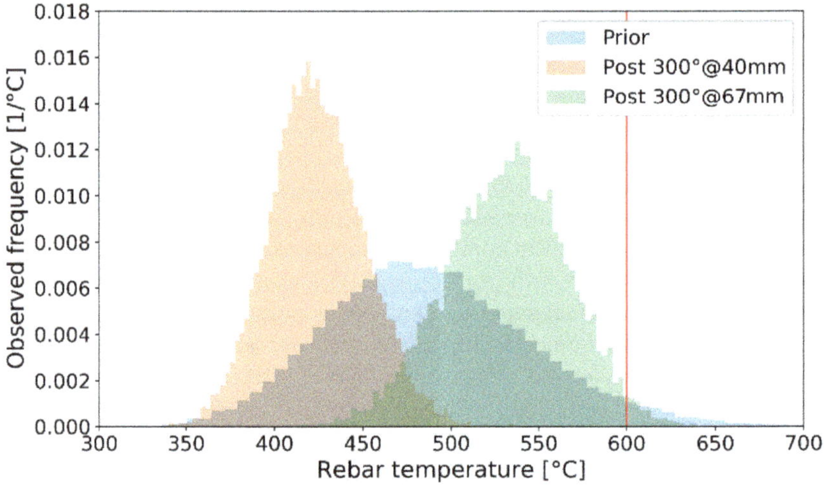

Fig. 3. Posterior and prior observed probability density for the maximum reinforcement temperature for select measurements using the discoloration method.

Table 3. Example posterior decision-making. Select measurements for the discoloration test (measurement of 300 °C isotherm depth)

Measurement of 300 °C depth [mm]	Probability of reinforcement temperature exceeding 600 °C [–]	Probability of reinforcement temperature remaining below 600 °C [–]	Repair decision	Expected cost
40	0.00	1.00	leave-as-is	0
67	0.02	0.98	leave-as-is	$0.2{\cdot}C_R$
71	0.06	0.94	leave-as-is	$0.6{\cdot}C_R$
75	0.13	0.87	repair	C_R
78	0.19	0.81	repair	C_R

3.4 VoI Evaluation and Conclusion on the Value of Performing a Test

Having evaluated the expected costs for all possible measurement outcomes as in Table 3, the pre-posterior assessment of the expected cost following application of respectively the rebound hammer test and the discoloration test are evaluated, i.e., Eq. (5). Since the range of possible measurement points was discretized in areas with equal probability for the purposes of this paper, within this paper the average expected cost is taken for both test methods. The results are listed in Table 4. Taking into account the starting consideration that the rebound hammer test is cheaper than the discoloration test, and recognizing that that the cost of doing such rebound hammer test will be significantly cheaper than $0.29 \cdot C_R$, these results lead to the conclusion that a rebound hammer should be done as it is the test method with the highest positive net VoI, see Eq. (2).

Table 4. Example posterior decision-making. Select measurements for the discoloration test (measurement of 300 °C isotherm depth)

Scenario	Expected cost	VoI	Cost of obtaining the measurement	Net VoI
Prior	$0.40 \cdot C_R$	–	–	0
Rebound hammer	$0.11 \cdot C_R$	$0.29 \cdot C_R$	$C_{hammer} < C_{color} < 0.25 \cdot C_R$	>>0
Discoloration	$0.15 \cdot C_R$	$0.25 \cdot C_R$		>0

4 Conclusion

The Value-of-Information (VoI) concept provides a decision framework which is readily applicable to the post-fire assessment of concrete structures. The results of such analyses can be contrary to intuition due to the large difference between repair costs and failure costs (i.e., costs incurred when deciding not to repair in cases where repair is actually necessary).

A simplified example application has been presented whereby two possible test methods (rebound hammer test and discoloration test) were evaluated and compared. Despite the high prior probability of 96% that no repair is needed, the VoI evaluation indicates a very clear net value for obtaining additional information. Of the compared test methods, the rebound hammer test is preferred. This outcome is subject to the simplifying assumptions adopted within this work and should not be generalized.

The work presented in this paper constitutes a proof-of-concept for the application of the VoI concept within Fire Safety Science and Engineering. Follow up work is foreseen to apply the concept amongst others to the design of fire safety experiments, maximizing the net information gain from performing fire tests.

Acknowledgments. Funded by the European Union (ERC, AFireTest, 101075556). Views and opinions expressed are however those of the author(s) only and do not necessarily reflect those of the European Union or the European Research Council Executive Agency. Neither the European Union nor the granting authority can be held responsible for them. The last author further thanks the Research Foundation of Flanders (FWO) for the financial support on the research project (Grant number 3G010220) "Vibration-based post-fire assessment of concrete structures using Bayesian updating techniques". Florian Put is funded by Research Foundation of Flanders (FWO) within the scope of the research project (Grant number 1137123N) "Characterization of the thermal exposure and material properties of concrete during the fire decay phase for performance-based structural fire engineering".

References

1. Jovanović, B., Caspeele, R., Lombaert, G., Reynders, E., Van Coile, R.: State-of-the-art review on the post-fire assessment of concrete structures. Struct. Concr. **24**(4), 5370–5387 (2023)
2. Straub, D.: Value of information analysis with structural reliability methods. Struct. Saf. **49**, 75–85 (2014)
3. Khoury, et al.: Fire design of concrete structures – materials, structures and modelling. fib bulletin 38. In: The International Federation for Structural Concrete, Lausanne, Switzerland (2007)
4. Howard, R.A.: Information value theory. IEEE Trans. Syst. Sci. cybern. **2**(1), 22–26 (1966)
5. Amacher, G.S., Malik, A.S., Haight, R.G.: Forest landowner decisions and the value of information under fire risk. Can. J. For. Res. **35**(11), 2603–2615 (2005)
6. Van Coile, R., Hopkin, D., Lange, D., Jomaas, G., Bisby, L.: The need for hierarchies of acceptance criteria for probabilistic risk assessments in fire engineering. Fire Technol. **55**, 1111–1146 (2019)
7. Van Coile, R., Caspeele, R., Taerwe, L.: Towards a reliability-based post-fire assessment method for concrete slabs incorporating information from inspection. Struct. Concr. **15**(3), 395–407 (2014)
8. Elghazouli, A.Y., Cashell, K.A., Izzuddin, B.A.: Experimental evaluation of the mechanical properties of steel reinforcement at elevated temperature. Fire Saf. J. **44**(6), 909–919 (2009)
9. Felicetti, R., Gambarova, P.G., Meda, A.: Residual behavior of steel rebars and R/C sections after a fire. Constr. Build. Mater. **23**(12), 3546–3555 (2009)
10. EN 1991-1-2:2002. Eurocode 1: Actions on structures – Part 1-2: General actions – Actions on structures exposed to fire. European Committee for Standardization, Brussels, Belgium
11. JCSS: JCSS Probabilistic Model Code. Joint Committee on Structural Safety (2015). https://www.jcss-lc.org/jcss-probabilistic-model-code/. Accessed 13 Jan 2024
12. EN 1992-1-2:2004. Eurocode 2: Design of concrete structures – Part 1–2: General rules – Structural fire design. European Committee for Standardization, Brussels, Belgium
13. Gernay, et al.: *fib* bulletin 108: performance-based fire design of concrete structures (state-of-the-art report). In: The International Federation for Structural Concrete, Lausanne, Switzerland (2023)
14. Jovanović, B., Caspeele, R., Lombaert, G., Reynders, E., Van Coile, R.: Bayesian updating methodology for the post-fire evaluation of the maximum temperature profile inside concrete elements. Fire Technol. (2024)

Author Index

J. C. Matos et al. (Eds.): IPW 2024, LNCE 494, pp. 529–530, 2024.
https://doi.org/10.1007/978-3-031-60271-9

www.ingramcontent.com/pod-product-compliance
Ingram Content Group UK Ltd.
Pitfield, Milton Keynes, MK11 3LW, UK
UKHW022217051026
15624UKWH00001B/206
9783031602702